Advanced Reinforced Concrete Design

P.C. VARGHESE

Honorary Professor, Anna University, Madras
Formerly
Professor and Head, Dept. of Civil Engineering, Indian Institute of Technology Madras
and
UNESCO Chief Technical Advisor, University of Moratuwa, Sri Lanka

Prentice-Hall of India Private Limited
New Delhi - 110 001
2002

ADVANCED REINFORCED CONCRETE DESIGN
by P.C. Varghese

© 2001 by Prentice-Hall of India Private Limited, New Delhi. All rights reserved. No part of this book may be reproduced in any form, by mimeograph or any other means, without permission in writing from the publisher.

ISBN-81-203-1733-5

The export rights of this book are vested solely with the publisher.

Second Printing ... ... **February, 2002**

Published by Asoke K. Ghosh, Prentice-Hall of India Private Limited, M-97, Connaught Circus, New Delhi-110001 and Printed by Jay Print Pack Private Limited, New Delhi-110015.

To
P.C. Mathew
and
Omana Mathew

Contents

Preface		xv
Acknowledgements		xvii
Introduction to IS 456 (2000)		xix

1. Deflection of Reinforced Concrete Beams and Slabs 1–23
  1.1 Introduction 1
  1.2 Short-term Deflection of Beams and Slabs 1
  1.3 Deflection due to Imposed Loads 2
  1.4 Short-term Deflection of Beams due to Applied Loads 3
  1.5 Calculation of Deflection by IS 456 5
  1.6 Calculation of Deflection by BS 8110 10
  1.7 Deflection Calculation by Eurocode 11
  1.8 ACI Simplified Method 14
  1.9 Deflection of Continuous Beams by IS 456 14
  1.10 Deflection of Cantilevers 15
  1.11 Deflection of Slabs 15
  References 23

2. Estimation of Crackwidth in Reinforced Concrete Members 24–38
  2.1 Introduction 24
  2.2 Factors affecting Crackwidth in Beams 24
  2.3 Mechanism of Flexural Cracking 25
  2.4 Calculation of Crackwidths 25
  2.5 Simple Empirical Method 26
  2.6 Estimation of Crackwidth in Beams by IS 456 and BS 8110 28
  2.7 Shrinkage and Thermal Cracking 30
  References 38

3. Redistribution of Moments in Reinforced Concrete Beams 39–49
  3.1 Introduction 39
  3.2 Redistribution of Moments in a Fixed Beam 39
  3.3 Positions of Points of Contraflexures 40
  3.4 Conditions for Moment Redistribution 41
  3.5 Final Shape of Redistributed Bending Moment Diagram 42
  3.6 Moment Redistribution for a Two-span Continuous Beam 42
  3.7 Advantages and Disadvantages of Moment Redistribution 43
  3.8 Modification of Clear Distance between Bars in Beams (for Limiting Crackwidth) with Redistribution 43
  3.9 Moment–Curvature (M–ψ) Relation of Reinforced Concrete Sections 44
  3.10 ACI Conditions for Redistribution of Negative Moments 45
  Conclusion 46
  References 49

4. Design of Reinforced Concrete Deep Beams — 50–72
- 4.1 Introduction — 50
- 4.2 Minimum Thickness — 51
- 4.3 Steps of Designing Deep Beams — 51
- 4.4 Design by IS 456 — 52
- 4.5 Design according to British Practice — 54
- 4.6 ACI Procedure for Design of Deep Beams — 59
- 4.7 Checking for Local Failures — 61
- 4.8 Detailing of Deep Beams — 61
- *References* — 72

5. Design of Ribbed (Voided) Slabs — 73–83
- 5.1 Introduction — 73
- 5.2 Specification regarding the Slabs — 74
- 5.3 Analysis of the Slabs for Moment and Shears — 75
- 5.4 Ultimate Moment of Resistance — 76
- 5.5 Design for Shear — 76
- 5.6 Deflection — 77
- 5.7 Arrangement of Reinforcements — 77
- 5.8 Corrosion of Steel with Clay Blocks — 77
- *References* — 83

6. Approximate Analysis of Grid Floors — 84–95
- 6.1 Introduction — 84
- 6.2 Analysis of Flat Grid Floors — 84
- 6.3 Analysis of Rectangular Grid Floors by Timoshenko's Plate Theory — 85
- 6.4 Analysis of Grid by Stiffness Matrix Method — 88
- 6.5 Analysis of Grid Floors by Equating Joint Deflections — 88
- 6.6 Comparison of Methods of Analysis — 88
- 6.7 Detailing of Steel in Flat Grids — 89
- *References* — 95

7. Design Loads other than Earthquake Loads — 96–128
- 7.1 Introduction — 96
- 7.2 Dead Loads — 96
- 7.3 Imposed Loads (IS 875 Part 2) — 97
- 7.4 Loads due to Imposed Deformations — 99
- 7.5 Characteristic Imposed Loads — 99
- 7.6 Partial Safety Factors for Loads — 99
- 7.7 Some General Provisions regarding Live Loads on Structures — 100
- 7.8 Wind Loads on Structures — 101
- 7.9 Indian Code for Wind Loads — 103
- 7.10 General Theory of Wind Effects of Structures — 108
- 7.11 Wind Force on Stiff Buildings (Quasi-static Method) — 112
- 7.12 Gust Factor (GF) Method (Dynamic Analysis) — 114
- 7.13 Wind Tunnel Tests — 115
- 7.14 Classification of Wind (Cyclones) — 115
- 7.15 Importance of Checking for Wind Loads at All Stages of Construction — 117
- 7.16 Construction Loads — 117
- 7.17 Joints in Concrete Construction — 118
- *References* — 128

CONTENTS

8. Analysis of Reinforced Concrete Frames for Vertical Loads by using Substitute Frames   129–146

- 8.1 Introduction   129
- 8.2 Distribution of Loads from Slabs to Supporting Beams   129
- 8.3 Other Methods for Distribution of Loads   132
- 8.4 Frames Analysis   132
- 8.5 Braced and Unbraced Frames   133
- 8.6 Analysis by Substitute Frames   133
- 8.7 Analysis by Continuous Beam Simplification   134
- 8.8 Use of Sub-frame for Analysis of Vertical Loads   135
- 8.9 Procedure for Calculation of (P and M) for Central Columns   138
- 8.10 Procedure for Calculation of (P and M) for External Columns   140
- 8.11 Reduction in Loads on Columns and Beams   140
- 8.12 Partial Restraint at End-supports   140
- 8.13 Analysis of Single-span Portals   140
- 8.14 Critical Section for Moment and Shear   141
- 8.15 Recommended Procedure   141
- 8.16 Formulae to Determine Span Moment   141
- *References*   *146*

9. Analysis of Frames under Horizontal Loads   147–157

- 9.1 Introduction   147
- 9.2 Effect of Lateral Loads   147
- 9.3 Methods of Analysis   147
- 9.4 Portal Method (Method of Proportional Shear)   148
- 9.5 Cantilever Method (Method of Proportional Axial Stresses)   150
- 9.6 Comparison of Results of Analysis   152
- 9.7 Analysis of Rigid Frames with Transfer Girders   152
- 9.8 Drift Limitation in Very Tall Buildings   152
- 9.9 Classification of Structural System for Tall Building   153
- *References*   *157*

10. Preliminary Design of Flat Slabs   158–170

- 10.1 Introduction   158
- 10.2 Advantages and Disadvantages of Flat Slabs   159
- 10.3 Historical Development   160
- 10.4 Action of Flat Slabs and Flat Plates   160
- 10.5 Preliminary Design of Flat Slab   161
- 10.6 Basic Action of Two-way Slabs   161
- 10.7 Determination of Minimum Thickness of Slab   162
- 10.8 Moment of Inertia of Flanged Beams   165
- *References*   *170*

11. Design of Two-way Slabs by Direct Design Method   171–210

- 11.1 Introduction   171
- 11.2 Limitations of Direct Design Method   173

11.3	Calculation of Total Static Moments for Interior and Exterior Frames	174
11.4	ACI 318 Formula for Moments	175
11.5	Distribution of Moments M_o	176
11.6	Longitudinal Distribution of M_o	177
11.7	Effect of Pattern Loading on Positive Moment (Minimum Stiffness of Columns)	180
11.8	Transverse Distribution of Moments and Factors affecting the Distribution	182
11.9	General Equations for Transverse Distribution of Moments to Column Strips	183
11.10	Allocation of Moments to Middle Strips	184
11.11	Analysis of Exterior Frame Supported on a Wall	184
11.12	Treatment of Unequal Moments in Adjacent Spans	185
11.13	Design Loads on the Beams	185
11.14	Design of Reinforcements	185
11.15	Moments in Columns in DDM	186
11.16	Moments and Shear Transfer from Slab to Columns	188
11.17	Detailing of Steel in Flat Slabs	189
11.18	Design of Two-way Slabs	191
	Conclusion	*192*
	References	*210*

12. Shear in Flat Slabs and Flat Plates   211–240

12.1	Introduction	211
12.2	Checking for One-way (Wide Beam) Shear	211
12.3	Two-way (Punching) Shear	212
12.4	Permissible Punching Shear	213
12.5	Shear due to Unbalanced Moment (Torsional Moments)	214
12.6	Calculation of J Values	218
12.7	Strengthening of Column Areas for Moment Transfer by Torsion which Produces Shear	219
12.8	Shear Reinforcement Design	219
12.9	Effect of Openings in Flat Slabs	227
12.10	Recent Revisions in ACI 318	228
12.11	Shear in Two-way Slabs with Beams	228
	References	*240*

13. Equivalent Frame Analysis of Flat Slabs   241–262

13.1	Introduction	241
13.2	Historical Development of the Concept of Equivalent Frame	241
13.3	Background of ACI (1977) Method—Equivalent Column Method	243
13.4	Summary of Provisions in ACI 318	248
13.5	Arrangement of Live Load	249
13.6	Reduction in Negative Moments	249
13.7	Generalised Space Frame Model	250
13.8	Lateral Loads and Two-way Slab Systems	251
13.9	Design Procedure	251
	References	*262*

CONTENTS

14. Design of Spandrel (or Edge) Beams   263–274
- 14.1 Introduction   263
- 14.2 Design Principles   264
- 14.3 Size of Beam to be Considered   264
- 14.4 Bending Moments and Shears in the Beam   264
- 14.5 Torsion to be Taken for Design   264
- 14.6 Designing for Torsion   266
- *Conclusion*   *267*
- *References*   *274*

15. Provision of Ties in Reinforced Concrete Slab—Frame System   275–281
- 15.1 Introduction   275
- 15.2 Design for Overall Stability (Robustness)   276
- 15.3 Design Procedure for Ties   276
- 15.4 Continuity and Anchoring of Ties   278
- *Conclusion*   *279*
- *References*   *281*

16. Design of Reinforced Concrete Members for Fire Resistance   282–294
- 16.1 Introduction   282
- 16.2 ISO 834 Standard Heating Conditions   282
- 16.3 Grading or Classifications   283
- 16.4 Effect of High Temperature on Steel and Concrete   284
- 16.5 Effect of High Temperatures on Different Types of Structural Members   285
- 16.6 Fire Resistance by Structural Detailing from Tabulated Data   287
- 16.7 Analytical Determination of the Ultimate Bending Moment Capacity of Reinforced Concrete Beams under Fire   288
- 16.8 Other Considerations   291
- *References*   *294*

17. Design of Plain Concrete Walls   295–309
- 17.1 Introduction   295
- 17.2 Braced and Unbraced Walls   296
- 17.3 Slenderness of Walls   297
- 17.4 Eccentricities of Vertical Loads at Right Angles to Wall   298
- 17.5 Empirical Design Method for Plane Concrete Walls Carrying Axial Load   299
- 17.6 Design of Walls for In-plane Horizontal Forces   301
- 17.7 Rules for Detailing of Steel in Concrete Walls   305
- *References*   *308*

18. Earthquake Forces and Structural Response   310–337
- 18.1 Introduction   310
- 18.2 Bureau of Indian Standards for Earthquake Designs   311
- 18.3 Earthquake Magnitude and Intensity   311
- 18.4 Historical Development   312
- 18.5 Basic Seismic Coefficients and Seismic Zone Factors   314

18.6	Determination of Design Forces	320
18.7	Choice of Method for Multi-storeyed Buildings	325
18.8	Difference between Wind and Earthquake Forces	326
18.9	Torsion in Buildings	326
18.10	Partial Safety Factors for Design	326
18.11	Distribution of Seismic Forces	327
18.12	Analysis of Structures other than Buildings	328
18.13	Ductile Detailing	328
18.14	Increased Values of Seismic Effect for Vertical and Horizontal Projections	328
18.15	Proposed Changes in IS 1893 (Fifth Revision)	328
	Conclusion	329
	References	337

19. Design of Shear Walls — 338–357

19.1	Introduction	338
19.2	Classification of Shear Walls	339
19.3	Classification according to Behaviour	341
19.4	Loads in Shear Walls	342
19.5	Design of Rectangular and Flanged Shear Walls	344
19.6	Derivation of Formula for Moment of Resistance of Rectangular Shear Walls	348
	References	357

20. Design of Cast *in situ* Beams—Column Joints — 358–377

20.1	Introduction	358
20.2	Types of Cast *in situ* Joints	358
20.3	Joints in Multi-storeyed Buildings	358
20.4	Forces acting on Joints	359
20.5	Strength Requirement of Columns	360
20.6	Forces Directly acting on Joints	361
20.7	Design of Joints for Strength	362
20.8	Anchorage	363
20.9	Confinement of Core of Joint	365
20.10	Shear Strength of the Joint	366
20.11	Corner (Knee) Joint	368
20.12	Detailing for Anchorage in Exterior Beam—Column Joint	370
20.13	Procedure for Design of Joints	371
	Conclusion	371
	References	377

21. Ductile Detailing of Frames for Seismic Forces — 378–391

21.1	Introduction	378
21.2	General Principles	380
21.3	Factors that Increase Ductility	380
21.4	Specifications of Materials for Ductility	381
21.5	Ductile Detailing of Beams—Requirements	381
21.6	Ductile Detailing of Columns and Frame Members with Axial—Load (P) and Moment (M)—Requirements	384

21.7	Shear Walls	387
21.8	Joints in Frames	387
	Conclusion	*387*
	References	*391*

22. Inelastic Analysis of Reinforced Concrete Beams and Frames   392–424

22.1	Introduction	392
22.2	Inelastic Behaviour of Reinforced Concrete	392
22.3	Stress–Strain Characteristics of Concrete	393
22.4	Stress–Strain Characteristics of Steels	396
22.5	Moment Curvature Relation (M–ϕ Curves)	397
22.6	Concept of Plastic Hinges (Moment Rotation at Plastic Hinges)	400
22.7	Effect of Shear on Rotation Capacity	401
22.8	Inelastic or Non-linear Analysis of Reinforce Concrete Beams	401
22.9	Allowable Rotation for Collapse Load Analysis	403
22.10	Baker's Method for Plastic Analysis of Beams and Frames	403
	References	*424*

23. Strip Method of Design of Reinforced Concrete Slabs   425–434

23.1	Introduction	425
23.2	Theory of Strip Method	426
23.3	Application to a Simply-supported Slab	426
23.4	Modified Strip Method	427
23.5	Points of Inflection for Fixed Ends	428
23.6	Use of Strong Band in Slab Design	428
23.7	Support Reaction	428
23.8	Method of Design	428
23.9	Design of Skew Slabs	428
23.10	Affinity Theorems	429
	References	*434*

24. Durability and Mix Design of Concrete   435–458

24.1	Introduction	435
24.2	Types of Cements Produced in India	435
24.3	Durability of Concrete	437
24.4	Carbonation and Atmospheric Corrosion of Steel	440
24.5	Chloride Penetration and Steel Corrosion	440
24.6	Presence of Salts and Steel Corrosion	441
24.7	Sulphates and Concrete Disintegration	441
24.8	Curing of Concrete (IS 456 Clause 13.5)	442
24.9	Summary of Recommendations for Durability of Concrete	442
24.10	Polymer Concrete	442
24.11	Mix Design of Concrete	443
24.12	Design of Ordinary Grade Concrete	444
24.13	Design of High Strength Concrete	450
24.14	Design of Very High Strength Concrete Mixes	452
24.15	Mix Design Method	452
	Conclusion	*457*
	References	*457*

25. Quality Control of Concrete in Construction — 459–472
    25.1 Introduction — 459
    25.2 Statistical Principles — 459
    25.3 Application of Statistics in Concrete Mix Design — 463
    25.4 Evaluation of Concrete at Construction Sites — 465
    25.5 Load Test of Structures — 469
    References — *472*

26. Design of Structures for Storage of Liquids — 473–502
    26.1 Introduction — 473
    26.2 Design of Water Tanks — 473
    26.3 Details of IS 3370 (1967) — 473
    26.4 Durability Requirements — 474
    26.5 Details of Design Methods — 475
    26.6 Limit State Design Procedure — 475
    26.7 Crackwidth Analysis in Limit State Method — 476
    26.8 Deflection Analysis — 480
    26.9 Strength Analysis by Ultimate Limit State — 482
    26.10 Design of Sections Subjected to Bending and Tension — 482
    26.11 Modification of Crackwidth Formula in Sections with Bending and Tension — 483
    26.12 Recommended Procedure of Design — 485
    26.13 Alternate Method of Design — 485
    26.14 Design of Section Subjected to Tension Only — 486
    26.15 Design of Section Subjected to Bending and Tension — 487
    26.16 Design of Plain Concrete Tanks — 487
    26.17 Structural Analysis of Tank Walls — 487
    26.18 Design of Ground Slabs — 489
    26.19 Minimum Steel provided for Shrinkage and Temperature — 489
    References — *501*

27. Historical Development of Reinforced Concrete — 503–506
    27.1 Concrete in the Roman Period — 503
    27.2 Later Development in Concrete — 503
    27.3 Birth of Reinforced Concrete — 504
    27.4 Recent Development in Reinforced Concrete Design — 505
    27.5 Developments in India — 505
    27.6 Evolution of an International Code — 506

Appendix A   Calculation of Bending and Torsional Stiffness of Flanged Beams — **507–508**

Appendix B   Durability of Structural Concrete — **509–512**

Index — *513–515*

Preface

This text is the outgrowth of the lectures I have delivered to the postgraduate students in advanced reinforced concrete design. It is a sequel to my first book *Limit State Design of Reinforced Concrete* which has been well received by students, teachers and professionals alike. These two books together explain the provisions of IS 456 (2000). Besides, the text analyzes the procedures specified in many other BIS codes such as those on winds, earthquakes, and ductile detailing.

The book also discusses such modern topics as design of shear walls, design of beam–column joints, and inelastic analysis of reinforced concrete. An analysis of these topics is so necessary because, during the last two decades, considerable advances have taken place in the theory of design of reinforced concrete members which form the building blocks for design of all specialized reinforced concrete structures, for instance, tall buildings, bridges, and chimneys. Even though a number of textbooks on advanced reinforced concrete dealing in detail with specialized structures like bridges are available in India, hardly any book, covering the fundamentals of the modern advancements in the theory of reinforced concrete design, can be found. This text strives to address this need.

The book is lecture based, with each chapter discussing only one topic, and an indepth coverage being given to the fundamentals. Such treatment has been found most suitable for self-study and class-room teaching.

Since the book is also addressed to the practising engineers, often the topics have been discussed in greater detail than found necessary for the students. For example, the chapters on deep beams and flat slabs give more information than is covered in IS 456 (2000). Teachers are urged to orient their courses accordingly. They may restrict their lectures to the necessary fundamentals as required by the prescribed course. But I do believe that the practising engineers will find the detailed analysis very useful.

The book has a number of distinguishing features. Codes of other countries, specially of USA and UK, have been presented and compared with the Indian code to explain the fundamentals and expose the reader to international practices. Also, a large number of worked-out examples have been provided to illustrate the theory and to demonstrate their use in practical designs. Finally, a number of typical detailing of reinforced concrete members have been included, which will be found useful in field applications.

I fervently hope that both students and professionals will find the text stimulating and useful.
Any constructive suggestions for improving the contents will be warmly appreciated.

P.C. VARGHESE

Acknowledgements

I wish to acknowledge the help I received from various individuals and institutions for the preparation of this book. I am indebted to Prof. P. Purushothaman, formerly professor of civil engineering, Anna University for helping me in many ways in bringing out this book. He gave me invaluable assistance at various stages of manuscript preparation. I am also thankful to Prof. Natarajan who, with the assistance of Mrs. Rajeswari Srinivasan, helped me immensely in word processing the manuscript.

I am indebted to Anna University and the Vice Chancellor, Dr. A. Kalanidhi for giving me the opportunity to continue as honorary professor of civil engineering at the university. It has been a pleasure to work with the students and my colleagues in the teaching profession.

This book is a compilation from various books, notes, codes and other publications. I am indebted to all these authors and publishers; they are too numerous to be mentioned individually.

Acknowledgement is also due, especially to the Bureau of Indian Standards (BIS) for granting me permission to reproduce from their publications. Finally, the Publishers, Prentice-Hall of India, specially the editorial and production team, deserve my sincere appreciation for the great care they have taken in processing the manuscript.

P.C. VARGHESE

Introduction to IS 456 (2000)

Since IS 456 (1978) has been in use in India for the past 22 years, teachers in various technical institutions as well as practising civil engineers in India are very familiar with the clauses and tables of that code. Now, it has been revised by the Bureau of Indian Standards (BIS) and published in July 2000 as IS 456 (2000). Till the next revision is made it is mandatory that teaching and practice of the subject in India should be based on this latest code of practice. An introduction to the important revisions made is presented here, which can be used to get a full understanding of the changes and its impact on future design practice.

1. Make-up of the Revision

Of the six sections into which IS 456 (1978) was divided, the first-five sections have been retained in IS 456 (2000) also. The working stress method which constituted the sixth ection of the old version has now been discontinued and it is presented as Annex B, of this code. Accordingly, the status of working stress method as an alternate method for design of reinforced concrete has now been removed.

However, due to the introduction of References as Clause 2 in Section 1 of the new code as well as due to revision of the various clauses concerning durability in Section 2, there has been a one-digit-forward shift of the various clauses in Sections 3–5. For example, whereas stability was dealt with as Clause 19 in the 1978 version it is dealt with as Clause 20 in IS 456 (2000). In fact, there are not many changes in the contents of the various clauses regarding limit state design presented in Sections 3–5 of the codes. As far as limit state design clauses are concerned, the main changes are only the one-digit forward shift of the clause numbers as stated above and a complete change in numbers of the tables and figures of IS 456 (1978). A list of the changes in various sections is given below. For changes in the numbers of tables and figures, reference should be made to the BIS publication IS 456 (2000).

2. Changes in Section 2 on Materials Workmanship, Inspection and Testing

Most of the changes that have been made in IS 456 (1978) are in Section 2 of the Code. The revised code has incorporated many changes to give importance to durability of concrete structures, which of late had been neglected at the expense of too much importance on strength. The following are the important changes:

1. Recognition of all the three grades of OPC cements along with other types of cements [Clause 5.1 of IS 456 (2000)].
2. Enumeration of allowable minerals admixtures (Clause 5.2) and the approval of the practice of chemical admixtures (Clause 5.5).
3. The method of testing water for concreting has been described (Clause 5.4).
4. Characteristic strength of steel has been defined as the minimum yield or 0.2% proof stress (Clause 5.6.3).

5. The minimum strength of concrete for structural purpose is specified as M20, with minimum cement content including admixtures of 300 kg/m^3 and maximum water–cement ratio of 0.55 (Clauses 6.2.1.1 and Table 5). Grade up to M 80 has been included in the code (Table 2). The maximum cement content (not including admixtures) has been specified as of 450 kg/m^3 (Clause 8.2.4.2).

6. The value of modulus of elasticity of concrete is to be taken as $E_c = 5000\sqrt{f_{ck}}$ (in N/mm^2) (Clause 6.2.3).

7. Workability has been tabulated in terms of slump (Table 7).

8. The factors affecting durability have been fully explained (Clause 8). The new five environmental classification has been described (Table 3); the requirements to withstand sulphate attack of concrete has been expanded (Table 4); the minimum cement contents and allowable maximum water–cement ratio for different exposures (Tables 5 and 6) and the limit of chlorides in concrete (Table 7) have also been specified. Recommendation for using a minimum grade of M30 in sea-water construction as well as other precautions to be taken to protect-steel in saline atmosphere have been laid down (Clause 8.2.8). Alkali aggregate reaction prevention has also been dealt with in this important clause (Clause 8.2.5.4).

9. The recommended value to be used for standard deviation for concrete mix design under Indian conditions have been revised (Table 8). Its value remains constant at 5 N/mm^2 from M30 to M50.

10. Quality assurance factors have been clearly defined (Clause 10.1)

11. Tolerance limits for cover to steel fabrication has been specified (Clause 12.3.2).

12. The accuracy of measuring equipment for weighing cement, aggregates, water and for batching has been laid down (Clause 10.2.2).

13. Recommendations for curing of concrete has been made (Clause 13.5). Whereas ordinary portland cement concrete (OPC) requires seven days of good moist curing, portland puzzolana cement (PPC) concrete requires at least ten days of good curing to attain full strength.

14. Requirements of concrete for under-water placement has been described (Clause 14.2.2) and placing concrete by pumps has been included (Clause 14.2.4).

15. Simple acceptance criteria have been introduced for acceptance and quality control of small batches of concrete production (Clause 16 and Table 11).

Thus a large number of additional information has been provided in Section 2 to ensure durability and control of quality of concrete.

3. Changes in Section 3 on General Design Considerations

The text of the main clauses remain more or less the same as in earlier version. However as pointed out earlier, the one digit forward shift of clause numbers as well as the changes in table and figure numbers are the main difference between the two versions in this section. Following are the main changes.

1. Bases for design (Clause 18), the status of working stress method without using load factor as an alternate method has been discontinued. Design should be normally made by limit

state method, and working stress method is to be used only where the former method is not applicable; like carrying out serviceability limit state of deflection cracking etc.

2. Factor for stability against overturning against dead load should be 1.2 or 0.9 depending on its action and that for imposed load 1.4. The lateral sway due to transient loads should be limited to H/500 (Clause 20).

3. Fire resistance requirements are important additions to the code in this section. Minimum dimensional requirements for slabs, beams columns and walls as well as minimum cover requirements for different fire ratings have been specified (Fig. 1 Clause 21 and Table 16A).

4. Effective length of cantilevers has been defined (Clause 22.2.6c.).

5. Bending moment coefficient at mid-point of interior spans has been increased from 1/24 to 1/16 and brings its value to 3/4 the value at the support (Table 12).

6. The curves for the modification factor for tension reinforcement for checking deflection requirements of beams and slabs has been changed to represent the actual steel stress at service loads instead of the old curves based on types of steel and allowable stresses (Fig. 4).

7. The method for adjusting differences in support moments obtained in restrained slabs by Table of coefficients for two-way slabs has been described (Clause 24.4.1).

8. The concept of determining effective length of columns by 'stability index' defined in Annexure E has been introduced (Clause 25.2).

9. When considering biaxial bending it is sufficient to ensure that eccentricity exceeds the minimum about only one axis at a time (Clause 25.4).

10. Specifications for lap slices have been modified (Clause 26.2.5.1).

11. Strength of welds has been modified so that for joints in tension values of 100% strength can be taken if there is strict supervision and not more than 20% of the bars is welded (Clause 26.2.5.2).

12. The maximum spacing of main steel in slabs has been limited to 300 mm (Clause 26.3.3b).

13. The term 'nominal cover' has been defined and the nominal cover as well as the strength of concrete necessary to satisfy durability conditions have to be satisfied for various exposures (Clause 26.4 and Table 16).

14. The minimum covers for various fire ratings have been specified (Table 16A).

4. Changes in Section 4 on Special Design Requirements for Structural Members and Systems

The main changes in this section are the addition of the following topics. The one-digit forward shift of the clause numbers of the old version of the code and changes in the numbers of the tables and figures continue in this section also.

1. A new clause on design of reinforced concrete corbels has been added (Clause 28).

2. The Clauses regarding design of plain concrete walls have been modified to includ design of walls in horizontal shear also (Clause 32).

3. Mention has been made of precast joints and filler block construction (Clause 30.8).

5. Changes in Section 5 on Structural Design by Limit State

There has not been many changes in the contents of the various clauses of this section. However, the one-digit forward shift in clause numbers as well as changes in table and figure numbers have been carried over to this section also. The following are the main changes introduced in IS 456 (2000) Section 5.

1. For limit state of cracking guidance regarding width of cracks allowed in different environments have been specified (Clause 35.3.2).
2. The design shear strength for tension reinforcement values equal or less than 0.15 per cent has been added (Table 19).
3. The most important revision has been the increased shear strength close to the supports and rules regarding detailing of steel close to the support (Clause 40.5).
4. The general approach to design of reinforced concrete members in torsion has been explained in detail (Clause 41.1).
5. A new clause on deflection of flexural members has been added (Clause 42). Formulae for calculation of deflections have been presented in Annex C.
6. A new clause an cracking of flexural members has been added. (Flexural members is defined as one subjected to axial loads lesser than $0.2f_{ck}A_c$) (Clause 43). Formulae for crackwidth calculation are given in Annex F.

6. Section 6 of IS 456 (1978) on Structural Design (Working Stress Method)

As already stated this section has been deleted as all future routine design are to be carried out by limit state method. However, as there are situations where the concepts have to be used it has been put as Annexure B.

Conclusion: As can be seen from the above list there has not been any radical changes in Sections 3–5 of the code but there are many additions. The major changes are in Section 2, in which the importance of durability has been fully dealt with in IS 456 (2000). It has also been enriched with additional clauses regarding design for fire resistance, design of corbels, design of plain concrete walls and the British practice of considering enhanced shear close to support in favourable situations. It also gives guidances for calculation of deflection and crackwidth in members under bendings. In general we may say IS 456 (2000) has kept abreast with the rapid developments in concrete technology and reinforced concrete design.

CHAPTER 1

Deflection of Reinforced Concrete Beams and Slabs

1.1 INTRODUCTION

Excessive deflection of beams and slabs causes cracking of finishes, loss of strength of members, improper drainage and unsightly appearance. IS 456 (2000) Clause 23.2 and BS 8110 limit allowable deflection under service loads as follows:

1. The deflection in the members due to all causes (namely, loads as well as effects of temperature, creep, shrinkage, etc.) should not exceed span/250.

2. The deflection which will take place after completion of the main construction (including erection of partitions and applications of finishes) due to long-term effects of the permanent loads (i.e. due to creep and shrinkage) together with the deflection due to the transient load (that part of applied load that is applied and removed intermittently) should not exceed span/350 according to IS 456 and span/500 according to BS 8110 or 20 mm whichever is less.

The first condition refers to the deflection that can be noticed by the eye and the second condition is to prevent damages to the finishes. The empirical method to limit deflection is to limit span/depth ratio as given in IS 456 Clause 23.2.1. However, in marginal cases and in the case of special structures, deflection may have to be calculated. The accepted methods of calculations are discussed in Section 1.5.5.

1.2 SHORT-TERM DEFLECTION OF BEAMS AND SLABS

The instantaneous load deflection curve of a beam with increasing load is a flat S curve as shown in Fig. 1.1. It has five parts:

1. An almost linear portion OA with cracking of the beam starting at A due to tension in concrete.

2. A curvilinear cracked-portion ABC with increasing loads leading to decrease in the value of the moment of inertia I and hence rigidity EI.

3. A linear part beyond C extending to D, where EI remains more or less constant (in normal beams this portion corresponds to the service loads where the stress in concrete is of the order of one-third cube strength).

4. A curvilinear part DE where further cracking of concrete occurs.

5. An almost horizontal part beyond E where a small increase in load produces a large increase in deflection. (We should be aware that in hot countries where beams and slabs are directly exposed to solar radiation, variation of thermal stresses can also cause additional cracking leading to enhanced deflection).

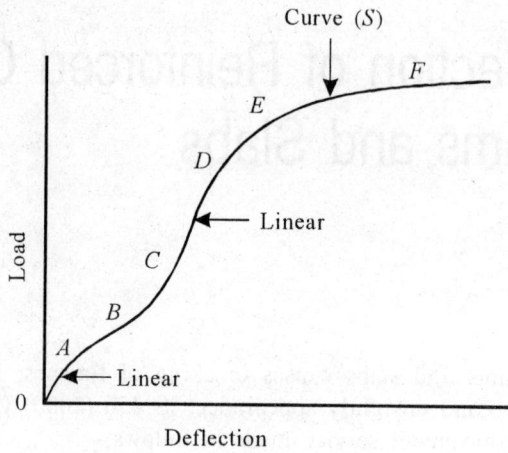

Fig. 1.1 Load deflection diagram of beams.

1.3 DEFLECTION DUE TO IMPOSED LOADS

The total load that comes on a structure is composed of *dead* and *applied* loads. The dead load and that part of the live load that always act on the structure is called the *permanent* load. The balance of the applied load that is applied off and on at short intervals of time is called the *transient* part of the applied load. The percentage of the applied load that may act as permanent load in buildings as recommended by the Australian code (AS 1988) are given in Table 1.1:

TABLE 1.1 LIVE LOAD FACTORS TO CALCULATE PERMANENT LOADS TO ESTIMATE DEFLECTION

Item	Long-term factor	Short-term factor
Roof with traffic	0.2	0.7
Floor—residential	0.3	0.7
Floor—office	0.2	0.5
Floor—retail	0.3	0.6
Floor—storage	0.5 to 0.8	1.0

With this information we can identify the following different aspects of deflection in beam and slabs due to applied loads:

1. Short-term deflection due to permanent load.
2. Short-term deflection due to varying part of applied load. Further, we have to add the effects of creep in increasing the deflection due to permanent load and also the deflection caused by shrinkage of concrete.

1.4 SHORT-TERM DEFLECTION OF BEAMS DUE TO APPLIED LOADS

The basic equation of deflection of beam is

$$a = \iint \frac{M}{EI} dx\, dx$$

The quantity $M/EI = 1/r = \psi$ is called the *curvature*. Deflection can be calculated either by moment area method or by the curvature method. The latter method has the following advantages:

1. The curvature-area theorem is equally applicable to elastic or plastic state as they are purely geometrical relationship between curvature, slope and deflection.
2. Unlike moment-area theorem it can be used also for deflections other than that due to bending, as for example, deflection due to shrinkage.

The formula for deflection of a simply supported beam due to a uniformly distributed load w is:

$$a = \frac{5}{384} \frac{wL^4}{EI}$$

Maximum bending moment,

$$M_{max} = \frac{wL^2}{8}$$

Therefore

$$a = \frac{5}{48} \frac{M_{max}}{EI} L^2$$

But $M/EI = 1/r = \psi$. Therefore,

$$a = \frac{5}{48} \psi_{max} L^2 = K \psi_{max} L^2 \qquad (1.1)$$

where

ψ_{max} = Maximum curvature in the beam = M_{max}/EI

E = Modulus of elasticity of concrete

I = Effective moment of inertia of the cracked section usually represented as I_{eff}

K = A constant depending on the type of beam and loading as given in Table 1.2.

TABLE 1.2 DEFLECTION COEFFICIENTS FOR SOME LOADINGS

S.No.	Loading	M_{max}	K
1	Beam with equal end moments M	M	0.125
2	Beam with moment M at one end	M	0.0625
3	UDL W on simply supported beam	$Wl^2/8$	0.104
4	UDL W with end moments M_A, M_B	M_A, M_C, M_B	$0.104[1 - (\beta/10)]$ $\beta = (M_A + M_B)/M_C$
5	Two point loads $W/1$, $W/2$ at αl	$\frac{W}{2}\alpha l$	$0.125 - (\alpha^2/6)$
6	Central point load with end moments	M_A, M_C, M_B	$0.083[1 - (\beta/4)]$ $\beta = (M_A + M_B)/M_C$
7	Point load W at αl	$W\alpha(1-\alpha)L$	$(3 - 4\alpha^2)/[48(1-\alpha)]$
8	Cantilever with point load W at αl	$W\alpha L$	$\alpha(3-\alpha)/16$
9	Cantilever with UDL W over αl	$\dfrac{W\alpha^2 L^2}{2}$	$\alpha(4-\alpha)/12$

There are two types of moment of inertia as under:

(i) The gross moment of inertia, $I_{gr} = bd^3/12$.

(ii) The moment of inertia of the transformed cracked section where concrete below the neutral axis is considered as ineffective = I_r (Table 87 of SP 16). I_{eff} will be less than I_{gr} but equal to and greater than I_r depending on the amount of cracking in the beam under the applied moment M.

1.5 CALCULATION OF DEFLECTION BY IS 456

The procedure recommended for calculation of deflection by IS 456 and SP 16 Section 7 is as under:

1.5.1 Calculation of Deflection Due to Load

As already derived, deflection is given by the formula

$$a = \frac{M_{max}}{EI_{eff}} KL^2$$

An empirical expression for the value of I_{eff} has been recommended by IS 456 Annexure C. This is similar but not the same as the expression in ACI. In BS 8110, $E_c I$ is calculated by a different approach.

$$I_{eff} = \frac{I_r}{1.2 - (M_r/M)(z/d)[1-(x/d)](b_w/b)} \quad (1.2)$$

where $I_r \leq I_{eff} < I_{gr}$ and

M_r = Cracking moment = $(I_{gr}/y_r)f_{cr}$

y_r = Distance of extreme tension fibre from controid of Section

f_{cr} = Modulus of rupture = $0.7\sqrt{f_{ck}}$

z = Lever arm in elastic theory = $d - x/3$

x = Depth of neutral axis (Tables 91–93 of SP 16)

d = Effective depth

b_w = Breadth of compression face

Figure 1.2 (which is Chart 89 of SP 16) is a readymade chart for reading of I_{eff}/I_r. As specified in the formula, the value of the applied moment M (under service load) should not be more than $4M_r$ as, otherwise, there will be considerable cracking and the equation will not be valid for calculation of deflections.

Accordingly, we can calculate the following:

Immediate deflection due to permanent load = a_{ip}
Immediate deflection due to total load = a_{it}

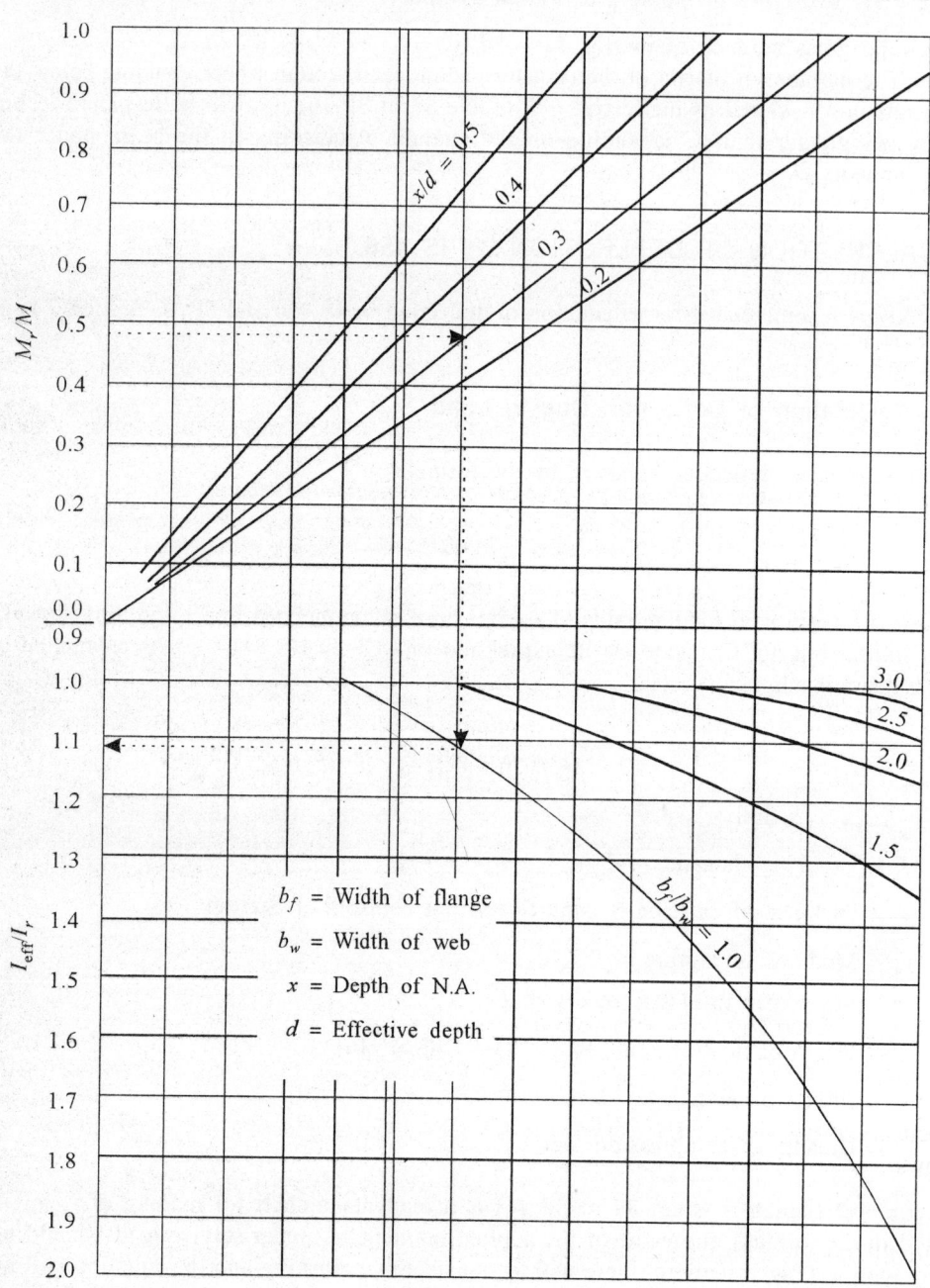

Fig. 1.2 Effective moment of inertia of section for calculation of deflection of beams. (Ref. SP 16 Chart 89)

1.5.2 Long-term Deflection Due to Creep by IS 456

Deflection caused by permanent loads goes on increasing with time due to creep of concrete. For taking the effect of creep of concrete, it is generally accepted that strain due to creep is proportional to the elastic strain (or stress) in concrete as long as the stress does not exceed one third the cube strength which is also the usual service stresses in reinforced concrete. We write the relationship as follows:

$$\text{Creep strain} = \text{Elastic strain} \times \text{Ultimate creep coefficient} = \varepsilon_c \theta$$

The ultimate creep coefficient, θ, is considered as the result of a number of factors at which the important factors are humidity, age of loading, composition of concrete, thickness of member and duration of loading. However, IS 456 takes into account only the effect of age in creep coefficient as shown in Table 1.3. Accordingly the long term strain $\varepsilon_{c\infty}$ can be expressed as follows:

$$\varepsilon_{c\infty} = \varepsilon_c + \varepsilon_c \theta = \varepsilon_c (1+\theta) \tag{1.3}$$

We modify the short-term value of the elastic modulus by taking

$$E_c = \frac{\text{Stress}}{\text{Strain}} = \frac{\text{Stress}}{\varepsilon_c}$$

We get

$$E_{c\infty} = \frac{\text{Stress}}{\varepsilon_c (1+\theta)} = \frac{E_c}{1+\theta}$$

TABLE 1.3 ULTIMATE CREEP COEFFICIENT θ
(IS 456 Clause 6.2.5.1)

Age at loading	Value of θ
7 days	2.2
28 days	1.6
1 year	1.1

As deflection is inversely proportional to creep, for calculation of long-term deflection with creep effects we simply multiply the short-term deflection by $(1 + \theta)$. Applying this principle to deflection due to permanent load, the long-term final deflection due to loads will be as follows if we consider I a constant value.

$$\text{Final deflection} = (\text{Initial deflection})(1 + \theta)$$

(Refer SP 16 amendment and Step 8, Example 1.1 for new value of I using $E_{c\infty}$.)

1.5.3 Graphical Representation of Deflection Due to Load

The various components of deflection due to the permanent loads and varying loads can be visualized as shown in Fig. 1.3.

8   ADVANCED REINFORCED CONCRETE DESIGN

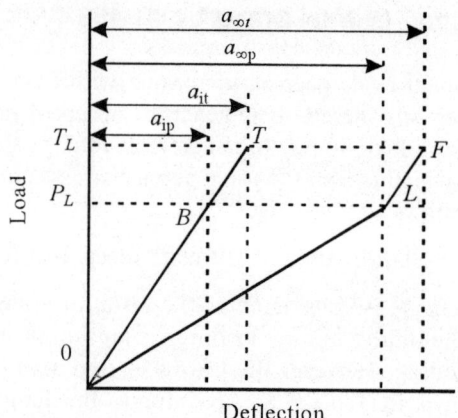

Fig. 1.3   **Approximate calculation of deflection of beams.**

1. The total elastic or immediate deflection due to all loads (dead and applied loads) is represented by OBT and designated as a_{it} (t for total). This is calculated by using E_c and I_{eff}.

2. The elastic deflection due to the permanent load is represented by point B. It is designated by a_{ip} (p for permanent).

3. The immediate deflection due to the varying part of the applied load then will be $(a_{it} - a_{ip})$ (i indicates immediate).

4. The long-term deflection including creep effects due to the permanent load acting on the beam will be represented by the line OL and the final deflection $a_{\infty p} = a_{ip}(1 + \theta)$ approximately.

5. Assuming that the deflection due to the varying part of the live load will be elastic, we have the total deflection due to all loads including the effect of creep as equal to $a_{\infty p} + (a_{it} - a_{ip})$. This is represented by point F in Fig. 1.3, LF being parallel to BT.

6. The long-term deflection due to creep will be equal to $(a_{\infty p} - a_{ip})$.

1.5.4. Deflection Due to Shrinkage

It is more appropriate to calculate shrinkage deflection by the curvature method. IS 456 Appendix B uses the following equation for shrinkage deflection:

$$a_{cs} = k_3 \psi_{cs} L^2 = k_3 \frac{k_4 \varepsilon_{cs}}{D} L^2 \qquad (1.4)$$

where

$k_4 = \varphi(p_t - p_c)/\sqrt{p_t} \leq 1.0$

$\varphi = 0.72$ for $(p_t - p_c) = 0.25$ to 1.0%

$\varphi = 0.65$ for $(p_t - p_c) > 1.0\%$

$k_3 =$ Constant depending on the support condition of the beams as given in Table 1.4.

$\varepsilon_{cs} =$ Shrinkage strain (average = 0.0003 and maximum = 0.0006)

$\psi_{cs} =$ Shrinkage curvature

$D =$ Total depth

$L =$ Span

TABLE 1.4 IN DEFLECTION DUE TO SHRINKAGE
(IS 456 Annexure C)

No.	Support condition	Coefficient k_3
1.	Cantilever	0.500 (1/2)
2.	Simple beams	0.125 (1/8)
3.	Continuous at one end only	0.086 (11/128)
4.	Continuous at both ends	0.063 (1/16)

k_3 can be derived from the curvature area theorem for deflection. For example, as shown in Fig. 1.4

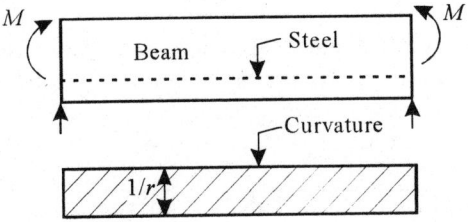

Fig. 1.4 Deflection of beams due to shrinkage.

for a simply supported beam undergoing deflection we use the theory of conjugate beam maximum deflection is bending moment at mid point.

$$a_{cs} = \frac{L}{2}\frac{M}{EI}\frac{L}{4} = \frac{1}{8}\psi L^2$$

Thus in Table 1.4, we have $k_3 = 1/8 = 0.125$ for a simply-supported beam.

1.5.5 Deflection Requirements According to IS 456

The two requirements in Section 1.1 can be expressed as follows:

1. (Long-term deflection due to permanent load) + (elastic deflection due to varying part of applied loads) + (shrinkage deflection) should not be more than $L/250$.

$$\alpha_{\infty p} + (\alpha_{it} - \alpha_{ip}) + \alpha_{cs} \not> \frac{L}{250} \qquad (1.5)$$

2. (Creep deflection) + (shrinkage deflection) + (elastic deflection due to varying part of applied load) should not be more than $L/350$.

$$(a_{\infty p} - a_{ip}) + a_{cs} + (a_{it} - a_{ip}) \not> \frac{L}{350} \qquad (1.6)$$

1.6 CALCULATION OF DEFLECTION BY BS 8110

The BS method is based on the recommendation of CEB-FIP model code using curvature area method. In this procedure, a reduction in the applied moment causing deflection is made, as in reality the concrete below the neutral can carry limited tension between the cracks. Its effect, called the *tension stiffening* can be looked upon as a reduction of moment causing deflection to $(M - \Delta M)$, where ΔM is the moment carried out by the tension in concrete. In addition, in BS method the curvature due to shrinkage is calculated more accurately than in IS method.

BS method does not assume I_{eff} as in IS 456 or ACI method. Instead it calculates $E_c I_r$ for short-term and long-term loadings separately by using the appropriate E_{ci} and $E_{c\infty}$ values.

1.6.1 Deflection due to Loads

As explained, in BS method, the concept of partially cracked section has been introduced. For convenience in calculation, the maximum tension (f_{ct}) at the C.G. of the steel of 1 N/mm² for short-term loading and 0.55 N/mm² for long-term loading independent of the applied moment is assumed in BS 8110. ΔM is the moment of these tensile forces about the neutral axis (Fig. 1.5).

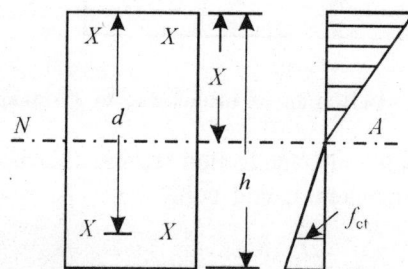

Fig. 1.5 Effect of tension in concrete on deflection of beams.

Assuming f_{ct} as tension at C.G. of steel we proceed as follows:

$$\text{Tension at the extreme tension fibre} = f_{ct} \frac{h-x}{d-x}$$

Taking moment of this triangular distribution of tension about N.A.

$$\Delta M = \frac{b(h-x)^3}{3(d-x)} f_{ct} \qquad (1.7)$$

$$\text{Moment causing deflection} = (M - \Delta M) \quad \text{and} \quad \frac{1}{r} = \frac{M - \Delta M}{EI}$$

$$\text{Deflection } a = K \frac{1}{r} L^2 = K \psi_{max} L^2$$

as in Section 1.4.

1.6.2 Deflection due to Shrinkage in BS 8110

The shrinkage curvature is to be separately calculated in BS 8110 by the following formula:

$$\psi_{cs} = \frac{m S_s \varepsilon_{cs}}{I_r}$$

where

ε_{cs} = Free shrinkage strain

m = Modular ratio = E_s/E_{eff}

E_{eff} = Long-term modulus, $E_c/(1 + \theta)$

I_r = Moment of inertia of cracked section using E_{eff}

S_s = First moment of area section using the centroid of the cracked section (neutral and axis) is equal to $\sum A_s e_s$, the moment of compression steel is taken as negative.

The above formula can be derived with reference to Fig. 1.6, Table 1.5 and Ref. 1.

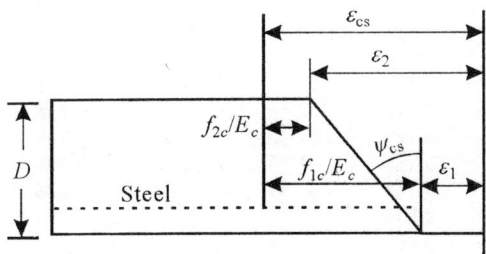

Fig. 1.6 Deflection due to shrinkage.

1.7 DEFLECTION CALCULATION BY EUROCODE

The Eurocode follows generally the same procedure as in BS 8110 except that the value of ΔM is modified by a correction factor to account for the fact that cracks are spaced at s and only portions between the cracks can carry the tension 2. Accordingly, the value of the curvature for deflection due to load ψ is modified as follows:

$$\psi \text{ (mean)} = \psi \text{ (cracked)} + (1 - \xi)\psi \text{ (uncracked)}$$

where

$\xi = 1 - \beta_1 \beta_2 (M_r/M)^2$

β_1 = 1 for HYD bars and 0.5 for plain bar

β_2 = 1.0 for short-term loads and 0.5 for sustained and repeated loads

Details of the method can be obtained in Ref. 2.

TABLE 1.5 MOMENT OF INERTIA OF CRACKED SECTION

(Ref: Table 87 of SP 16) $\left[d'/d = 0.05, I_r / \left(\dfrac{bd^3}{12} \right) \right]$

$p_t \times m$	\multicolumn{8}{c}{$p_c(m-1)/(p_t m)$}
	0.0	0.1	0.2	0.3	0.4	0.6	0.8	1.0
1.0	0.100	0.100	0.100	0.100	0.100	0.100	0.100	0.100
1.5	0.143	0.144	0.144	0.144	0.144	0.145	0.145	0.145
2.0	0.185	0.185	0.186	0.186	0.186	0.187	0.188	0.188
2.5	0.224	0.225	0.225	0.226	0.227	0.228	0.229	0.230
3.0	0.262	0.263	0.264	0.264	0.265	0.267	0.269	0.270
3.5	0.298	0.299	0.300	0.302	0.303	0.305	0.308	0.310
4.0	0.332	0.334	0.336	0.338	0.339	0.343	0.346	0.348
4.5	0.366	0.368	0.371	0.373	0.375	0.379	0.383	0.386
5.0	0.398	0.401	0.404	0.407	0.409	0.414	0.419	0.424
5.5	0.430	0.433	0.437	0.440	0.443	0.449	0.455	0.460
6.0	0.460	0.465	0.469	0.472	0.476	0.483	0.490	0.496
6.5	0.490	0.495	0.500	0.504	0.509	0.517	0.525	0.532
7.0	0.519	0.525	0.530	0.535	0.540	0.550	0.559	0.567
7.5	0.547	0.554	0.560	0.566	0.571	0.582	0.592	0.602
8.0	0.575	0.582	0.589	0.596	0.602	0.614	0.626	0.636
8.5	0.602	0.610	0.617	0.625	0.632	0.646	0.659	0.670
9.0	0.628	0.637	0.645	0.654	0.662	0.677	0.691	0.704
9.5	0.653	0.663	0.673	0.682	0.691	0.708	0.723	0.738
10.0	0.678	0.689	0.700	0.710	0.720	0.738	0.755	0.771
10.5	0.703	0.715	0.727	0.738	0.748	0.769	0.787	0.804
11.0	0.727	0.740	0.753	0.765	0.777	0.798	0.818	0.837
11.5	0.750	0.764	0.778	0.792	0.804	0.828	0.850	0.869
12.0	0.773	0.789	0.804	0.818	0.832	0.857	0.880	0.902
12.5	0.795	0.812	0.829	0.844	0.859	0.866	0.911	0.934
13.0	0.818	0.836	0.853	0.870	0.885	0.915	0.942	0.966
13.5	0.839	0.859	0.877	0.895	0.912	0.943	0.972	0.998
14.0	0.860	0.881	0.901	0.920	0.938	0.972	1.002	1.030
14.5	0.881	0.904	0.925	0.945	0.964	1.000	1.032	1.061
15.0	0.902	0.926	0.948	0.969	0.990	1.027	1.062	1.093
15.5	0.922	0.947	0.971	0.994	1.045	1.055	1.091	1.124
16.0	0.942	0.968	0.994	1.018	1.040	1.083	1.121	1.155
17.0	0.980	1.010	1.038	1.065	1.090	1.137	1.179	1.217
18.0	1.018	1.051	1.082	1.111	1.139	1.191	1.237	1.278
19.0	1.054	1.090	1.125	1.157	1.188	1.244	1.294	1.340
20.0	1.089	1.129	1.166	1.202	1.235	1.296	1.351	1.400
21.0	1.123	1.167	1.207	1.246	1.282	1.348	1.408	1.461
22.0	1.156	1.203	1.248	1.289	1.328	1.400	1.464	1.521
23.0	1.188	1.239	1.287	1.332	1.374	1.451	1.519	1.581
24.0	1.220	1.274	1.326	1.374	1.419	1.502	1.575	1.640
25.0	1.250	1.309	1.364	1.415	1.464	1.552	1.630	1.699
26.0	1.280	1.342	1.401	1.456	1.508	1.602	1.685	1.758
27.0	1.308	1.376	1.438	1.497	1.552	1.651	1.739	1.817
28.0	1.337	1.408	1.475	1.537	1.595	1.701	1.794	1.876
29.0	1.364	1.440	1.510	1.576	1.638	1.750	1.848	1.934
30.0	1.391	1.471	1.546	1.615	1.681	1.798	1.902	1.993

*Similar tables are available in SP 16 for $d'/d = 0.10, 0.15$ and 0.20.

TABLE 1.6 DEPTH OF NEUTRAL AXES—VALUES OF x/d BY ELASTIC THEORY
(Ref. SP 16 Table 91*) ($d'/d = 0.05$)

$p_t \times m$	\multicolumn{8}{c}{$p_c(m-1)/p_t m$}
	0.0	0.1	0.2	0.3	0.4	0.6	0.8	1.0
1.0	0.132	0.131	0.131	0.130	0.130	0.130	0.128	0.126
1.5	0.159	0.158	0.157	0.156	0.155	0.153	0.152	0.150
2.0	0.181	0.180	0.178	0.177	0.176	0.173	0.171	0.169
2.5	0.200	0.198	0.197	0.195	0.194	0.190	0.187	0.185
3.0	0.217	0.215	0.213	0.211	0.209	0.205	0.202	0.198
3.5	0.232	0.230	0.227	0.211	0.209	0.233	0.218	0.214
4.0	0.246	0.243	0.240	0.238	0.235	0.230	0.225	0.210
4.5	0.258	0.255	0.252	0.249	0.246	0.241	0.235	0.230
5.0	0.270	0.267	0.263	0.260	0.257	0.251	0.245	0.239
5.5	0.281	0.277	0.274	0.270	0.267	0.260	0.253	0.247
6.0	0.292	0.287	0.284	0.280	0.276	0.268	0.261	0.255
6.5	0.301	0.297	0.293	0.288	0.284	0.276	0.269	0.262
7.0	0.311	0.306	0.301	0.297	0.292	0.284	0.276	0.268
7.5	0.319	0.314	0.309	0.305	0.300	0.291	0.282	0.274
8.0	0.328	0.323	0.317	0.312	0.307	0.298	0.289	0.280
8.5	0.336	0.330	0.325	0.319	0.314	0.304	0.294	0.285
9.0	0.344	0.338	0.332	0.326	0.321	0.310	0.300	0.291
9.5	0.351	0.345	0.339	0.333	0.327	0.316	0.305	0.295
10.0	0.358	0.352	0.345	0.339	0.333	0.321	0.310	0.300
10.5	0.365	0.358	0.351	0.335	0.339	0.326	0.315	0.304
11.0	0.372	0.365	0.358	0.351	0.344	0.332	0.320	0.309
11.5	0.378	0.371	0.363	0.356	0.349	0.336	0.324	0.313
12.0	0.384	0.377	0.369	0.362	0.355	0.341	0.328	0.316
12.5	0.390	0.382	0.374	0.367	0.359	0.345	0.332	0.320
13.0	0.396	0.388	0.380	0.372	0.364	0.350	0.336	0.324
13.5	0.402	0.393	0.385	0.377	0.369	0.354	0.340	0.327
14.0	0.407	0.398	0.390	0.381	0.373	0.358	0.344	0.330
14.5	0.413	0.403	0.394	0.386	0.378	0.362	0.347	0.333
15.0	0.418	0.408	0.399	0.390	0.383	0.365	0.354	0.336
15.5	0.423	0.413	0.404	0.395	0.386	0.369	0.354	0.339
16.0	0.428	0.418	0.408	0.399	0.390	0.373	0.357	0.342
17.0	0.437	0.427	0.416	0.407	0.397	0.379	0.363	0.347
18.0	0.446	0.435	0.425	0.414	0.404	0.386	0.368	0.352
19.0	0.455	0.443	0.432	0.421	0.411	0.392	0.374	0.357
20.0	0.463	0.451	0.439	0.428	0.417	0.397	0.379	0.362
21.0	0.471	0.459	0.446	0.435	0.424	0.403	0.383	0.366
22.0	0.479	0.466	0.453	0.441	0.429	0.408	0.388	0.370
23.0	0.486	0.472	0.459	0.447	0.435	0.413	0.392	0.373
24.0	0.493	0.479	0.465	0.453	0.440	0.417	0.396	0.377
25.0	0.500	0.485	0.471	0.458	0.445	0.422	0.400	0.380
26.0	0.507	0.491	0.477	0.463	0.450	0.426	0.404	0.384
27.0	0.513	0.497	0.482	0.468	0.455	0.430	0.407	0.387
28.0	0.519	0.503	0.488	0.473	0.459	0.434	0.411	0.390
29.0	0.525	0.508	0.493	0.478	0.464	0.437	0.414	0.392
30.0	0.531	0.514	0.498	0.482	0.468	0.441	0.417	0.395

*Similar tables are available in SP 16 for $d'/d = 0.01$, 0.15 and 0.20.

1.8 ACI SIMPLIFIED METHOD

The ACI method uses the simple approach suggested in 1960 by Yu and Winter. It also assumes an effective moment of inertia proposed by Branson in 1971 [3]. It is based on a study of beams with M/M_r values varying from 2.2 to 4.0. and I_{gr}/I_r varying from 1.3 to 3.5. The equation for I_{eff} is given in Clauses 9.5 of ACI 318 as follows:

$$I_{eff} = \left(\frac{M_r}{M}\right)^3 I_{gr} + \left[1 - \left(\frac{M_r}{M}\right)^3\right] I_r$$

Its value should not be more than I_{gr}. Calculations for deflection can be made as in the IS 456 method using the above formula for I_{eff}.

1.8.1 Long-term Deflection in ACI Method

Additional long-term deflection due to combined effects of creep and shrinkage is obtained in ACI method by multiplying the short-term deflection by the following factor:

$$\lambda = \frac{t}{1+50\rho} \qquad (1.8)$$

where

ρ = ratio of compression steel A_{sc}/bd at midspan of simply supported or continuous beams. and at supports in cantilevers

t = time dependent factor taken as follows: 3 months = 1.0; 6 months = 1.2; 1 year = 1.4; and 5 or more years = 2.0

1.9 DEFLECTION OF CONTINUOUS BEAMS BY IS 456

IS 456 Annexure C specifies that deflection of continuous beams should be calculated by modified value of I_{gr}, I_r, M_r. The modification is given by the formula

$$X_e = k_1 \frac{X_1 + X_2}{2} + (1 - k_1) X_0 \qquad (1.9)$$

where

X_e = Modified value of X (I_{gr}, I_r, M_r)

X_1, X_2 = Values at support

X_0 = Value of the mid span.

k_1 = A coefficient given in Table 1.7 for values of k_2

$k_2 = (M_1 + M_2)/(M_{F1} + M_{F2})$

M_1, M_2 = Actual support moments

M_{F1}, M_{F2} = Theoretical fixed-end moments if the loads are placed only in the span with its ends fixed, this condition being considered as the standard to which all the others are considered relative.

TABLE 1.7 VALUES OF k_1 FOR CONTINUOUS BEAMS
(IS 456 Table 25)

k_2	0.5 or less	0.6	0.7	0.8	0.9	1.0	1.1	1.2	1.3	1.4
k_1	0	0.03	0.08	0.16	0.30	0.50	0.73	0.91	0.97	1.0

where

$$k_2 = \frac{M_1 + M_2}{M_{F1} + M_{F2}}$$

1.10 DEFLECTION OF CANTILEVERS

For deflection in cantilevers, effective span is taken as clear span plus one half effective depth of the cantilever. Deflection in cantilevers also depends on the rotation of its base. This depends on many factors especially when the cantilever forms part of a framed member. This effect is shown in Fig. 1.7.

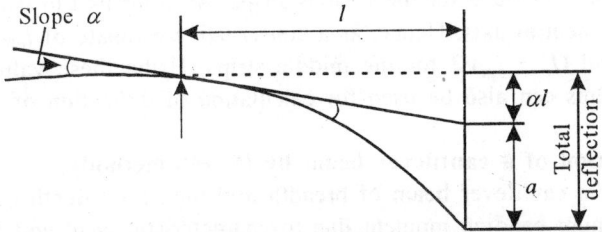

Fig. 1.7 Deflection of cantilevers.

1.11 DEFLECTION OF SLABS

The present empirical method calculating deflection of slabs is shown in Fig. 1.8.

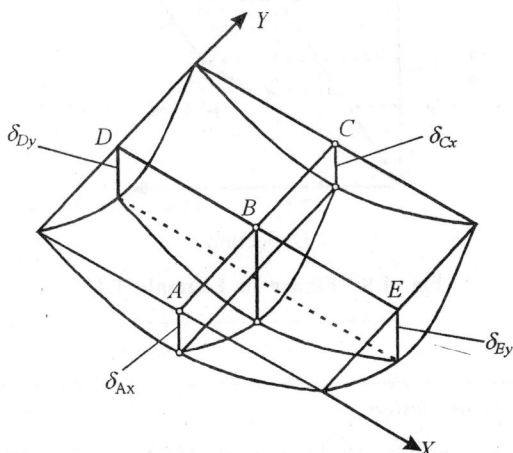

Fig. 1.8 Deflection of slabs: δ_{Bx} and δ_{By} are deflections of middle strips in X and Y directions respectively.

Maximum deflection of two-way slab = Mean mid-span deflection of column strip in the long direction + mean mid-span deflection of middle strip in the short direction.

Referring to Fig. 1.8 we take deflections δ in both directions along the strips. Average deflections are obtained as follows:

$$\delta_{1B} = \frac{1}{2}(\delta_{Ax} + \delta_{Cx}) + \delta_{By}$$

$$\delta_{2B} = \frac{1}{2}(\delta_{Dy} + \delta_{Ey}) + \delta_{Bx} \quad (1.10)$$

$$\delta = \frac{1}{2}(\delta_{1B} + \delta_{2B})$$

As deflection of column strips contribute more to the total deflection (as much as 75%) a proper estimate of I_{eff} should be made for the various strips. As compared to beams, slabs do not crack much below the neutral axis. Hence, as a conservative estimate of I_{eff}, we can use I_r for the column strip and $(I_r + I_{gr})/2$ for the middle strips. Using these values, the same procedure as used for beams can also be used for calculation of deflection of slabs.

EXAMPLE 1.1 (Deflection of a cantilever beam by IS 456 method)

Estimate the deflection of a cantilever beam of breadth 300 mm, total depth 600 mm, span 4 m subjected to a maximum bending moment due to characteristic dead and live loads of 210 kNm of which 60% is due to permanent loads.

Assume tension steel is 1.17%, compression steel 0.418% cover to centre of steel is 37.5 mm; creep factor = 1.6; shrinkage strain = 0.003; grade of concrete M, 15; grade of steel Fe, 415 (Example SP 16 (1980), p. 212)(see Fig. 1.9). [Example 12 of SP 16 (1980) of BIS].

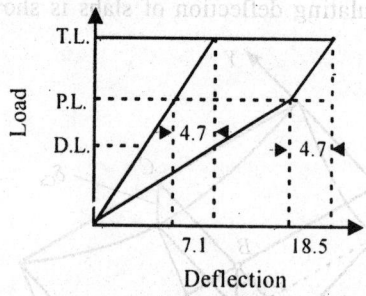

Fig. 1.9 Figure in Example 1.1.

Reference	Step	Calculations
	1	*Data for design* $E_c = 5700\sqrt{15} = 22.1 \text{ kN/mm}^2$ (IS 456(2000); $E_c = 5000\sqrt{f_{ck}}$) $E_s = 200 \text{ kN/mm}^2$

DEFLECTION OF REINFORCED CONCRETE BEAMS AND SLABS

Reference	Step	Calculations
IS Cl. 6.2.2	2	$\dfrac{E_s}{E_c} = \dfrac{200}{22.1} = 9.05$ Calculate cracking moment M_r $f_r = 0.7\sqrt{15} = 2.71 \text{ N/mm}^2$ $y_t = \dfrac{600}{2} = 300 \text{ mm}$ $I_{gr} = \dfrac{bd^3}{12} = \dfrac{300(600)^3}{12} = 5.4 \times 10^9 \text{ mm}^4$ $M_r = I_{gr}\dfrac{f_r}{y_t} = (5.4 \times 10^9)(2.71/300)$ $\quad = 48.8 \times 10^6 \text{ Nmm} = 48.8 \text{ kNm}$
	3	Check value of M/M_r $\dfrac{M}{M_r} = \dfrac{210}{48.8} = 4.30$ ACI limits application of the theory to ratios equal or less than 4. For values more than 4, cracking will be more than allowable. Let us assume that the theory is applicable in this case.
	4	Find x/d and z/d m short-term $\dfrac{E_s}{E_c} = \dfrac{200}{22.1} = 9.05$ $\dfrac{p_c(m-1)}{p_t(m)} = \dfrac{0.418 \times 8.05}{1.117 \times 9.05} = 0.333$ $d = 600 - 30 - \dfrac{15}{2} = 562.5 \text{ mm}$ $p_t(m) = 10.10;$ $\dfrac{d'}{d} = \dfrac{37.5}{562.5} = 0.06 \text{ (use } 0.05)$
Table 1.6		$\dfrac{x}{d} = 0.338$ $\dfrac{z}{d} = 1 - \dfrac{x}{3d} = 0.887$

18 ADVANCED REINFORCED CONCRETE DESIGN

Reference	Step	Calculations
Table 1.5	5	*Obtain* I_r $\dfrac{I_r}{bd^3/12} = 0.720$ $I_r = \dfrac{0.720 \times 300 \times (562.5)^3}{12} = 32.04 \times 10^8 \text{ mm}^4$
SP 16 Chart 89 Fig. 1.2	6	*Find* I_{eff} Either calculate by Eq. (1.2) or use SP 16 $\dfrac{I_{\text{eff}}}{I_r} = 1.0$ $I_{\text{eff}} = 3.204 \times 10^9 \text{ mm}^4$
Table 1.2	7	*Calculate immediate deflections* Cantilever deflection = $0.25(4)^2 \dfrac{M_{\max}}{EI}$ (i) M (total) = 210 kNm (ii) M due to permanent load = $0.6 \times$ Total load $\qquad\qquad\qquad\qquad\qquad = 0.6 \times 210 = 126$ kNm Immediate deflection: due to total load $a_{\text{it}} = \dfrac{0.25(4000)^2}{22.1 \times 10^3} \dfrac{210 \times 10^6}{3.204 \times 10^9} = 11.86$ mm Immediate deflection: due to permanent load $a_{\text{ip}} = \dfrac{11.86 \times 126}{210} = 7.12$ mm
Table 1.3 Step 3	8	*Determine total deflection due to permanent load* $1 + \theta = 1 + 1.6 = 2.6$ Long-term deflection due to permanent load = $(7.12)(2.6) = 18.51$ mm (*Note:* For more exact values, we recalculate x/d and I_{eff} using $E_{c\infty}$ to find $a_{\infty p}$ as in B.S. practice shown in Example 1.2 Step 3. This procedure is adopted in later editions of SP 16 as an amendment.)
	9	*Find deflection due to varying part of load* $a_{\text{it}} - a_{\text{ip}} = 11.86 - 7.12 = 4.74$ mm

DEFLECTION OF REINFORCED CONCERETE BEAMS AND SLABS 19

Reference	Step	Calculations
	10	Obtain deflection due to shrinkage
Sec. 1.5.4		$a_{cs} = k_3 \psi_{cs} L^2$
Table 1.4		$k_3 = 0.5$ for cantilever
		$p_t - p_c = 1.117 - 0.418 = 0.699 < 1.0$
		$k_4 = \dfrac{(0.72)(0.699)}{\sqrt{1.117}} = 0.476$
		assume $\varepsilon_c = 0.0003$
		$\psi_{cs} = \dfrac{k_4 \varepsilon_c}{D} = \dfrac{0.476 \times 0.0003}{600} = 2.38 \times 10^{-7}$
		$a_{cs} = (0.5)(2.38 \times 10^{-7})(4000)^2 = 1.90$ mm
		Assume one half occurs after 28 days = 0.95 mm
	11	Check deflection requirements—Condition 1
		Total deflection due to all causes $L/250$ = 16 mm
		Maximum deflection components
		1. Permanent load with creep = 18.51 mm
		2. Varying load with no creep = 4.74 mm
		3. Shrinkage (total) = 1.90 mm
		Total = 25.15 mm
	12	Check deflection requirement—Condition 2
		Total deflection after all construction $\not> L/350$
		$\dfrac{L}{350} = \dfrac{4000}{350} = 11.42$ mm
		Component of deflection after construction.
		1. Creep = (18.51 − 7.12) = 11.39 mm
		2. Shrinkage = 1/2 total = 0.95 mm
		3. Varying part of load = 4.74 mm
		Total = 17.08 mm
		Actual deflection is more than that is allowed.
		Note: Possible remedial measures are as follows:

Reference	Step	Calculations
		1. Revise the section by increasing depth
		2. Provide an upward camber during construction (to correct the total deflection) of 10 mm at the end.
		3. Reduce long-term effect by delaying application of applied loads that act as permanent loads (partitions and other non-varying applied loads).

EXAMPLE 1.2 (Deflection by BS 8110 method)
Estimate deflection of cantilever in Example 1.1 by the method recommended in BS 8110.

Reference	Step	Calculations
	1	*Design data* $$E_c = 22.1 \text{ kN/mm}^2; \quad m = \frac{200}{22.1} = 9.05$$
Table 1.3		$$E_{c\infty} = \frac{E_c}{1+\theta} = \frac{22.1}{1+1.6} = 8.5 \text{ kN/mm}^2$$ $$m_\infty = 200/8.5 = 23.5$$
	2	*Calculate EI for instantaneous deflection*
Ex. 1.1		$E_c I_r = (22.10)(32.04)(10^{11}) = 70.80 \times 10^{12}$
Step 5	3	*Determine EI for long-term deflection* $$m_\infty = 23.5 \text{ and } \frac{d'}{d} = 0.05$$ $$p_c(m-1) = 0.418 \times 22.5 = 9.405$$ $$p_t(m) = 1.17 \times 23.5 = 27.495$$ Ratio of $p_c(m-1)$ and $p_t(m) = \frac{9.405}{27.495} = 0.342$
Table 1.5		$$\frac{I_{r\infty}}{bd^3/12} = 1.50 \text{ (approx.) (refer step 8, Ex. 1.1)}$$ $$I_{r\infty} = 150 \frac{bd^3}{12} = 6.67 \times 10^9 \text{ mm}$$ $$E_{c\infty} I_{r\infty} = 8.5 \times 6.67 \times 10^{12} = 56.69 \times 10^{12} \text{ Nmm}^2$$ (*Note*: This value is less than $E_c I_r$ of Step 2.)

DEFLECTION OF REINFORCED CONCRETE BEAMS AND SLABS

Reference	Step	Calculations
Table 1.6	4	Find ΔM taken by concrete in short-term deflection $\dfrac{x}{d} = 0.388$ $x = 0.388 \times 489.7 = 190$ mm
Eq. 1.7		$\Delta M = \dfrac{b(h-x)^3}{3(d-x)} f_{ct}$ where $f_{ct} = 1$ N/mm^2 $= \dfrac{300(600-190)^3}{3(562.5-190)} = 18.5 \times 10^6$ Nmm $= 18.5$ kNm
Table 1.6	5	Calculate ΔM taken by concrete on long-term effects With values from Step 3 and assuming $d'/d = 0.05$ $\dfrac{x}{d} = 0.455$ $x = 0.455 \times 562.5 = 256$ mm $F_t = 0.55$ N/mm^2 (for long-term effects) $\Delta M = \dfrac{300(600-256)^3 \times 0.55}{3(562.5-256)} = 7.3$ kNm
Table 1.2	6	Determine short-term deflection due to permanent load $\dfrac{1}{r} = \dfrac{M - \Delta M}{EI} = \dfrac{(126-18.5) \times 10^6}{70.80 \times 10^{12}} = \dfrac{107.5 \times 10^6}{70.80 \times 10^{12}} = 1.52 \times 10^{-6}$ mm^{-1} $a_{ip} = k \dfrac{1}{r} L^2 = 0.25(1.52 \times 10^{-6})(4000)^2 = 6.08$ mm
Step 7, Ex. 1.1	7	Find short-term deflection due to varying load $M = 210 - 126 = 84$ kNm By proportion $a_{iv} = 6.08 \dfrac{84}{107.5} = 4.75$
	8	Obtain long-term deflection due to permanent load $M - \Delta M = 126 - 7.3 = 118.7$ kNm $\dfrac{1}{r} = \dfrac{M}{EI} = \dfrac{118.7 \times 10^6}{56.67 \times 10^{12}} = 2.09 \times 10^{-6}$ mm^{-1}

Reference	Step	Calculations
		Deflection $a_{\infty p} = 0.25 \dfrac{1}{r} L^2$
		$= 0.25(2.09 \times 10^{-6})(4000)^2 = 8.36$ mm
		(This is less than that got by IS method.)
	9	*Calculate deflection due to shrinkage*
		Shrinkage curvature $= \dfrac{1}{r} = \dfrac{E_{cs} m S_s}{I_{r\infty}}$
Sec. 1.6.2		$I_{r\infty} = 6.674 \times 10^9$ S_s and moment area of reinforcement about the natural axis
Step 3		$d = 600 - 37.5 = 562.5$ mm
		$A_{st} = \dfrac{1.117(300 \times 562.5)}{100} = 18.85$ mm^2 at top
		$A_{sc} = \dfrac{0.416(300 \times 562.5)}{100} = 702$ mm^2 at bottom
		Position of both steels = 37.5 mm from surface
		Depth N.A. (long-term effects) = 256 from bottom and 344 from top of beam.
		$S_s = 1885 \times (344 - 37.5) - 702(256 - 37.5)$
		$= 4.24 \times 10^5$ mm^3 (moment for comp. steel is $-$ve)
		$\psi = \dfrac{1}{r} = \dfrac{\varepsilon_{cs} m s_s}{I_{r\infty}}$
		$= \dfrac{0.003 \times 23.5 \times 4.24 \times 10^5}{6.67 \times 10^9} = 4.48 \times 10^{-6}$ mm^{-1}
		Deflection $= k_3 \psi_{cs} L^2 = 0.05(4.48 \times 10^{-6})(4000)^2 = 3.58$ mm
		(This is much larger than that got by IS method.)
	10	Check deflection requirements—Condition 1
		Total deflection: $\not> \dfrac{L}{250}$, i.e. 16 mm
		1. Long-term deflection: due to permanent load = 8.36 mm
		2. Varying load with no creep = 4.75 mm
		3. Shrinkage deflection (total) = 3.58 mm
		Total deflection = 16.69 mm
		Deflection is satisfactory.

Reference	Step	Calculations
	11	*Check deflection requirements—Condition 2* Deflection after construction $\not> \dfrac{L}{350}$, i.e. 11.42 mm 1. Deflection due to creep = 8.36 − 6.08 = 2.28 mm 2. One-half shrinkage deflection = 3.52/2 = 1.76 mm 3. Varying part of load = 4.75 mm Total = 8.82 mm Deflection is satisfactory.

REFERENCES

1. Hughes B.P. *Limits State Theory of Reinforced Concrete Design*, Pitman's Book Company, London, 1981.

2. Joseph Eibl, Ernst and Sahn (Eds.), *Concrete Structures,* Eurodesign Handbook, Berlin, 1994.

3. Branson D.E. *Deformation of Concrete Structures*, McGraw Hill, New York, 1977.

4. SP 16 (1990) and (Amendment), *Design Aids for Reinforced Concrete,* Bureau of Indian Standards, New Delhi.

CHAPTER 2

Estimation of Crackwidth in Reinforced Concrete Members

2.1 INTRODUCTION

The allowable crackwidth in reinforced concrete members are given in Table 2.1. The empirical methods to limit crackwidth in normal concrete structures due to normal loads are the following:

(i) Limit the allowable steel stresses in the members as is done in the design of water tanks.

(ii) Detail the steel in the members according to the rules laid down in the Code of Practice as given in IS 456 Clause 26.

However, under exceptional cases such as severe exposure conditions and in water-retaining structures we may check the crackwidth by theoretical calculations. Cracks can be produced also by shrinkage and temperature variations. Extra reinforcements to reduce such cracks are always necessary in reinforced concrete. The methods for calculation of widths of cracks due to loads as well as those due to shrinkage and temperature changes are discussed in this chapter. Calculation of crackwidth in water tanks recommended by BS 5337 is dealt with in Chapter 26.

TABLE 2.1 ALLOWABLE CRACKWIDTHS FOR BUILDINGS

Exposure condition	IS 456	BS 8110	ACI 318
1. Normal (in mm)	0.3	0.3	0.33 (outside)
			0.44 (inside)
2. Aggressive (in mm)	0.1	0.1	0.1

2.2 FACTORS AFFECTING CRACKWIDTH IN BEAMS

The main factors that affect crackwidth at a given point in a reinforced concrete beam are the following:

1. Distance of the point from the surface of the nearest main reinforcing bar a_{cr} (Fig. 2.1)

2. Spacing of the tension reinforcement (s)
3. Cover to steel (c). (Greater the cover the larger the crackwidth. Under normal condition, crackwidth is said to be 0.004 times the nominal cover.)
4. Stress level in steel and hence the tensile strain in concrete (ε_c) and in steel (ε_s).

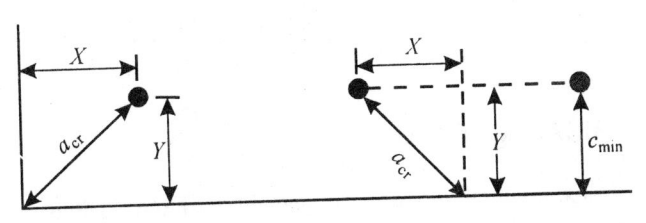

Fig. 2.1 Factors affecting crackwidth in beams.

2.3 MECHANISM OF FLEXURAL CRACKING

On gradual loading of a beam, the first cracks are formed when the tensile strain in concrete reaches the limiting value. We assume that the reinforcement and the concrete are bonded together and these cracks are formed at a regular spacing which will release the tensile strain in concrete on both sides of the cracks as shown in Fig. 2.2. Further loading increases

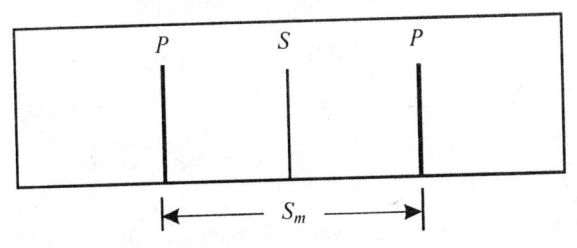

Fig. 2.2 Crack spacing in beams (P, Primary and S, Secondary cracks).

the width of the cracks. With bars having good bond characteristics there will be no slip between the bar and the concrete. Hence subsequent secondary cracks are formed to release the strain on further loading. The width of these cracks will be finer than that of the primary cracks. From statistical point of view we may use the following relation for the mean width of cracks W_{mean}.

$$W_{mean} = \text{(Tensile strain in concrete at the level considered)} \times \text{(Mean spacing of cracks)}$$
$$= \varepsilon_c \times S_m \tag{2.1}$$

Taking the maximum crackwidth will be twice the mean

$$W_{mean} = 2\varepsilon_c \times S_m \tag{2.2}$$

2.4 CALCULATION OF CRACKWIDTHS

The following are some of the proposed methods to estimate width of cracks in beams:

1. Simple empirical methods
2. IS 456 and BS 8110 methods [1, 2]
3. ACI method [3] [4]
4. CEB-FIP method modified as Eurocode method [5]

We shall restrict ourselves to the first two methods in this book and references 3, 4 and 5 can be used to study the other methods.

2.5 SIMPLE EMPIRICAL METHOD

The study of the empirical method gives us an insight into the various factors that affect crackwidth. In this method we arrive at suitable values of ε_c in Eq. (2.2) to arrive at the crackwidth.

2.5.1 Empirical Value of Mean Spacing of Cracks (S_m)

Test results show that spacing of cracks at a point depends on (a_{cr}) the distance of the point from the outside of the nearest tension bar. Hence,

$$S_m = (\text{Constant}) (a_{cr})$$

$$= 1.67 a_{cr} \text{ for deformed bars}$$

$$= 2.00 a_{cr} \text{ for plain bars}$$

As shown in Fig. 2.1

$$a_{cr} = \sqrt{x^2 + y^2} - \frac{\phi}{2} \qquad (2.3)$$

Where x and y are the horizontal and vertical distances of the point from the nearest bar surfaces and ϕ the diameter of the bar.

2.5.2 Mean Strain in Concrete ε_m

As plane sections remain plane even after bending the strains at a point of concrete and steel at the same level are the same.

$$E_c = \frac{\varepsilon_s \bar{y}}{d - x} \qquad (2.4)$$

where $\bar{y}$ is the distance of point from the neutral axis, ε_s is the steel strain in the tension steel and ε_c the concrete strain at $\bar{y}$. $(d - x)$ is the distance of steel from the neutral axis. Therefore,

$$\varepsilon_s = \frac{f_s}{E_s}$$

where f_s is the stress level in steel and $Es = 200 \times 10^3$ N/mm².

2.5.3 Expressions for Crackwidth

Using Eq. (2.1), we have

$$W_{mean} = S_m \varepsilon_c$$
$$= 1.67 a_{cr} \varepsilon_c \text{ for HYD bars}$$
$$= 2.00 a_{cr} \varepsilon_c \text{ for smooth bars}$$

with

$$W_{max} = 2 W_{mean}$$
$$= 3.33 a_{cr} \varepsilon_c \text{ for HYD bar}$$
$$= 4.00 a_{cr} \varepsilon_c \text{ for smooth bars}$$

Using this formula crackwidth at different positions can be estimated as follows:

1. Crackwidth at the surface of steel is zero as $a_{cr} = 0$.
2. Crackwidth at the neutral axis is zero as $\varepsilon_c = 0$.
3. Along the bottom of the beam, where ε_c is constant, crack will be the widest and a_{cr} is the largest at the point midway from the steel Fig. (2.1).
4. Cracks at the corners, position 2 in Fig. 2.1, a_{cr} is large and so cracks will also be large.
5. Along the side of the beam towards the bottom, ε_c is large but a_{cr} is not large. Towards the top ε_c is small but a_{cr} is large. It can be found that the widest crack occurs at about one-third the distance from the neutral axis and the level of the reinforcement.

The importance of the following empirical rules in IS 456 for crack control is evident from the above discussions.

(i) Rule regarding spacing of steel: If the spacing is too large, large cracks can occur between them. Smaller diameter bars at closer spacing is better than larger diameter bar further apart.

(ii) Rule regarding side reinforcement for beams of large depth greater than 750 mm.

From the above empirical procedure we can also deduce that cracks in concrete beams are diamond shaped as shown in Fig. 2.3 with no width at the level of the steel.

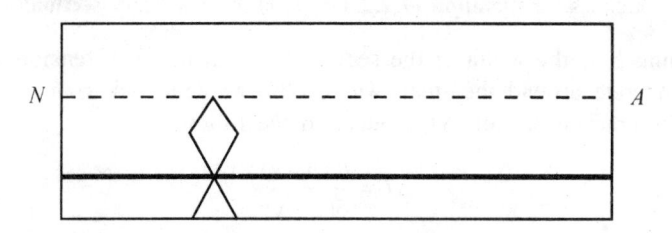

Fig. 2.3 Theoretical shape of crack in beams.

2.6 ESTIMATION OF CRACKWIDTH IN BEAMS BY IS 456 AND BS 8110

IS 456, Annexure F and BS 8110 specify the same formula for estimation of crackwidth in beams. The BS method is based on the work of Beeby [6] and is the best fit of extensive experimental data, where the stress in tension steel is limited to 0.8 f_y. It is based on the following assumptions:

1. $S_m = 1.5 a_{cr}$ (is a function of a_{cr})
2. w_{mean} is a function of S_m and ε_m
3. $w_{max} = 2 w_{mean}$

$$\text{Crackwidth } w_d = \frac{3 a_{cr} \varepsilon_m}{1 + 2(a_{cr} - c_{min})/(h-x)} \tag{2.5}$$

where

c_{min} = Minimum cover of tension steel
h = Overall depth of the member
x = Depth of neutral axis

The value of ε_m is the average strain at the level where the cracking is being considered. An approximate value for ε_m can be obtained by first calculating the steel strain on the basis of cracked section and then reduce it for the stiffening effect of concrete around the steel. Assuming that the concrete below the neutral axis can take a tension equal to 1N/mm² as shown in Fig. 2.4.

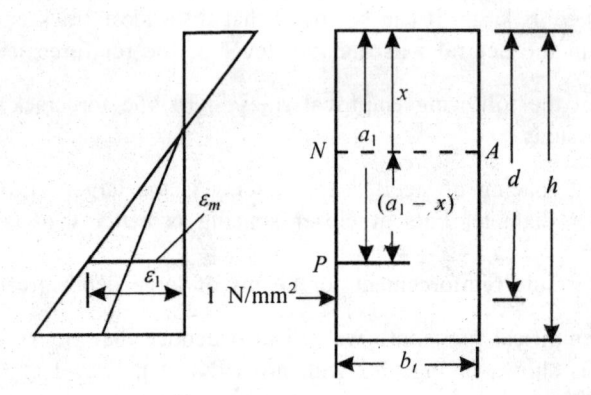

Fig. 2.4 Estimation of E_m for point P of a beam section.

Also, assuming b_t is the width of the section at the centroid of tension steel and ΔT the tension in the concrete around the steel, which is to be deducted from the tension in steel obtained from the cracked section ΔT reduces to the following:

$$\Delta T = \frac{b_t (h-x)}{3(d-x)}$$

[assumed]
$$\text{Strain (due to } \Delta T) = \frac{\Delta T}{A_s E_s} \tag{2.6}$$

Based on the above derivation the value of the ε_m at depth a_1, in Eq. (2.6), will be given by the following equation (Fig. 2.4).

$$\varepsilon_m = \varepsilon_1 - \frac{b_t(h-x)(a_1-x)}{3E_s A_s(d-x)} \quad (2.7)$$

where

ε_m = The average strain in concrete at the level where crackwidth is required. In cases, where it is expected that the concrete may be subjected to abnormally high shrinkage strain as much as 0.006, ε_m may be increased by adding 50% of the expected shrinkage strain in it. With good curing, the effect of shrinkage may be ignored

ε_1 = Strain at the level considered of a cracked section ignoring the stiffening effect of concrete in the tension zone and effective E_c taken as one-half the value to allow for creep effect (see Chapter 1)

b = Width of section at the centroid of tension steel

a_1 = Distance of compression face to point considered for crackwidth calculations

$$\varepsilon_1 = \frac{Mx_1}{I_c(0.5E_c)} \quad (2.8)$$

where E of concrete is taken as $0.5E_c$. x_1 the distance from the neutral axis to the point where strain is required. I_c is the moment of inertia of the cracked section.

It can be seen from Fig. 2.4 that the second term in Eq. (2.7) represents 'the stiffening effect of concrete' assuming concrete at level of steel can take a tension of 1N/mm^2. A negative value of ε_m denotes that the section is uncracked. For a section which is subjected to bending and tension, the value of $h - x$, may be interpolated by the following conditions:

1. When the neutral axis is at the least compressed face then x is zero, so $h - x = h$
2. For axial tension where the whole section is in tension, $h - x$ is taken as $2h$.

2.6.1 Specifying Service Stress for Limiting Crackwidth

Reducing the stress level in steel is one of the easiest methods of reducing crackwidths. Let f_s be the steel stress. The strain ε_1 at point P in Fig. 2.4 is given by

$$\varepsilon_1 = \frac{f_s(h-x)}{E_s(d-x)} \quad (2.9)$$

Substituting this value in Eq. (2.7) the value of ε_m can be found in terms of f_s. By specifying the value of the design crackwidth at the point where the crackwidth will be maximum the value of the limiting f_s 'the stress for the designed crackwidth' can be determined. This method is usually used for limit state design of liquid retaining structures based on BS 5337 and is dealt with in Chapter 26. Both BS 8110 and BS 5337 give similar formulae for crackwidth. However, there are slight differences in the expressions due to the following reasons:

1. BS 8110 allows crackwidth up to 0.3 mm whereas in BS 5337 it is to be limited to lower values (see Chapter 26).

2. In BS 8110, the probability of the crackwidth exceeding the mean design crack is taken as 20% whereas in BS 5337 it is taken as only 5%.

2.6.2 Crackwidth under Long-term Loading

The effect of sustained loading (creep effect) on crackwidth is at present only qualitatively understood. At present, the effect is taken in calculation by the reduced value of E_c to find the depth of the neutral axis and by adding a percentage of the excepted shrinkage strain into ε_m.

2.7 SHRINKAGE AND THERMAL CRACKING

Shrinkage and temperature changes produce strains in concrete. Extra reinforcements are to be provided in concrete to take care of these strains.

2.7.1 Shrinkage Strains in Concrete

The final shrinkage strain in concrete is of the order of 3×10^{-4}. If concrete is allowed to shrink freely there will be no cracking but restrains produce tension in concrete if reinforcements are introduced, it goes into compression (IS 456 Section 6.2.4).

Total shrinkage = (Tensile strain concrete) + (Compression strain in steel)

$$\varepsilon_{cs} = \varepsilon_{ct} + \varepsilon_{sc} = \frac{f_{ct}}{E_s} + \frac{f_{sc}}{E_s}$$

From force equilibrium of tension and compression

$$A_c f_{ct} = A_s f_{sc} = \rho A_c f_{sc}$$

where

ρ = Steel ratio A_s/A_c

f_{ct} = Concrete stress in tension

f_{sc} = Steel stress compression

If we designate the steel ratio at which f_{ct} and f_{sc} simultaneously reach their maximum allowed values. ρ is the critical steel ratio. Then, taking f_{ct} as the tensile strength of concrete. We get

$$\rho_c = \frac{f_{ct}}{f_y} = \frac{0.24\sqrt{f_{ck}}}{f_y} \qquad (2.10)$$

(Another expression for $f_{ct} = 0.12(f_{ck})^{0.7}$ and in any case f_{ct} is less than the modules of rupture f_{cr} given in IS 456). If ρ is less than ρ_c, the steel will reach yield and the cracking will be random. So that it is always better to have rightly more than the critical steel ratio in all structures.

with

$$f_{ck} = 25 \quad \text{and} \quad f_y = 415$$

$$\rho_c = \frac{1.25}{415} = 0.003$$

Hence the steel should not be less than 0.3%. Values of f_{sc}, f_{ct} and ε_{ct} are given by the following expression:

$$f_{sc} = \frac{\varepsilon_{cs} E_s}{1+mp}, \quad f_{ct} = \rho f_{sc}, \quad \varepsilon_{ct} = \frac{f_{ct}}{E_c} \tag{2.11}$$

The value of ε_{ct} varies with the spacing of cracks but the mean value can be taken as $0.5\varepsilon_{ct}$.

2.7.2 Combined Thermal and Shrinkage Strains

Temperature changes in concrete can be divided into two parts:

1. Temperature drop of concrete from its maximum value during hydration to the ambient temperature T_1
2. The seasonal variation to $\pm T$ degrees after hardening. In this case,

$$\varepsilon_{temp} = \alpha TR$$

where

α = Co-efficient of thermal expansion (Table 2.2)

T = Variation of temperature (drop or rise in temperature)

R = Restraint factor usually taken as 0.1 to 0.4 for ordinary cases and 0.8 to 1.0 i rigid restraints

$$\text{Total strain } \varepsilon = \varepsilon_{cs} + \varepsilon_{temp} \frac{-\varepsilon_{ct}}{2} \tag{2.12}$$

where ε_{cs} and ε_{ct} are strains due to shrinkage and tension in concrete, respectively.

Many authors express the capacity of concrete to withstand cracking in terms of the maximum strain depending on the type of concrete as given in Table 2.2.

TABLE 2.2 THERMAL EXPANSION OF CONCRETE
(for concrete with normal cement contents)

Type of aggregate in concrete	Coefficient of expansion (°C)	Tensile strain capacity
Limestone	8.0×10^{-6}	90×10^{-6}
Granite, basalt	10.0×10^{-6}	80×10^{-6}
Gravel, quartzite	12.0×10^{-6}	70×10^{-6}

32   ADVANCED REINFORCED CONCRETE DESIGN

2.7.3 Spacing of Cracks due to Shrinkage and Temperature

Method 1

Using the assumption that at the crack there is no stress, and between the cracks there is no slip between concrete and steel with bars diameter ϕ spaced at s, we get

$f_{ct} A_c$ = (Bond) (Perimeter of bars) (Crack spacing)

$\Sigma 0$ = Total perimeter = $4\rho A_c/\phi$

$f_{ct} A_c = (f_b)(4\rho A_c/\phi)(s)$

$s = (f_{ct}/f_b)(\phi/4\rho)$

where

ρ = Steel ratio

s = Spacing of cracks

f_b = The bond of strength

ϕ = Diameter of bar

Experiments show that $f_b/f_{ct} = 1$ for plain bars and 1.5 for deformed bars. Accordingly, $s = \phi/(4\rho)$ for plain bars and it is $\phi/(6\rho)$ for deformed bars. (Mean value = $\phi/5\rho$)   (2.13)

Method 2

The B.S. crackwidth formula can also be used to calculate the width of these cracks as shown in Example 2.1.

EXAMPLE 2.1   (Calculation of crackwidth by IS 456 and BS 8110)
A beam of width 450 mm, depth 750 mm and cover of reinforcement 40 mm is reinforced with 3 rods of 40 mm (3780 mm^2) diameter. Calculate the crackwidth when the section is subjected to a bending moment of 490 kNm at the following points.

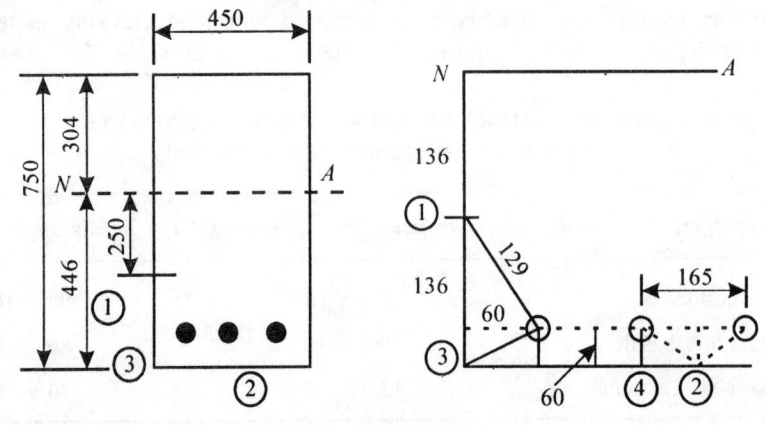

Fig. 2.5   Example 2.1.

ESTIMATION OF CRACKWIDTH IN REINFORCED CONCRETE MEMBERS

1. At a point on the side of the beam 250 mm below the neutral axis
2. At a point midway between bars at the tension face
3. At the bottom corner
4. At the tension face directly under the bar

Assume f_{ck} = 25 N/mm² and f_y = 415 N/mm². Use the method recommended by IS 456 and BS 8110.

Reference	Step	Calculations
	1	*Calculate the position of the N.A.* $b = 450; h = 750; d = 690; f_{ck} = 25;$ $E_c = 5000\sqrt{f_{ck}} = 25 \times 10^3$ N/mm² $E_{c\infty} = \dfrac{E_c}{2} = 12.5$ kN/mm² $m = \dfrac{E_s}{E_c} = \dfrac{200}{12.5} = 16$ $p_t = \dfrac{3780 \times 100}{450 \times 690} = 1.21$ $mp_t = 1.21 \times 14.29 = 17.3$
SP 16 Table 91 (Table 1.6 of Text)		$\dfrac{x}{d} = 0.44$ $x = 0.44 \times 690 = 304$ mm $d - x = 386$ mm
	2	*Determine strain in concrete* *By method 1 from stress in concrete* Find I_r and f_c
SP 16 Table 88 (Table 1.5)		$\dfrac{d'}{d} = \dfrac{60}{690} = 0.08$ (use 0.1) $p_t m = 17.3$; coefficient = 0.991 $I_{cracked} = \dfrac{0.991 \times 450 \times 690^3}{12} = 1.22 \times 10^{10}$ mm⁴ $\varepsilon_1 = \dfrac{f_c}{E_c} = \dfrac{M}{I} \dfrac{y}{E_c} = \dfrac{490 \times 10^6 \times 250}{1.22 \times 10^{10} \times 14 \times 10^3} = 7.2 \times 10^{-4}$ *By method 2 from stress in steel*

Reference	Step	Calculations
		$f_s = \dfrac{M}{(d-x/3)A_s} = \dfrac{490 \times 10^6}{589 \times 3780} = 220 \text{ N/mm}^2$
		Strain in steel $= \dfrac{220}{220 \times 10^3} = 1.0 \times 10^{-3}$
		$\varepsilon_1 =$ strain in concrete at 250 mm
		$= \dfrac{1.0 \times 10^{-3} \times 250}{386} = 6.4 \times 10^{-4}$
	3	Calculate crackwidth at side 250 mm below N.A.
		$h - x = 750 - 304 = 446 \text{ mm}$
		$\varepsilon' = \dfrac{b(h-x)}{3E_s A_s} \dfrac{a-x}{d-x}$
		$= \dfrac{450 \times 446 \times 250}{3 \times 200 \times 10^3 \times 3780 \times 386} = 0.6 \times 10^{-4}$
		$\varepsilon_m = \varepsilon_1 - \varepsilon' = (7.2 - 0.6) \times 10^{-4} = 6.6 \times 10^{-4}$
		$a_{cr} = \sqrt{136^2 + 160^2} - 20 = 190 \text{ mm}$
		$w = \dfrac{3 \times 190 \times 6.6 \times 10^{-4}}{1 + 2(190-40)/446} = 0.225 \text{ mm}$
	4	Calculate crackwidth midway between bars at tension face
		$\varepsilon_1 = \dfrac{1.0 \times 10^{-3} \times 446}{386} = 11.5 \times 10^{-4}$
		$\varepsilon' = \dfrac{b(h-x)}{3E_s A_s} \dfrac{446}{386} = 1 \times 10^{-6}$
		$\varepsilon_m = \varepsilon_1 - \varepsilon' = 11.7 \times 10^{-6}$
		$a_{cr} = \sqrt{82.5^2 + 60^2} - 20 = 82 \text{ mm}$
Eq. (2.5)		$w = \dfrac{3 \times 82 \times 1.7 \times 10^{-4}}{1 + (2 \times 42)/446} = 0.035 \text{ mm}$
	5	Calculate crackwidth at bottom corner
		$\varepsilon_m = 11.7 \times 10^{-4}$
		$a_{cr} = \sqrt{60^2 + 60^2} - 20 = 64.85 \text{ mm}$

ESTIMATION OF CRACKWIDTH IN REINFORCED CONCRETE MEMBERS 35

Reference	Step	Calculations
	6	$w = \dfrac{3 \times 64.85 \times 11.7 \times 10^{-4}}{1 + 2(64.85 - 40)/446} = 0.20$ mm Calculate crackwidth under the bar on tension face $\varepsilon_m = 11.7 \times 10^{-4}$; $a_{cr} = 40$ mm
Eq. (2.5)		$w = 3 a_{cr} \varepsilon_m = 3 \times 40 \times 11.7 \times 10^{-4} = 0.14$ mm

EXAMPLE 2.2 (Critical steel ratio to avoid cracking of walls)
A reinforced concrete wall is 100 mm thick. Determine the critical ratio of horizontal steel required to control shrinkage cracks with grade 25 concrete and grade 415 steel. If 10 mm grade 415 steel at 300 mm spacing is provided on both faces of the wall estimate the spacing of the cracks and its width due to a drop in temperature of 20°C.
  Assume the following data:

1. Drying shrinkage of concrete = 100×10^{-6}
2. Coefficient of thermal expansion = 10 μ/°C
3. Modulus of concrete in tension $E_c = 10 \times 10^3$ N/mm²
4. Strength of concrete in tension = 1.5 N/mm²

Reference	Step	Calculations
Text Sec. 2.7.1	1	Determine critical steel ratio ρ_c $\rho_c = \dfrac{f_{ct}}{f_y} = \dfrac{1.5}{415} = 0.0036$
	2	Calculate spacing of cracks with 10 mm @ 300 10 mm @ 300 on both faces gives $\rho = \dfrac{2 \times 262}{100 \times 1000} = 0.0052 > 0.0036$
Eq. 2.13		Maximum spacing $= s = \dfrac{\phi}{5\rho}$ $s = \dfrac{10}{(5)(0.0052)} = 385$ mm
Sec. 2.73		Secondary cracks at $s/2$, i.e. 190 mm (say)
	3	Determine contraction strain ε = Due to shrinkage + Due to drop in temperature − Due to elongation of concrete because of tension 1. Due to shrinkage = 100×10^{-6} (given)

36 ADVANCED REINFORCED CONCRETE DESIGN

Reference	Step	Calculations
		2. Drop in temperature = $10 \times 10^{-6} \times 20 = 200 \times 10^{-6}$
		3. Due to elongation of concrete = f_{ct}/E_{ct}
		$$= -\frac{1.5}{10 \times 10^3} = -150 \times 10^{-6}$$
		Item 3 varies from zero at crack to maximum at the middle between cracks. Hence the average strain = -75×10^{-6}
		$\varepsilon = (100 + 200 - 75) \times 10^{-6} = 225 \times 10^{-6}$
	4	*Estimate the crackwidth*
		w = Spacing × Strain = $385 \times 225 \times 10^{-6}$ = 0.087 mm
	5	*Calculate crackwidth from B.S. formula*
		Assume clear cover of 25 mm and steel as 10 mm at 300 spacing.
		$a_{cr} = \sqrt{150^2 + 30^2} - 5 = 148$ mm
	6	*Spacing of cracks*
		$s = 3a_{cr} = 3 \times 148 = 444$ mm
	7	*Crackwidth*
Eq. (2.5)		$$w = \frac{3a_{cr}\varepsilon_m}{1 + 2(a_{cr} - c_{min})/2h}$$
		$$= \frac{3 \times 148 \times 225 \times 10^{-6}}{1 + 2(148 - 25)/200} = 0.045 \text{ mm @ 445 mm spacing}$$
		Note: The second method gives smaller crackwidth than the first empirical method. All these methods give only an estimate of the magnitude of crackwidth and its spacing. In actual practice the degree of restraint also affect the results.

EXAMPLE 2.3 (Early thermal cracking)

A concrete wall 600 mm thick is reinforced with 10 mm bars at 100 mm centres in the horizontal direction and 20 mm at 250 mm in the vertical direction on both the faces. The vertical steel in the outer layer with clear cover of 25 mm so that the clear cover to horizontal steel is 40 mm. Check the wall for cracks due to early thermal cracking for a change in temperature of 32°C, assuming $\alpha = 12 \times 10^{-6}$°C and that the net strain ε_r for early thermal cracking can be expressed by the following formula: $\varepsilon_r = 0.8T\alpha R$. where T is the rise in temperature and R is the restraining factor. Use the IS formula,

$$w = \frac{3a_{cr}\,\varepsilon_r}{1 + 2(a_{cr} - c_{min})/(h - x)}$$

ESTIMATION OF CRACKWIDTH IN REINFORCED CONCRETE MEMBERS 37

Reference	Step	Calculations
	1	*Value of R* Let us consider the base of the wall where the restraining factor will be high 0.8 (on the top, the restraining factor will be small of the order of 0.2 to 0.4).
	2	*Calculate strain for 20°C temperature* $\varepsilon_r = 0.8 T \alpha R$ $= (0.8)(32)(12 \times 10^{-6})(0.8)$ $= 2.46 \times 10^{-4}$
	3	*Check horizontal bars for vertical cracks* Horizontal steel spacing = 10 mm at 100 mm centres Cover to centre of steel = 50 mm Crackwidth is maximum midway between rods. $a_{cr} = \sqrt{50^2 + 50^2} - 5 = 65.7$
Eq. (2.5)	4	*Calculate crackwidth* Assume $x = h/2$; $h-x = 300$ mm $w = \dfrac{3 a_{cr} \varepsilon_r}{1 + 2(a_{cr} - C_{min})/(h-x)}$ $= \dfrac{(3)(65.7)(2.46 \times 10^{-4})}{1 + 2(65.7 - 45)/300}$ $= 0.04$ mm < 0.3 mm *Note:* Horizontal steel controls the cracks normal to that steel, i.e. vertical cracks. Similarly, we check vertical bars for horizontal cracks.
	5	*Check vertical bars for horizontal cracks* Assume 20 mm bars at 250 mm in the vertical directions as outer layer. Let the clear cover be 25 mm. $a_{cr} = \sqrt{125^2 + (25+10)^2} - 10$ mm $= 120$ mm $w = \dfrac{(3)(120)(2.46 \times 10^{-4})}{1 + 2(120 - 25)/300}$ $= 0.05$ mm < 0.3 mm This is allowable.

REFERENCES

1. IS 456 *Code, Practice for Plain and Reinforced Concrete* (4th revision), Bureau of Indian Standards, 2000).

2. BS : 8110 British Standard, *Structural Use of Concrete*, Part 2, British Standards Institution, London, 1989.

3. ACI 318, *Building Code Requirements for Reinforced Concrete*, American Concrete Institute, Detroit, 1989.

4. Ferguson P.M., Breen J.E. and Jirsa J.O. *Reinforced Concrete Fundamentals*, 5th ed., John Wiley & Sons, New York.

5. Eibl J., *Concrete Structure—Eurodesign*, Handbook, Ernst & Sohn, Berlin, 1996.

6. Beeby A.W. and Beeby J.R. *Proposals for the Control of Deflection in the New Unified Code for Structural Concrete*, Vol. 3, No. 3, March 1969.

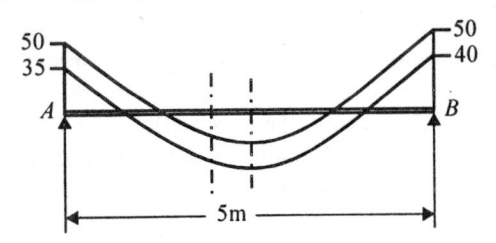

CHAPTER **3**

Redistribution of Moments in Reinforced Concrete Beams

3.1 INTRODUCTION

Laboratory tests on fixed and continuous R.C. beams have conclusively proved that the indeterminate structures redistribution of moments takes place in R.C., as that happen in steel. Because of the limited rotation capacity of R.C. sections, the allowable redistributions in the moments are not as much as in steel structures. More so, these limited allowable redistributions can improve the steel placement in R.C. beams and slabs. Hence the subject of redistribution of moments has received much attention and rules have been laid down in R.C. codes for the redistribution of moments.

It should be remembered, at the outset the redistribution of moments is allowed only when the moments have been obtained by theoretical methods of elastic analysis of structures such as moment distribution, slope-deflection analysis. Moments obtained by bending moment coefficients, as in the case of direct design method or tables of coefficients of two-way slabs, are not allowed to be redistributed.

This chapter examines the general requirements laid down by IS, BS and ACI codes for normal redistribution of moments in R.C. members, and explains how the design calculations can be carried out. For larger redistribution of moments we can use inelastic methods of analysis discussed in Chapter 22.

3.2 REDISTRIBUTION OF MOMENTS IN A FIXED BEAM

Let us take the example of a fixed beam as shown in Fig. 3.1 of length L and subjected to a uniformly-distributed load w per unit length. Let $w = 24$ kN/m and $L = 5$ m.

Fig. 3.1 Redistribution of moments in beams.

The bending moment diagram in elastic behaviour is shown in the figure and the values of the salient moments are given below:

$$M_A = M_B = \frac{wL^2}{12} = \frac{(24)(5)(5)}{12} = 50 \text{ kNm}$$

$$M_{\text{centre}} = \frac{wL^2}{24} = \frac{(24)(5)(5)}{24} = 25 \text{ kNm}$$

The points of contra-flexure are at $0.21L$ (1.05 m) from either ends.

It is required to reduce the value of the fixing moments M_A by 30% (i.e. $M_A = 0.7 \times 50 = 35$ kNm) and M_B by 20% (i.e. $M_B = 50 \times 0.8 = 40$ kNm) and plot the resultant bending-moment diagram.

By finding the point where the shear force is zero we can calculate the distance x_m of the point of maximum positive moment from A.

$$V_A = \frac{wL}{2} + \frac{M_A - M_B}{L}$$

$$= \frac{(24)(5)}{2} + \frac{35 - 40}{4} = 58.75 \text{ kN}$$

$$x_m = \frac{58.75}{24} = 2.45 \text{ m}$$

We obtain the maximum positive moment by Eq. (8.7).

$$M_{\max} = \frac{0.5(V_A)^2}{w} - M_A = \frac{0.5(58.75)^2}{24} - 35 = 37 \text{ kNm}$$

The points of contraflexure are found by equating the bending moment at x_0 to zero.

$$\frac{24x_0^2}{2 - 59x_0 + 35} = 0$$

$x_0 = 0.69$ m from A and 0.78 m from B.

3.3 POSITIONS OF POINTS OF CONTRAFLEXURES

The following points should be noted from Section 3.2. The position of the points of contraflexure of the elastic bending-moment diagram and the one obtained after redistribution of moments are not the same. The actual point of contraflexure in the structure can be anywhere between the corresponding points of the diagram before and after the redistribution. So, one should carefully follow the second rule given in the next section (condition for moment redistribution) that the reinforcement provided at all points in a structure should not be less than 70% of the maximum elastic moment at the point. This will ensure that the structure is safe both in elastic and plastic stages.

3.4 CONDITIONS FOR MOMENT REDISTRIBUTION

IS 456 (2000) Clause 37.1.1 and BS 8110 give the following five conditions to be satisfied for moment re-distribution in reinforced concrete structures:

Condition 1: Equilibrium must always be maintained at all times between the internal forces and external loads applied.

Condition 2: The ultimate moment of resistance provided for at any section should not be less than 70% of the moment at the section obtained from the elastic analysis for obtaining the maximum moment diagram covering all load combinations.

Condition 3: When using limit state design of R.C. section moments obtained at any section by elastic analyses should not be reduced by more than 30% of the numerically largest moment (i.e. the redistribution should not be more than 30%) covering all load combinations. It may be noted that with working stress design, the redistribution allowed is only 15% (IS 456 Clause 37.1.1 and Section B.1.2). ACI code does not allow any redistribution when members are designed by the working-stress method.

Condition 4: At sections where the moment capacity provided is less than the moment obtained from the elastic analysis of the section (i.e. locations from which moments are to be shed and which have to undergo large rotations), the following relationship should be satisfied:

$$\frac{x}{d} + \frac{\delta M}{100} \leq 0.6$$

which may also be stated as

$$\frac{x}{d} \leq 0.6 - \frac{\Delta M}{M} \quad (3.1)$$

where

x = Depth of neutral axis

d = Effective depth and

δM = Percentage reduction in moment $= \dfrac{\Delta M}{M} 100$ and $\dfrac{\delta M}{100} = \dfrac{\Delta M}{M}$

In BS 8110 the same condition is stated by a simple rule that the depth of neutral axis should obey the following relationship:

$$\frac{x}{d} \leq \beta_b - 0.4 \quad (3.2)$$

where

β_b = Ratio of moments = $\dfrac{\text{Moment at section after redistribution}}{\text{Moment at section before redistribution}}$

$= \dfrac{M - \Delta M}{M} = \dfrac{1 - \Delta M}{M}$

Hence Eq. (3.2) reduces to Eq. (3.1).

Condition 5: For structural frames over 4 storeys in height and which provide lateral stability also, the reduction of moments should be restricted to 10%.

Additional condition: Even though, IS 456 specifies only the above conditions, one should in addition check that the sections of the maximum moments are provided with enough lateral binders to help the necessary rotation for redistribution, by increasing the allowable value of ε_c, the strain of concrete at failures. (See Section 3.9 and Chapter 22.)

3.4.1 Influence of x/d Condition on Redistribution

Condition 4 in Section 3.4 puts severe limitations on the allowable redistribution moments in structures. By Eq. (3.2), in design of beams for 30% of redistribution the depth of neutral axis should be restricted to

$$\frac{x}{d} \leq 0.6 - 0.3 = 0.3 \tag{3.3}$$

Most of the formulae for design have been worked out on the assumption that the limiting value of $x/d = 0.5$ (for Fe 415 steel it is 0.479). To obey Condition 4, the real redistribution that can be allowed works out to approximately only 10 to 12%. In columns, x/d is very large so that no redistribution of moment can take place in columns. Because of the rotations that are necessary for the redistribution, it has to be very limited in the case of columns.

These discussions show that redistribution of moments is not as easy as it may look at first glance. Larger redistribution should be carried out with care. However, redistribution of moments up to 10 to 30% of the beams can be accounted for in routine designs without any difficulty. (The Australian code allows redistribution of $75(0.4 - x/d)$ for x/d more than 0.2 and 30% for x/d less than or equal to 0.2.)

3.5 FINAL SHAPE OF REDISTRIBUTED BENDING MOMENT DIAGRAM

It should be noted that two conditions that (1) the elastic moment should not be reduced by more than 30% and that (2) the moment value at any section should not be less than 70% of the original elastic value, are not complementary. If these are to be satisfied simultaneously the point of contra-flexure should be assumed (as already pointed out in Section 3.2) to lie in a zone and not at a particular point on the beam.

3.6 MOMENT REDISTRIBUTION FOR A TWO-SPAN CONTINUOUS BEAM

3.6.1 Theoretical Method for Redistribution of Moments

To consider the redistribution in a two-span continuous beam, we first draw the elastic bending moment diagrams by considering the various cases of loading (IS 456 Clause 22.4.1).

Case 1: Both spans loaded with dead load and left span only with live load.

Case 2: Both spans loaded with dead load and right span only with live load.

Case 3: Both spans loaded with dead load and live load.

For a purely theoretical solution of redistribution we have to first choose a suitable value for the support as the base value. The B.M. diagram for each of the three cases are then drawn with redistribution carried out separately for the chosen value of the fixed support moment. Finally, the envelopes for positive and negative bending moments for each section are drawn as the design envelope.

3.6.2 Shortened Graphical Method

However, a quicker solution which is valid for all practical purposes can be made by drawing the envelope of the un-distributed moments for the three cases and working the redistribution on their envelope as indicated below:

Step 1: Draw the individual elastic bending-moment diagrams for the loading cases 1, 2 and 3 as explained in Section 3.6.1.

Step 2: Draw the elastic moment envelope considering all the three loading cases 1, 2 and 3 (envelope curve *A*)

Step 3: Draw the 70% value of the above envelope to confirm the specification that the chosen value should be at least 70% of the elastic value (envelope curve *B*).

Step 4: As the aim of redistribution is to reduce the difference between span and support moments (the latter is usually the larger of two), choose a suitable moment from the envelope (curve *A*) as the base value. This value should not be less than 70% of the maximum elastic moment at the support.

Step 5: Join these support moment values at the various supports to form the base line. On this base line draw the free bending-moment diagrams for the three cases of dead and full design combinations already obtained in Step 1.

Step 6: Draw the envelope of the diagrams thus obtained (envelope curve *C*).

Step 7: The conditions satisfying the values of envelopes *B* and *C* give the distributed design-moment diagram for the continuous beam (envelope curve *D*).

3.7 ADVANTAGES AND DISADVANTAGES OF MOMENT REDISTRIBUTION

The main advantage of redistribution is that the maximum positive moment can be increased and the maximum negative moments at supports can be reduced, thus, resulting in lesser congestion of steel at the supports. On the other hand, without redistribution there is always an advantage that the maximum moments at various points are not reached simultaneously so that there is reserve of strength in the structure even if one of the points of maximum moment reached the failure stage. However, the redistribution of 30% of the larger moments in limit state design and 15% in working stress design allowed in IS 456 are quite reasonable and safe for all types of R.C. structures.

3.8 MODIFICATION OF CLEAR DISTANCE BETWEEN BARS IN BEAMS (FOR LIMITING CRACKWIDTH) WITH REDISTRIBUTION

BS 8110 (1985) uses the following expression for modifying spacing of steel with redistribution:

$$\text{Clear spacing of steel} = 75000\frac{\beta_b}{f_y} \not> 300 \text{ mm}$$

where

$$\beta_b = \frac{\text{Moment after redistribution}}{\text{Moment before redistribution}}$$

Thus, if the redistribution is negative, i.e. the design moment is less than the theoretical curve, the spacing of bars has to be decreased. IS 456 Clause 26.3.2 Table 15 gives the clear distance between the bars to be provided for crack control in beams when redistribution of moments is also made in the analysis. It is to be noted that using Fe 415 steel with no redistribution the maximum spacing is 180 mm whereas with the moments reduced by 30% the spacing is to be reduced to 125 mm (70% of 180 mm) and when the moments are increased by 30% the spacing can be increased to 235 mm (130% of 180 mm). Thus proper spacing and adjustments in maximum spacing of bars for crack control should be incorporated in design when using redistribution of moments.

3.9 MOMENT–CURVATURE ($M - \psi$) RELATION OF REINFORCED CONCRETE SECTIONS

The redistribution of moments and collapse analysis of reinforced concrete indeterminate structures involve rotation characteristics of R.C. sections. Hence, we will examine briefly this property of R.C. beams. More details of elastic analysis is given in Chapter 22.

From theory of simple bending we know that the rotation of a section undergoes when a bending moment is applied. The rotation is given by the following relation:

$$\frac{M}{I} = \frac{E}{R} = \frac{f}{y}$$

Taking x as the depth of neutral axis (Fig. 3.2)

$$\frac{M}{EI} = \frac{1}{R} = \frac{\varepsilon_c}{x} = \frac{f_c}{E_c x} = \psi$$

where

$$\text{Strain in concrete } \frac{f_c}{E_c} = \varepsilon_c$$

$$\psi = \frac{\text{Strain in extreme fibre}}{\text{Depth of neutral axis}} = \text{Curvature}$$

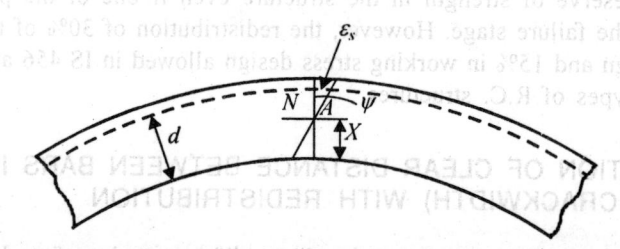

Fig. 3.2 Calculation of curvature in reinforced concrete beams.

The above expression means that the moment-curvature relation is related to the distribution of moments and the maximum value of curvature is related to the maximum value of ε_c at failure, which is usually assumed as 0.0035. It is also inversely proportional to x, the depth of neutral axis. The rotation capacity of reinforced concrete sections is directly related to ε_c. Investigations show that the value of ε_c can be increased very much by provision of steel binders. Thus, as pointed out in Section 3.4, it is also obligatory to provide adequate lateral steel in the form of binders at closer spacing at sections of maximum moment when redistributions of moments are assumed to take place in our designs. This capacity for large deformation in structure is called *ductility*. It can be ensured by paying attention to the following factors:

1. Structural design of beams and columns in frames should be such that the columns remain elastic and redistribution of moments takes place only in beams. This is sometimes referred to as adoption of strong column vs weak beam design.

2. The amount of tensile steel in beams should be restricted so that the value of x, the depth of neutral axis in them will allow suitable redistribution of moments. Provision of compression reinforcements also improves ductility of the section.

3. Stirrups and spirals to enclose the longitudinal steel and concrete should be provided at sufficiently close intervals at sections of maximum moments. This should also be provided at the upper and lower ends of columns. This will increase the failure strain of concrete at these places.

4. Detailing reinforced concrete members for ductility is a major requirement for earthquake resistant structures and is dealt with in Chapter 21.

3.10 ACI CONDITIONS FOR REDISTRIBUTION OF NEGATIVE MOMENTS

The concept of redistribution of moments was recognised in ACI codes in 1963. The conditions for redistribution of moments in beams and slabs according to the present ACI code Section 8.4 are the following:

1. Bending moments obtained by theoretical methods of elastic analysis with factored load only are allowed to be redistributed. (No redistribution is allowed in elastic design using working loads.)

2. The reinforcement at the sections, where the moment is to be analysed should not exceed one half the balanced steel ratio. This ensures that there will be enough ductility in the section. If ρ and ρ' are the tension and compression steel ratios then

$$\rho \text{ or } (\rho - \rho') \not> 0.50\, \rho_b$$

where ρ_b is balanced steel for failure strains.

3. The maximum increase or decrease of negative moment that can be distributed is given by the expression:

$$\frac{\Delta M}{M} 100 = 20\left(1 - \frac{\rho - \rho'}{\rho_b}\right)$$

This equation is represented by Fig. 3.3.

4. Adjustment of negative moment is to be made for each configuration of loading and members should be proportioned for the maximum adjusted envelope developed from all loading conditions.

5. Static equilibrium must always be maintained: increase or decrease in moments is to be accompanied by corresponding decrease or increase in other moments.

6. In the case of unequal moments occuring on the two sides of a fixed support (which happen when the adjacent spans are unequal) the difference between the two moments is taken into the support.

7. The resultant redistribution will generally amount to only 10 to 20% depending on the reinforcement provided.

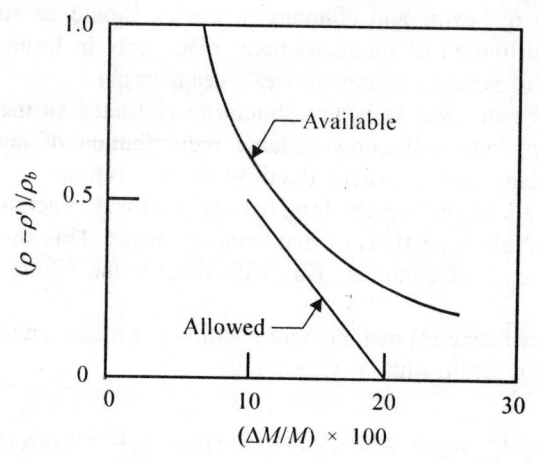

Fig. 3.3 ACI conditions for redistribution of negative moment.

CONCLUSION

Redistribution of moments obtained from elastic analysis of structures is allowed in IS:456 Clause 37 when the structure is designed by limit-state methods. By using I.S. and B.S. codes of practice, a redistribution upto 30% can be achieved. ACI code permits a redistribution varying from 10 to 20% depending on the reinforcement ratios. If higher amounts of redistribution has to be made, this can be achieved only by using inelastic methods of analysis of the structures as described in Chapter 22.

EXAMPLE 3.1 (Bending moment of a simply-supported beam)
Mark out the ordinates of the bending moment diagram of a beam subjected to the following loads and end conditions at every one-eighth points along its length.

1. A simply supported beam with a uniformly-distributed load.
2. A beam fully fixed at both ends with a uniformly-distributed load.

EXAMPLE 3.2 (Bending moments of beams with specified fixing moments)
Determine the ordinates of the bending moment diagram at every one-eighth point of beam AB of span 20 m with a uniformly-distributed load of 30 kN/m if the fixing moments at A is 1000 kNm and that at B is 500 kNm.

REDISTRIBUTION OF MOMENTS IN REINFORCED CONCRETE BEAMS 47

(a) Simply supported beam

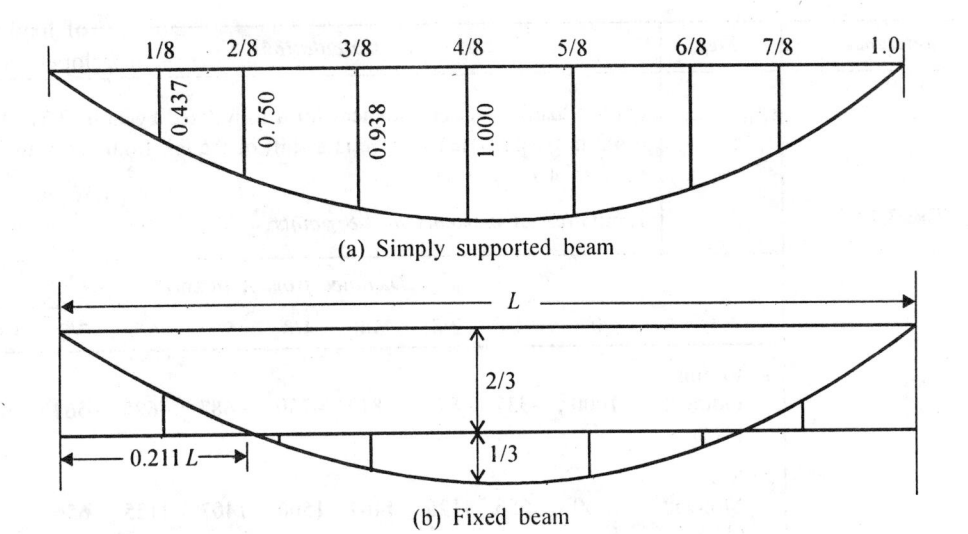

(b) Fixed beam

Fig. 3.4 Example 3.1.

Reference	Step	Calculations for Example 3.2
	1	*Data* $M_A = 1000$ kNm; $M_B = 500$ kNm; $l = 20$ m; $w = 30$ kN/m $$\frac{wl^2}{8} = 1500 > 1000 \text{ kNm (given)}$$
	2	*Find* R_A $$R_A = \frac{wl}{2} + \frac{M_A - M_B}{L} = 300 + \frac{500}{20} = 325 \text{ kN}$$
	3	*Find position of the maximum moment (i.e. point where S.F. = 0)* $$x = \frac{R_A}{w} = \frac{325}{30} = 10.83 \text{ from } A$$
	4	*Calculate the maximum bending moment* $$M' = R_A \frac{R_A}{w} - \frac{w}{2}\left(\frac{R_A}{w}\right)^2 = \frac{0.5 R_A^2}{w}$$ $$= \frac{0.5(325)^2}{30} = 1760.4 \text{ kNm}$$ $M_{max} = 1760 - 1000 = 760$ kNm
	5	*Find position of zero moment (m from A)* $$1000 - 325 m + \frac{30 m^2}{2} = 0$$ $m = 3.71$ m from A and 17.95 m from B.

Reference	Step	Calculations
Ex. 3.1	6	(*Note:* Position of zero moment for a fully-fixed beam of 0.211 × 20 = 4.2 m from *A*. Hence there is a shift of the P.I. from 4.2 m to 3.71 m from *A*.) *Tabulation of moments at 1/8 points*

	Distance from *A* in (m)								
	0	1/8	2/8	3/8	4/8	5/8	6/8	7/8	1.0L
1. Fixing moment (M_f)	−1000	−937	−875	−812	−750	−687	−625	−562	−500
2. S.S. Moment (M_0)	0	656	1125	1407	1500	1407	1125	656	0
3. Final moment (*M*)	−1000	−281	250	593	750 Max (760)	720	500	−94	−500

EXAMPLE 3.3 (**Conditions for redistribution of moments from one section to another section**)

A tee beam *ABC* is continuous over two spans of 8 m each and it carries uniformly distributed factored load of 75 kN/m. Assuming $f_{ck} = 25$ and $f_y = 415$ (with bilinear stress strains curve) check whether we can reduce the maximum moment by 30% and redistribute to the spans. Width of flange 1000 mm, width of web = 300 mm, thickness of slab = 150 mm, *D* = 820 mm, *d* = 770 mm are given.

Reference	Step	Calculations
	1	*Determine elastic bending moment diagram* Beam *ABC*. M_{max} at support $M_B = \dfrac{wL^2}{8} = \dfrac{75 \times 8 \times 8}{8} = 600$ kNm At support the beam is rectangular
Step 4 Ex. 3.2	2	*Calculate the maximum moment in span* Shear at support $A = R_A = \dfrac{75 \times 8}{2} - \dfrac{600}{8} = 225$ kN As $M_A = 0$, the maximum positive moment $= \dfrac{R_A^2}{2w}$ $= \dfrac{225^2}{2 \times 75} = 337.5$ kNm In the span, the beam acts as a *T*-beam.

Reference	Step	Calculations
	3	*Design the section at B for 30% redistribution and find x/d*
		Design moment at section B (centre of beam) (30% redistribution and minimum 70% M_B) = 0.7 × 600 = 420 kNm.
SP 16 Table 3		$$\frac{M}{bd^2} = \frac{420 \times 10^6}{300 \times 770^2} = 2.36$$
		for f_{ck} = 25 and Fe 415 steel singly-reinforced percentage steel = 0.748
		For rectangular sections, assuming the design stress strain curve of concrete is valid.
IS 456 Clause 37		$$0.87 f_y \frac{0.748}{100} bd = 0.36 f_{ck} bx$$
		$$\frac{x}{d} = \frac{0.748}{100} \frac{0.87 f_y}{0.36 f_{ck}} = \frac{0.748}{100} \frac{0.87 \times 415}{0.36 \times 25} = 0.3$$
	4	Check the condition for 30% redistribution
		$$\frac{x}{d} \leq 0.60 - \frac{\Delta M}{M} = 0.60 - 0.30 = 0.30$$
		As x/d < 0.30, we can redistribute 30% of the moments into the span.
	5	*Determine the maximum span moment*
		Shear at $A = \frac{75 \times 8}{2} - \frac{420}{8} = 247.5$ kN
		Point of maximum moment x from A where shear is zero
		$$x = \frac{247.5}{75} = 3.3 \text{ m}; \quad M = 247.5 \times 3.3 - \frac{75(3.3)^2}{2} = 408.38 \text{ kNm}$$
	6	*Design the section for the enhanced moment value*
		As the section is a T beam, the failure by yielding of steel will be
		$$A_s = \frac{M}{0.87 f_y L_A} = \frac{408.38 \times 10^6}{(0.87 \times 415)(770 - 75)} \quad \text{(approx.)}$$
		$= 1627$ mm^2
		Provide four rods of 25 mm (1963 mm^2).

REFERENCES

1. IS 456 2000, *Code of Practice for Plain and Reinforced Concrete*, Bureau of Indian Standards, New Delhi, 2000.

2. BS 8110 1985, *Structural Use of Concrete*, British Standards Institution, London, 1985.

3. ACI 318 1989, *Building Code Requirements for R.C.*, American Concrete Institute, Detroit, 1989.

CHAPTER 4

Design of Reinforced Concrete Deep Beams

4.1 INTRODUCTION

Beams with large depths in relation to spans are called *deep beams*. In IS 456 (2000) Clause 29, a simply-supported beam is classified as *deep* when the ratio of its effective span L to overall depth D is less than 2. Continuous beams are considered as deep when the ratio L/D is less than 2.5. The *effective span* is defined as the centre-to-centre distance between the supports or 1.15 times the clear span whichever is less.

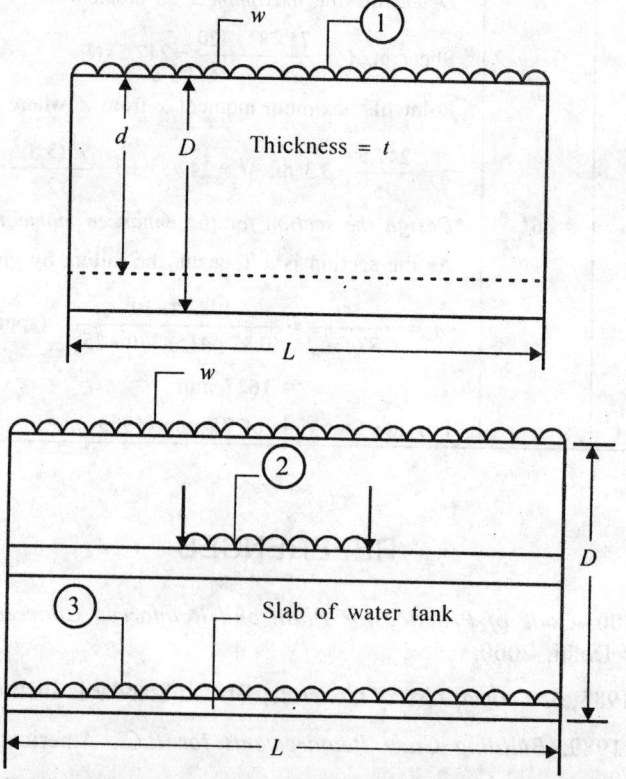

Fig. 4.1(A) Types of loadings in deep beams: 1. top loading; 2. side loading; 3. bottom loading.

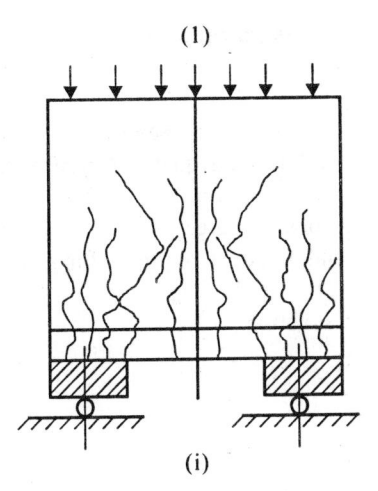

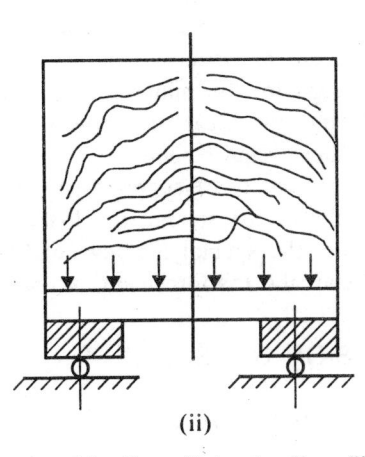

Fig. 4.1(B) Variation of cracking of deep beams with type of loading: (i) top loading, (ii) bottom loading.

The strain distribution across the section of a deep beam will not be linear as in a shallow beam. It can be determined by the theory of elasticity. Before 1970, designs of deep beams were based on this elastic strain distribution. However, experimental investigations by Leonhardt and Walter around 1966 [1] showed that such designs are not valid for reinforced concrete sections which crack under low value of tensile stresses. These tests also showed the importance of correct detailing of steel in deep beams. The present-day design practices are mainly based on these tests and some other tests. As shown in Fig. 4.1, behaviour of a deep beam depends also on how they are loaded and special considerations should be given to this aspect in our designs. This chapter deals with the commonly accepted methods of design and detailing of these deep beams.

4.2 MINIMUM THICKNESS

The minimum thickness of deep beams should be based on two considerations. Firstly, it should be thick enough to prevent buckling with respect to its span and also its height. The empirical requirement to prevent bulking can be expressed as follows:

$$\frac{D}{t} < 25 \quad \text{and} \quad \frac{L}{t} < 50 \tag{4.1}$$

where t is thickness of the beam. Secondly, the thickness should be such that the concrete itself should be able to carry a good amount of the shear force that acts in the beam without the assistance of any steel. This aspect is dealt with Section 4.5.3.

4.3 STEPS OF DESIGNING DEEP BEAMS

The important steps in the design of R.C. deep beams are the following:

   1. Determine whether the given beam is deep according to the definition.

2. Check its thickness with respect to buckling as well as its capacity to carry the major part of the shear force by the concrete itself.
3. Design for flexure.
4. Design for the minimum web steel and its distribution in the beam.
5. Design for shear. If the web steel already provided is inadequate, design additional steel for shear requirements.
6. Check safety of supports and loading points for local failure.
7. If the beams are not top loaded design the special features required for deep beam action under the special loading conditions.
8. Detail the reinforcements according to accepted practice.

4.4 DESIGN BY IS 456

4.4.1 Determination of Design Bending Moment

The design bending moments are calculated as follows:
In a simply supported beam, the bending moment is calculated as in ordinary beams. For a total load w uniformly distributed on the beam

$$M_{(max)} = \frac{wL^2}{8} \tag{4.2a}$$

In a continuous beam the bending moment, according to American practice [2] for a uniformly distributed load, w per unit length is as follows:

$$\text{At mid span, } M_{max} = \frac{wL^2}{24}(1-e^2) \quad \text{positive} \tag{4.2b}$$

$$\text{At face of support, } M_{max} = \frac{wL^2}{24}(1-e)(2-e) \quad \text{negative} \tag{4.2c}$$

where e is the ratio of width of support to effective span.

4.4.2 Check for Compression in Concrete

Eventhough stresses in compression in concrete in deep beams are always low, a routine check should be made to estimate the maximum compression in concrete by the standard beam formula.

4.4.3 Determination of Area of Tension Steel

The area of steel to carry the tension is determined by the empirical method of assuming a value for the lever arm. IS 456 Clause 29.2 [3] follows the CEB (Committee Euro-International du Beton) and gives the following values for z, the lever arm length.
For simply-supported beams,

$$z = 0.2(L+2D) \quad \text{when } L/D \text{ is between 1 and 2} \tag{4.3a}$$

$$= 0.6L \quad \text{when } L/D \text{ is less than 1}$$

For continuous beams

$$z = 0.5L \quad \text{when } L/D \text{ is less than 1}$$
$$= 0.2(L + 1.5D) \quad \text{when } L/D \text{ is between 1 and 2.5} \quad (4.3b)$$

From these values

$$M_u = A_s f_s z \quad (4.4)$$

where $f_s = 0.87 f_y$ in limit-state design.

4.4.4 Detailing of Tension Steel

IS 456 Clause 29.3 gives specific rules for detailing of the tension steel in simple and continuous beams. They can be summarised as follows

1. In simply-supported beams, the tension steel should develop tension as in a tied arch as shown in Fig. 4.2. They should consist of many bars preferably placed in a zone of depth equal to $(0.25D - 0.05L)$ adjacent to the tension face of the beam. All the bars should extend through the span without curtailment beyond the supports and preferably bent upwards at the ends to obtain adequate embedment and anchorage.

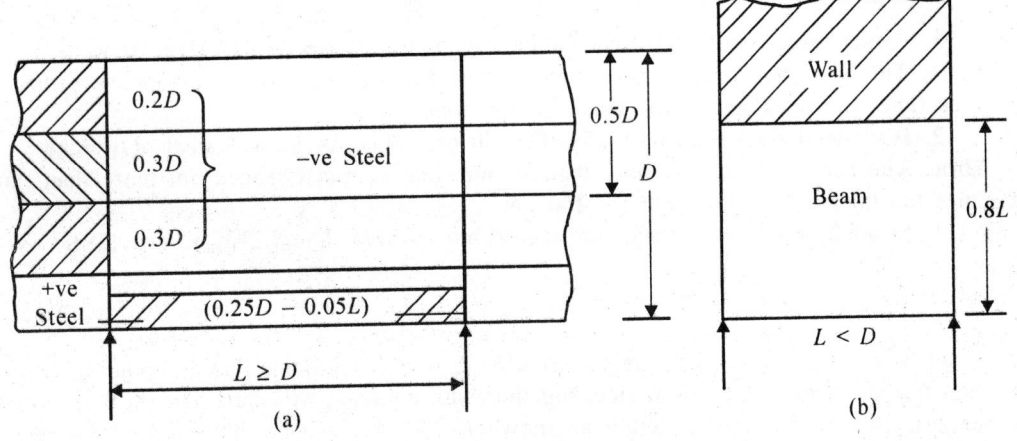

Fig. 4.2 Placement of tension steel in deep beams.

2. In continuous beams there will be positive steel for the positive moments and the negative steel over the supports for the negative moments. Termination and detailing of negative steel is dealt with in Clause 29.3.2. of IS 456 and are as follows:

(a) Termination of steel. Only one-half of the negative reinforcements are allowed to be terminated at a distance of $0.5D$ from face of support. All the remainder shall extend over the full span.

(b) Distribution of negative reinforcement.

(i) When the span depth ratio is less than unity the negative steel should be evenly placed over a depth of $0.8D$ measured from the tension face.

(ii) When the ratio of L'/D is in the range of 1.0 to 2.5 the steel over the support of a deep beam shall be placed in two following zones:

1. A zone of depth of 0.2D adjacent to tension face which shall contain a proportion of the tension steel given by

$$A_{st} = 0.5\left(\frac{L'}{D} - 0.5\right) \tag{4.5}$$

where L' is the clear span.

2. A zone measured $0.3D$ on either side of the mid-depth of the beam which shall contain the remainder of the tension steel.

As shown in Fig. 4.2 (b), when the depth of the beam is much larger than the span the portion above a depth equal to 0.8 times the span can be designed as a wall. Beam action is present only in the lower portion of such a member,

4.4.5 Detailing of Web Steel

All specifications require minimum amount of vertical steel A_v. Horizontal steel A_h in the form of U bars to be placed on both faces of deep beams. They not only overcome the affects of shrinkage and temperature but also act as sheer reinforcement. The amounts specified in IS 456 Clause 32.5 are as follows:

1. Vertical steel shall be 0.15% for Fe 250 or 0.12% for Fe 415 steel or welded fabric. The bars shall not be more than 14 mm diameter and spaced at not more than three times the thickness of the beam or 450 mm.

2. Horizontal steel shall be 0.25% Fe 250 or 0.20% for Fe 415 steel or welded fabric. The bars shall not be more than 16 mm diameter and spaced not more than three times the thickness of the beam or 450 mm.

3. In addition, the necessary side face reinforcement should also be provided.

4.4.6 Design for Shear in IS 456

No separate checking for shear is specified in IS 456. We assume that arching action of the main tension steel and the web steel together with concrete will carry the shear. In simply-supported beam the arching action as shown in Fig. 4.3 can be depended on if the main tension steel is properly detailed. However, in continuous beams, this arch action is not present and ACI recommends that we should design them as in ordinary beam.

4.5 DESIGN ACCORDING TO BRITISH PRACTICE

The British practice is adopted from the Guide published by CIRIA (Construction Industry Research and Information Association) and also the recommendations of Kong Robinson and Sharp [4,5,6].

4.5.1 Design for Flexure

The value recommended as the value of z by Kong is the lesser of the two values given by the following equations:

$$z = 0.6L \text{ or } 0.6D \tag{4.6}$$

It will be $z = 0.6L$ when $L/D < 1$ and $z = 0.6D$ when $L/D \geq 1$.

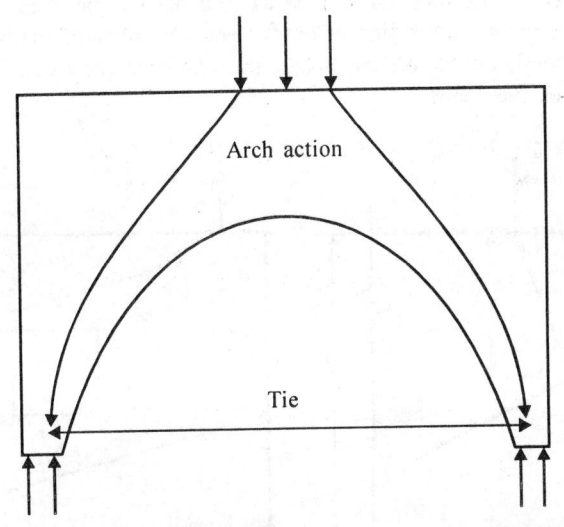

Fig. 4.3 Arch action in simply-supported deep beams.

The CIRIA Guide, however, recommends the CEB values given in IS 456. The values of steel area obtained from Eq. (4.4) are kept very conservative as the tension steel is also assumed to contribute to the shear resistance of the beam in British practice. Using Eqs. (4.4) and (4.6), we get

$$A_s = \frac{M_u}{(0.87 f_y)(0.6D)} = \frac{1.9 M_u}{f_y D}$$

$$A_s = \frac{M_u}{(0.87 f_y)(0.6L)} = \frac{1.9 M_u}{f_y L} \tag{4.7}$$

The greater value of A_s is taken as the tension steel.

4.5.2 Minimum Steel for Crack Control

We should also check whether the tension steel provided is enough for control of tension cracks due to beam action which should be atleast not less than 0.2 to 0.3% of the area of concrete.

4.5.3 Designing for Shear

Unlike the IS code, the British practice requires numerical calculations for design of deep beams for shear [6]. The design is based on the results of research carried out by Kong and others. It is applicable only to simply-supported beams of span depth ratio not exceeding two. The shear analysis is carried out by assuming a structural idealisation of 'critical diagonal tension failure line' along the natural load path which in the case of concentrated loads is taken as the line joining the load and the support as shown in Fig. 4.4. A uniformly distributed load is replaced by two equal concentrated loads acting at one-fourth the span from the supports [Fig. 4.4(c)]. The total shear V is carried out by the shear strength of concrete and steel cutting across the assumed line of crack. For a number of loads (none of which is more than 50% of the total) acting on the beam, the resultant load can be assumed to act at the centre of gravity of the loads.

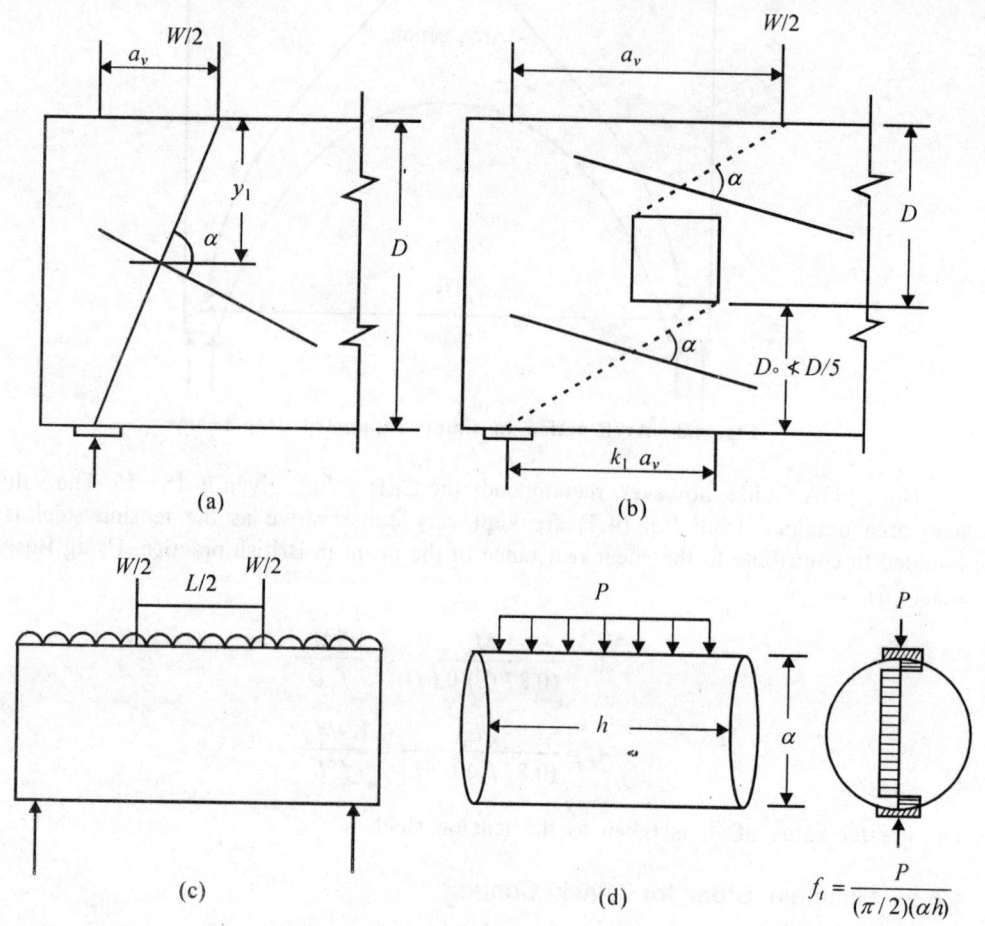

Fig. 4.4 Design of deep beams for shear: (a) deep beam without holes, (b) deep beams with holes, (c) action of UDL equated to concentrated loads, (d) split cylinder test (compressed between 25 mm wide 3 mm thick plywood) for tensile strength.

The procedure of design is as follows:

Step 1: Determine the nominal shear in concrete. It should not exceed the maximum allowable shear.

$$v = \frac{V_u}{tD_u}, \quad \text{where } v \not> \tau_{max}$$

DESIGN OF REINFORCED CONCRETE DEEP BEAMS   57

(Table 20 of IS 456 may also be used to find τ_{max} or we may use $\tau_{max} = 0.8 \sqrt{f_{ck}}$.

Step 2: $V_u = V_c + V_s$, where

V_c = Shear that can be carried by concrete

V_s = Shear that can be carried by steel

Step 3: The shear capacity of concrete V_c is given by the following formula:

$$V_c = C_1\left(1 - 0.35\frac{a_v}{D}\right)f_t t D \qquad (4.8)$$

where

C_1 = A coefficient equal to 0.72 for normal weight concrete

$\dfrac{a_v}{D} = \dfrac{\text{Shear span}}{\text{Depth ratio}}$

$f_t = 0.5\sqrt{f_{ck}}$, the tensile strength or the cylinder splitting strength

Hence Eq. (4.8) becames,

$$V_c = 0.72(D - 0.35a_v) f_t t \qquad (4.9)$$

When designing for shear it is also recommended that concrete itself should carry atleast 65% of the ultimate shear. This can be ensured by choosing a suitable thickness of the beam given by the following formula:

$$t = \frac{0.65 V_u}{0.72(D - 0.35a_v)f_t} \qquad (4.10)$$

Step 4: The shear capacities of the tension steel and the nominal web steel should also be taken into account in resisting the shear. Their shear capacity can be calculated from the following empirical formula:

$$V_s = C_2 \sum_1^n A_1 \frac{y_1}{D} \sin^2 \alpha \qquad (4.11)$$

where

C_2 = 225 N/mm² for Fe 415 steel and 100 N/mm² for Fe 250 steel

A_1 to A_2 = Areas of reinforcements (horizontal and vertical) cut by the assumed crack (The areas should also include the tension steel, and each bar is to be taken separately.)

y_1 = Depth from the top of the beam to the point where the bar intersects the critical diagonal crack line.

α = The angle between the bar considered and the critical diagonal crack (Fig. 4.4)

n = The number of bars including the tension steel cut by the assumed crack line

D = Total depth of beam

Step 5: Checking for shear. If V_u is less than $(V_c + V_s)$ the beam is safe in shear with the minimum web steel provided. If $(V_c + V_s)$ is less than V_u then additional steel should be provided as horizontal or vertical steel to satisfy the value of V_u. Provision of horizontal steel will be more efficient than the vertical steel.

4.5.4 Design of Deep Beams with Holes

References [4,5,6] give guidelines for design of deep beams with openings. The shear diagonal or line of crack is taken as shown in Fig. 4.4. The design steps can be as follows:

1. For determining the tension steel the lever arm depth $0.6D$ should be replaced by $0.75D_0$ where D_0 is the distance of bottom of the hole from the soffit of the beam as shown in Fig. 4.4 Accordingly the equations for tension steel can be expressed as follows:

$$A_s = \frac{1.9M_w}{f_y L} = \frac{1.55M_w}{f_y D_0} \tag{4.12}$$

2. For determining the shear strength of beams with openings, the effect of openings can be completely discarded if they are reasonably clear of the critical diagonal crack. If any of the openings is within the line of crack the value of V_c is obtained by modifying the equation for V_c as follows:

   (i) a_v is replaced by $k_1 a_v$, the distance of the edge of the hole to the face of support.
   (ii) D is replaced by D_0

Thus, we get

$$V_c = C_1(D_0 - 0.35k_1 a_v) f_t t$$

$$V_s = \lambda C_2 \sum A_1 \frac{y_1}{D} \sin^2 \alpha \tag{4.13}$$

where $\lambda = 1.5$ for web bars and 1.0 for main bars (or assume $\lambda = 1$ for all bars).
$C_2 = 225$ for Fe 415 and 100 for Fe 250
$C_1 = 0.72$

It is also *recommeded* that

$$V_u \not> 0.75 t D_0 \tau_{max} \tag{4.14}$$

and atleast 55% of the shear should be carried our by the strength of concrete V_c. We may also use the 65% rule here as well. The procedure of designing is shown in Table 4.1.

TABLE 4.1 DESIGN OF DEEP BEAMS WITHOUT HOLES

1. Design thickness: minimum thickness	$t = \dfrac{0.65 V_u}{0.72(D - 0.35 a_v) f_t}$ where $f_t = 0.5\sqrt{f_{ck}}$
2. Design for flexure: area of tension steel A_s (a) A_s got from IS rules for z $\quad A_s \not< \dfrac{1.9M_u}{f_y D}$ and $\not< \dfrac{1.9M_u}{f_y L}$ (b) $A_s \not< 0.3\%$ of gross area	

3. Provide web steel	Minimum $A_v \nless 0.12\%$; $A_h \nless 0.20\%$ for Fe 415 steel
4. Check for shear capacity along shear diagonal	$V_1 = 0.72 \, (D - 0.35a_v) \, f_t t$ $+ 225 \, \Sigma \, A_s(Y_1/D)\sin^2\alpha$ If $V_1 > V_u$ no addition shear steel required
5. Dsign for shear if $V_u > V_1$	$V_u - V_1 = 225 \, \Sigma \, A(y_1/D)\sin^2\alpha$ Provide steel area A, preferably as horizontal or horizontal plus vertical steel
6. Check bearing pressure, see Section 4.7	
7. Provide special steel if beam is side or bottom loaded and design the special steel required	
8. Detail the steel, Section 4.4.4	

(For beams with holes, modify D and a_v as given in Section 4.5.4.)

4.6 ACI PROCEDURE FOR DESIGN OF DEEP BEAMS

ACI 318 defines simple beams with ratio L/D less than 1.25 as deep beams. Continuous beams are considered as deep when the clear span/depth ratio (i.e. L/D) is less than 2.5.

4.6.1 Design for Flexure

The ACI code does not give any specific recommendation but PCA publication [2] gives a lever arm factor as follows:

$$z = 0.9d \tag{4.15}$$

where d is the effective depth. It also recommends to check the amount of tension steel obtained for crack control as given in Section 4.5.2.

4.6.2 Design for Shear by ACI Code

ACI 318 (89) Clause 11.8 gives the procedure for shear design of beams when clear span/effective depth ratio is less than 5 and which are loaded from the top so that compression struts can develop between the load and the support. The design procedure is as follows:

Step 1: The critical section for shear is taken as shown in Fig. 4.5 and as follows:

(a) $0.15L_n$ for uniformly distributed load and

(b) $0.5a$ for concentrated loads placed at not greater than the effective depth from the face of the support. Let the factored shear be V_u.

60 ADVANCED REINFORCED CONCRETE DESIGN

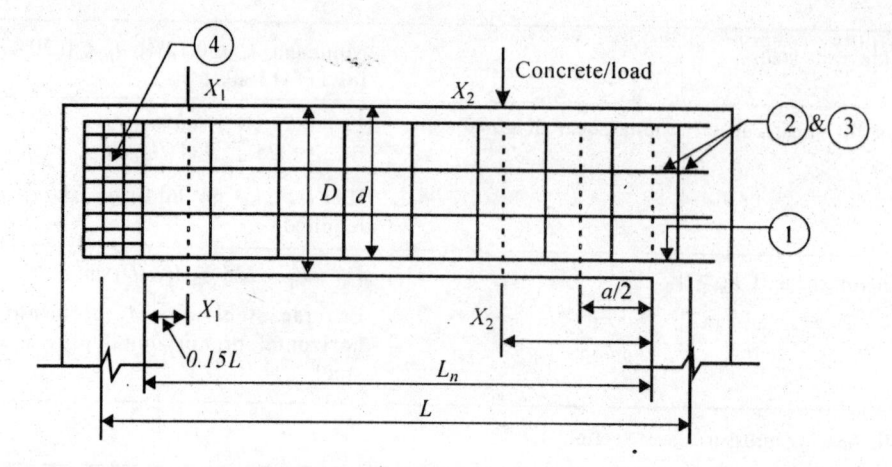

Fig. 4.5 Design for shear in deep beams by ACI method (L_n = clear span; a = shear span).

Step 2: Calculation of allowable shear (ACI Clause 11.8.4). The maximum allowable shear is given by the following expressions:
For L_n/d less than 2 (with reduction factor 0.85)

$$\tau_{\max} = 0.53\sqrt{f_{ck}} \tag{4.16}$$

For L_n/d between 2 and 5 taking $L_n/d = f$,

$$\tau_{\max} = 0.044(10+f)\sqrt{f_{ck}} \tag{4.17}$$

Hence

$$V_u < td\tau_{\max}$$

where t = thickness of the beam and d = effective depth.

Step 3: The shear that can be resisted by concrete is given by the formula (with reduction factor 0.85)

$$V_c = 0.13\sqrt{f_{ck}}\,td \tag{4.18}$$

This gives a very conservative value and more exact value can be obtained by Eq. (11.30) of ACI 318 (89).

Step 4: Minimum specified values of steel as web steel in the form of horizontal and vertical hoops should be provided in deep beams. The shear resistance of web steel can be expressed by the following equation using a reduction factor 0.85.

$$V_s = 0.85\left(\frac{A_v}{s_1}\frac{1+f}{12} + \frac{A_h}{s_2}\frac{11-f}{12}\right)f_y d \tag{4.19}$$

where

A_v = Area of vertical steel at spacing s_1

A_h = Area of horizontal steel at spacing s_2

$f = L_n/d$

ACI requires 0.15% vertical steel and 0.25% horizontal steel for 415 steel (IS 456 stipulates only 0.12% and 0.20%, respectively) putting $A_s/st = \rho$, the steel ratio, Eq. (4.19) is reduced as

$$V_s = 0.07 \, [d(\rho_v + 11\rho_h) - L_n(\rho_h - \rho_v)] \, tf_y$$

where s is the spacing and t is the thickness. Putting $\rho_v = 0.0012$ and $\rho_h = 0.002$ in the equation, we get

$$V_s = 0.07(0.0232d - 0.0008ln)tf_y \qquad (4.20)$$

This is the shear that the minimum steel can carry

Step 5: Check for shear. The beam is safe in shear if the area of concrete and the minimum steel can carry the shear. That is $V_c + V_s = V_u$. Otherwise, put additional A_h or A_v steel or A_h and A_v steel so that the shear capacity is satisfied by Eq. (4.19).

4.6.3 Design for Shear in Continuous Beams and under Special Loading Conditions

Design of continuous deep members in shear should be based on the design procedure for ordinary continuous beams. Similarly, side and bottom loaded deep members should be designed for shear by conservative principles, as in the case of regular beams.

4.7 CHECKING FOR LOCAL FAILURES

Special care should be taken to check the bearing stresses at the supports as well as at the loading points. In IS 456 Clause 33.4, the permissible ultimate stress is limited to $0.45f_{ck}$. The area of support should be strengthened by vertical steel and spiral binders to avoid brittle or premature failure at supports. BS 8110 Clause 5.2.3.4 allows the following bearing stresses:

1. Dry bearing on concrete = $0.4f_{ck}$ (no intermediate padding)
2. Bedded bearing on concrete = $0.6f_{ck}$ (with cementitious padding)
3. Concrete face of steel bearing plate casts into member or support and not exceeding 40% of concrete dimension = $0.8f_{ck}$.

4.8 DETAILING OF DEEP BEAMS

The methods of detailing of reinforcement for various types of deep beams are shown in Fig. 4.6. The following points should be noted:

1. When beams are bottom loaded the vertical bars have to carry the load to the arching zones so that the central vertical bars should be longer. Mechanical anchorage such as hooks should be provided at their ends.
2. When beams are side loaded as when a cross wall of a silo meets the main wall or column, suspender bars should be used as shown in Fig. 4.6(d)
3. Under special conditions as in bottom loaded beams it is recommended that the actual shear the beam has to carry should not exceed 75% of the capacity of the beam in shear.

Fig. 4.6 Detailing of steel in deep beams: (a) top loaded simply-supported, (b) bottom loaded simply-supported, (c) continuous, (d) side loaded through cross walls.

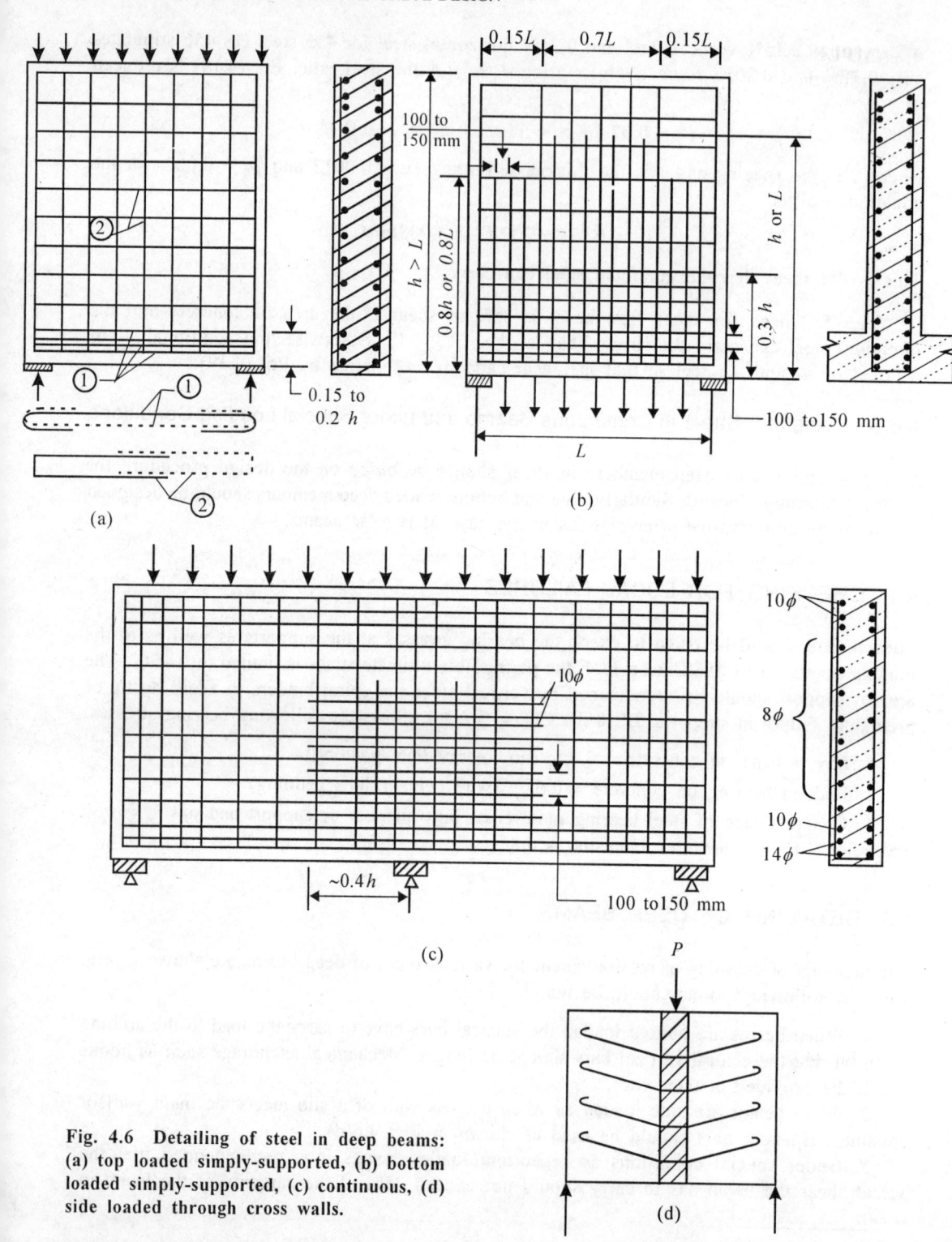

DESIGN OF REINFORCED CONCRETE DEEP BEAMS

EXAMPLE 4.1 (Design of simply-supported deep beam)
Determine the thickness and reinforcements for a simply-supported transfer girder of length 5.25 m loaded from two columns at 1.75 m from each end with 3750 kN (see Fig. 4.7). The total depth of the beam is 4.2 m and the width of supports is 520 mm. Assume grade 40 concrete and Fe 415 steel.

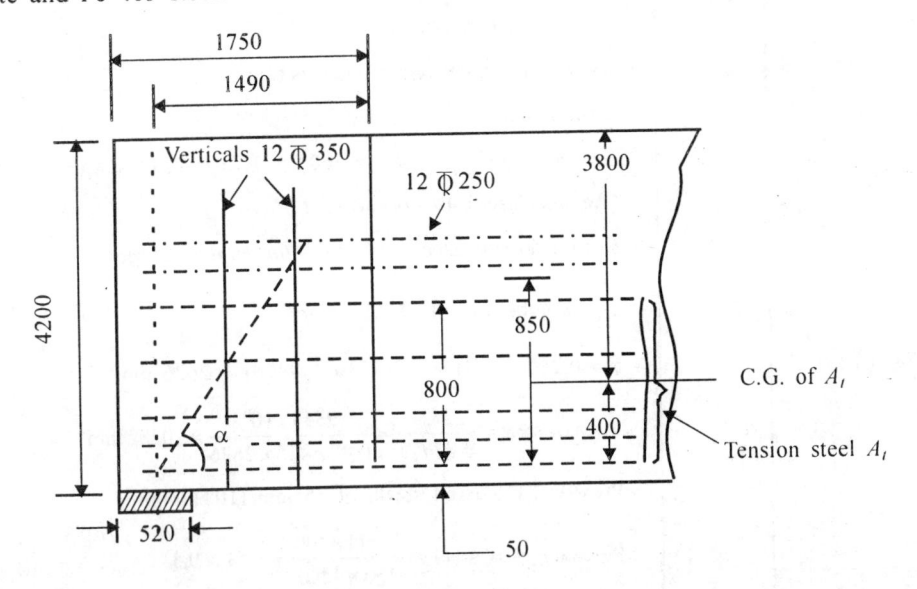

Fig. 4.7 Example 4.1.

Reference	Step	Calculations
	1	*Determine beam thickness*
		Condition 65% of shear to be taken by concrete
		Factored shear
		$V_u = 1.5 \times 3750 = 5625$ kN
Eq. (4.9)		$V_c = 0.72(D - 0.35a_u)f_t t$
		$a_u = 1750 - (520/2) = 1490$ mm
		$f_t = 0.5 \sqrt{f_{ck}} = 0.5 \sqrt{40} = 3.16$ N/mm^2
Eq. (4.10)		$0.72(4200 - 0.35 \times 1490)(3.16t) = 0.65 \times 5625 \times 1000$
		$t = 437$ mm
		Assume $t = 450$ mm
	2	*Check for deep beam action*
		Effective span $L = 5250 - 520 = 4730$

Reference	Step	Calculations
		$L/D = 4730/4200 = 1.13 < 2$
		$L/t = 4730/450 = 10.5 < 50$
		$D/t = 4200/450 = 9.33 < 25$
		Dimension satisfy deep beam action
	3	*Design loads*
		$V_u = 5625$ kN $M_u = 5625 \times 1.75 = 9844$ kNm
		(We may use 1.49 instead of 1.75.)
	4	*Design tension steel and check for minimum steel*
		$\dfrac{L}{D}$ is between 1 and 2.
Eq. (4.3a)		$z = 0.2(L + 2D) = 0.2(4730 + 8400) = 2626$ mm
		Area of steel $= \dfrac{M_u}{0.87 f_y z} = \dfrac{9844 \times 10^6}{0.87 \times 415 \times 2626} = 10382$ mm^2
		Provide 21 tension steels of 25 mm (10311 mm²)
		Percentage of steel $= \dfrac{10311 \times 100}{450 \times 4200} = 0.54 > 0.3$
		The steel provided is more than the minimum specified.
	5	*Detail of tension steel*
Eq. (4.3)		Zone (depth) of placement $= 0.25D - 0.05L$
		$= 0.25(4200 - 0.05 \times 4730) = 991$ mm
		Distribute steel in this area with cover say 40 mm.
	6	*Calculate the nominal steel areas required*
Sec. 4.4.5		(a) *Vertical steel*
		$A_v = \dfrac{0.12}{100} 450 \times 1000 = 540$ mm²/m length
		That is 270 mm² on both faces.
		Spacing not to exceed 3 times of 450 mm (i.e. 1350 mm).
		Provide 12 mm at 350 mm (given 323 mm²/m).
		(b) *Calculate nominal horizontal steel required*
		Length $A_h = \dfrac{0.20}{100}(450 \times 1000) = 900$ mm²/m
		That is 450 mm² on both faces.
		Provide 12 mm at 250 mm spacing (gives 452 mm²/m).

DESIGN OF REINFORCED CONCRETE DEEP BEAMS

Reference	Step	Calculations
IS 456 Table 20	7	Check $v < \tau_{max}$ $u = \dfrac{V}{tD} = \dfrac{5625 \times 10^3}{450 \times 4200} = 2.98 \text{ N/mm}^2$ τ_{max} for M40 = 4.0 N/mm^2
Eq. (4.9)	8	*Design for shear* Shear taken by concrete = $0.72(D - 0.35a_v)f_t t$ $V_c = 0.72(4200 - 0.35 \times 1490)(3.16)(450) = 3766$ kN Shear to be taken by steel = $V_u - V_c$ $V_s = 5625 - 3766 = 1859$ kN
Eq. (4.11)	9	*Determine the shear taken by tension steel in arch action V_{st}* $V_{st} = C_2 A_s \dfrac{y_1}{D} \sin^2 \alpha; C_2 = 225 \text{ N/mm}^2$ For horizontal steel $\tan \alpha = \dfrac{D}{a_v} = \dfrac{4200}{1490} = 2.82$ $\sin \alpha = 0.942$ and $\sin^2 \alpha = 0.89$ A_s = Area of tension steel provided $V_{st} = \dfrac{225 \times 10311 \times 3800 \times 0.89}{4200}$ = 1868 kN and V_s required = 1859 kN The arch action can take the balance of shear; however we will also calculate the capacity of web steel for resisting shear.
	10	*Find the shear taken by horizontal web steel V_{sh}* Web steel has to be provided in orthogonal directions. The shear resistance of horizontal and vertical steels can be read off from CIRA manual (Ref. 7) Table 4. We may also proceed from fundamentals as follows. For the horizontal steel, we draw bars and find y_1. The depth at which the bar intersects the critical diagonal crack. Area of 2-12 mm bars (one on each face) at each level = 226 mm^2 These bars are spaced at 250 mm. y of each layer (mm) = 850, 1100, 1350, 1600, 1850, 2100, 2350, 2600, 2850, 3100, 3350, 3600, 3850, 4100

Reference	Step	Calculations
Eq. (4.11)		$V_{sh} = C_2 \sum A_1 \dfrac{y_1}{D} \sin^2 \alpha$
		$= \dfrac{(225)(226)(0.89)}{4200}(850 + 1100 + \cdots + 4100) = 373.37 \text{ kN}$
	11	Calculate the shear taken by vertical steel V_{sv}
		Proceed as in Step 9 after finding θ.
	12	Check the bearing capacity at supports
		Width of support $S = 520$ mm
		(Maximum width that can be considered $= 0.2 L_0 = 946$ mm)
		Adopt lesser value = 520 mm
		$\text{Bearing stress} = \dfrac{v}{St} = \dfrac{5625 \times 10^3}{520 \times 450} = 24 \text{ N/mm}^2$
IS 456 Cl 34.4		Allowable stress $= 0.45 f_{ck} = 0.45 \times 40 = 18 \text{ N/mm}^2$
		Actual stress is larger than the allowable stress.
		The following remedies are possible:
		1. Provide reinforcement for the excess stress
		2. Provide better support like bearing on steel plate with allowable bearing $0.8 f_{ck}$.
		3. Increase breadth to decrease bearing stress.
	13	Determine the design shear by ACI code (as different from Step 8)
		(a) In ACI, tension steel does not contribute to shear as no arch action is assumed. Total shear = 5625 kN.
		(b) Shear taken by concrete with reduction factor
Eq. (4.18)		$= 0.13 \sqrt{f_{ck}}\, td = 0.13 \sqrt{40}\,(450)(4125) = 1526 \text{ kN}$
		and $d = 4200 - 75 = 4125$ mm
		This value of reduction factor is much less than that obtained by the British practice.
		(c) The shear taken by the minimum web steel
Eq. (4.19)		$V_w = 0.85 \left(\dfrac{A_v}{s_1} \dfrac{1+f}{12} + \dfrac{A_h}{s_2} \dfrac{11-f}{12} \right) f_y d$
		$\dfrac{A_v}{s_1 t} = 0.0015$ and $\dfrac{A_h}{s_2 t} = 0.0025$
		This reduces to

Reference	Step	Calculations
Eq. (4.20)		$V_w = 0.07(0.029d - 0.001L_n)tf_y$
		$= 0.07[0.029(4125) - 0.001(4730)](450 \times 415)$
		$= 1502 \text{ kN}$
		$V_c + V_w = 1526 + 1502 = 3028 \text{ kN}$
		Balance shear = 5625 − 3028 = 2597 kN
		This is taken by extra shear steel to be designed by Eq. (4.19). As can be seen from Eq. (4.19), the horizontal steel is more efficient than the vertical steel.

EXAMPLE 4.2 (Design of bottom loaded deep beam)

What extra precautions would you take if the beam in Example 4.1 is bottom loaded?

Reference	Step	Calculations
	1	*Shear taken by concrete*
		The thickness should be such that the total design shear should never exceed 75% of the capacity of solid beam. That is
		$V < 0.75tD\tau_{max}$
		(τ_{max} from Table 14 of IS 456)
	2	*Provision of larger bars*
		Provide larger bars on both faces to support the bottom loads. These bars should be well anchored in the top compression zone.
	3	*Extra precaution for horizontal steel*
		Provide horizontal web steel particularly in the lower half of the depth preferably over the whole length or at least for a length of span 0.4D measured from each support as shown in Fig. 4.6. Area of this steel should satisfy the minimum tension steel and bar spacing rules.

EXAMPLE 4.3 (Design of continuous deep beam)

A reinforced concrete deep girder is continuous over spans of 9 m apart from centre to centre. It is 4.5 m deep, 300 mm thick and the supports are columns 900 mm in width. If the girder supports a uniformly distributed load of 200 kN/m including its own weight, design the necessary steel assuming grade 20 concrete and Fe 415 steel.

Reference	Step	Calculations
	1	*Check deep beam action*
		$\dfrac{L}{D} = \dfrac{9000}{4500} = 2; \quad \dfrac{L}{t} = \dfrac{9000}{300} = 30$
		$\dfrac{D}{t} = \dfrac{4500}{300} = 15$
	2	*Determine bending moments and points of inflection*
		Let $r = \dfrac{\text{Width of support}}{c/c \text{ span}} = \dfrac{90}{900} = 0.1$
		Factored load $= 1.5 \times 200 = 300$ kN/m)
Eq. 4.2(b)		$M \text{ (positive in span)} = \dfrac{wL^2}{24}(1 - r^2)$
		$= \dfrac{300(9)^2(1 - 0.01)}{24}$
		$= 1002.4$ kNm
Eq. 4.2(c)		$M \text{ (negative at support)} = \dfrac{wL^2}{24}(1 - r)(2 - r)$ at face of support
		$= \dfrac{(300)(9)^2}{24}(1 - 0.1)(2 - 0.1)$
		$= 1731.4$ kNm
		Let the inside distance between points of inflection be x.
		$\dfrac{300x^2}{8} = 1002.4$ gives $x = 5.2$ m
		Clear span $= 9 - 0.9 = 8.1$ m
		Distance of P.I. from the face of support $= \dfrac{8.1 - 5.2}{2} = 1.5$ m (say)
	3	*Design of tension steels for moments*
Eq. (4.3b)		$\dfrac{L}{D} = 2;$ Lever arm $z = 0.2(L + 1.5D)$
		$= 0.2[900 + (1.5)(4500)] = 315$ cm
		By British practice,
Sec. 4.5.1		$z = 0.6D = 0.6 \times 450 = 270$ cm
		Assume lesser of the values

DESIGN OF REINFORCED CONCRETE DEEP BEAMS 69

Reference	Step	Calculations
		A_s at supports $= \dfrac{1725 \times 10^6}{(2700)(0.87 \times 415)} = 1769.5 \text{ mm}^2$
		A_s at mid span $= \dfrac{1002.4 \times 10^6}{2700(0.87 \times 415)} = 1028.3 \text{ mm}^2$
		Minimum steel of 0.3% $A_s = 0.003 \times 300 \times 1500 = 1350 \text{ mm}^2$
		Provide minimum steel of 4050 mm² at supports and spans.
	4	*Detail the steel*
Sec. 4.4.4		(a) Place of the positive steel in $(0.25D - 0.05L)$ region of the depth of the beam $= (0.25 \times 4500 - 0.05 \times 9000) = 675$ mm.
		Provide cover of 50 mm. All positive steel should be continued to the supports.
		(b) Negative steel to be placed in two groups as $\dfrac{L}{D} > 1$.
		Group A: $A_{st} = 0.5\left(\dfrac{L}{D} - 0.5\right) = 0.75$
		75% of the steel to be placed on $0.2D$ ($= 900$ mm) adjacent to tension face steel.
		Group B: The remaining 25% of the steel to be placed on $0.3D$ ($= 1350$ mm) on either side of the mid depth.
		Extend negative steel beyond face of supports as follows:
		Not more than 50% to be stopped at $0.5D$ from face of support and the balance should extend beyond the point of inflection by D, 12ϕ or 1/16th clear span, whichever is larger.
	5	*Design for shear*
Sec. 4.6.3		In continuous beam we cannot depend on arch action as in simple beams. Hence the design for shear as in ordinary beams.
		Shear at edge of support $= \dfrac{1}{2}$(Clean span)(Load)
		$= \dfrac{1}{2}(8.1)(300) = 1215$ kN
		Take critical section for shear as recommended by Portland Cement Association, U.S.A.
		At 0.15 of clear span from the face of support
		$0.15 \times 8.1 = 1.21$ m from the edge of support

Reference	Step	Calculations
		Design shear = 1215 − (300 × 1.21) = 852 kN
		Therefore, the design for this shear as in an ordinary beam is
		$V_c + V_s$ = 852 kN

EXAMPLE 4.4 (Design of deep beams with openings)

Design a symmetrical deep beam one-half of which is shown in Fig. 4.8. It supports two loads 415 kN at 400 mm from supports on each side. Assume f_y = 415 N/mm², grade 25 concrete and effective span 1650 mm.

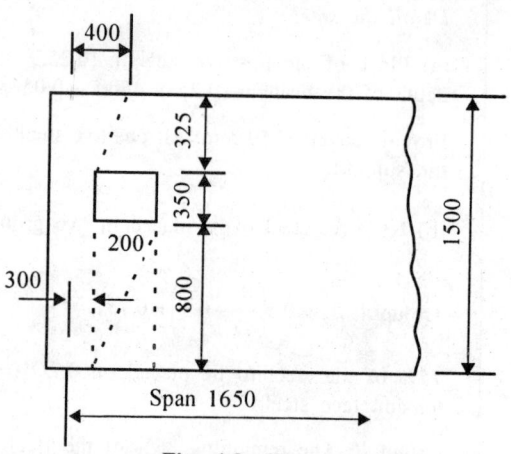

Fig. 4.8 Example 4.4.

Reference	Step	Calculations
	1	Check for the deep beam action
		$\dfrac{L}{D} = \dfrac{1650}{1500} = 1.1 < 2$, assume t = 300 mm
	2	Calculate the loads on the beam
		$P_w = 415 \times 1.5 = 622.5$ kN (say 625 kN)
		$M_u = 625 \times 0.4 = 250$ kNm
Eq. (4.14)		$V_w = 625$ kN ≯ $0.75 \times 800 \times 300 \times 3.1 = 558$ kN
		Assume this is marginally admissible.
	3	Determine the minimum thickness required
Eq. (4.14)		$t = \dfrac{0.55 \times 625 \times 10^3}{0.7(D_0 - 0.35 k_1 a_v) f_t} = \dfrac{0.55 \times 625 \times 10^3}{0.7(800-105) \times 2.5} = 283$ mm < 300
		where $k_1 a_v = 300$; $f_t = 0.5\sqrt{f_{ck}} = 2.5$ N/mm².

DESIGN OF REINFORCED CONCRETE DEEP BEAMS 71

Reference	Step	Calculations
	4	*Calculate the area of tension steel needed and check for the minimum* L.A. = Lesser of $0.75D_0$ or $0.6L$ $0.75 \times 800 = 600$; $0.6 \times 1650 = 990$ $A_s = \dfrac{1.55 M_u}{f_y D_0} = \dfrac{1.55 \times 250}{415 \times 800} = 1167 \text{ mm}^2$ 4 bars of 20 mm gives 1257 mm². Steel percentage $= \dfrac{1257 \times 100}{300 \times 1500} = 0.28\%$ is very low. Provided the minimum steel percentage required = 0.3%.
	5	*Detailing of tension steel* Detail tension steel in two layers with sufficient cover and with their ends bent upwards to ensure arch action so that the shear contributed by tension steel can also be taken into account as in B.S.
	6	*Calculate the web steel required* $A_v = (0.12\%) = \dfrac{0.12 \times 300 \times 1000}{100}$ per metre $= 360 \text{ mm}^2 \ (= 180 \text{ mm}^2)$ on each face Provide 12 mm spaced 150 mm (754 mm²).
Eq. (4.3)	7	*Determine the capacity of concrete in shear V_c* $V_c = 0.7(D_0 - 0.35 k_1 a_v) t f_t$ $k_1 a_v = 300$; $D_0 = 800$ $f_t = 0.5\sqrt{f_{ck}} = 2.5$ $V_c = 0.7[800 - (0.35 \times 300)] \, (2.5 \times 300) = 365 \text{ kN}$
Sec. 4.5.4 Eq. (4.13)	8	*Obtain the value of the shear taken by the tension steel* $V_s = C_2 A_s \dfrac{d}{D} \sin^2 \alpha$ $C_2 = 225$; $d = 1500 - 75 = 1425$ $\tan \alpha = \dfrac{800}{300}$ and $\sin^2 \alpha = 0.877$ $V_s = 225 \times 1257 \dfrac{1425}{1500} 0.877 = 236 \text{ kN}$
	9	*Determine the shear to be taken by web steel* $V_w = 625 - (365 + 236) = 24 \text{ kN}$

Reference	Step	Calculations
		Assume the contribution by the vertical steel is zero. Find y_1 values of the horizontal steel at 150 mm spacing. Let the first bar be at 150 mm from the bottom $y = 1350$. y_1 of other horizontal bars of web steel 1200 mm, 1050 mm, 900 mm, 750 mm, and (hole) 350 mm by 200 mm. Shear from these bars $= C_2 A_s \dfrac{\sin^2\alpha}{D} \Sigma y_1$ A_s of 12 mm bar = 113.1 $C_2 = 225$ for HYSD bars; $A_s = 113 \text{ mm}^2$ $\Sigma y_1 = 1350 + 1200 + 1050 + 900 + 750 + 350 + 200 = 5800$ $V_w = \dfrac{225(113) \times 0.877}{1500} 5800$ $= 86 \text{ kN} > 24 \text{ kN}$ (required)
	10	Check for bearing Continue as in Example 4.1.

REFERENCES

1. Leonhardt F. and Walter R., *Deep Beams,* Bulletin 178, Deutcher Ausschuss fur Stahlbeton, Berlin, 1966.

2. Notes on ACI 318 (1989), *Building Code Requirements for Reinforced Concrete—Design Applications*, Portland Cement Association, Chicago, 1996.

3. IS 456, *Indian Standard Code of Practice for Plain and Reinforced Concrete*, Bureau of Indian Standards, New Delhi, 2000.

4. Ove Arup and Partners, *The Design of Deep Beams in Reinforced Concrete*, CIRIA Guide 2, London, 1977.

5. Kong F.K., Robins P.J. and Sharp G.R., *The Design of Reinforced Concrete Deep Beams in Current Practice*, Vol. 53, No. 4, The Structural Engineer, London, April, 1975.

6. Reynolds C.E. and Steedman J.C., *Reinforced Concrete Designer's Handbook*, Cement and Concrete Association, London, 1997.

CHAPTER 5

Design of Ribbed (Voided) Slabs

5.1 INTRODUCTION

In IS 456 Clause 30, the terms *ribbed slabs* and *voided slabs* are used for the slabs constructed *in situ* by one of the following ways:

(i) As a series of concrete ribs with topping cast on forms which are removed after the concrete has set. The ribs are usually given a small taper for easy removal of shuttering.

(ii) As a series of concrete ribs cast between clay or concrete blocks which remain in the structure and the top of the ribs reconnected by a topping of concrete of the same strength as that used in the ribs.

(iii) As a series of concrete ribs with continuous top and bottom faces cast *in situ* but containing voids of any shape inside (Fig. 5.1).

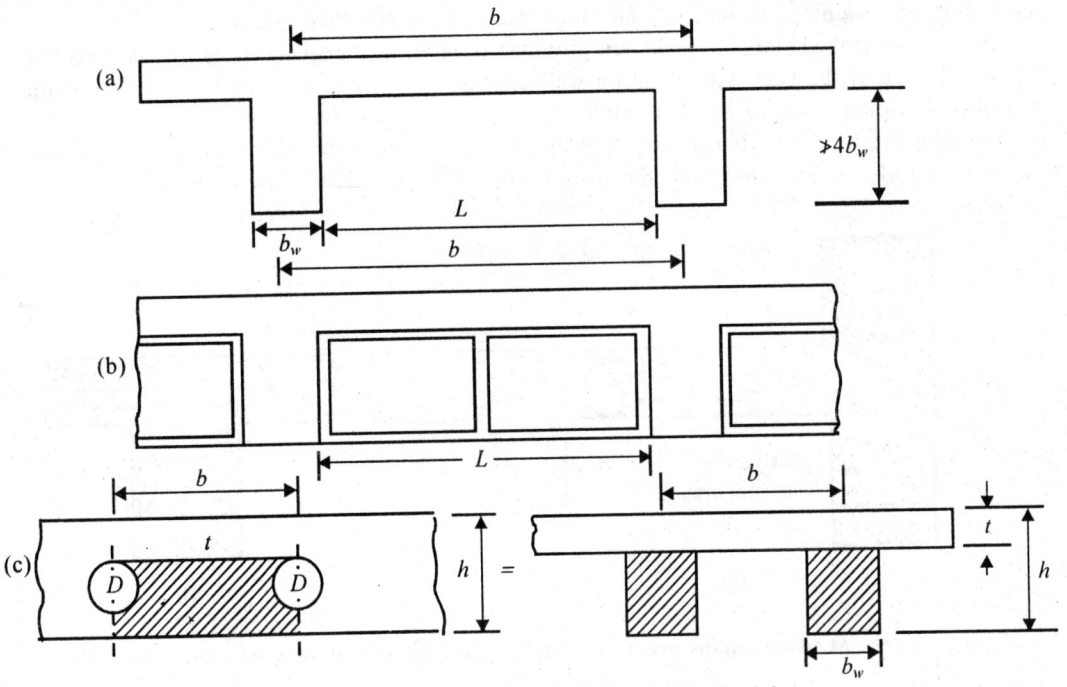

Fig. 5.1 Three types of ribbed (voided) slabs.

(Slabs made from precast concrete funicular shells with cast *in situ* ribs between them called *waffle shells* developed by Structural Engineering Research Centre, Madras, can be considered to belong to this system of ribbed slabs.)

In BS 8110, these slabs are allowed to be constructed as a series of ribs cast *in situ* with left in blocks and top of the ribs connected by a concrete of lower strength, the topping not considered to contribute to the structural strength. Thus, certain types of grid slabs can also be considered as voided slabs. These slabs are also called *coffered slabs*.

The principal advantage of these types of constructions is the reduction in weight as much as 20 to 30% achieved by introduction of voids in the slab. Such floors are economical for long spans and moderate live loads. They also reduce loads on foundations allowing more storeys to be built on the same foundations. According to IS:456, it is important that in these slabs the ribs and topping are both of *in situ* construction with the topping placed along with the ribs. In BS 8110, the topping can be non-structural.

There are also special conditions regarding sizes of ribs, thickness of topping, etc. specified in the code so that the conventional methods of analysis of slabs are applicable only to these slabs. (IS 6061 Parts I and II dealing with the construction of floors with filler blocks should be consulted for details of construction of these voided slabs.)

5.2 SPECIFICATION REGARDING THE SLABS

5.2.1 Size and Position of Ribs

IS 456 Clause 30.5 requires the *in situ* ribs for ribbed slabs without infills should not be less than 65 mm wide and space between the centres not greater than 1.5 m. The depth of rib (excluding any topping) should not be more than four times their width.

In one way ribbed slabs (the ribs are provided parallel to the span) are at regular spacing. When the edge of the slab is built into a wall or supported on beams, the bearings are made solid by providing a rib of suitable width (at least as wide as the bearing) along the edges as shown in Fig. 5.2. The portion of the slab which are under the concentrated loads inside the span are also made solid with the provision of ribs of suitable widths.

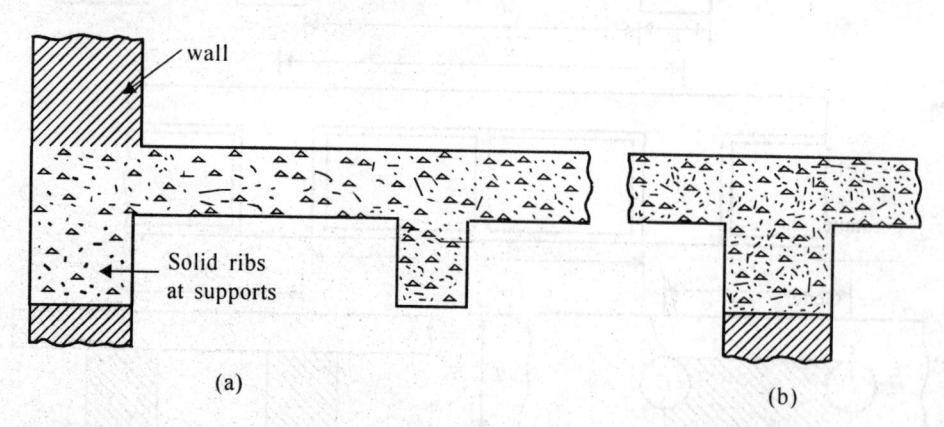

Fig. 5.2 Edges of voided slabs are to be made solid: (a) at end supports, (b) over walls.

5.2.2 Thickness of Topping

For slabs without permanent infills (of blocks or tiles) the minimum thickness of topping is generally recommended by BS 8110 to be 50 mm or one-tenth distance between ribs, whichever is greater. This ensures that the load can be transferred by arching action without the aid of reinforcement in the topping. However, a single layer of mesh having a cross sectional area not less than 0.12% of the area of the topping should be provided in each direction as nominal steel in the topping. The spacing between wires should not be greater than one-half the distance between the ribs. With block infilling, the thickness of topping may be reduced to 30 mm. Infills like funicular shells which are made of structural grade concrete can be assumed to act integral with the *in situ* top concrete even when it may be as little as 25 mm.

5.2.3 Hollow or Solid Blocks as Formers

BS 8110 specifies that if the strength of the formers are also be taken into account, it should be made of concrete or burnt clay of characteristic strength not less than 14 N/mm^2. The minimum thickness of slabs and the clear spacing of ribs should be as in Table 5.1 and Fig. 5.1.

TABLE 5.1 RIBBED SLAB DIMENSION (BS 8110)

Type of ribbed slab	Maximum rib spacing (mm)	Rib width	Maximum rib height excluding topping	Minimum topping thickness (mm)
Slabs with structural blocks jointed in cement mortar in 1:3 ratio (11 N/mm^2) with structural topping	500 (clear L)	b_w	$4b_w$	25 usually 40
As above but not jointed	500 (clear L)	b_w	$4b_w$	30
All other slabs with permanent blocks	500 (centre-to-centre)	b_w	$4b_w$	40 or $L/10$
All slabs without permanent blocks	500 (clear L)	b_w	$4b_w$	40 or $L/10$

Note: Generally clay blocks are 300 × 300 mm in plan and of depth varying from 75 mm to 250 mm.

5.3 ANALYSIS OF THE SLABS FOR MOMENT AND SHEARS

The moment and shear forces in ribbed slabs due to ultimate loads are calculated as in the case of solid slabs, depending on whether it is one-way, two-way or flat-slab construction.

Continuous slabs may also be designed as series of simply-supported slabs provided they are protected from corrosion caused by the tendency for these slabs to crack at the supports. This cracking should always be controlled by providing negative steel at the support. The practice is to provide negative steel equal to at least one-fourth the steel required at the middle of the adjoining spans. These bars should extend at least a distance of one-tenth the clear span on either side from the face of the support (BS specifies this extension to be 15%).

5.4 ULTIMATE MOMENT OF RESISTANCE

The theory of ordinary beams and slabs is also used for the calculation of the ultimate resisting moment of the slab. For analysing a solid slab, we take unit width and get the positive and negative design moments per metre width of the slab. The design moments and shears for each rib are taken as those induced in a slab of width equal to the spacing of the ribs. Where the topping concrete helps to the strength of the system, the ribs and topping are assumed to act as 'T' beams. The width of the slab taken is the lesser of the two following dimensions:

1. The distance from centre-to-centre or ribs
2. $(b + kL/5)$, where b is the breath of the web and L the span with $k = 1$ for simply-supported and 0.7 for continuous beams.

For resisting the moments of slabs with burnt clay blocks, strength of clay blocks in compression may also be taken into account. This strength may be assumed as 0.25 to 0.3 times the compressive strength of the block. The steel required in the slab for a distance equal to the spacing of the ribs is determined and this amount of steel is provided in each rib.

The compression due to the negative moment at supports (tension at top and compression at the bottom) can be resisted only by the concrete in the ribs. In cases where the rib is insufficient in compression, additional compression steel is provided at the bottom of ribs by extending the tension steel at the end mid-span to the supports or by providing separate compression steel at the supports.

5.5 DESIGN FOR SHEAR

The shear stress V is calculated from the expression:

$$v = \frac{V}{bd}$$

where

V = The shear force due to ultimate loads on a width of slab equal to the centre distance between ribs as calculated from the slab action

b = The width of the rib

d = The effective depth

In cases where structural hollow clay tiles or hollow blocks are used for infillng, IS 456 Clause 30.3 states that the value of rib for resisting shear may be increased by taking it as the width of the concrete rib plus one wall thickness of the block. An increased width should be used for precast elements like funicular shell made of structural concrete as well. The design shear value of Table 19 of IS 456 may be used for the design calculations.

However, as in the case of solid slabs it is better that these slabs are made safe in shear without shear reinforcements. The shear is unlikely to be critical in ribbed slabs unless the applied live loads are heavy and the dead load is light. Even when the special shear reinforcement is provided, the shear stress should not exceed the maximum permissible in Table 20 of IS 456. If shear reinforcement are to be provided they should be by means of links as in the case of beams. According to BS 8110, ribs reinforced by single bars do not require links. Where two or more bars are used links are required when v exceeds one-half the allowable stress.

5.6 DEFLECTION

The recommendation for deflection control (by limiting the span effective depth ratio) with respect to 'T' beams is also applicable to ribbed, hollow block or voided slabs. However, for calculating the reduction factor for 'T' beam the rib width should include the walls of the blocks on both sides of the rib.

For voided slabs constructed box or I section units, the effective rib-width calculated on the assumption flange of the unit is concentrated as a rectangular rib having the same depth and equivalent cross-sectional area. This is shown in Fig. 5.1(c), where

$$b_w(h-t) = b(h-t) - \frac{\pi D^2}{4} \tag{5.1}$$

5.7 ARRANGEMENT OF REINFORCEMENTS

The following rules given in IS 456 Clause 30.7 apply to detailng of steel:

1. Half of the total positive steel should be carried out through the bottom on the bearing and anchored for a length equal L_d, where L_d is equal to the full development length of the bar.
2. If continuous slabs are designed as simply-supported, a minimum negative steel or at least one-fourth of the positive steel required at the mid-span should be provided at the supports. These should be extended at least one-tenth clear span into adjoining spans.
3. In the topping it is advisable to provide 0.12 % of steel in both directions.

5.8 CORROSION OF STEEL WITH CLAY BLOCKS

Great care should be taken when slabs made from clay blocks are used in roofs or in the places where they get wet. Observations have shown that steel tends to corrode very fast when it is in contact with bricks or other clay products because of the salts present in them. Hence, we should always ensure that the steel placed in ribbed slabs with clay blocks have enough concrete cover. The steel rods should be so placed during construction that under no conditions it will come in contact with the clay products. We should also not allow water to seep through the slab and start corrosion of steel. This type of corrosion has been the cause of failure or reinforced brickwork which was once popular in India. In slabs made with left-in-blocks inside them, the inside cover to reinforcement should not be less than 10 mm. All other covers should satisfy the general rules for cover for the given exposure conditions.

EXAMPLE 5.1 (Design of waffle slabs)

Design a waffle slab 3.6 × 3.9 m continuous over two adjacent sides and simply-supported on the other two sides if it is made of precast funicular shells so that ribs are spaced at 1.2 m × 1.2 m as shown in Fig. 5.3. Assume factored U.D.L;w = 10 kN/m^2, f_{ck} = 25 N/mm^2 and f_y = 41.5 N/mm^2.

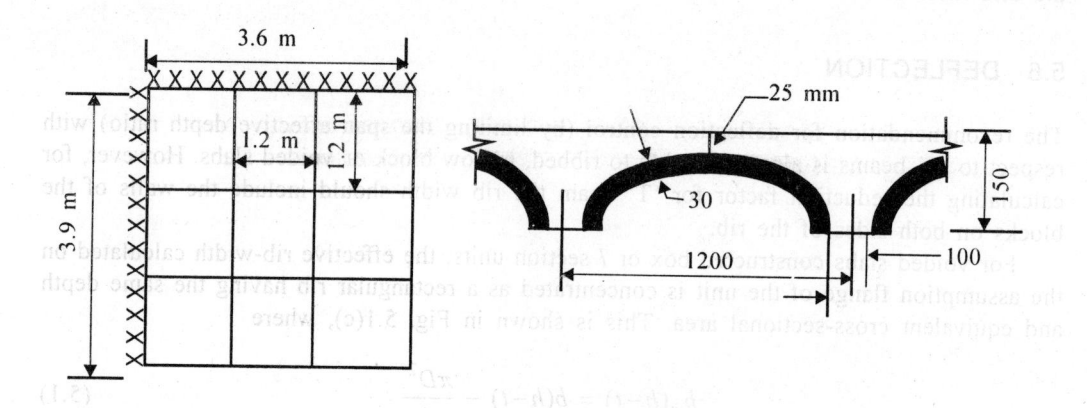

Fig. 5.3   Example 5.1.

Reference	Step	Calculations	
	1	*Check dimension for code requirements*	
IS 456 (2000)		Rib width = 100 m > 65 m	
Clause 30.5		Rib spacing = 1.2 m < 1.5 m	
		Topping 25 mm allowed for funicular shells.	
		$d = D -$ cover = 150 − 20 = 130 mm	
		The slab can be designed as two-way slabs.	
	2	*Calculate design moments*	
		$\dfrac{L_y}{L_x} = \dfrac{3.9}{3.6} = 1.08$ (two - way slab)	
IS 456			
Table 26		Two adjacent sides are continuous (case 4)	
		Coefficient Short span α_x	Long span α_y
		−ve 0.052	0.047
		+ve 0.039	0.035

Reference	Step	Calculations
		Assume same rib size in x and y directions
		Maximum $-$ve moment $= 0.052 w L_x^2$
		$\qquad\qquad\qquad\qquad = 0.052 \times 10 \times 3.6^2 = 6.74$ kNm/m
		Maximum $+$ve moment $= 0.039 w L_x^2$
		$\qquad\qquad\qquad\qquad = 0.039 \times 10 \times 3.6^2 = 5.05$ kNm/m
		Maximum moment one rib has to carry = (moment per merte) (Rib spacing)
		Maximum $-$ve B.M. $= 1.2 \times 6.74 = -8.1$ kNm
		Maximum $+$ve B.M. $= 1.2 \times 5.05 = +6.1$ kNm
	3	*Design section at support*
		Only the rib will be effective. Compression steel is provided at the bottom and additional tension steel on the top.
		$M_u = 8.1$ kNm; $d = 130$ mm; $b = 100$ mm
		$\dfrac{d^1}{d} = \dfrac{20}{130} = 0.15$
		$\dfrac{M_u}{bd^2} = \dfrac{8.1 \times 10^6}{100(130)^2} = 4.79$
SP 16 Table 51		Tension steel (top) = 1.634% p_t
		Compression steel (bottom) = 0.478% p_c
		$A_{st} = \dfrac{1.634}{100} \times 130 \times 100 = 212$ mm^2
		Provide 3T10 (235 mm^2)
		$A_{sc} = \dfrac{0.478}{100} \times 130 \times 100 = 62$ mm^2
		Provide 1T10 (79 mm^2)
	4	*Design section at mid-span*
		Section acts with top concrete in compression (minimum thickness of topping is 25 mm.)
		$A_{st} = \dfrac{M_u}{0.87 f_y \left(d - \dfrac{D_f}{2}\right)} = \dfrac{6.1 \times 10^6}{(0.87 \times 415)(130 - 12.5)}$
		$= 143.8$ mm^2 (2T10 gives 157 mm^2)

Reference	Step	Calculations
		Percentage steel provided $= \dfrac{157 \times 100}{130 \times 1200} = 0.1\%$
		Provide minimum steel for T beams (=0.2%)
		$= \dfrac{0.2 \times 130 \times 1200}{100} = 312 \text{ mm}^2$
		2T16 will give 402 mm^2.
	5	Check for shear (case 4)
		Maximum shear $= 0.43\, wL_x$
		$= 0.43 \times 10 \times 3.6 = 15.5 \text{ kN/m}$
		$V_v =$ Shear $\times$ Rib spacing $= 15.5 \times 1.2 = 18.6 \text{ kN}$
		Breadth of rib (precast unit thickness 30 mm)
		$b = 100 + 30 + 30 = 160 \text{ mm};\quad d = 130 \text{ mm}$
		$v = \dfrac{18.6 \times 10^3}{160 \times 130} = 0.89 \text{ N/mm}^2$
IS 456		For 1.25% steel assumed at support with enhancement factor 1.3 and depth of slab 150 mm., the shear
Table 19		$\tau_c = 0.70 \times 1.30 = 0.91 \text{ N/mm}^2$
		The shear is less than the allowable value. This shear limit is safe without extra shear reinforcement.
	6	*Check deflection by empirical procedure. Consider as T beam*
		Depth of beam = 150 mm
		Basic *L/D* ratio (continuous spans) = 26 (satisfactory)
	7	*Detail the steel*
		Special care should be taken to provide the necessary corner steel. The slab should be made solid at support as in all voided slabs.

EXAMPLE 5.2 (Layout of an infilled slab)

Find suitable dimension of a simply-supported slab of span 6.5 to be made from structural hollow clay blocks 300 × 300 × 250 mm height with 20 m wall thickness. Determine the reinforcements required if the slab is to carry an imposed load of 4.0 kN/m^2.

Reference	Step	Calculations
Table 5.1	1	*Determine limiting dimensions*
		(a) Minimum top slab depth = 40 mm

DESIGN OF RIBBED (VOIDED) SLABS   81

Reference	Step	Calculations

Text
Table 5.1
 	
Sec. 5.2.1
 	
Sec. 5.6
 	
Sec. 5.4
 	
Sec. 5.6
Sec. 5.5
		(b) Total depth = depth of (block + slab) = 250 +40 = 290 mm
		(We may adopt a plaster thickness of 10 mm on the underside of tiles to cover up the tiles thus making total depth of 300 mm.)
		(c) Spacing of ribs is to be not more than 500 mm.
		   Adopt spacing = 400 mm.
		(d) Width of rib not less than 290/4 = 73 mm.
		   and less than 65 mm
		   Actual width of concrete = 400 − 300 = 100 mm
		   Add thickness of block for deflection is
		   100 + 2(20) = 140 mm
		(e) Flange width $[b + (kL/5)]$ is much more than 400.
		(f) Adopt the following dimensions:
		   Flange width = 400 mm
		   Slab thickness = 40 mm
		   Depth = 290 mm
		   b for deflection = 140 mm
		   b for shear = 120 mm

	2	*Calculate load on floor* Dead load = 3.5 kN/m² Plaster = 0.25 kN/m² Imposed load = 4.0 kN/m² Total factored load = 1.5(7.75) = 11.63 kN/m² Factored load per rib = 11.63 × 0.4 = 4.65 kN/m²
	3	*Calculate steel required for one T beam (Rib spacing = 400 mm)* $M = \dfrac{4.65 \times 6.5^2}{8} = 24.6$ kNm/m; For 400 mm = 24.6 × 0.4 = 9.84 kNm Assume 20 mm bars are used. $d = 290 - 20 - \dfrac{20}{2} = 260$ mm $A_s = \dfrac{M}{0.87 f_y [d - (t/2)]}$

Reference	Step	Calculations
Sec. 5.5		Assuming, $f_y = 415$ and $f_{ck} = 20$ $$A_s = \frac{9.84 \times 10^6}{0.87 \times 415(260-20)} = 113.6 \text{ mm}^2$$ Use 1T20 (314 mm^2) (provide one rod)
IS Cl. 26.5.1	4	Ckeck for minimum steel (based on rib width) Required $p = \dfrac{0.85}{f_y} = 0.2\%$ (approx.) p provided $= \dfrac{100 A_s}{b_w h} = \dfrac{100 \times 314}{100 \times 260} = 1.0\%$
Step 3 SP 456 Fig. 4	5	Check for deflection (T beam) A_s (required) $= \dfrac{113.6 \times 100}{400 \times 260} = 0.11\%$ Basic L/D ratio for simply-supported beam = 20 Modification factor F1, for tension steel 0.11% = 2.0 Modification factor F2, for compression steel = 1.0 Modification factor for T beam $\dfrac{b_w}{b_f} = \dfrac{140}{400} = 0.35; \quad F_3 = 0.8$ Allowable: $\dfrac{L}{D} = 20 \times 2.0 \times 0.8 = 32.0$ Actual: $\dfrac{L}{D} = \dfrac{6500}{260} = 25.0$ The design is satisfactory.
IS Cl. 30.3 Sec. 5.5 IS 456 Table 19	6	Check for shear without shear steel b for the shear = 100 + 20 = 120 mm $V = \dfrac{4.65 \times 6.5}{2} = 15.1 \text{ kN}$ $v = \dfrac{V}{bd} = \dfrac{15.1 \times 10^3}{120 \times 260} = 0.48 \text{ kN/mm}^2$ % steel provided $= \dfrac{314 \times 100}{120 \times 260} = 1\%$ V allowed = 0.62 N/mm^2 (without enhancement) The design is satisfactory (without shear steel).

Reference	Step	Calculations
	7	*Detailing of steel*
		These slabs are made solid near the beaming. Then the shear at supports is much less
		$$v = \frac{15.1 \times 10^3}{400 \times 160} = 0.23 \text{ N/mm}^2$$
		The effective anchorage of steel at support is provided by extending the steel beyond the center line of support by 300 mm or one-third the support width.
Sec. 5.3	8	*Design for continuity* (negative steel at supports)
		Negative steel = $\frac{1}{4}$ (positive steel) = $\frac{314}{4} = 78.5$ mm^2
		Provide 1T10 giving 78.5 mm^2 steel. This steel should extend 10% by IS: 456 and 15% by BS 8110 into span.
		Extension into span from face of support = $0.15 \times 6500 = 976$ mm

REFERENCES

1. BS 8110 1985, *Structural Use of Concrete*, Part 1, Code of practice for design and construction, British Standards Institution, London, 1985.

2. IS 456 2000, *Code of Practice for Plain and Reinforced Concrete*, Bureau of Indian Standards, New Delhi, 2000.

CHAPTER 6

Approximate Analysis of Grid Floors

6.1 INTRODUCTION

There are many approximate methods available for the analysis and design of different types of R.C. structures. Even though more and more structures are being designed by computers, analysis of structures by approximate methods should also be encouraged as they are useful in developing an engineering sense and can also be used as checks on the computer outputs. These types of checks are always necessary for successful design of complex structures. This chapter illustrates the use of approximate methods for analysis of grid floors.

6.2 ANALYSIS OF FLAT GRID FLOORS

For enclosing large rooms such as theatre halls, auditoriums, where column free space is the main requirement, grid floors are used. The types of commonly-used grids are shown in Fig. 6.1. Grids with diagonal members are called *diagrids*. We have seen in Ch. 5 that grid floors

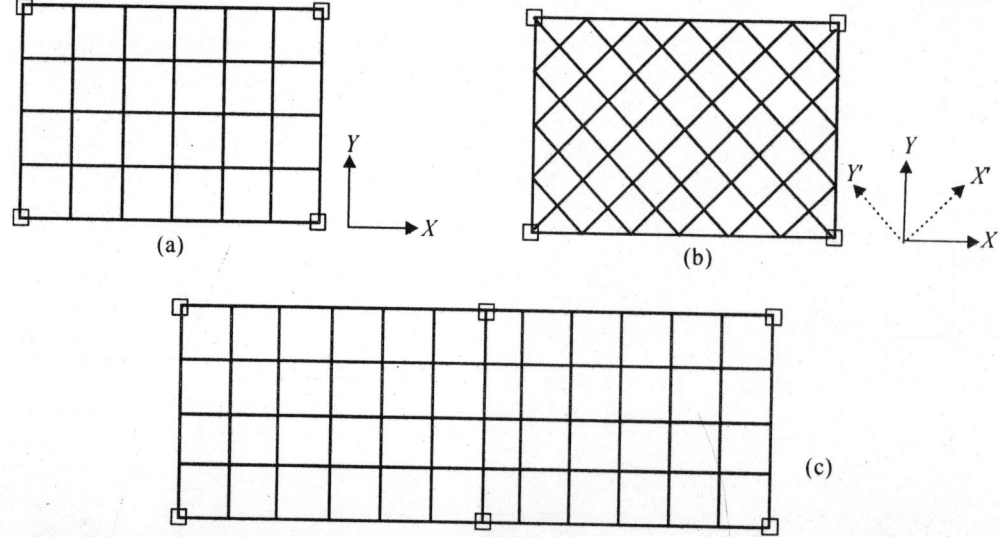

Fig. 6.1 Layout of grid floors: (a) rectangular grid, (b) diagrid, (c) continuous grid.

with restricted layout of beams, thickness of slab and edge beams can be analysed by the conventional methods as in ribbed slabs. Large grid floors which do not follow these restricted layouts are analysed by other methods. These methods can be divided into the three following groups:

1. Method based on plate theory (approximate method)
2. Stiffness matrix method using computer
3. Equating deflection method of each intersecting node by simultaneous equations.

In the following sections, a brief discussion of the method based on the plate theory and reference to other two methods are given.

6.3 ANALYSIS OF RECTANGULAR GRID FLOORS BY TIMOSHENKO'S PLATE THEORY

Many variations in the plate theory by different authors are available for analysis of grid floors. The most popular one is that given by Timoshenko and Krieger [1]. They have shown that the moments and shears in an *anisotropic* plate, freely supported on four sides, depend on the deflection surface. The vertical deflection ω at any point of a symmetrical grid shown in Fig. 6.2 is given by

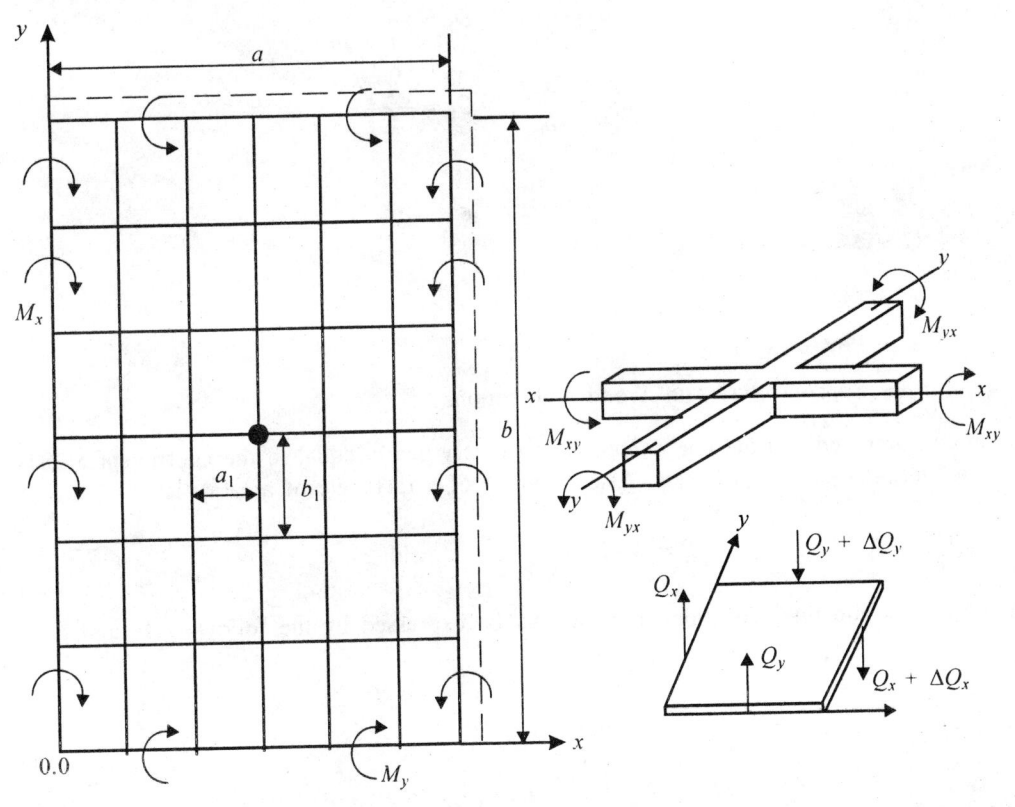

Fig. 6.2 Analysis of flat grid floors (See also Fig. 6.3).

$$\omega = \frac{16q}{\pi^6 \left(\dfrac{D_X}{a^4} + \dfrac{2H}{a^2 b^2} + \dfrac{D_Y}{b^4} \right)} \sin \frac{\pi x}{a} \sin \frac{\pi y}{b} \tag{6.1}$$

where

$\pi^6 = 960$ (approx.)

q = The load per unit area

a, b = Lengths of plate in X and Y directions respectively (a_1 and b_1 are spacing of ribs along X any Y directions.)

D_X, D_Y = Flexural rigidity EI of beams in X and Y directions 'per unit width'

C_X, C_Y = Torsional rigidity of beams parallel to X and Y directions per unit width

$C_X, C_Y = 2H$

Taking I_1 and I_2 as the moment of inertia of the beams in X and Y directions, C_1 and C_2 as the torsion constants or the polar (torsional) section modulus of the beams in X and Y directions, the values of D_X, D_Y, C_X and C_Y can be calculated by Eq. 6.2. If a_1 and b_1 are the spacing of the beams in X and Y directions, then

$$D_X = \frac{EI_1}{b_1}, \quad D_Y = \frac{EI_2}{a_1}$$

$$C_X = \frac{GC_1}{b_1}, \quad C_Y = \frac{GC_2}{a_1} \tag{6.2}$$

where

G = Modulus of shear = $\dfrac{E}{2(1+\mu)}$

E = Modulus of elasticity

μ = Poisson's ratio

a_1, b_1 = Spacing of ribs in X and Y directions

It should be noted that the coefficient $16/\pi^6$ is very nearly equal to the coefficient $5/384$ of the deflection equation of beams. Also, C value of a rectangle of area xy is

$$C = \left(1 - 0.63 \frac{x}{y}\right) \frac{x^2 y}{3} \quad \text{with } x < y$$

The bending and torsional moments can then be expressed by the following formulae:

$$M_X = D_X \frac{\partial^2 \omega}{\partial x^2} \quad \text{and} \quad M_Y = D_Y \frac{\partial^2 \omega}{\partial y^2}$$

$$M_{XY} = C_X \frac{\partial^2 \omega}{\partial x \partial y} \quad \text{and} \quad M_{YX} = -C_Y \frac{\partial^2 \omega}{\partial x \partial y} \tag{6.3}$$

$$Q_X = \frac{\partial}{\partial x}\left[D_X \frac{\partial^2 \omega}{\partial x^2} + C_Y \frac{\partial^2 \omega}{\partial y^2}\right]$$

$$Q_Y = \frac{\partial}{\partial y}\left[D_Y \frac{\partial^2 \omega}{\partial y^2} + C_X \frac{\partial^2 \omega}{\partial x^2}\right]$$

The maximum bending moments are at the centre of the span while the maximum torsional moments are at the corners. The maximum shears develop at mid points of the longer and shorter side supports.

Taking the first term of the series in equation (6.1) and putting,

$$\omega_1 = \frac{16q/\pi^6}{(D_X/a^4) + [2H/(a^2 b^2)] + (D_Y/b^4)} \tag{6.4}$$

The expression for the stress-resultants along the grid can be written down by using Eqs. (6.3) and (6.4) as follows:

1. *Values of M_X and M_Y (bending moments)*

$$M_X = D_X \left(\frac{\pi}{a}\right)^2 \omega_1 \sin\frac{\pi x}{a} \sin\frac{\pi y}{b} \tag{6.5}$$

$$M_Y = -D_Y \left(\frac{\pi}{b}\right)^2 \omega_1 \sin\frac{\pi x}{a} \sin\frac{\pi y}{b} \tag{6.6}$$

These are maximum at the centre where the second term becomes equal to unity.

2. *Values of M_{XY} and M_{YX} (torsional moments)*

$$M_{XY} = C_X \left(\frac{\pi^2}{ab}\right) \omega_1 \cos\frac{\pi x}{a} \cos\frac{\pi y}{b} \tag{6.7}$$

$$M_{YX} = -C_Y \left(\frac{\pi^2}{ab}\right) \omega_1 \cos\frac{\pi x}{a} \cos\frac{\pi y}{b} \tag{6.8}$$

3. *Values of Q_X and Q_Y (shears)*

$$Q_X = -\omega_1 \left[D_X \left(\frac{\pi}{a}\right)^3 + C_Y \left(\frac{\pi^3}{ab^2}\right)\right] \cos\frac{\pi x}{a} \sin\frac{\pi y}{b} \tag{6.9}$$

$$Q_Y = -\omega_1 \left[D_Y \left(\frac{\pi}{b}\right)^3 + C_X \left(\frac{\pi^3}{a^2 b}\right)\right] \sin\frac{\pi x}{a} \cos\frac{\pi y}{b} \tag{6.10}$$

Using these relationships the bending moments, shears and the twisting moments at the salient points of the grid can be calculated.

4. *Value of I and C for analysis*

The question arises as to what the values of I_1 and I_2 and C_1 and C_2 are to be taken for analysing Eqs. (6.5)–(6.10). The most common method is to take these values corresponding to the beam with the attached slabs on either side. Some designers however assume the values for *I* and *C* as those due to the beam portions of the grids only neglecting the slab. The use of this method for approximate analysis of a grid floor is shown in Example 6.1.

6.4 ANALYSIS OF GRID BY STIFFNESS MATRIX METHOD

This is an exact method and needs a computer for its application. It assumes the frame coordinates system X and Y and the member coordinate system X'-Y' as shown in Fig. 6.1(b). The active joints are identified first and then the force displacement equations are written for the various active member ends. The member stiffnesses are then transferred to the frame coordinate system. From the stiffness matrix and joint displacement equations the force at each member joint can be obtained by use of computer programme. Readymade programmes for the various types of grids are also available for quick analysis in design offices. If a computer and the relevant software programmes are available, the analysis can be carried out quickly. However, the results of the computer analysis should be checked with values got from approximate methods.

6.5 ANALYSIS OF GRID FLOORS BY EQUATING JOINT DEFLECTIONS

In this method deflection of the joints and the loads on each of the beams are determined. As the number of resulting equations are large, a computer will be required for its easy solution. *Analysis of Grid Floors* [2] by the National Buildings Organisation, New Delhi gives tables and charts which may be conveniently used for the analysis of rectangular and diagonal grid floors. In this method, torsion component is generally neglected.

6.6 COMPARISON OF METHODS OF ANALYSIS

No torsion analysis is suitable for the preliminary analysis and design. Torsion analysis method gives fairly good results for rectangular grids. The handbook method gives best results for square grids. However complex grids, like diagonal grids, are best solved by matrix method for final designs. Even though the slab analogy of voided slabs is straightforward, its use should be restricted with the following limitations:

1. Spacing of ribs should not be greater than 1.5 m or twelve times the flange thickness (the clear spacing should not be more than ten times the thickness).
2. The depth of the rib should not be more than four times the width of the ribs.

With large spacing of the beams, grids do not act as a plate but as individual units. Accordingly, with the above spacing restrictions the plate analogy can be used for approximate analysis of rectangular grids. For grids not obeying the above spacing rules, the approximate methods can be used for the preliminary analysis only. In all the cases, the matrix method gives more realistic values and should be used for the final analysis and design.

6.7 DETAILING OF STEEL IN FLAT GRIDS

Detailing of slabs and beams in flat grids is shown in Fig. 6.3. Steel is provided in the slabs at the top and bottom for convenience. Beams are provided with steel as in conventional beams so as to satisfy the B.M. and S.F. requirements along its length. To take care of concentrated reactions at beam junctions, additional stirrups are also provided at the grid beam junctions. The edge beams should also be properly reinforced depending on the way they are supported.

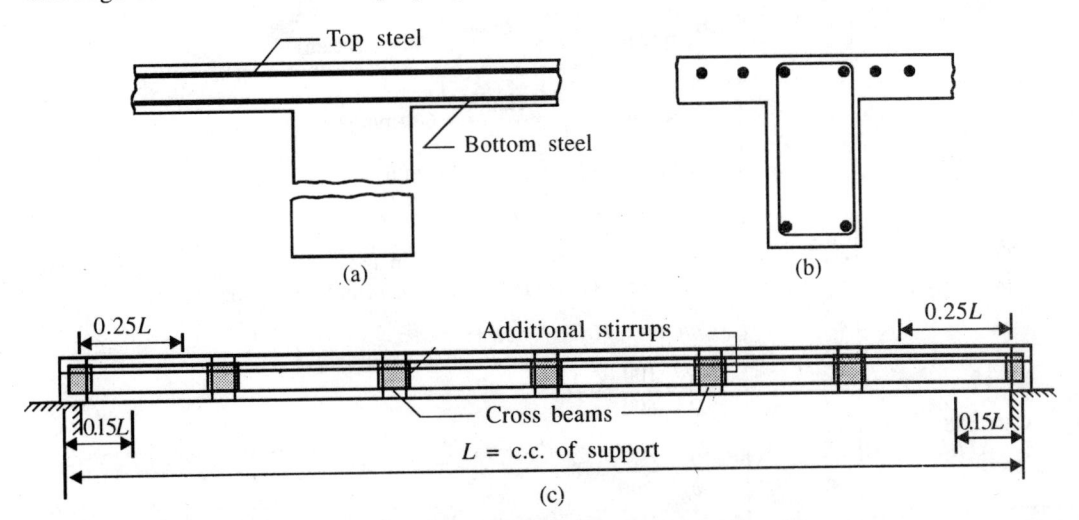

Fig. 6.3 Detailing of steel in grid floors.

EXAMPLE 6.1 (Approximate analysis of grid floors)

A reinforced rectangular grid floor is 12 m × 16 m with the centre-to-centre spacing of ribs at 2 m both ways, as shown in Fig. 6.4. Determine the bending moments and shears at the

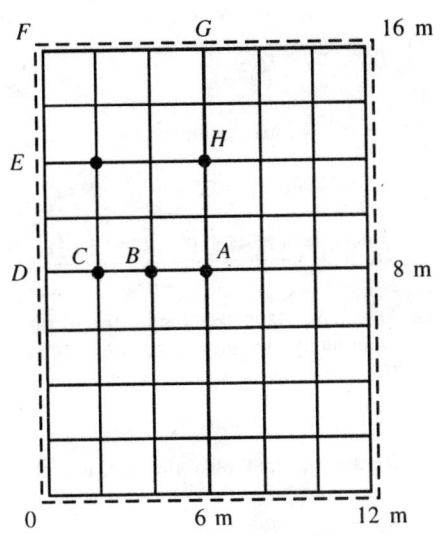

Fig. 6.4 Example 6.1.

ADVANCED REINFORCED CONCRETE DESIGN

salient points. Assume slab thickness is approximately 1/20th span, total load including self-weight is 6.5 kN/m², f_{ck}, = 20 N/mm² and it is simply-supported on all the four sides.

Reference	Step	Calculations
	1	Estimate necessary thickness of beams and floor slab
		Assume spacing of ribs 2 m both ways
		$L_x = 12 \text{ m}, \quad D = \dfrac{12{,}000}{20} = 600 \text{ mm(beam)}$
		$L_x = 2 \text{ m}, \quad D_f = \dfrac{12{,}000}{20} = 600 \text{ mm(slab)}$
		Assume width of rib = 200 mm
Sec. 5.2 IS 456 Clause 30.5	2	Check dimensions for design as voided slab As spacing of ribs is more than the allowed we cannot design the structure as a voided slab. Use orthotropic plate theory.
	3	Calculate M.I. of ribs (T beams)
SP 16 Chart 88		$\dfrac{D_f}{D} = \dfrac{100}{600} = 0.166$
		$\dfrac{b_f}{b} = \dfrac{2000}{200} = 10$
		$k = \dfrac{2.3}{I} = \dfrac{Kb_w D^3}{12}$
		$= \dfrac{2.3 \times 0.20 (0.6)^3}{12} = 82.6 \times 10^{-4} \text{ m}^4$
	4	Calculate rigidity of ribs per unit width
IS 456 Clause 6.2.3.1		$D_X = \dfrac{EI}{a_1} \quad D_Y = \dfrac{EI}{b_1}$
		$a_1 = b_1 = 2 \text{ m (spacing of ribs)}$
		$E = 5700\sqrt{f_{ck}} = 25.49 \times 10^6 \text{ kN/m}^2$
		$D_X = D_Y = \dfrac{(25.49 \times 10^6)(82.8 \times 10^{-4})}{2} = 1.055 \times 10^5 \text{ m unit}$
Text Appendix (A)		Note: We have assumed, the whole slab is acting with the rib. Alternately we may take into account, b_f is less than four times $(D - D_f)$, i.e. b_f = 1000 mm.
	5	Calculate torsional rigidity
		We assume that only the ribs are effective (200 × 600)
		$C_X = \dfrac{GC_1}{b_1} \text{ and } C_Y = \dfrac{GC_2}{a_1}$

Reference	Step	Calculations
Step 4		Also assume $\mu = 0.15$, then $$G = \frac{E}{2(1+\mu)} = 0.435E = 0.435 \times 25.49 \times 10^6 = 11.09 \times 10^6 \text{ kN/m}^2$$ $$C_1 = C_2 = \left(1 - 0.63\frac{x}{y}\right)\frac{x^3 y}{3}$$ $$= \left(1 - \frac{0.63 \times 0.2}{0.6}\right)\frac{(0.2)^3(0.6)}{3} = 1.264 \times 10^{-3} \text{ m}^3$$ $$C_X = C_Y = \frac{GC_1}{b_1} = \frac{GC_2}{a_2} = \frac{(11.09 \times 10^6)(1.264 \times 10^{-3})}{2}$$ $$= 7.0 \times 10^3$$ $$2H = C_X + C_Y = 14.0 \times 10^3 \text{ m unit}$$
	6	Calculate central deflection ($a = 12$ m, $b = 16$ m) $$\frac{D_X}{a^4} = \frac{1.055 \times 10^5}{(12)^4} = 5.09 \quad \text{(say } A\text{)}$$ $$\frac{D_Y}{b^4} = \frac{1.055 \times 10^5}{(16)^4} = 1.61 \quad \text{(say } B\text{)}$$ $$\frac{2H}{a^2 b^2} = \frac{14.0 \times 10^3}{(12)^2(16)^2} = 0.38 \quad \text{(say } C\text{)}$$ $$A + B + C = 7.08$$ Central deflection $\omega_1 = \dfrac{16q}{7.08\pi^6}$ $$= \frac{16 \times 6.5}{960 \times 7.08} = 0.015 \text{ m} = 15 \text{ mm}$$ Long-term deflection = 2 to 3 times elastic deflection depending on the date of the removal of supports. $$\frac{\text{Span}}{\text{Deflection}} \doteq \frac{12000}{2.5 \times 15} = 320 < 250 \text{ (allowed)}$$ (Otherwise calculate long-term effects separately as in Chapter 1.)
	7	Find trignometric ratios of salient points X-axis along 12 m = a; spacing of ribs = $a_1 = 2$ m Y-axis along 16 m = b; spacing of ribs = $b_1 = 2$ m

92   ADVANCED REINFORCED CONCRETE DESIGN

Reference	Step	Calculations
		TABLE 1 (EXAMPLE 6.1) CALCULATION OF TRIGONOMETRIC RATIOS (STEP 7)

Points	x, y	$\dfrac{\pi x}{a}$	$\sin\dfrac{\pi x}{a}$	$\cos\dfrac{\pi x}{a}$	$\dfrac{\pi y}{b}$	$\sin\dfrac{\pi y}{b}$	$\cos\dfrac{\pi y}{b}$
A	6, 8	$\pi/2$	1	0	$\pi/2$	1	0
B	4, 8	$\pi/3$	$\sqrt{3}/2$	1/2	$\pi/2$	1	0
C	2, 8	$\pi/6$	1/2	$\sqrt{3}/2$	$\pi/2$	1	0
D	0, 8	0	0	1	$\pi/2$	1	0
E	0, 12	0	0	1	$3\pi/4$	$1/\sqrt{2}$	$-1/\sqrt{2}$
F	0, 16	0	0	1	π	0	1
G	6, 16	$\pi/2$	1	0	π	0	1
H	6, 12	$\pi/2$	1	0	$3\pi/4$	$1/\sqrt{2}$	$-1/\sqrt{2}$

Reference: Text

Step 8

Calculate M_x at A to H

M_X maximum at A $= D_x\left(\dfrac{\pi}{a}\right)^2 \omega_1$

$= 1.055 \times 10^5 \times \left(\dfrac{\pi}{12}\right)^2 \times 0.015 = 108 \text{ kNm}$

It varies as $\left(\sin\dfrac{\pi x}{a} \sin\dfrac{\pi y}{b}\right)$

Using Table 1 we can tabulate M_x at other points.

Step 9

Calculate M_y at A where it is maximum

$M_{Y(\max)} = D_Y\left(\dfrac{\pi}{b}\right)^2 \omega_1 = (1.055 \times 10^5)\left(\dfrac{\pi}{16}\right)^2 (0.015) = 61 \text{ kNm}$

It varies as $\left(\sin\dfrac{\pi x}{a} \sin\dfrac{\pi y}{b}\right)$ and can be calculated as in Table 2.

TABLE 2 (EXAMPLE 6.1) CALCULATION OF M_x (STEP 8)

Points	$\sin\dfrac{\pi x}{a} \sin\dfrac{\pi y}{b}$	M_x (kNm)
A	1	108.0
B	$\sqrt{3}/2$	93.5
C	1/2	54.0
D	0	0
E	0	0
F	0	0
G	0	0
H	$1/\sqrt{2}$	77.0

APPROXIMATE ANALYSIS OF GRID FLOORS 93

Reference	Step	Calculations
	10	Calculate M_{xy} and Q_x, Q_y $$M_{XY} = C_X \frac{\pi^2}{12 \times 16} \omega_1$$ it varies as $\left(\cos\dfrac{\pi x}{a} \cos\dfrac{\pi y}{b}\right)$ $$Q_X = \left[D_X\left(\frac{\pi}{a}\right)^3 + C_Y\left(\frac{\pi^3}{ab^2}\right)\right]\omega_1$$ It varies as $\left(\cos\dfrac{\pi x}{a} \sin\dfrac{\pi y}{b}\right)$ $$Q_Y = \left[D_X\left(\frac{\pi}{b}\right)^3 + C_X\left(\frac{\pi^3}{ba^2}\right)\right]\omega_1$$ It vaires as $\left(\sin\dfrac{\pi x}{a} \cos\dfrac{\pi y}{b}\right)$ These can also be tabulated as in Table 3.

TABLE 3 (EXAMPLE 6.1) VALUES OF MOMENTS AND SHEAR (Step 10)

Point	Moments, kN/m			Shear, kN/m		
	M_X	M_Y	M_{XY}	M_{YX}	Q_X	Q_Y
A	108	61	0	0	0	0
B	94	53	0	0	14.7	0
C	54	30.5	0	0	25.5	0
D	0	0	0	0	29.5	0
E	0	0	3.74	3.74	20.9	0
F	0	0	5.30	5.30	0	0
G	0	0	0	0	0	13.4
H	77	43	0	0	0	9.4

Notes: (1) The moments and shears given above are for unit lengths. To get design values multiply the above by their respective spacing of ribs (i.e. b, for M_x and V_x and a_1 for M_y and V_y).

(2) Multiply values by $(4/\pi)$ to campensate for the sinusoidal loading used.

(3) Side beams of the grid should be disigned for the canditions in which they are supported, i.e. on walls, on columns, etc.

EXAMPLE 6.2 (Design of a cantilever grid) [4]

A cantilever canopy for a residential building consists of a grid slab as shown in Fig. 6.5. Assuming that the total load on the slab is a UDL of 3.5 kN/m² indicate how an approximate design of the structure can be made. Neglect torsional effects in the slab. $a = 4.83$ m; $b = 2.95$ m; $a_1 = b_1 = 0.94$ m.

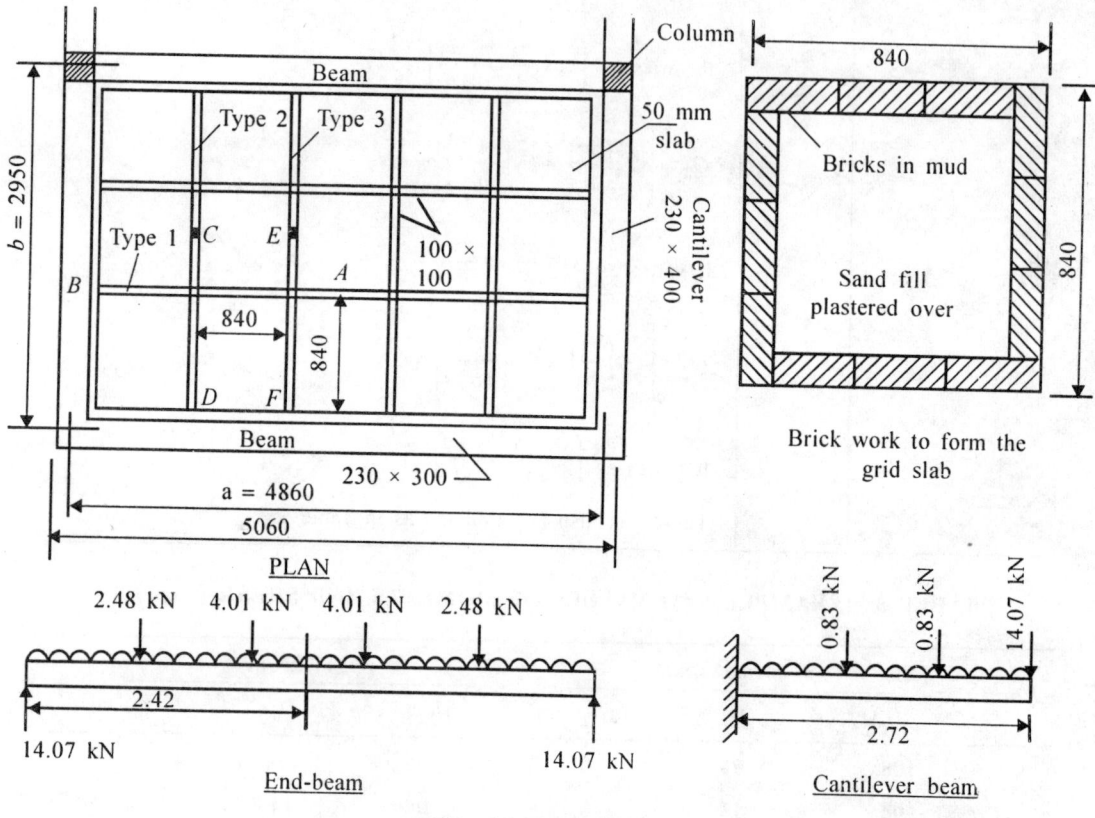

Fig. 6.5 Example 6.2.

Reference	Step	Calculations
	1	Write down the formulae for M_x, Q_x, M_y, Q_y (neglecting torsion) per unit width
Eqs. 6.1 6.5, 6.9		$M_X = \dfrac{16q_0}{\pi^4} \dfrac{a^2 b}{1+(D_y/D_x)(a/b)^4} \sin\dfrac{\pi x}{a} \sin\dfrac{\pi y}{b}$ $Q_X = \dfrac{16q_0}{\pi^3} \dfrac{ab_1}{1+(D_y/D_x)(a/b)^4} \cos\dfrac{\pi x}{a} \cos\dfrac{\pi y}{b}$ Similarly we can find M_y and Q_y.
	2	Find design moments and shears for longitudinal and transverse ribs (Fig. 6.5).

APPROXIMATE ANALYSIS OF GRID FLOORS 95

Reference	Step	Calculations
		There are only three types of ribs as under:
		Type 1 ribs:
		M_X at $A(x = a/2, y = b/3)$; Q_X at $B(x = 0, y = b/3)$
		Type 2 ribs:
		M_Y at $C(x = a/5, y = b/2)$; Q_Y at $D(x = a/5, y = 0)$
		Type 3 ribs:
		M_Y at $E(x = 2a/5, y = b/2)$; Q_Y at $F(x = 2a/5, y = 0)$
		3 *Design of longitudinal beam at the end of grid* (*simply- supported on the side of cantilevers*).
		Load on beam (230 × 300) = (UDL from part of slab acting with beam) + (UDL from self-weight) + (load due to end shears from the transverse Types 2 and 3 ribs)
		Find end-reactions on cantilevers as well as the designs of BM and SF from the above loadings.
	4	*Design of side cantilever beams* (230 × 400)
		Loads = (UDL from slab acting with the beam) + (UDL due to self-weight) + (Load from end-reaction from end-beams got in Step 3) + (Load from shear of the longitudinal type 1 ribs)
		Determine the design BM and SM from above loading.
	5	*Design of longitudinal beam near the wall between columns*
		Assume this beam as fixed at the two ends and acted on by the loads as in Step 3.

Notes: 1. Although we have omitted torsional effects, all beams should be liberally provided with stirrups for torsion effects.
2. The approximate values of loads acting on the end and cantilever are shown in Fig. E. 6.2.
3. The inner size of grid of 840 mm can be easily obtained with $3\frac{1}{2}$ brick formwork in mud mortar and plastering it with (mud + cement) mortar as indicated in Fig. 6.5.

REFERENCES

1. Timoshenko S. and Krieger S.W. *Theory of Plates and Shells*, 2nd ed., McGraw-Hill, New York, 1953.

2. *Analysis of Grids*, National Building Organisation, New Delhi, 1968.

3. Krishna Raju, *Advanced Reinforced Concrete Design—S.I. Units,* C.B.S., New Delhi, 1986.

4. Tamil Nadu P.W.D. publication, 1986.

CHAPTER 7

Design Loads other than Earthquake Loads

7.1 INTRODUCTION

IS 875 (1987) code of practice for design loads (other than earthquake loads for buildings and structures) deals with the magnitude of such loads that being used for designs in India. It was published in 1957 and was first revised in 1964. In the second revision of the code in 1987, it was expanded and the relevant specification for each type of load has been put in the following five parts.

Part 1: Dead loads

Part 2: Imposed loads

Part 3: Wind loads

Part 4: Snow loads

Part 5: Special loads and load combinations

This chapter deals with a brief description of various loads for design of buildings, a more detailed discussion on wind loads and an indication of the loads that may come on buildings during construction. The original publications should be consulted for accurate and up-to-date information regarding loadings to be used in designs. Earthquake loads are dealt with in a separate code, IS:1893 (1984) and is discussed in Chapter 18 of this book.

7.2 DEAD LOADS

IS 875 Part 1 gives the dead loads that can be taken for design. The most important dead load is the self-weight of the structure. According to IS 456 Clause 19.2.1, the unit weight of plain concrete is generally taken as 24 kN per cubic metre and that of reinforced concrete as 25kN per cubic metre. The weight of finishes will depend on the types of floor finish.

Terrazzo floor tiles are 225 × 225 × 18 mm or 300 × 300 × 25 mm and are usually laid in cement slurry on 1.3 cement mortar at least 18 mm thick. Hence, the dead load of this finish will be about 0.72 to 0.75 kN/m^2, and the nominal weight of light partitions will be 1 kN/m^2 (minimum).

Thus, the characteristic self-weight of a 100 mm slab will work out as follows:

Self-weight of 100 mm slab = 2.50 kN/m^2

Finishes (with mosaic floor) = 0.72 kN/m^2

Partitions (minimum) = 1.00 kN/m^2

Total = 4.22 kN/m^2

To this, we have to add the characteristic imposed load. Roof finish with about 125 mm weathering course in brickjelly concrete is generally taken as 2.25 kN/m^2, 20 mm asphalt as 0.48 kN/m^2, 50 mm insulating screed as 0.72 kN/m^2 and special ceiling finish as 0.25 kN/m^2.

7.3 IMPOSED LOADS (IS 875 PART 2)

These are loads which are not permanently acting on the structure. They can be classified into the following categories:

7.3.1 Live Loads

These are service loads or working loads caused by superimposed loads which vary only in a gradual manner. Loads due to furniture, stored materials and persons occupying or moving into the room come in this category. In the case of buildings, for simplicity in design calculations, these loads are assumed as uniformly distributed on the area considered. In designing bridges, they are generally considered as 'specified moving concentrated loads'. They may also include a dynamic coefficient which depends on the type of structure. They are also called *characteristic live loads*.

7.3.2 Climatic Loads

Loads due to the effect of wind, snow, etc., are known as *climatic loads*. The wind loads that are prescribed for the given region should be used for the design of structures in a given location. The loads induced in a restrained structure due to likely temperature changes come under this category. These loads are to be taken into account in designs. This depend on the maximum and minimum temperature experienced by that region. *Climatological and Solar Data for India* published by Central Building Research Institute, Roorkee, gives the values of maximum and minimum temperatures for various parts of India. Some of these values are given in Table 7.1 (see also IS 875 Part 5). Expansion joints should be planned on this data. The temperature variation in South India is much smaller than that in North India.

7.3.3 Exceptional Overloads

These are the loads due to cyclones, earthquakes, etc., applicable to the concerned region. Values assessed from the past records and available in the codes of practice are to be used for this purpose.

TABLE 7.1 BASIC WIND SPEED AND APPROXIMATE MINIMUM AND MAXIMUM TEMPERATURES FOR SOME TOWNS IN INDIA (REF. 5)

S.No.	Town	Wind speed (m/s)	Range of temperature °C Minimum	Maximum
1	Agra	47	−2	48
2	Ahmedabad	39	2	48
3	Asansol	47	5	47
4	Bangalore	33	8	39
5	Bhopal	39	0	44
6	Bikaner	47	−1	49
7	Calcutta	50	6	44
8	Chennai	50	13	45
9	Coimbatore	39	11	40
10	Cuttack	50	7	47
11	Darjeeling	47	−5	27
12	Delhi	47	−1	46
13	Guwahati	50	2	40
14	Jabalpur	47	0	46
15	Jamshedpur	47	4	47
16	Kanpur	47	0	47
17	Kochi	39	17	34
18	Madurai	39	15	42
19	Mangalore	39	16	38
20	Mumbai	44	11	38
21	Mysore	44	10	38
22	Patna	47	2	46
23	Pondicherry	50	18	40
24	Srinagar	39	−2	50
25	Thiruvananthapuram	39	18	34
26	Tiruchirapalli	39	14	44
27	Varanasi	47	1	47
28	Visakhapatnam	50	14	44

7.3.4 Dynamic Superimposed Loads

These loads are produced on structures by the dynamic effects, vibration and impact caused by machinery or mobile equipment such as overhead crane, etc. The generally accepted values are usually given in codes of practice. Thus the impact factors used for design of road bridges is given in the Indian Roads Congress (IRC) Bridge Code.

7.3.5 Construction Loads

These are the loads caused during the construction of buildings etc. They include the effect of forces due to prestressing the order of construction of the slabs of multistoreyed building, etc. These loads should be estimated by the engineer depending on the order and method of construction used at the site. This is dealt with briefly in Section 7.16.

7.4 LOADS DUE TO IMPOSED DEFORMATIONS

These loads may be due to one or more of the following effects:

1. Shrinkage, temperature or prestressing
2. Movement of supports

They are estimated from the knowledge of materials, climate, soil properties, erection methods, etc., used for the structure under consideration.

7.5 CHARACTERISTIC IMPOSED LOADS

The loads which have the 'accepted probability' of not being exceeded throughout the specified lifespan of the structure are known as the *characteristic imposed* or *live loads*. Due to the difficulty of correctly assessing their values on statistical basis, live loads at present are taken simply as the *design loads* that are being used with safety in the various codes for elastic design. Value of the common live loads are given in the revised IS:875 code. Wind loads are separately dealt with in Section 7.8.

7.6 PARTIAL SAFETY FACTORS FOR LOADS

In limit state design, the isolated design load for each of the limit states is got by the formula:

Factored design load = (Partial safety factor) × (Characteristic load)

The partial safety factors for the various limits states is given in IS 456 Table 18. The above partial safety factors should cover the variation that may occur in loading and depend on the type of load. For example, as the dead loads can be calculated more accurately than the live loads only a smaller partial safety factor needs be prescribed for dead loads rather than live loads. This is the practice in BS 8110 where a value of 1.4 is given for DL and a factor 1.6 for LL. In IS:456, however, the same partial safety factor is used for DL and LL to facilitate the use of the result of structural analysis to both limit state and working stress designs. Similarly, the partial safety factor will also vary with the limit state that is being considered.

Usually, a factor equal to 1.5 is used for the ultimate limit state condition and the unity for the serviceability limit state.

7.7 SOME GENERAL PROVISIONS REGARDING LIVE LOADS ON STRUCTURES

In many cases, where effects of moving concentrated loads are considered, simplified specifications can be evolved for complicated cases of live loadings. Thus, for design of bridges, many codes replace the series of concentrated wheel loads that should be used for design by a uniformly distributed load. These simplified loads should be evolved from theoretical calculations and experience. The following are some of the practices that are generally recommended for design of buildings:

1. The effect of lightweight movable partitions in buildings is estimated by assuming an additional superimposed load for the effect of these partitions. A minimum partitions loadings of 1.0 kN square metre is usually specified for lightweight partitions. However, detailed analysis should be made for fixed heavy partitions (IS 875 Part 2 Clause 3.1.2).

2. In multistoreyed buildings a progressive reduction of live loads (but not dead load) is allowed for designs of columns and foundations. The British and Indian Practices (IS 875 Clause 3.2) allow reduction of live load for buildings like flats and offices, as indicated in Table 7.2. This reduction is not allowed for buildings for storage such as warehouses and stores. However, such reduction is also applicable for factories or workshops that are designed for live loads more than 5 kN/m^2, provided that the loading assumed for any column is not less than the load, as if all floors are designed for a live load of 5 kN/m^2.

3. According to IS 875 (1987) Part 2 Clause 3.2.2, the loads in beams can be reduced by 5% for every 5 m^2 it supports subject to a maximum of 25%.

4. The code also gives the magnitude of horizontal loads for handrails and increase in impact and vibrations to be taken in designs.

TABLE 7.2 REDUCTION IN FLOOR LIVE LOAD IN BUILDINGS ON COLUMNS AND FOUNDATIONS [IS 875 (PART 2) AND BS 6399]

No. of floors (including roof) supported by column	Reduction in load on all floor (percentage)
1	0
2	10
3	20
4	30
5 to 10	40
over 10	50

Notes: (1) No reduction in loads from ground floor to basement.
(2) No reduction is made in any of the dead loads.

7.8 WIND LOADS ON STRUCTURES

Wind produces three different types of effects on structures: static, dynamic and aerodynamic interference effects in tall chimneys.

The response of loads depends on the type of the structure. When the structure deflects in response to the wind load then the dynamic and aerodynamic effects should be analysed in addition to static effects. The concepts given in IS 875 published in 1964 regarding wind effects have been fully revised in the code published in 1987. The Australian code AS 1170 Part 2 (1983) and the British code CP 3 Chapter V revised in 1972 are some of the other modern codes on the subject. As the IS code is based on the concepts of the BS code it is interesting to trace the development of the British code to get an understanding of wind action and the development of the modern codes on wind effects.

7.8.1 History of Development of Specification for Wind Load

The first attempt to study wind effects in Britain was made in the eighteenth century by the engineers who were interested in wind mills, where wind is used as a source of power. Their aim was to calculate the areas of sails required for the grinding capacity of wind mills. In 1756, John Smeaton studied the effect of wind on building and he recommended the following wind loads for design of structures:

1. High winds (up to 45 mph) 0.28 kN/m^2
2. Very high winds (45–60 mph) 0.43 kN/m^2
3. Storms (over 60 mph) 0.56 kN/m^2

After 1880, great interest was shown in U.K. on wind effects on structures when the Tay bridge in Britain failed in a gale. Conflicting evidences were presented during the Board of Trade Enquiry of that failure. Those who had taken measurements of windpressure opined that wind pressures as high as 4.3 or 4.7 kN/m^2 (430–480 kg/m^2) were quite frequent in the U.K. However, engineers led by Benjamin Baker maintained that pressure cannot be greater than 1.2 or 1.4 kN/m^2 as otherwise most of the structures that survived the storm would not have done so. After the enquiry they recommended a value of 2.7 kN/m^2 applied on the projected area. This was based on typical values that were used in the neighbouring countries. These values are as under:

1. In France, 2.63 kN/m^2
2. In the U.S.A., 2.39 kN/m^2 on unloaded bridges, 1.45 kN/m^2 on train surfaces and 2.90 kN/m^2 on loaded bridges.

These values led to the actual measurements of wind pressure by tube anemographs on which the highest gust velocities could be recorded. Baker, who was co-designer of the Forth Railway Bridge in Scotland, made independent studies on an island on the Firth of Forth. He erected two boards one of the area 27.87 sq.m., (300 sq. ft) and other of 0.14 sq. m (1.5 sq.ft). Readings on these boards between 1883–1890 showed that whereas the larger board shows pressure of only 0.19 kN/m^2 the smaller board shows up to 1.48 kN/m^2. It was then considered to have been caused by the dimensional effect. We now know that this is due to turbulent nature of the wind which causes fluctuations in velocity ranging from short duration gusts lasting a few seconds to squalls lasting several minutes. Higher the velocity shorter will be the duration of gusts. Thus, the wind pressures are very high at a point during a gust, while the average pressure on the structure can remain low. This explains the measured differences in pressures experienced by different areas to the same wind. Larger areas will not be affected by winds of shorter duration. In early 1900, T.E. Stantion conducted a series of experiments at the National Physical Laboratory in England and showed that pressure p on plates obeys the law of hydraulics (Bernoulli's equation)

$$p = \frac{1}{2}\rho v^2 \qquad (7.1)$$

where ρ is the density (taken as 1.2 kg/m^3 for Indian conditions) and v the steady velocity of air. ρ is assumed as 1.226 in U.K. and 1.293 kg/m^3 in Canadian codes, respectively.

Pressure p is independent of the size of the plate for areas more than about two sq. ft so that the idea of dimensional effect on buildings were proved to be not correct. This discrepancy between pressures, found in larger and smaller areas, was due to something else, which we now know was the gust effect.

Thus for a cyclonic wind speed of 180 km/hr (50 m/s) the pressure is

$$p = 0.6v^2 = 0.6(50)^2 = 1.50 \text{ kN/m}^2$$

7.8.2 Gust Effect

The wind velocities at any locality vary considerably with time. In addition to a steady wind, there are effects of gusts which last only a few seconds. The modern trend is the adoption of 'peak gust' loading, as the basis for design, in place of mean load averaged over one minute which was until recently used. It is considered to yield a more realistic assessment of wind load. In practice, the peak gust likely to be observed over an average time of 3, 5 or 15 seconds (depending on the location and size of the structure, as explained in Section 7.8.3, is nowadays used in design structures for wind effects).

In U.S.A., the wind speed is expressed as the velocity of the fastest mile of wind to pass through the instrument in T second. That is,

$$\text{Average velocity} = \frac{3600}{T} \text{ mph}$$

Designs are made assuming this mean speed and gust factor as given in Section 7.12.

7.8.3 Structure of Wind

It is now clear that the difference between the laboratory experiments by Stanton and the natural wind effects on varying areas is due to properties of wind turbulence. Stanton attempted to study the effect of wind on different sizes of structure but he terminated it rather inconclusively. In 1930, A. Bailey and N.D.C. Vincent extended the work of Stanton and studied the distribution of pressure over the surface of common building shapes. This resulted in many of the present day recommendations of the British code. The First Birtish code of wind loads was issued in 1944 as the British Standard code of practice CP 3 (1994) (Code for Functional Requirements of Buildings, Chapter V, 'Loading'). This specified the worst gust speed in various parts of Britain as unit pressure for the combined effect of pressure and suction for buildings located in various terrain conditions such as: sheltered condition, open country subjected to maximum exposure to sea coast and high altitude inland.

Normal internal pressure and pressure on roofs, etc., were recommended to be obtained by the use of coefficients. Local effects due to winds were indicated by increased coefficients to be used for such situations. The 1944 Code was revised in 1952 as CP 3 (1952), Chapter V. It adopted one minute mean wind speed rather than the gust speed previously advocated to allow for reduction in large areas. Four categories of exposure to wind were defined with average one minute wind speed of 45, 54, 63 and 73 mph.

Basic wind pressure was recommended, as before, for combined effect of pressure and suction at different heights above ground level. Coefficients similar to 1944 Code were used for pressures on roof and wall structures. However, many cladding failures occurred in Britain between 1950 and 1960. Hence, when the scheduled revision of the 1952 Code in 1958, code markers introduced many new features (as indicated further below) in the revised code.

The revised code was finally published as the 1970 revision of the code of wind and was called British Code of Practice CP-3 (1970), *Code of Basic Data for Design of Buildings*, Chapter V, 'Loadings', Part 2: Wind Loads. Pressure coefficients and force coefficients for limited range of building shapes were given. This 1970 version was revised again in 1972 as CP 3 Chapter V Part 2 (1972), and this code is currently in practice in U.K.

As compared with the 1952 edition the present code considers wind loading in detail based on many ideas developed after 1952. One of the very important changes is the adoption of 'gust loading' as the basic for design in place of the 'mean load' averaged over one minute. Investigation of numerous cases of building damages due to wind action has shown that many cases of damage was due to underestimation of wind forces due to gust action. It is found that gust action is more important than the formerly assumed one. The important principle that evolved for selecting the design wind speed for a structure was that only those gusts which are of size large enough to envelope the structure are to be considered. The size effect can be accommodated by varying gusts. Three second gusts are taken for small buildings of size up to 20 m (class A), 5-second gusts for building sized 20–50 m (class B) and 15-second gust for buildings greater than 50 m (class C). The 1972 BS code specifies basic winds from 85 to 125 mph (38 to 56 m/s in U.K. as compared to 33 to 55 m/s in IS) depending on locations. They are much more than those specified in old code. The intensity of gust is also related to duration of the gust that affects buildings. Larger structures will be affected only by gust of larger duration and thus subjected to smaller overall pressures compared to smaller structures.

7.9 INDIAN CODE FOR WIND LOADS

The wind maps given in IS 875 (1964) give two wind pressures (as different from wind speed) maps. One map shows the basic pressures excluding winds of short duration and the other basic pressures including winds of short duration (less than 5 minutes). The whole of India was divided into three regions. In the present revision of the code, IS 875 (1987), Part 3, consistent with other modern codes, we have a single wind map of India giving the basic maximum wind speed as peak gust velocity averaged over a short time interval of 3 s duration. The Map of India is divided into six zones with basic wind speeds of 33, 39, 44, 47, 50 and 55 m/s corresponding to 120, 140, 160, 170, 180 and 200 km/h. Many other recommendations, common to other International codes have been incorporated in the code.

In many respects the IS code is very similar to the BS code. The highest peak wind speed of 55 m/s is assigned to the hilly regions around Silchar and Ladakh. For most parts of North India, the assigned wind speed is 47 m/s (170 km/h) and the wind speeds for India are shown in Fig. 7.1. The wind speed assigned to the East coast of India and the coastal regions of Gujarat where cyclones are possible in 50 m/s (180 km/h). We should clearly remember that the above values are the basic wind speeds. The actual design wind speed will depend on many other factors as given in the next section.

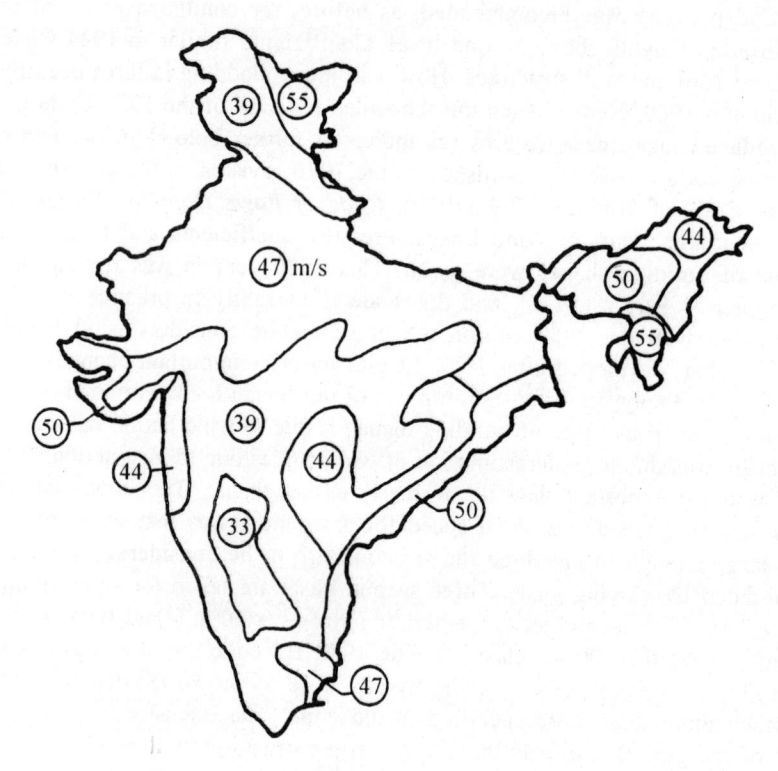

Fig. 7.1 Basic wind speeds in India (in m/s) based on 50 years return period.

7.9.1 Effects of Wind on Structures

There are two approaches to determine the affects of wind on structures:

1. The peak wind approach with gust related factors (PWA)
2. The mean wind approach with gust response factor (MWA)

For ordinary structures the Indian, British and Australian codes recommend the peak wind approach. For tall and flexible structures, the Indian and Australian codes recommend the use of the *gust response factor* approach known as the *dynamic load method* (Section 7.12). In this chapter we shall deal in detail with the peak wind approach only.

7.9.2 Design Wind Speed

The design wind speed for a place is obtained from the basic wind speed at the place by using the formula:

$$V_z = V_b \, k_1 \, k_2 \, k_3 \qquad (7.2)$$

where

V_z = Design wind speed (in m/s) at height z

V_b = Basic wind speed for the site (33 to 55 m/s) as shown on the map of India (Fig. 7.1). It is the peak gust speed averaged over 3 s for terrain category 2 (open level country with scattered obstructions) at 10 m above ground level on 50 year mean return period

k_1 = Probability factor or risk coefficient with return periods

k_2 = Factor for the combined effects of terrain (ground roughness) height and size of the component on structure

k_3 = Factor for local topography (hills, valleys, cliffs, etc.)

The value of these different are found as follows:

(i) Values of probability factor (risk coefficient) k_1. As the basic wind speeds are given for terrain category 2 based on a return period of 50 years and the return period that is taken for design of different types of structures will be different, a corresponding reduction in basic wind speeds has to be made for lesser return periods and an increase in basic wind speeds for higher return period. This probability factor also depends on the basic wind speed. The values of k_1 recommended and used in IS 875 (Table 1) are given in Table 7.3.

TABLE 7.3   RISK FACTOR k_1
(IS 875 Part 3 Table 1)

No.	Class of structure	Design life (in years)	33	39	44	47	50	55
1	General buildings	50	1.0	1.0	1.0	1.0	1.0	1.0
2	Temporary sheds	5	0.82	0.76	0.73	0.71	0.70	0.67
3	Structures with hazards in the event of failure	25	0.94	0.92	0.91	0.90	0.90	0.89
4	Important buildings	100	1.05	1.06	1.07	1.07	1.08	1.08

(Basic wind speed (in m/s))

(ii) Value of type of terrain (category), height and structure size factor k_2. k_2 is a function of three parameters, namely, the type of terrain, the height of the building and the size of the structure. In IS 875, this factor for terrain category 2, type of structure class A (3 s gust speed) is by definition 1.0 at 10 m level. (The correspondence is BS for category 1, class *A* and 10 m.) Wind speed increases from zero near the ground level to full streaming speed at a certain height above the ground surface. The height at which the wind speed reaches 99% of full streaming speed is called the gradient height Z_g and the corresponding wind as the gradient wind. This height is less in open areas and more in areas crowded with tall building. The terrain classification takes this aspect into account. In IS code, gradient height is assumed 500 m as the same for all categories. Accordingly, terrains are classified into the following four categories:

Category 1: Exposed open terrain with no obstructions

Category 2: Open terrain with scattered obstructions of height between 1.5 and 10 m

Category 3: Terrain with numerous low-rise (10 m) obstructions (outskirts of large cities)

Category 4: Terrain with numerous high-rise obstructions (city centres)

Again, the size of the structure is important as small duration gust has less influence on large structures. To take this aspect into account, structures are grouped into 'classes' depending on the largest (horizontal or vertical) dimension.

Class A: Structures and components such as cladding, glazing, roofing, etc., having maximum dimension less than 20 m. (Here, the 3-second gust speed governs.)

Class B: Structures and components as above with maximum dimension 20 to 50 m. (Here the 5-second gust speed governs.)

Class C: Structures and components as above with maximum dimension greater than 50 m. (Here the 15-second speed govern.)

The value of k_2 which considers all the various factors like height, terrain category and class are given in IS 875 Part 3 (1987). A part of the table is reproduced as Table 7.4 for illustration. It should be noted that the factor k_2 for terrain category 2 at 10 m for class *A* type structure should be obviously 1.0.

TABLE 7.4 TERRAIN CATEGORY CLASS AND HEIGHT FACTORS k_2
[IS 875 (1987) Part 3 Table 2]

Height (m)	Terrain 1 Class			Terrain 2 Class			Terrain 3 Class			Terrain 4 Class		
	A	B	C	A	B	C	A	B	C	A	B	C
10	1.05	1.03	0.99	1.00	0.98	0.93	0.91	0.88	0.82	0.80	0.76	0.67
15	1.09	1.07	1.03	1.05	1.02	0.97	0.97	0.94	0.87	0.80	0.76	0.67
20	1.12	1.10	1.06	1.07	1.05	1.00	1.01	0.98	0.91	0.80	0.76	0.67
30	1.15	1.13	1.19	1.12	1.10	1.04	1.06	1.03	0.96	0.97	0.93	0.83
50	1.20	1.18	1.14	1.17	1.15	1.10	1.12	1.09	1.02	1.10	1.05	0.95
100	1.26	1.24	1.26	1.24	1.22	1.17	1.20	1.17	1.10	1.20	1.15	1.05
250	1.34	1.32	1.28	1.32	1.31	1.26	1.29	1.26	1.20	1.28	1.24	1.16
500	1.40	1.38	1.34	1.39	1.37	1.32	1.36	1.33	1.28	1.32	1.30	1.22

DESIGN LOADS OTHER THAN EARTHQUAKE LOADS 107

We assume that the velocity at a height h is related to the velocity at the ground at a height h_0 by formula. The following power law is used in the USA and the UK.

$$V_h = V_g \left(\frac{h}{h_o}\right)^a \tag{7.3}$$

Whereas the Indian Standards use a log formula, which is more involved.

Assuming that wind is measured at a standard height of 10 m, Eq. (7.3) gives:

$$V_h = V_g \left(\frac{h}{10}\right)^a \tag{7.3 a}$$

where a is usually taken as 1/10 to 1/12 depending on site conditions.

The relationship between k_2 factor and height for each class of structure depends on the terrain category. For example, this relationship for class A type structure and terrain categories 2 and 4 can be taken from Table 7.4 and the velocity profile can be plotted as shown in Fig. 7.2. The gradual decrease in velocity as we approach the ground can be considered as retardation of wind near the grounds. Similarly, the velocity profile for a given terrain category does not develop immediately at the start of the terrain. The height at which it will be fully developed depends on the fetch or distance through which the wind blows. This fetch-height relationship for each category is given in Table 3 of IS 875 and is shown in Fig. 7.3. These two relationships can be used to find the velocity, values for a given fetch and transition from one category to another. This is shown in Fig. 7.2 and is given in IS 875 (Part 3) Appendix B.

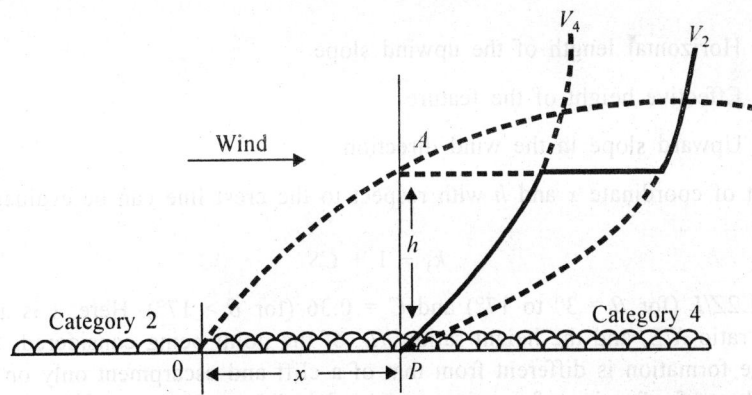

Fig 7.2 Velocity change with height: change of terrain from less rough to more rough (V_2 is height – velocity profile for terrain category 2 and V_4 that for the category 4. Curve OA is the fetch developed height relationship for category 4 at point P).

(iii) *Values of topography factor k_3*. The local topological features such as hills, valleys, cliffs, significantly alter the speed of wind. As a matter of general observation, we know that winds are higher near the summit of hills and lower in valleys. This effect becomes important only on slopes higher than 3° below which the value of k_3 is to be taken as unity. Above 3°, the value of k_3 can vary from 1.0 to 1.36. This factor becomes important in hilly regions in

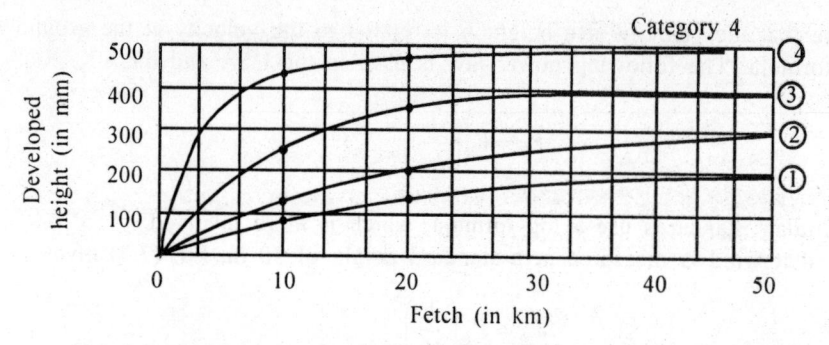

Fig. 7.3 Fetch-developed height—relationship for terrain category 1 to 4.

South India and the northern boundary states of India and it should be given its due importance. IS 875 Part 3 (1987) Appendix C gives details of method of computation of k_3 factor. A brief account of the recommended method is given below:

Regions affected by the topographic factor and the value k_3 are shown in Fig. 7.4 when the downward slope of a hill is less than 3° it is called a *escarpment*. The influence of the topographic factor can be considered to extend $1.5L_e$ upwind and $2.5L_e$ downwind of the summit or the crest. L_e is known as the effective horizontal length. It is taken as:

$$L_e = L, \text{ when the slope } \theta = 3° \text{ to } 17°$$

$$L_e = \frac{Z}{0.3} = 3.3Z, \text{ when the slope } \theta > 17° \text{(or } \tan \theta = 0.3\text{)}$$

where

L = Horizontal length of the upwind slope

Z = Effective height of the feature

θ = Upward slope in the wind direction

k_3 of a point of coordinate x and h with respect to the crest line can be evaluated from the expression:

$$k_3 = 1 + CS \tag{7.4}$$

where $C = 1.2Z/L$ (for $\theta = 3°$ to $17°$) and $C = 0.36$ (for $\theta > 17°$). Here, s is a function of the distance ratio X/L_e and the height ratio H/L_e of the point being considered. Its value for hill and ridge formation is different from that of a cliff and escarpment only on the leeward side. The values of s for these formations can be found from Fig 7.5 which is taken from IS 875. The final value of k_2 is given by Eq. 7.4.

7.10 GENERAL THEORY OF WIND EFFECTS OF STRUCTURES

Section 7.9 with the method of arriving at the peak gust value of design wind speed used for quasi-static calculation of wind forces. In this section, we shall look in more detail into the general effects of wind on different types of structures, which have been briefly mentioned in Section 7.8.

DESIGN LOADS OTHER THAN EARTHQUAKE LOADS 109

Fig. 7.4 Topological dimensions.

(a) General definition

(b) Cliff and escarpment

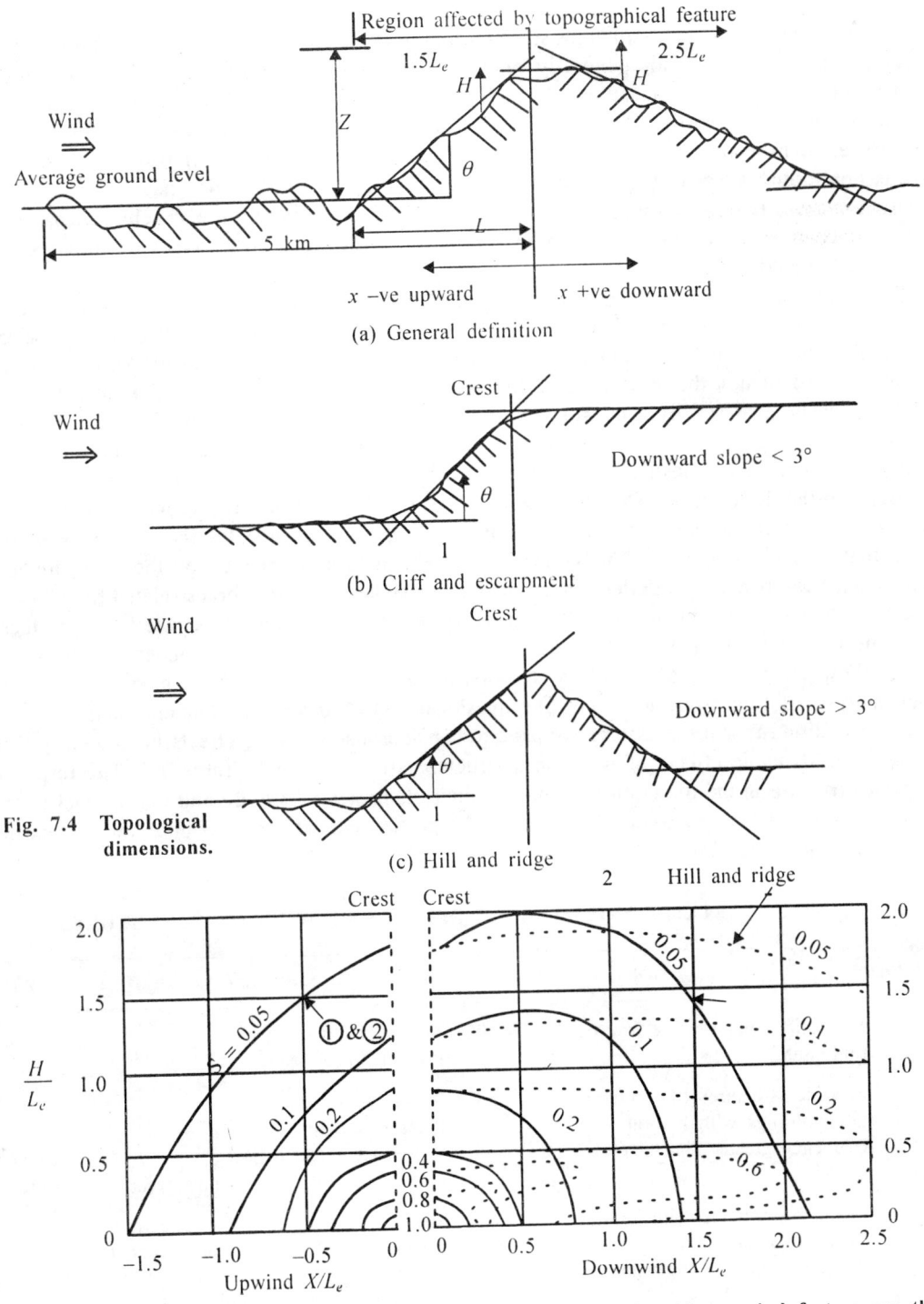

(c) Hill and ridge

Fig. 7.5 Factors for (1) Cliff and escarpment, (2) hill and ridge (the upwind factors are the same for both types of slopes).

Most of the man-made structures stand bluff (perpendicular) and the winds blow horizontally. As the structures are not generally made streamlined or smooth, wind-flow over these obstructions tends to be turbulent. In addition, the velocity of wind itself varies with time. The stiffness of structures also varies from structure to structure. The stiffness of a building is different from that of a tall chimney. As the theory of the action of winds in structures was not fully known, old specifications for calculation of wind loads on different types if structures like buildings, bridges, towers, chimneys treated each of these structures separately. It is only in recent times that a unified approach to wind load on all types of structure has been attempted.

As already explained in Section 7.8.2., the wind speed can be considered as fluctuating with a steady component and a fluctuating component superimposed on it. As regards the structure itself, its stiffness can vary from a very stiff structure to a flexible structure. With a very stiff structure the movement of the structure with wind will be negligible and the wind effect can be analysed by a quasi-static approach. However, as the stiffness of the structure decreases, the deflection, natural frequency and damping characteristics of the building have also to be considered in the calculation of the effect of the winds on the structure. The wind forces on the deflected structure will be different from that of the stiff structure. In addition, if the natural frequency of the structure is low there is a chance for the frequency of the fluctuating part of the wind to coincide with the natural frequency of the structure and resonance can result. The deflection can become very large and the forces related to mass and acceleration of the structure can be much greater than the wind force itself. Under these conditions natural frequency, structural damping, etc., are of great importance. Analysis that take these effects into consideration is called *dynamic analysis*. Thus we have two methods of analysis of forces due to winds: the quasi-static method and the dynamic analysis.

In addition to static pressures we have to consider also dynamic effects in structures. The static and dynamic effects of wind on a structure are as shown in Table 7.5. The response of the structure in the direction of wind is called *along-wind response* and that normal to the direction of wind as *across-wind response*. The portion behind the bluff body away from the wind is called its *wake*.

TABLE 7.5 STATIC AND DYNAMIC EFFECTS OF WIND ON STRUCTURES

S.No.	Types of structure	Effects of wind
1	Very stiff structure like low rise buildings	1. No deflection 2. Pressure quasi-static
2	Flexible structures like chimneys, tall buildings with natural frequency greater than that of wind	1. Structure deflects 2. Pressure dynamic 3. Resonance effect (nil) Both along and across wind effects are possible
3	Flexible structures with natural frequency equal to that of the fluctuating part of the wind.	1. Structure undergoes large deflection 2. Effects of deflection can be larger than wind effects 3. Resonance with damping Both along and across wind effects are possible

When a bluff body is exposed to wind, vortices tend to be shed from the sides of the body creating a pattern in its wake called the *vortex shedding* as shown in Fig. 7.6. The frequency of shedding depends mainly on the shape and size of the body, the velocity of flow, to a lesser degree on the surface roughness and the turbulence of the flow. This can cause 'across-wind effects' such as transverse vibrations in lightly-damped structures such as chimneys and towers.

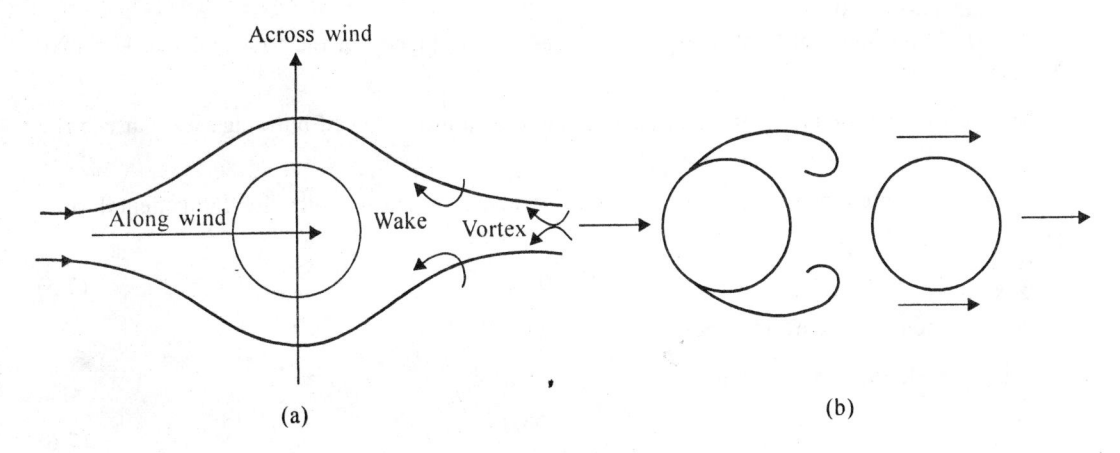

Fig. 7.6 Wind effects on tall structures: (a) Vortex shedding (b) interference.

In addition to the static and dynamic effects of wind on individual structures discussed above, we have to recognise the effect of adjacent structures on the pattern of flow of the wind itself and on each other. These are known as *aerodynamic interference effects* referred to in Section 7.8. The venturi effect and the reverse-flow effect as shown in Fig. 7.7 belong to this category. If one structure is located in the wake of another the vortices shed from the upstream structure may cause the oscillations of the downstream structure. This is called

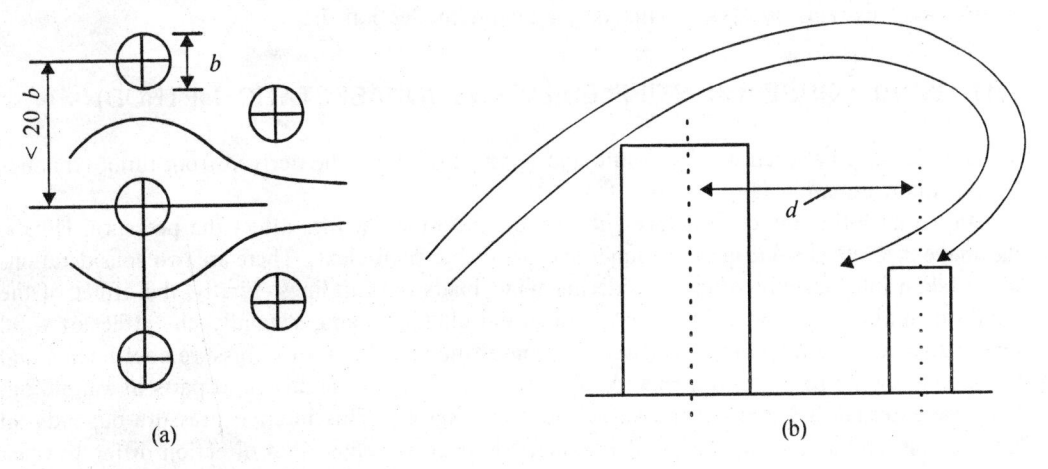

Fig. 7.7 Wind Effects on tall structures: (a) interference as in Ferry bridge cooling towers (b) reverse wind flow.

wake buffeting. These interference effects are best studied by 'wind tunnel tests' than by calculations. Wind tunnel tests is a third method that analyse wind effects on structures.

The rough guide given in IS 875 (Part 3) 1987 Clause 7.1 states that normally only building or structure which has either of the following conditions needs to be examined for dynamic effects of wind:

1. Buildings and closed structures with a height to minimum lateral dimension ratio of more than about 5.0 and

2. Building and closed structures whose natural frequency in the first mode is less than 1.0 Hz.

For a rough estimate of the natural frequency of multi-storeyed buildings we can use the following formulae (IS 875 Clause 7):

(a) For moment resisting frames without bracing or shear walls the fundamental time period T is given by:

$$T = 0.1n \tag{7.5}$$

where n = number of storeys including basement.

(b) For all other structures:

$$T = \frac{0.09H}{\sqrt{d}} \tag{7.6}$$

where

H = Total height of main structure

d = Maximum dimension of building (in metre) in the direction parallel to the wind force

As most cases of low-rise buildings can be considered as rigid the pressures on them can be determined by the tabulated values of pressure and force coefficients given in the codes without any dynamic analysis. This is dealt with in Section 7.11.

7.11 WIND FORCE ON STIFF BUILDINGS (QUASI-STATIC METHOD)

The relationship between wind pressure and wind velocity can be derived from fluid dynamics by Bernoulli equation [Eq. (7.1)].

In actual buildings or structures, the shape and other factors affect the pressure. Hence, the above theoretical value has to be modified by suitable coefficients. There are two considerations to be taken into account when considering wind loads on buildings. Firstly, the effect of the wind on 'building components' such as walls and claddings and, secondly, the effect of wind on the structure 'as a whole'. The first is accomplished by the use of pressure coefficients and the second by the force coefficients. As shown in Fig. 7.8 for the design of parts of a building, both the external and internal pressures should be known. This internal pressure depends on the location and size of openings. If the internal space is vented in a direction different from which the inlet windows are situated (as in corner rooms and corridors) large internal pressures can develop.

DESIGN LOADS OTHER THAN EARTHQUAKE LOADS 113

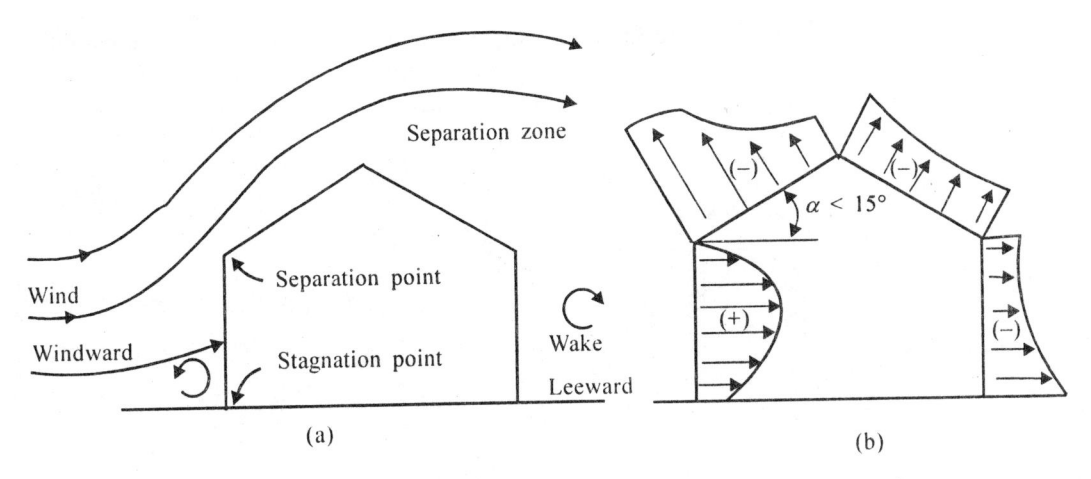

Fig. 7.8 Effect of wind on buildings: (a) external pressure (b) internal pressure.

A clear demonstration of the effects of internal pressure can be seen in roofs covered with tiles such as Mangalore tiles. There is considerable difference in the magnitude of the pressure under the tile with and without an underlay or 'ceiling'. With the underlay, the pressure difference between the inside acts only on a small part (the lower part of the tiles that rest on the underlay) whereas without the underlay the whole tile, will be under the action of the internal pressure. Because of this effect it has been observed in houses in the eastern coast in India that with cyclone winds the tiled roof with some sort of underlay has been found to be more stable than tiled roof without the underlay.

Thus, two types of coefficients—pressure coefficients for components and force coefficients for the structure as a whole are given in the wind code to calculate wind forces in buildings. These coefficients are based on a limited number of actual tests.

7.11.1 Pressure Coefficients for Components

Tables 4 to 22 of IS 875 Part 3 give the pressure coefficients to be used for calculating forces on component or various parts of buildings. Certain local parts of the buildings (like overhangs and eves of a slopped roof) will be subjected to larger pressures than other parts. These are shown hatched in Fig. 7.9. Corresponding higher values are also given in these tables.

Having obtained the internal and external pressure coefficients from Figs. 7.8 and 7.9, the force on a component of the building is calculated from the expression:

$$\text{Force} = (C_{pe} - C_{pi})pA \tag{7.7}$$

where

C_{pe} = External pressure coefficient

C_{pi} = Internal pressure coefficient

p = Pressure due to wind speed

A = Area of the component

Investigations have shown that breakage of windows in the windward side of a building during a storm can cause variation in internal pressure and can have serious effects on safety of building. This effect should be taken into account in the design of buildings in cyclonic areas as explained in the code.

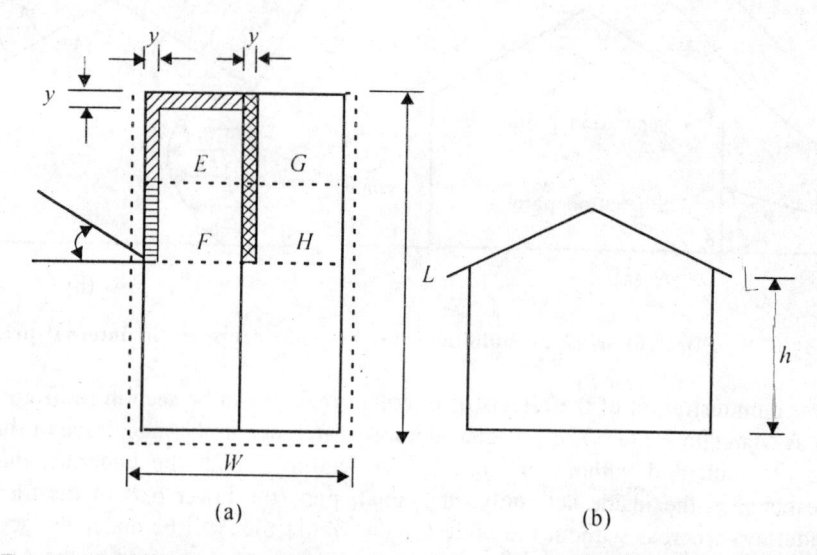

Fig. 7.9 External pressure coefficients for pitched roofs: (a) plan of building (b) section of building (hatched portions show regions of increased pressures).

7.11.2 Force Coefficients for Structure as a Whole

The force coefficients given in the wind codes are used for determining the overall wind load on the structure taken as a whole without considering the other parts. The expression for the force is given in the equation

$$\text{Force} = C_f p A_e$$

where
  C_f = Force coefficient
  p = Pressure
  A_e = Effective area of the building

The values of force coefficients used for various shapes of buildings are given in Table 23 of IS 875 (Part 3).

7.12 GUST FACTOR (GF) METHOD (DYNAMIC ANALYSIS)

Calculation of the along-the-wind loads on a structure was so far considered by the quasi-static method using peak-gust values and the design wind speed V_z given by Eq. (7.2) However wind can also be looked upon as a steady wind, say, mean hourly speed over which is varying speed is superimposed. The second method of analysis is known as 'dynamic analysis'. Using a simple model of atmospheric wind (it is likely that the real values at the site may be different) mean hourly wind speed is taken as

$$\overline{V}_z = k_1 k'_2 k_3 V_b \qquad (7.8)$$

where k_1 and k_3 are discussed in Section 7.9.2 but k'_2 is given in Table 33 IS 875 Part 3. It can be seen that k'_2 makes V_z less than V_b. k'_2 depends on the height and the category of the terrain as explained in Section 7.9.1.

According to the gust factor method, the along-wind F_z is given by the expressions

$$F_z = \text{Mean load due to } V_z \times G$$

where

$$\text{Gust factor } G = \frac{\text{Peak load}}{\text{Mean load}}$$

The data and procedure for calculation of G is given in Clause 8 of IS 875 Part 3 (1987).

We may say that generally (but not always, as in the case of earthquake analysis) the dynamic analysis based on mean wind speed gives lower values of load than the elastic analysis using peak-gust method. (For more details on dynamic analysis of wind specialized literature on wind effect on structure should be consulted).

7.13 WIND TUNNEL TESTS

The third method for evaluating wind loads on tall structures is by use of properly scaled wind tunnel tests. A good wind tunnel test will give results, which are nearer to the true values. In many cases, these values can also be very much less than those given by the other two methods of analysis. In very important cases of slender structures such as very tall buildings, or tall chimneys, it is advisable to make such an analysis for the wind forces.

7.14 CLASSIFICATION OF WIND (CYCLONES)

Wind usually denotes the horizontal wind. If vertical winds are referred they should be identified as such.

Winds with velocities higher than 80 km/h are termed as very strong winds. Such high-speed winds with circular motion are generally associated with cyclonic storms. A typical feature of cyclonic storm is that, whereas they grow in magnitude and speed on the high seas (picking up energy over the seas due to latent heat of varporisation of water) they loose their source of energy and rapidity weaken due to the resistance of trees, landscape, etc., after crossing the coast. Cyclones generally do not penetrate more than 40 to 60 km inland from the sea. These cyclones very rarely affect structures 120 km away from the coast.

Cyclonic winds rotate (anti-clockwise in the northern hemisphere) at very high speeds and the cyclone itself move from east to west at low speeds of about 12 to 16 km/h. The eastern coast in India (the coast around the Bay of Bengal) is very often subjected to cyclones between the months of October to December during the North-East monsoon. These storms can occasionally also cross over peninsular India and again gather speed and strike the coasts of Gujarat. But the latter are of less intensity than those on the west coasts of India. Buildings in these regions adjacent to the seacoast should be designed not only for very high winds but also for the special effects of cyclonic winds. The coastal regions of Gujarat may also be subjected to cyclones during the South-West monsoon.

The Australian code for wind forces recommends that the regional basic design wind velocities should be multiplied by a factor varying from 1.5 to 3.3 for cyclonic effects. In addition, design of buildings in tropical cyclonic areas should take into account repeated loading criteria wherever they are relevant. However, in the Indian code, the effect of cyclone on land is only reflected in the basic wind speeds specified in that code. But many authorities feel that the speed provided in IS code for the eastern coast are rather on the low side. Preliminary analysis of available data show that for a return period of 50 years the velocity of cyclone winds can be as high as 70 m/s in Tamilnadu, 73 m/s on the coasts of Andhra Pradesh to West Bengal and 60 m/s on the coast of Gujarat. Sriharikota near Madras has recorded speeds up to 200 km/h (55 m/s) and Paradeep in Orissa has record speeds up to 260 km/h (72 m/s). (According to IS:875, the wind speed to be used off-shore structures should be 1.15 times the value near the coast in absence of other wind data.) As against these observed data, the wind speed indicated in the wind map of India for these places is 50 m/s only. The philosophy of the design of structures in cyclonic areas is to divide them into three types depending on their importance:

1. Very low-cost structures
2. Normal structures
3. Post-disaster structures

Post-disaster structures are those which should survive a cyclone. Hospitals, telecommunication buildings, community centres, power stations, meteorological stations, police stations, air-traffic control buildings, fire stations, etc., are classified as post-disaster structures. They should be designed and constructed to survive the most servere cyclone. A return period of 100 year is usually adopted for them. On the other hand, whenever normal structures, undergo limited damage during a severe cyclone they can be repaired at nominal cost after the disaster to make them economical in construction. A return period of only 50 year is used for these structures. In low-cost structure, no particular care is taken to design them against wind, but detailing of construction can be adopted to make it as much resistant to cyclone as possible. Details of good construction practice of 'anchoring', 'bracing' and providing 'continuity' from roof to foundation are adopted for them. (The above three factors are sometimes referred to as the ABC of cyclone resistant construction.)

Structures can also be classified as wind *sensitive* and wind *insensitive* types. Sloped roofs with A.C. sheets or tile claddings are wind sensitive whereas flat roof in R.C. construction or a circular shaped R.C. building with domed roof are not wind sensitive. Wind sensitive structures should be totally avoided in cyclonic area and if adopted for any reason they should be carefully planned. Funnel-shaped hyperbolic water towers have been found to suffer heavily in cyclonic areas as their shape is not suitable to such regions. In tall structures, like hollow shafts of tall water tanks, special care should be taken during the construction to join each lift of construction with the ones below. Defective joints tend to separate during severe cyclones. For towers used to support water tanks, slip-form technique without joints will be the most suitable method of construction in cyclonic regions.

DESIGN LOADS OTHER THAN EARTHQUAKE LOADS 117

It should also be noted that, in general, the use of IS 875 (1987) gives higher pressure of wind than the British code CP 3 (1972). In addition in IS the minimum wind speed to be used is that pertaining to a height of 10 m above the ground level. In CP 3 reduced values of wind speeds are allowed for heights less than 10 m as shown in Fig. 7.10.

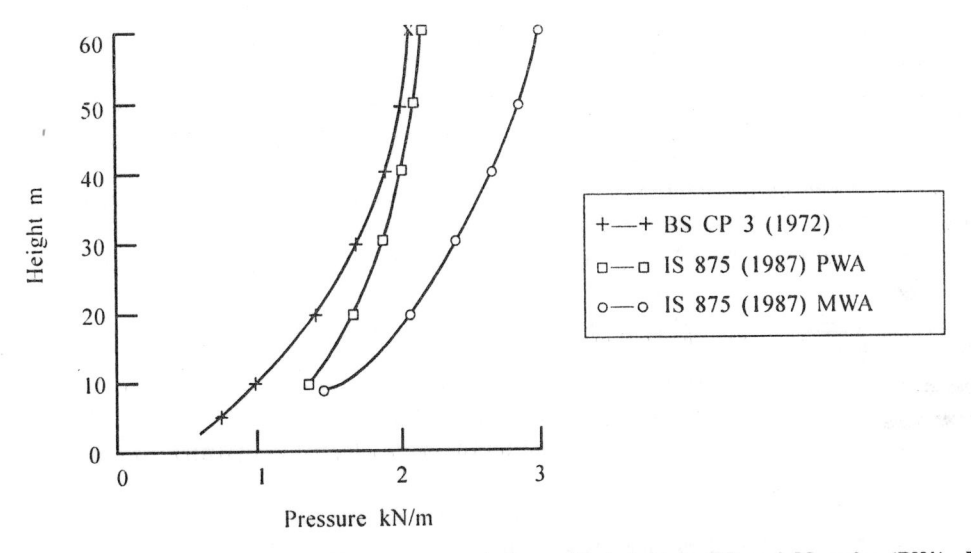

Fig. 7.10 Comparison of overall pressure variation with height in BS and IS codes (PWA, Peak wind approach; MWA, Mean wind approach).

7.15 IMPORTANCE OF CHECKING FOR WIND LOADS AT ALL STAGES OF CONSTRUCTION

There has been many cases of failure of structures due to high winds during construction of structures even though the completed structure would have been safe against these winds. The reason for such failures is that effect of wind on the unbraced incomplete structure is more severe than that on the braced completed structures. It is thus necessary that in zones of high winds, especially along hilly tracts and the sea-coasts, stability of structures should be checked and temporary supports should be provided against wind for all stages of construction. Thus the walls of a building without the roof is more unstable against the wind than the completed building.

7.16 CONSTRUCTION LOADS

There are many types of construction loads. In this section, we shall consider loads only due to the construction sequence of a multi-storeyed building. Nowadays we do not wait for 28 days for the concrete to attain its full strength before the next floor is cast and supported on that floor. In all such cases, the maximum load that can come on each floor and its capacity to carry the load at the stage of its strength development should be investigated. Floors are usually laid in a cyclic period (say one floor every one week) with one, two or more floors below fully shored. Alternately, the bottom floors may be reshored with the others shored as shown in Example 7.3.

We can calculate the maximum load that come on each floor for a given method of construction by making the following assumptions regarding sharing of loads between slabs and shores as illustrated by Examples 7.5 and 7.6.

1. The weight of each freshly-cast-floor is equal to the full self-weight of the slab plus the formwork (about 10% of the slab) as well as the weight of the workers, equipment, etc. The total load may amount to as much as 1.4 times the dead weight of the slab alone during construction. After construction, part of the load due to equipment and people is removed and the load on each floor is taken as unity.

2. Shoring can be considered as very stiff and is placed in such a way that it produces a uniformly-distributed load on the slabs below, and on which it rests.

3. When a new slab is laid the concrete is assumed to have no strength and all the weight on the slab is fully carried by the shores immediately below that slab.

4. This weight is also assumed to be equally distributed to all the slabs supporting it through the shores.

5. When shoring is removed (say 5 days after the floor is cast) from any of the bottom floor, the load that was being carried by that shoring is thrown back and evenly distributed to all the slabs above, which were connected to the shores including the slab which was cast most recently.

6. Slab loads are first calculated using the above assumptions and shore loads are then computed by equilibrium of the downward forces.

The loads on slabs should be calculated and checked against the strength of the concrete attained by the slabs from the time it has been cast. Lack of knowledge about construction over-loads has caused many failures especially in cantilever slabs.

7.17 JOINTS IN CONCRETE CONSTRUCTION

IS 11817 (1986) gives the classification of joints used in building construction. The principal type of joints used in reinforced concrete construction can be classified as follows:

1. Construction joint (joints left for longer period)
2. Cold joints (joints made temporarily for short period)
3. Expansion joints (joints with specified gaps)

IS 456 (2000) Clause 13.4 gives details of construction and cold joints. IS 456 (2000) Clause 27 recommends that structures exceeding 45 m in length should be normally be designed with one or more expansion joints.

EXAMPLE 7.1 (Wind load on buildings)
A reinforced, framed building is 45 × 15 m in plan and 60 m in height consisting of storeys 4 m in height. It is braced in the longitudinal direction by rigid frame action and by a reinforced concrete infill wall in the transverse direction. Determine the design wind force on the framed building.

Assume that the building is situated in terrain category 3 with basic wind speed of 50 m/s in a fully-developed velocity profile.

DESIGN LOADS OTHER THAN EARTHQUAKE LOADS 119

Reference	Step	Calculations
	1	*Data* Plan of building 45 × 15 m; height, 60 m Basic wind speed V_b = 50 m/s
	2	*Find design wind speed V_z* $V_z = k_1 k_2 k_3 V_b$ Risk coefficient $k_1 = 1.0$ Topography factor $k_3 = 1.0$. k_2 factor depends on the following
IS 875 (part 3) Table 2		(i) Terrain category 3, (ii) structural size factor—greater horizontal or vertical dimension is larger than 50 m, 15 s gust size is appropriate. Hence class C. Read off k_2 for terrain category 3 and class C for varying heights, the value of k_2. For convenience, we divide the heights into three divisions and use the greater value for each division, i.e. (0–20 m, 20–40 m and 40–60 m).
IS 875 (5.4)	3	*Determine dynamic pressure for different heights* V_b = 50 m/s; $V_z = k_1 k_2 k_3 V_b$ Pressure $p = 0.6 V_z^2$ N/m²

TABLE 1 (From Table 2 of IS : 875)

Reference		Interval	k_2 (*max*)	V_z (m/s)	p (N/m²)
IS 875 Table 2 (Table 7.4)		0–20	0.91	45.5	1242
		20–40	1.00	50.0	1500
		40–60	1.04	52.0	1622

| | 4 | *Calculate the force coefficients C_f*
 Let wind be normal to 45 m base = wind at 0°
 Wind normal to 15 m base = wind at 90°
 a = Depth of plan dimension = 15 m
 b = Dimension on which wind at 0° acts = 45 m
 h = Height = 60 m |

Reference	Step	Calculations
IS 875 Cl. 6.3.2.1 and Fig. 4		**TABLE 2** (From IS 875 Fig. 4 Page 39) | Dimension | Wind at 0° | Wind at 90° | |---|---|---| | a/b | 15/45 = 0.33 | 45/15 = 3.0 | | h/b | 60/45 = 1.33 | 60/15 = 4.0 | | C_f | 1.3 | 1.1 |
	5	*Determine forces of wind at 0°* Wind force = pC_f (Area) Area = 45 (Floor height of 4 m) U.D.L. on floor slab = $\dfrac{1.3p(45 \times 4)}{45} = 5.2p$ = 5.2p for each metre of length Tabulating for the intervals adopted.
Step 3		**TABLE 3** | Intervals | p (N/m^2) | Force (kN/m) | |---|---|---| | 0–20 | 1242 | 5.2p = 6.46 | | 20–40 | 1500 | 5.2p = 7.80 | | 40–60 | 1622 | 5.2p = 8.43 |
	6	*Determine wind forces in 90° direction* As above = $\dfrac{1.1 \times p \times 45 \times 4}{45} = 4.4p$ for each metre of length
	7	*Nature of action of wind forces* These uniformly-distributed forces from the edge of the slab are transferred to the frames and shear walls. The floor slab acts as rigid diaphragms between the walls and the frames.

EXAMPLE 7.2 (Wind load on a sloped roof building)
A single-storey shed is as shown in Fig. 7.11. It is 20 m wide and 30 m in length. It is situated in terrain category 3 and basic wind speed for its location is 47 m/s. The frames are spaced at 5 m centres. Determine the pressure for which the walls and the roof are to be designed.

DESIGN LOADS OTHER THAN EARTHQUAKE LOADS   121

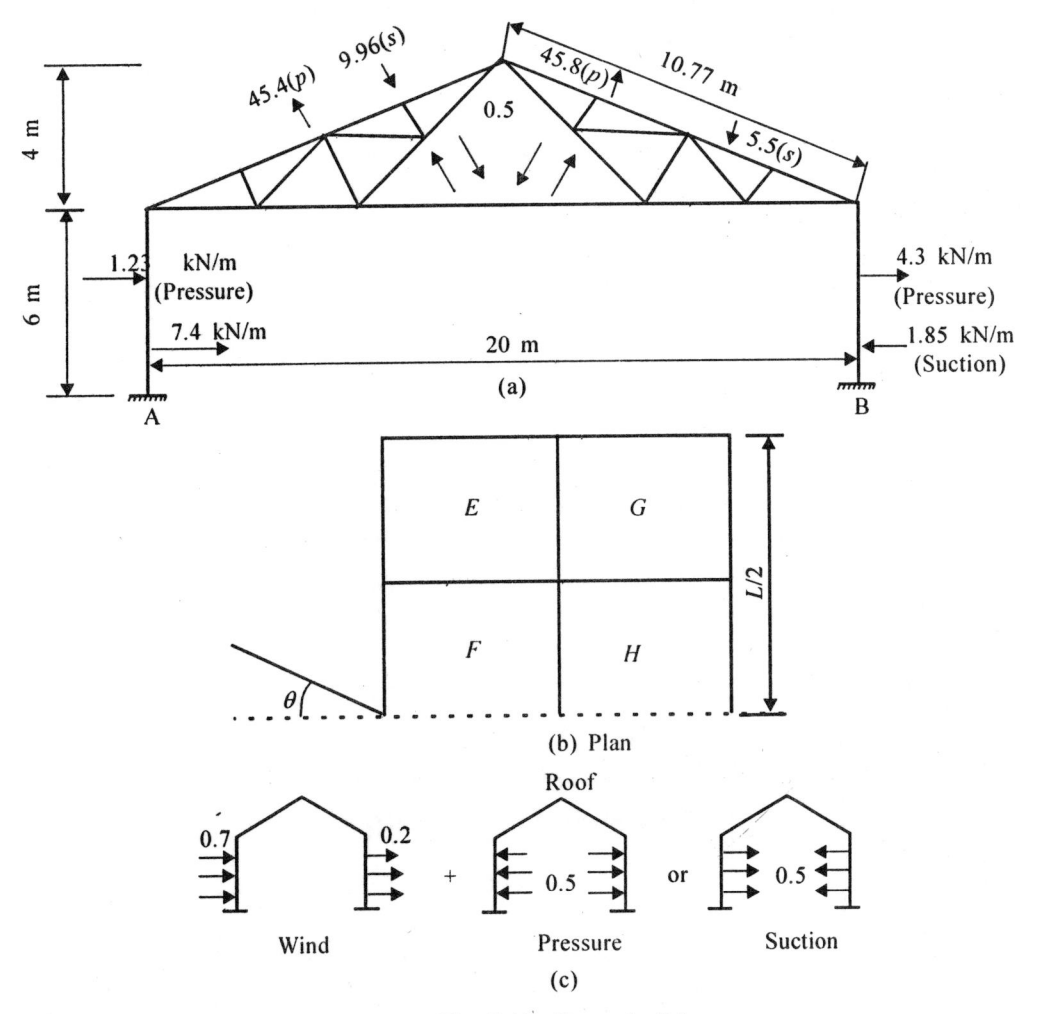

Fig. 7.11 Example 7.2

Reference	Step	Calculations
	1	*Data*
		Basic wind speed V_b = 47 m/s
		Plan of building = 20 × 30 m
		Height to eves-6 m heights of roof-4 m
	2	*Determine design wind speed V_z*
		$V_z = k_1 k_2 k_3 V_b$
		where k_1 = 1.0; k_3 = 1.0
		k_2 factor depends on the following:

Reference	Step	Calculations
		(1) Terrain category—3
		(2) Structural size factor—as the maximum vertical or horizontal dimension does not exceed 50 m, the 5 s speed governs. Hence class B.
		(3) Height—assume 10 m as IS does not recommend values lower than 10 m.
IS 875 Table 2		Read off from Table $2k_2 = 0.88$ for roof and walls < 10 m in height. $V_z = 0.88 \times 47 = 41.4$ m/s
	3	Calculate dynamic pressure $P = 0.6 V_z^2 = 0.6(41.4)^2 = 1028$ N/m^2
		(*Note:* The corresponding value in BS for walls will be only 550 N/m^2 as it gives values of k_2 below also 10 m.)
	4	Calculate pressure coefficient for walls (C_{pe} and C_{pi})
IS 875 Table 4 Page 14		$\dfrac{h}{w} = \dfrac{6}{20} = 0.3$ less than 0.5 $\dfrac{L}{w} = \dfrac{30}{20} = 1.5$
IS 875 (6.2.3.2)		A. windward wall. B leeward wall. For C_{pi} assume 5 to 20% opening.

TABLE 1 (IS 875 Table 4 Page 14)

	Wind angle	External pressure C_{pe}		Internal pressure C_{pi}		Action
		A	B	A	B	
Fig. 7.11	0°	$\overrightarrow{0.7}$	$\overrightarrow{0.2}$	$\overleftarrow{-0.5}$	$\overrightarrow{0.5}$	Pressure
IS 875 Cl. 6.2.3.2 Page 36	0°			$\overrightarrow{0.5}$	$\overleftarrow{-0.5}$	Suction

| | 5 | Determine pressure on walls $P = pA(C_{pe} - C_{pi})$ |

TABLE 2

Wall	Internal action	Wind pressure (for metre length)
A	Pressure	$P = (1028 \times 6 \times 1)(0.7 - 0.5)N = 1.23$ kN
	Suction	$P = (1028 \times 6 \times 1)(0.7 + 0.5)N = 7.40$ kN

DESIGN LOADS OTHER THAN EARTHQUAKE LOADS 123

Reference	Step	Calculations

		Wall	Internal action	Wind pressure (for metre length)
		B	Pressure	$P = (1028 \times 6 \times 1)(0.7) N = 4.3$ kN
			Suction	$P = (1028 \times 6 \times 1)(-0.3) N = -1.85$ kN

| | 6 | Calculate pressure on truss |

Spacing = 5 m

Height/width ratio = 6/20 = 0.3 < 0.5

Roof angle = 22°

Consider wind angle $\theta = 0°$

TABLE 3

External pressure C_{pe}		Internal pressure C_{pi}		Action
EF	GH	EF	GH	
−0.32	−0.4	−0.5	−0.5	Pressure ↑
		+0.5	+0.5	Suction ↓

Fig. 7.11
IS 875
Table 5
page 16

TABLE 4 [Fig. 7.3 (b)]

Portion of truss	Internal action	Load on truss ($L = 10.77$ and $s = 5$ m)

EF—Windward side Pressure $P = pA (C_{pe} - C_{pi})$
$= (1028 \times 10.77 \times 5)(-0.32 - 0.5)$
$= -45.4$ kN

Suction $P = 55.357(-0.32 + 0.5)$
$= 9.96$ kN

GH—Leeward side Pressure $P = 55.357(-0.4 - 0.5)$
$= -49.8$ kN

Suction $P = 55.357(-0.4 + 0.5)$
$= 5.54$ kN

Comments: These values are much higher than those got by BS code, due to the following reason:

(1) BS value starts from 3 m level and there is a reduction in velocity of wind between 10 m and 3 m. IS:875 recommends 10 m values to those below 10 m.

(2) In BS code, the coefficient for internal pressure is + 0.20 (pressure) and −0.3 (suction). In IS, the values are +0.5 and −0.5, respectively.

EXAMPLE 7.3 (Calculation of k_3 for escarpment)

Calculate k_3 for a point in a crest and escarpment with the following data:

Horizontal length of upwind slope in wind direction L = 35 m
Effective height of topographic feature Z = 10 m
Distance of point x = 5 m down the crest on the windward line
Height of point from the ground H = 20 m

Reference	Step	Calculations
Fig 7.4	1	*Data* slope of angle $= \dfrac{H}{L} = \dfrac{10}{35}; \theta = 15.9 < 17°$ L_e for slopes less than 17° = 35 m $\dfrac{x}{L_e} = \dfrac{5}{35} = 0.14$ $\dfrac{H}{L_e} = \dfrac{20}{35} = 0.57$
IS 875 Fig. 14	2	*Find s for cliff and escarpment* From Fig. 14, s = 0.35
(Fig. 7.5)	3	*Find the value of C* $C = \dfrac{1.2Z}{L} = \dfrac{1.2 \times 10}{30} = 0.4$
IS 875 Part 3 Page 55	4	*Find the value of k_3* $k_3 = 1 + CS = 1 + (0.4 \times 0.35) = 1.14$ (*Note:* Escarpment is one in which the downslope is flat (<3°) and hence titles in IS:456 Fig. 13 are not correct.)

EXAMPLE 7.4 (Correction for change of category)

A class C building 60 m in height is situated 1 km inside a category 2 area which is on a sea coast (Fig. 7.12). What will be effect of change of category from 1 to 2 in the estimation of design speed for this building?

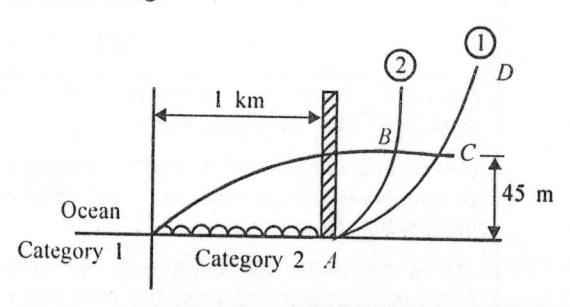

Fig. 7.12 Example 7.4.

DESIGN LOADS OTHER THAN EARTHQUAKE LOADS   125

Reference	Step	Calculations
IS 875 Table 3 App: B	1	*Estimate the fetch development height* **For category 2 (page 12 of code)** 	Fetch x (in km)	Height h (in m)
---	---
0.2	20
0.3	30
1.0	45	
IS 875 Table 2	2	*Determine the variation of velocity with height* **For categories 1 and 2 area and class C building** | Height h (in m)	k_3 Category 1	k_3 Category 2
10	0.93	0.99
15	0.97	1.03
20	1.00	1.06
30	1.04	1.09
50	1.10	1.14	
	3	*Estimation of effects* From step 1 it is seen that for a building distant 1 km from the boundary of category 2 area, for a height up to 45 m wind effects will be controlled by category 2. But beyond a height of 45 m the wind from category 1 will govern i.e. curve *ABCD* governs. (*Note:* This aspect is fully dealt with in the Australian code AS (1170) (1989) Part 2.

EXAMPLE 7.5 (Construction loads with two shores)
A reinforced concrete high-rise building is built with each floor being completed in *T* days (say one week) with two levels of shores and no reshores. Assuming it takes 5 days to set up the shores make a diagrammatic representation of the operation to determine the loads to be carried by the slabs and shores at each of the levels of construction (refer ACI Journal December 1963) (refer Fig. 7.13).

126 ADVANCED REINFORCED CONCRETE DESIGN

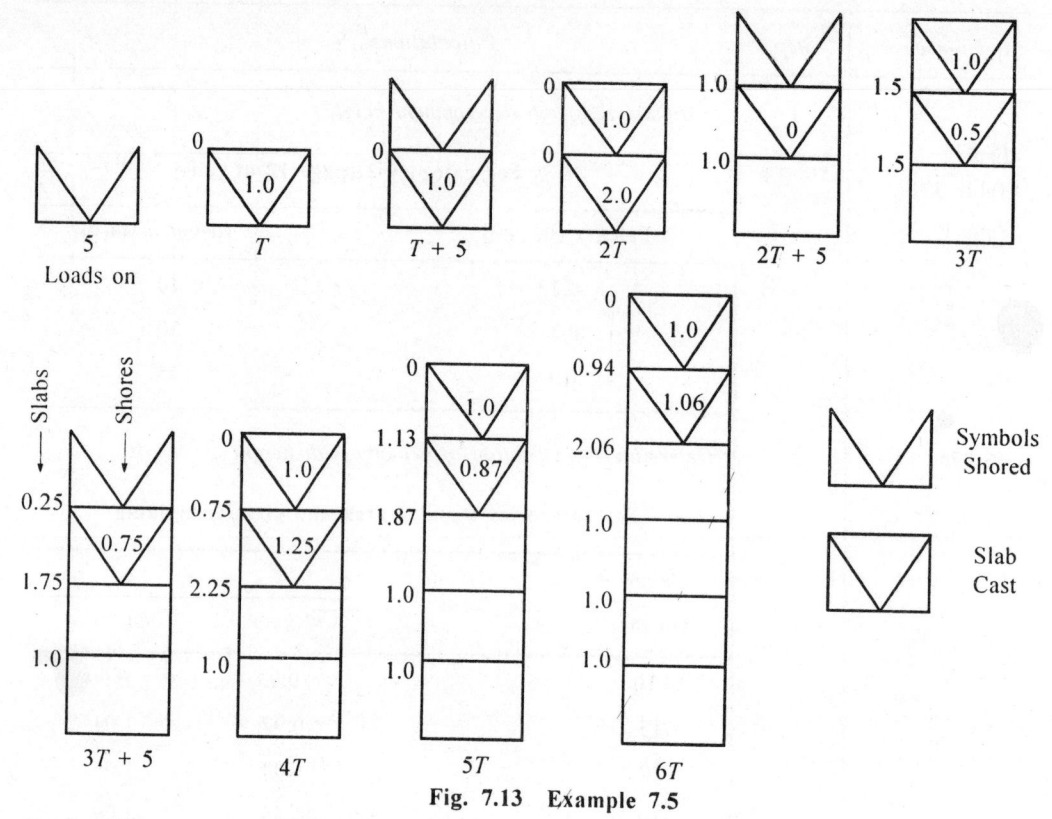

Fig. 7.13 Example 7.5

Explanation

1. In 5 days, the framework is set.
2. In T days, the slab is cast-whole load of the slab is carried by shores.
3. In $2T$ days, the second slab is cast. The bottom shore carries all the loads. The slab is cured for 5 days.
4. In $(2T + 5)$ days, bottom shore is removed and its load 2.0 is distributed among shored slabs at levels 1 and 2.
5. In $3T$ days, third level slab is cast. Its load is shared by the two slabs through the shores as shown.
6. In $(3T + 5)$ days, the lower shore is removed. Its load of 0.25 is shared by the two shored slabs 0.25 each.
7. In $4T$ days the new slab weight is distributed through the shores on shored slabs 2 and 3. The second level slab takes a load of 2.25.

Note: In the final analysis, the maximum value of load will converge to approximately 2.06, occuring at definite cycles. The initial maximum value is 2.25.

EXAMPLE 7.6 (Construction loads with two shores and one reshore)
Determine the loads on slabs if two levels of shoring and one level of reshoring after removal of shoring is used for construction of a multi-storeyed building. (Refer ACI Journal November 1974 and Concrete International, July 83).

DESIGN LOADS OTHER THAN EARTHQUAKE LOADS   127

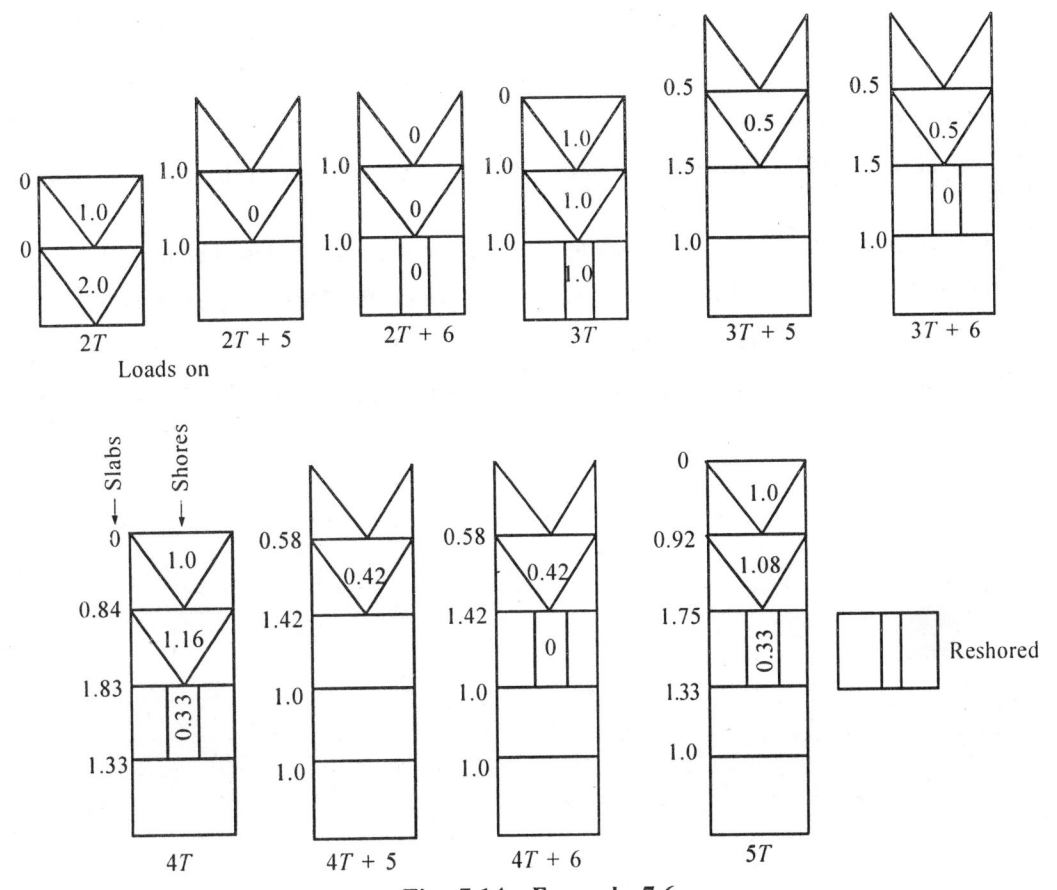

Fig. 7.14 Example 7.6

A. Assumptions

Same as in Example 7.5. Reshoring is carried out on the next day after removal of shores.

B. Explanation

1. In (2T + 6) days, reshoring is carried out but no load is taken by it.
2. In 3T days, the slab is cast and the entire load is taken through the shores by the reshore.
3. In (3T + 5) days, we remove reshore and shore for level 2 slab. The loads get redistributed as shown.
4. In 4T days, the weight of the fourth level slabs gets distributed among two shored and one reshored (total 3) slabs.

Conclusion: 1. Reshoring reduces the maximum load on slab from 2.25 to 1.83.
2. It can be shown that increasing the number of shores (to say three) does not decrease the maximum load ratio, but it increases the age and hence the strength of the slab at the time of application of the maximum load.

REFERENCES

1. IS 875 Parts 1 to 5, *Code of Practice for Design Loads (other than Earthquake) for Buildings and Structures*, B.I.S., New Delhi, 1987.

2. AS 1170.2 (1989), *Australian Standard,* Part 2, Windloads Standards Australia, North Sydney.

3. CP 3 Chapter V Part 2, *Wind Loads.* British Standards Institution, London, 1973.

4. *Wind Loading Hand Book*, Report Building Research Establishment, 1974.

5. Climatic and Solar Data for India, CBRI, Roorkee, 1973.

6. MacGinley T.J., *Structural Steelwork Calculations and Detailing*, Butterworth, 1973.

CHAPTER 8

Analysis of Reinforced Concrete Frames for Vertical Loads by Using Substitute Frames

8.1 INTRODUCTION

For analysing of R.C.—framed buildings, the loads from slabs are first transferred to the beams. The beam-column system is then analysed as frames. Layout of columns along a rectangular grid in plan is very convenient as the two resultant frames in two perpendicular directions can be analysed separately. In irregular layout of columns and beams, as happens in many residential flat constructions, exact analysis of the frames is difficult without complex calculations but approximate methods can be used for all practical purposes by subdividing the frame into smaller frames and analysing them individually.

IS 456 Clause 22 deals with analysis of R.C. framed structures, Clause 22.4.2 allows approximate analysis for gravity loads by means of substitute frames. This chapter first deals with the method of distributing the loads from the slab to the beams and then analysing the beams and columns by using substitute frames.

8.2 DISTRIBUTION OF LOADS FROM SLABS TO SUPPORTING BEAMS

The transfer of loads from one-way slabs to the supports is simple. However, in the case of the two-way slabs, there are several accepted methods for calculating the loads on the beams. IS 456 Clause 24.5 recommends the trapezium method for this purpose. According to the trapezium method, for a two-way rectangular slab the division of load occurs along a 45° line drawn at the corners as shown in Fig. 8.1.

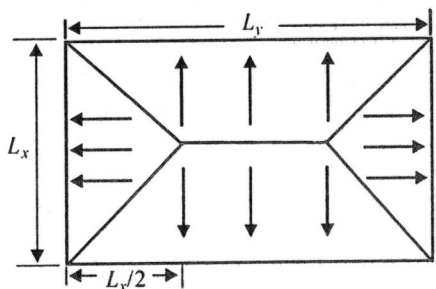

Fig. 8.1 Distribution of loads from slab to beams.

The loads from the area bounded by the beams and 45° lines are assumed to be transferred to the corresponding beam. This total load which will be triangular on the short beam and trapezoidal on the long beam can be found from the following formulae. (For design purpose, the dead load per unit area from slab is multiplied by factor 1.15 to account for weight of beams.) The following formulae for determining the equivalent loads on beams in two-way slab are very useful (Fig. 8.2).

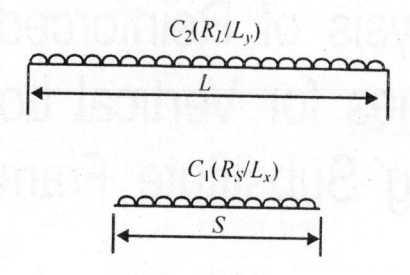

Fig. 8.2 Equivalent Loads on long (L) and short (S) beams.

Case 1: Loads for calculation of shear forces

Let L_x = Shorter span and L_y = Longer span. Then,

$$k = \frac{L_y}{L_x} \ (\geq 1)$$

Also if w = Load per unit area.

Total load $R = wL_x L_y$

Hence,

$$\text{Load on each short panel } R_x = \frac{wL_x^2}{4} \qquad (8.1)$$

Alternately, the percentage of load on each short side is given by

$$R_x = \frac{100 w L_x^2}{4 w L_x L_y} = \frac{25}{k} \qquad (8.2)$$

Thus the percentage of load coming on a short beam is $25/k$, where $k = L_y/L_x$. For a square slab where $k = 1$ the load is obviously 25%. The total load on the two short beams will be double the amount and the balance of the load is distributed to the long beams.

Case 2: Equivalent load for bending moments in simply-supported short span beams

The load on the short beam is triangular in shape and the maximum ordinate $m = wL_x/2$. Maximum B.M. at the centre $= mL_x^2/12 = wL_x^3/24$. If this is replaced by a UDL w_{es} so that the B.Ms. are equal. Then

$$\frac{w_{es} L^2}{8} = \frac{wL_x^3}{24}$$

$$\text{Equivalent UDL on } L_x = w_{es} = \frac{wL_x}{3}$$

$$\text{Total load on } L_x = W_{es} = \frac{wL_x^2}{3} \tag{8.3}$$

$$\text{Using Eq. (8.1) } W_{es} = \frac{4R_s}{3}$$

$$W_{es} = C_1 \times R_s$$

where $C_1 = 1.33$ (see Table 8.1).

Case 3: Equivalent load for bending moment in simply-supported-long span beams.

By similar calculation we can find the equivalent load for a trapezoidal loading on the long beam.

$$\text{Average load on long beam } W_{el} = C_2$$

$$C_2 = \frac{2k(3k^2 - 1)}{3k^2(2k - 1)}$$

$C_2 = 1.33$ when $k = 1$ (square slab) and is about 1.25 when $k = 2$.

Case 4: Fixed beams (continuous beams)

Similar values can be worked out for fixed short and long beams for calculating bending moments at the support and in the span. The various values can be tabulated as Table 8.1.

Theoretically, the 45° distribution is less accurate for slabs which are supported on one edge and continuous over the other. More load is likely to be supported by the beam over which the slab is continuous than the beam over which it is discontinuous. An estimate of such loads is usually made by calculating the load by trapezium method and adding 20% of the load on the beam over which the slab is continuous and deducting the same amount from the discontinuous side [1].

TABLE 8.1 FACTOR C FOR EQUIVALENT LOADS FOR BENDING MOMENTS IN BEAMS UNDER TWO-WAY SLABS [1]

Types of beam	Support B.M.		Span B.M.	
	$k = 1$	$k = 2$	$k = 1$	$k = 2$
(A) In terms of total load on beams				
1. *Simply supported*				
Short span (C_1)	–	–	1.33	1.33
Long span (C_2)	–	–	1.33	1.25 (approx.)
2. *Continuous spans*				
Short span (C_1)	1.25	1.25	1.50	1.50
Long span (C_2)	1.25	1.25	1.50	1.29

TABLE 8.1 Contd.

Types of beam	Support B.M.		Span B.M.	
	$k = 1$	$k = 2$	$k = 1$	$k = 2$

(B) As UDL on the beams
UDL on span = $(1/2)wl_x$(factor)

1. Simply supported				
Short span (C_1)	–	–	2/3	2/3
Long span (C_2)	–	–	2/3	15/16
2. Continuous spans				
Short span (C_1)	5/8	5/8	3/4	3/4
Long span (C_2)	5/8	0.90	3/4	0.96

Approximate formulae for simply supported cases (loads for B.M.). The following formulae are also used for calculation of loads for bending moments:

$$\text{UDL on short span} = \frac{1}{2}\frac{2}{3}wL_x \tag{8.4}$$

$$\text{UDL on long span} = \frac{1}{2}wL_x\left(1 - \frac{1}{3k^2}\right) \tag{8.5}$$

8.3 OTHER METHODS FOR DISTRIBUTION OF LOADS

Following methods can also be used to determine the loading on the beam:

1. By using the tables in the handbooks like Table 62 in *Reynold's Concrete Designers Handbook*. It gives tables for the trapezoidal loads that come on each beam (R_1 to R_4) for different boundary conditions.

2. By using the tables given in BS 8110 [Table 12.3 in Ref. 2] gives the shear in two-way slabs. The shear at the ends of the slab in L_x direction can be taken as the load on beam in the L_y direction. The load is assumed to act only on the middle three-fourth length of the beam so that the simply-supported maximum bending moment becomes $0.117\ W_e\ L_y^2$.

3. By using yield lines: in all cases a distribution of the load based on the yield line patterns for the slabs gives a fair estimate of the load on the beams.

8.4 FRAME ANALYSIS

The beams and the columns in the layout of buildings form frames. Generally for two way slabs two sets of frames in the perpendicular direction (one in the E-W and the other in the N-S direction) can be identified. Having obtained the loads that act on the beams of these frames from the slabs by one of the above methods, the frame can be analysed for the loads. This is the procedure recommended in the IS and BS codes for the analysis of the building frames.

However, for the analysis of two-way slab frames ACI advocates a different approach. The frames are analysed by incorporating the slab also as a part of the frames in the two perpendicular directions. This type of analysis is called the "equivalent frame method" and it is explained in Chapter 13 of this book along with the analysis of flat slabs.

8.5 BRACED AND UNBRACED FRAMES

Braced frames are those whose sway deflection is reduced considerably by the provision of shear walls or stiff cores carrying services, lifts, staircases, etc. Generally, reinforced concrete frames up to 12 storeys can be designed as unbraced frames and the vertical and horizontal loads can be taken by frame action. The method of analysis for gravity (vertical) loads are given in this chapter and the approximate methods of analysis for horizontal loads (due to wind or earthquakes) are given in Chapter 9.

8.6 ANALYSIS BY SUBSTITUTE FRAMES

Analysis of concrete structures for gravity loads are made by methods of elastic analysis. The complete elastic analysis of a large structure even with the availability of a computer package is expensive, time consuming and not necessary except under special circumstances. In order to simplify the analysis, IS 456 Clause 22.4 and BS 8110 Clause 3.2.1 allow the use of substitute frames which are also called as sub-frames. Their analysis is made by conventional elastic methods like slope deflection, moment distribution, etc.

8.6.1 Arrangement of Live Loads

IS Clause 21.4.1 recommends the following combination of gravity loads:

1. Design dead load (1.5DL) on all spans with full design live load (1.5LL) on two adjacent spans.
2. Design dead loads on all spans with full design live load on alternate spans.
3. When design LL does not exceed 3/4 design DL the load arrangement can be 1.5 (LL + DL) on all the spans.

(*Note:* It should be noted that BS 8110 stipulates the analysis with combinations of (DL) only and (1.4DL + 1.6LL) which is the theoretically correct procedure. IS recommendations however facilitates analysis for elastic design to be used also for limit state design by multiplying it by the factor 1.5. The IS procedure tends to produce less moments at the support and the columns than the BS practice.)

8.6.2 Relative Stiffness for Elastic Analysis

According to IS Clause 22.3 and BS, any one of the following methods can be used for the same frame analysis:

1. Cross-section method: using only concrete section neglecting the steel in the section.
2. Transformed-section method: using the whole concrete section plus the equivalent steel area using modular ratio.
3. Cracked-section method: using only the uncracked compression part of concrete plus the equivalent area of steel using modular ratio.

Usually the first method is used because of its simplicity. The next question is whether the moment of inertia of beams is that of the beam at the supports where it is rectangular beam or that of the beam in the span, where it acts as a T beam. ACI code 318 Clause 13.2.4 allows the use of a uniform T or L section with the projection of the slab to the sides of the beam equal to the depth of the beam under the slab or four-times the slab thickness whichever is less as shown in Fig. 8.3. This is a good practice to follow.

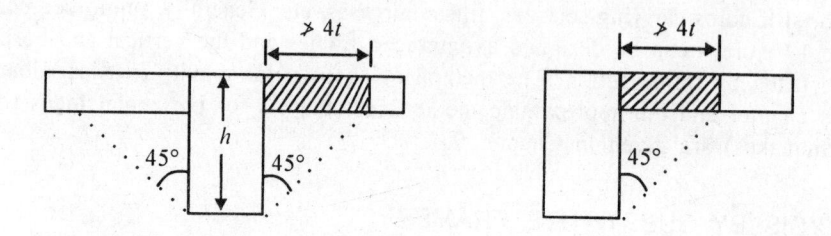

Fig. 8.3 Portions of slab that act with the beams: (a) T beam (b) L beam.

For calculating the moment of inertia of I and L beams, we can use SP 16 Chart 88 and the formula:

$$I_{gr} = \frac{Kb_w D^3}{12}$$

where K is the constant to be read off from the chart.

Another commonly-used approximation is to take the inertia of T beams as 1.5 and that of the L beams as 1.25 times the moment of inertia of rectangular part of those beams. These values are based on experimental evidence that the variation of the actual width of T beam along the length of the T beam is 1 to 2 times that of the rectangular part depending on the dimensions of the T beam.

8.6.3 Effective Spans of Frames

IS 456 Clause 22.3 gives rules for effective span of slabs. For the frame analysis centre-to-centre distances are taken and final corrections for moments and shears are made at the face of supports during the design stage.

8.7 ANALYSIS BY CONTINUOUS BEAM SIMPLIFICATION

The 1964 version of IS 456 and the present BS 8110 Clause 3.3.1.2.4 allow the beams in building frames to be separately analysed by the continuous beam simplification. This enables us to use formulae with equivalent loads for bending moments and shears. The bending moments in columns are then separately analysed by the approximate method given in Section 8.8.2. This procedure gives conservative design for both the beams and columns. However, this is not allowed in the present IS 456 code. According to Is 456 (2000) Clause 22.4.2, building frames are to be analysed as a complete frame or by the use of one or more substitute frames.

(However, many other structures, like bridges, can be analysed as continuous beams. For such continuous beams over three or more spans which do not vary more than 15% of the largest span the S.F. and B.M. coefficients are given in Tables 12 and 13 of IS 456. As maximum moments at supports and spans as well as the maximum shear is obtained with different dispositions of loads, separate coefficients are given for dead load and live load in these tables. As these coefficients have been arrived at after some amount of redistribution, no further redistribution of moments is allowed.)

8.8 USE OF SUB-FRAME FOR ANALYSIS OF VERTICAL LOADS

The types of sub-frames used for analysis are shown in Figs. 8.4 and 8.5. Although IS 456 does not specify the frames to be used, they are discussed in the explanatory handbook SP 24.

1. Type 1 sub-frames are for the analysis of beams and columns (allowed by IS 456 and SP 24).

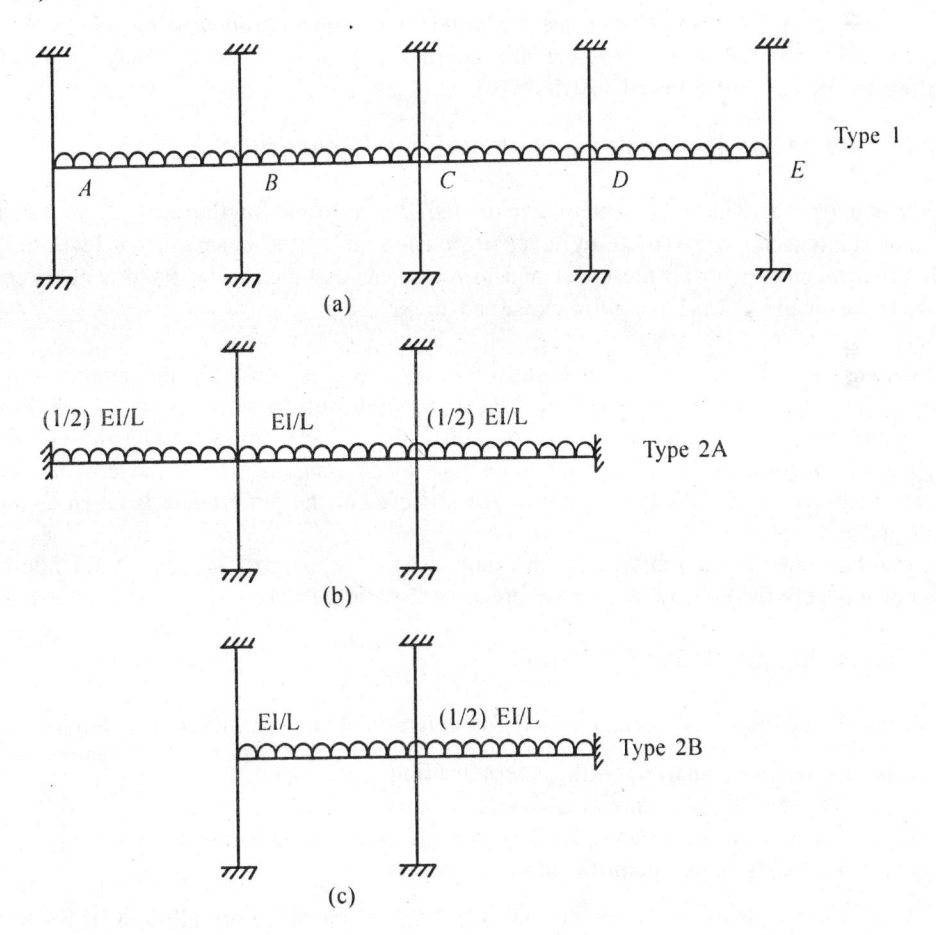

Fig. 8.4 Substitute frames for the analysis of beams and columns in frames specified by IS:456 for gravity loads.

136   ADVANCED REINFORCED CONCRETE DESIGN

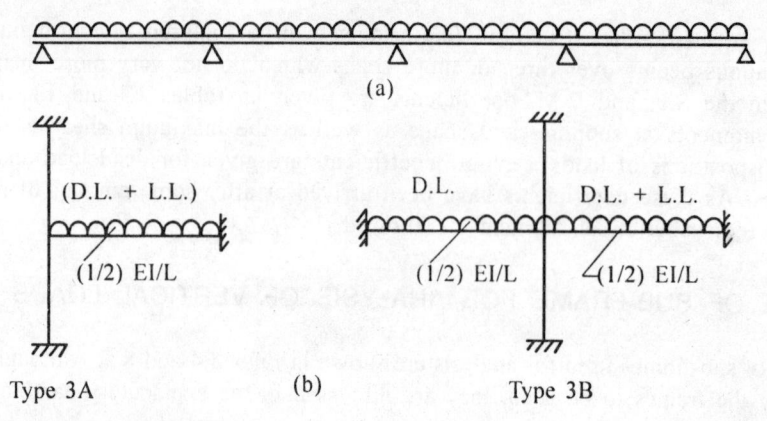

Fig. 8.5 Analysis allowed for beams and columns in BS:8110—(a) 'continuous beam simplification' for beams (b) sub-frames for columns.

2. Types 2A and 2B sub-frames are for beams and columns (allowed by SP 24).

3. Type 3A and 3B sub-frames are for approximate analysis of columns. (not fully recognised by IS 456 but allowed in BS 8110).

8.8.1 Sub-frame for the Analysis of Beams and Columns

1. Sub-frame type 1. The sub-frame shown in Fig. 8.4 is used for the analysis of all the beams and columns for a given storey level. It consists of all the beams at the level under consideration together with all the upper and lower beams assumed to be fixed at their ends away from the beams. These are fully explained in SP 24.

2. Sub-frame type 2. The sub-frames shown in Fig. 8.4 is used for the analysis of a particular beam and associated columns. This simplified sub-frame consists of the beam under consideration along with the beams on either side, if any, together with the columns attached to the beam under consideration. The ends of columns as well as those of the side beams are assumed to be fixed. In addition, the stiffness of the outer beam is taken as only one-half its real value.

We can have two layouts in type 2 sub-frames which we can call as type 2A for internal beams and type 2B for end beams. These are described in SP 24.

8.8.2 Sub-frame for Column Analysis

Three following methods are recommended for determining the moments in columns:

 1. Type 1 sub-frame analysis with pattern loading
 2. Types 2A and 2B sub-frame analysis for beams
 3. Types 3A and 3B sub-frames for approximate analysis of moments in columns, along with 'continuous beam simplification' for beams.

Eventhough IS 456 does not recommend the type 3 sub-frames they are allowed in BS 8110 and are used by many designers for quick design columns along with the continuous beam simplification for beams. Type 3 sub-frames consists of the column under consideration with the upper and lower columns fixed at their ends as shown in Fig. 8.5. The beams framing into these columns are also considered but fixed at their ends. The full stiffness of the columns are taken into account but the beams are assumed to have only one-half of its stiffness.

Accordingly, there will be two types of sub-frames: type 3A corresponding to internal columns and type 3B corresponding to external columns. The distribution of loads in the beams should be considered for maximum moment in the columns.

Eventhough theoretically redistribution of moments obtained from the first and second methods are possible, it can be seen from Chapter 3 that the restriction imposed on x/d values do not allow any redistribution of columns moments. The third method is a continuation of the methods used in olden codes. Being an approximate method no redistribution of moments is allowed by the codes. We should also remember that this approximate method overestimates the moments in the columns, especially, in the external columns.

8.8.3 Loading for Maximum (P and M) in Columns

Column sections at each floor are designed for direct load (P) and also for the bending moment (M) produced at the foot of upper columns as well as at the head of the lower column.

Usually the foot of upper column is referred to as the topside of the floor (T.S.) and the head of lower columns as the underside of the floor (U.S.). The simple formulae in Table 8.2 can be used for determining the end moments of the columns by the approximate method. In these tables, M_e is the unbalanced moment in the beams for the external column and M_{es} the unbalanced moment for the internal column. The loadings on the beams for calculating M_e and M_{es} should be such that these unbalanced moments are a maximum. Accordingly, for maximum M_e of external column, the end beam should be fully loaded with the dead and live load. For maximum M_{es} of internal columns, the beams should be so loaded that the differential moments is maximum, i.e. the longer beam should be fully loaded with dead and live loads and the shorter beam only with the dead load.

TABLE 8.2 DISTRIBUTION OF UNBALANCED MOMENTS IN COLUMNS

Column considered	Moments	
External columns:		
1. At the foot of upper column (T.S.)	$\dfrac{M_e K_u}{K_u + K_L + 0.5 K_b}$	(8.6)
2. At the head of lower column (U.S.)	$\dfrac{M_e K_L}{K_u + K_L + 0.5 K_b}$	
Internal columns:		
1. At foot of upper column (T.S.)	$\dfrac{M_{es} K_L}{K_u + K_L + 0.5 K_{b1} + 0.5 K_{b2}}$	(8.7)
2. At head of lower column (U.S.)	$\dfrac{M_{es} K_L}{K_u + K_L + 0.5 K_{b1} + 0.5 K_{b2}}$	

where

M_e = Unbalanced moment in the beam for external column

M_{es} = Unbalanced moment in the beams for internal column

K_u = Stiffness of upper column

K_L = Stiffness of lower column

K_{b1} = Stiffness of first beam framing into the column

K_{b2} = Stiffness of second beam framing into the column

In theory, the maximum moment in the column will not be associated with the maximum vertical load in the column. The two cases require two different loading patterns. However, for design purpose, the maximum P and M are calculated separately and the column designed for the combination of P and M [1].

8.9 PROCEDURE FOR CALCULATION OF (P AND M) FOR CENTRAL COLUMNS

The following procedure can be used for the estimation of the maximum axial load and moment in columns. The loads and moments are calculated from top downwards. Taking for example the frame in Fig. 8.6 of an internal column, the following steps can be followed.

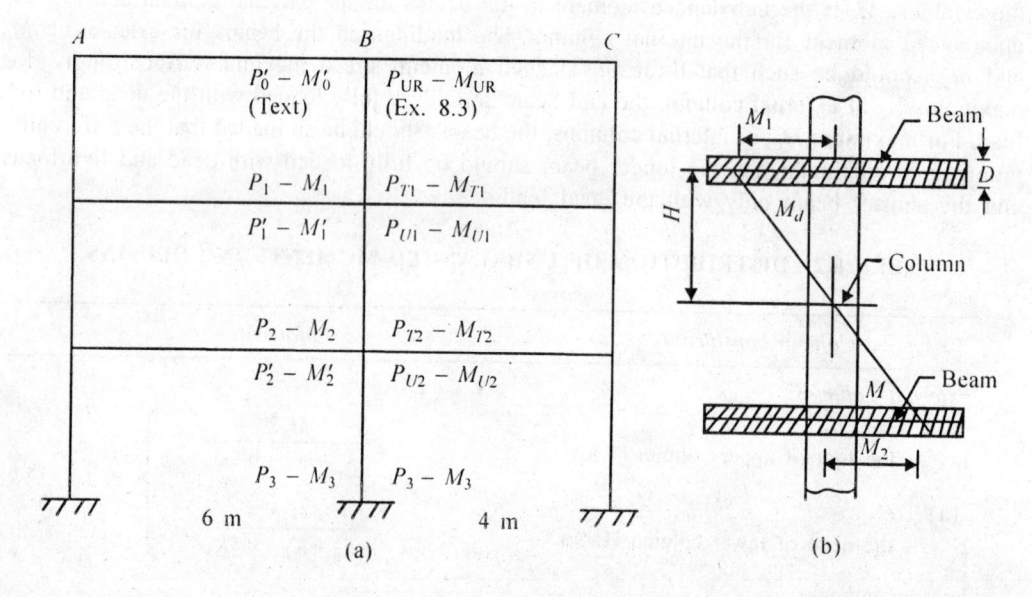

Fig. 8.6 Determination of design moments in columns. (a) M and P for internal columns (c) reduction in moments at the bottom of beams in upper floor and top of beams in lower floor for design.

where P', M' denote quantities on the underside of the floors and P_1, M_1 on the topside of the floors.

Step 1: First consider the U.S. (underside) of roof. Loading of sub-frame should produce maximum M (i.e. full dead and live loads from longer beam and dead load on the other beam). Also,

P'_0 = One-half of the loads from each beam with above load

$$M'_0 = \frac{M_{es}K_L}{K_u + K_L + 0.5K_{b1} + 0.5K_{b2}}$$

Note: 1. In this case, K_u = (stiffness of upper column) is zero.

2. M_{es} is determined by assuming the beams fixed at their ends. For UDL, the B.M. at support = $wL^2/12$ (Coefficients for other configurations are readily available.)

Step 2: Next consider the floor, one floor down from roof

(a) T.S. (Topside) of floor: foot of upper column

P_1 = (One half of maximum loads from roof above, i.e. with full dead and live loads on the beams above) + (Weight of column above)

$$M_1 = \frac{M_{es}K_u}{K_u + K_L + 0.5K_{b1} + 0.5K_{b2}}$$

where

M_{es} = corresponding unbalanced moment on the substitute frame at the joint being considered (i.e. when longer beam carries full (DL + LL) and the other one only dead load).

(b) U.S. of floor: head of lower column

$P'_1 = P_1$ + one-half load from the beam of sub-frame to produce M_{es}

$$M'_1 = \frac{M_{es}K_L}{K_u + K_L + 0.5K_{b1} + 0.5K_{b2}}$$

where M_{es} is the corresponding unbalanced moment on the frame at the joint being considered (and as M'_1 is the moment at the head of the lower column).

Step 3: Consider the two floors down from roof.

(a) T.S. of floor: foot of upper column

P_2 = (One-half of full dead and live loads on beams from roof) + (One-half of full dead and live loads on beams one floor below) + (weight of columns above)

$$M_2 = \frac{M_{es2}K_u}{K_u + K_L + 0.5K_{b1} + 0.5K_{b2}}$$

(b) U.S. of floor: head of lower column

$$P'_2 = P_2 + \text{(One-half of loads on beams of sub-frame to produce } M_{es2})$$

$$M'_2 = \frac{M_{es2} K_L}{K_u + K_L + 0.5 K_{b1} + 0.5 K_{b2}}$$

where M_{es2} = corresponding unbalanced moment on substitute frame.

This calculation for T.S. and U.S. of floors is repeated till the foundation is reached. This procedure is illustrated by Example 8.2.

8.10 PROCEDURE FOR CALCULATION OF (P AND M) EXTERNAL COLUMNS

The principle is same for the enternal column as is used in the central column. It should be noted that M_e is maximum, when full dead and live loads are superposed on the beam of sub-frame.

8.11 REDUCTION IN LOADS ON COLUMNS AND BEAMS

As it is very unlikely that full loading will happen on all the floors of a multi-storeyed frame simultaneously it is usual to allow a reduction in the imposed loads when several floor loads are carried by a single column or a large area of floor is carried by a single beam. This is explained in Section 7.7 of Chapter 7. This factor should also be considered in calculation of loads on columns and beams.

8.12 PARTIAL RESTRAINT AT END-SUPPORTS

When ends of beams are built into the masonry or when full fixity is not provided at the ends of beams and slabs, only partial restraint will develop at these ends. Depending on the degree of fixity, the B.M. at these ends can vary from WL/120 to WL/12. In normal cases, it is considered sufficient to assume the negative bending moment produced as WL/24. Adjustment may also be made in the remaining portion of the beam for this end-moment (IS:456 Clause 22.5.2).

8.13 ANALYSIS OF SINGLE-SPAN PORTALS

For single-span portal frames where the ratio of span of the beam to height of column is more than 2 and the reinforcement in the beam is less than 1%, it is customary to carry out distribution of moments in the elastic analysis with cracked EI value for the beam so that the columns are designed for comparatively larger moments. Alternately, the bending moment transmitted to the column can be limited by proper detailing of the corner. In the later case, the top reinforcement in the beam at the support under overload should be made to yield before the concrete on the inside of the column reaches ultimate stress, so that large moments are not transferred to the columns. Detailing of corner should comply with the assumption of moment transfer to columns.

ANALYSIS OF R.C. FRAMES FOR VERTICAL LOAD USING SUBSTITUTE FRAMES

8.14 CRITICAL SECTION FOR MOMENT AND SHEAR

IS 456 Clause 22.2 allows analysis of monolithic frames by taking the centre-to-centre distances. As the variation of bending diagram of beams with uniformly supported load is very large near the supports, some saving can be made in using the bending moment values at the faces of the supports. However, such reduction will not be applicable to simply supported beams. Suitable modification as given in IS 456 Clause 22.6 regarding critical section for shear can also be made when designing for shear. Similar reduction in moments can be made for columns moments as stated in Example 8.3.

8.15 RECOMMENDED PROCEDURE

IS 456 stipulates that for final design the analysis of braced frames should be made either by complete analysis of the frame (nowadays made by computer software) or by the stipulated sub-frame analysis with redistribution of moments. This will lead to economical designs of the structure. However, for preliminary analysis of frames with equal spans much economy will not be lost if the coefficient method or the continuous beam simplification without redistribution, as allowed by BS 8110, is followed. These approximate methods can also be used as a quick check of the calculations made by the more elaborate computer methods.

It should be remembered that the only sub-frame that is theoretically applicable to unbraced frames is the type 1 sub-frame. Others are approximations.

8.16 FORMULAE TO DETERMINE SPAN MOMENT

The analysis of frames gives the moments at the end of the beams. It will be necessary to calculate the span moments from these end moments and the loading on the beam. The following procedure may be used for the case of a beam with UDL:

Let AB in Fig. 8.7 be a fixed beam of length L loaded with a uniformly distributed load w. Let the fixing moments at A be M_A and at B be M_B. It is required to determine the maximum positive span moment M_{AOB} and its location at a point O.

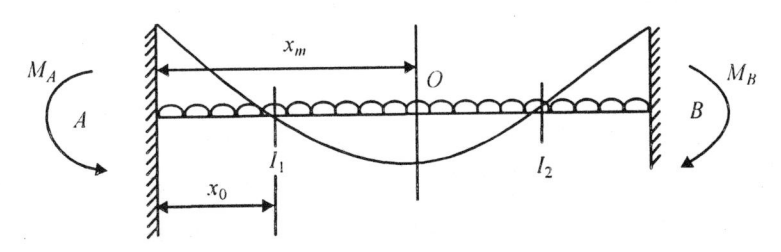

Fig. 8.7 Position of maximum moment and point of inflection in beams.

It is known the span moment will be a maximum where the shear force is zero. Let the point of maximum moment be x_m from the left-hand support A and the point be designed as O. Assuming, M_A is greater than M_B the value of x_m from A is given by the point of zero shear.

$$x_m = \frac{L}{2} + \frac{M_A - M_B}{wL} \quad \text{and} \quad V_A = \frac{wL}{2} + \frac{M_A - M_B}{L}$$

The resultant shear force at support A is given by:

$$V_A = wx_m \tag{8.8}$$

Taking moments about O, we get the maximum bending moment

$$M_{max} = V_A x_m - \frac{1}{2} w x_m^2 - M_A$$

$$= \frac{V_A V_A}{w} - \frac{1}{2} w \left(\frac{V_A}{w}\right)^2 - M_A = \frac{V_A^2}{2w} - M_A \tag{8.9}$$

where $x_m = V_A/w$. Also, when M_A and M_B are not very different, one may use the approximation:

$$M_{max} = \frac{M_A + M_B}{2} \quad \text{and} \quad x_m = \frac{L}{2}$$

The maximum bending moment at mid-point is

$$M_{AOB} = \frac{wl^2}{8} - \frac{M_A + M_B}{2} \quad \text{(approx.)}$$

The points of inflection x_0 can be obtained by solution of the following quadratic equation:

$$\left(\frac{wL}{2} + \frac{M_A - M_B}{L}\right) x_0 - \frac{w x_0^2}{2} - M_A = 0 \tag{8.10}$$

These data give the distribution of the bending moment along the length of the beam so that the beam can be properly designed and detailed.

EXAMPLE 8.1 (Estimation of loads on roof and floor slabs)
A reinforced concrete building on beams and columns has its slabs in panel of 6 × 5 m. The thickness of the roof and floor slabs are 150 mm. Estimate the design load for the roof slab and the floor slab.

Reference	Step	Calculations	
	1	*Roof Panels*	
	1.1	Calculate the design load	
Text		D.L. of slab 0.15 × 25	= 3.75 kN/m²
Sec. 7.2		20 mm asphalt (water proof)	= 0.48 kN/m²
		50 mm screed	= 0.72 kN/m²
		special ceiling finish	= 0.24 kN/m²
		Total	= 5.19 kN/m²
IS 875	1.2	Estimate imposed load	
(Part 1)		L.L. on roof = 1.5 kN/m²	

ANALYSIS OF R.C. FRAMES FOR VERTICAL LOAD USING SUBSTITUTE FRAMES 143

Reference	Step	Calculations
	1.3	Determine the design load
		I.S. Design load = 1.5(1.5 + 5.19) = 10.0 kN/m^2
Text	2	Floor panels
Sec. 7.2	2.1	Calculate the dead load
		D.L. of 150 mm slab = 3.75 kN/m^2
		Floor finish (Terazzo) = 0.72 kN/m^2
		Ceiling finish = 0.24 kN/m^2
		Partitions = 1.00 kN/m^2
		Total = 5.71 kN/m^2
IS 875	2.2	Estimate imposed load
		L.L. on floor = 4.0 kN/m^2
	2.3	Determine the design load
		I.S. design load = 1.5(5.71 + 4.0) = 14.6 kN/m^2

EXAMPLE 8.2 (Estimation of equivalent loads on beams)
A reinforced concrete slab 6 × 5 m is discontinuous on one of the longer sides and is continuous on all other sides. If it carries a factored load of 9 kN/m^2 find (a) the loads supported by the beams (b) the equivalent load for the bending moment, for the continuous long beam.

Reference	Step	Calculations
	1	Dimension of beams
		$L_x = 5$ m; $L_y = 6$ m; $k = \dfrac{6}{5} = 1.2$
	2	Estimate the loads on long beams from trapezoidal distribution
		(Method 1)
		Total load = wL_xL_y = 9 × 6 × 5 = 270 kN
Eq. (8.2)		% of load on short span = $\dfrac{25}{k} = \dfrac{25}{1.2} = 20.8\%$ (each beam)
		% of load on long span = $\dfrac{100 - 41.6}{2} = 29.2\%$ (say 29%)
		(Correction for load on beam with continuous slab: add 20% of reaction.)
		Load on the long beam with slab continuous over it
		= 1.2 (0.29 × 270) = 1.2 × 78.3 = 94 kN
		Add 15% for weight of beam = 0.15 × 78.3 = 11.7 kN

144   ADVANCED REINFORCED CONCRETE DESIGN

Reference	Step	Calculations
Table 8.1	3	Total load on beam = (94 + 11.7) kN = 105.7 (say 106 kN) Calculate the bending moments in long beam *Method 1: Trapezoidal distribution (S.S.)* Equivalent load = $C_2 W$ where $C_2 = 1.3$ $$\text{B.M.} = \frac{WL}{8} = \frac{1.3 \times 106 \times 6}{8} = 103.35 \quad (\text{say } 104 \text{ kNm})$$
	4	*Method 2: Reynold's Hand book method* Formulae for the trapezoidal distribution for various cases given in the handbook (Table 62) can be used to determine the loads on the beams.
Table 8.1 or Eq. (8.2)	5	*Method 3: Using approximate formulae (Table 8.1)* Short span (15% self wt.) $$w_c = \frac{wL_x}{3} = \frac{1 \times 9 \times 5}{3} = 15 \text{ kN/m}$$ $$M_{max} \text{ (with self weight)} = \frac{1.15 \times 15 \times 5 \times 5}{8} = 53.9 \text{ kNm}$$
Eq. (8.3) Table 8.1	6	*Long span using formula* $$w_c = \frac{1}{2} wL_x \left(1 - \frac{1}{3k^2}\right) = 17.29 \text{ kN/m}$$ Factor for continuity and self wt. = 1.35 $w = 1.35 \times 17.29 = 23.34$ kN/m $$M = \frac{23.34 \times 6 \times 6}{8} = 105 \text{ kNm}$$ (Values are same order as in step 3.)

EXAMPLE 8.3 (Calculation of *P* and *M* for column analysis)

A reinforced concrete frame as shown in Fig. 8.6 has beams of 300 × 700 and column 300 × 400. The columns are 3.5 m in height. The characteristic dead and live loads are 40 kN/m and 60 kN/m on the beam for the roof and floors. Determine the direct load and bending moments for which the internal column has to be designed at each level, using the approximate method. Assume there are three floors only.

Reference	Step	Calculations
	1	Data of loads 1.5(DL) = 1.5 × 40 = 60 kN 1.5(DL + LL) = 1.5 (100) = 150 kN

ANALYSIS OF R.C. FRAMES FOR VERTICAL LOAD USING SUBSTITUTE FRAMES 145

Reference	Step	Calculations
	2	Design DL of columns = $1.5(0.3 \times 0.4 \times 25 \times 3.1)$ = 13.75 kN (say 14 kN) *Calculate the relative stiffness* $k_c = \dfrac{I}{L}$ of column $= \dfrac{0.3(0.4)^3}{12 \times 3.5} = 0.46 \times 10^{-3}$ $K_A = \dfrac{I}{L}$ of beam $AB = \dfrac{0.3(0.7)^3}{12 \times 6} = 1.42 \times 10^{-3}$ $K_B = \dfrac{I}{L}$ of beam $BC = \dfrac{0.3(0.7)^3}{12 \times 4} = 2.14 \times 10^{-3}$ Assume E is constant
Fig. 8.6		(Next we have to determine the loads and moments at the top side (T.S.) and underside (U.S.) of floor and each roof level.)
	3	*Determine P and M on the underside of roof* To produce maximum moment in column we put 150 kN on 6 m span and 60 kN on 4 m span $M_{es} = \dfrac{w_1 l_1^2}{12} - \dfrac{w_2 l_2^2}{12} = \dfrac{160 \times 6^2}{12} - \dfrac{60 \times 4^2}{12} = 400 \text{ kNm}$ M_{UR} and P_{UR} are moments and load with maximum moment in the column at bottom of roof. $M_{UR} = \dfrac{M_{es} K_{CL}}{K_{CL} + 0.5 K_A + 0.5 K_B}$ $= \dfrac{370 \times 0.46}{0.46 + 0.71 + 1.07} = 76 \text{ kNm}$ $P_{UR} = \dfrac{(150 \times 6) + (60 \times 4)}{2} = 570 \text{ kN}$ Mamimum load on underside of roof $P'_{UR} = \dfrac{150(6+4)}{2} = 750 \text{ kN}$
	4 (a)	*Calculate P and M on the first floor below roof* *Topside of the first floor from the roof (TI)* P_{TI} = Maximum load at T_{II} = Maximum load on roof + Weight of column $P_{TI} = 750 + 16 = 766$ kN M_{es} (as in step 3) = 370 kNm $M_{TI} = \dfrac{M_{es} K_{CU}}{K_{CU} + K_{CL} + 0.5 K_A + 0.5 K_B}$ $= \dfrac{370 \times 0.46}{0.46 + 0.46 + 0.71 + 1.07} = 63 \text{ kNm}$

Reference	Step	Calculations
	(b)	Underside of first floor roof (UI)
		$P_{UI} = 766 + 570 = 1336$ kN
		$M_{UI} = 63$ kNm
		$P'_{UI}(\text{max}) = 766 + 750 = 1516$ kN
	5	Calculate P and M for the second floor from the roof
		Topside of first floor from roof (T2)
		$P(T2) = 1561 + 16 = 1532$ kN
		$M(T2) = 63$ kNm
		Underside of the second floor from roof
		$P_{U2} = 1532 + 570 = 2102$ kN
		$M_{U2} = 63$ kNm
		$P'_{U2}(\text{max.}) = 1532 + 750 = 2282$ kN
	6	Estimate load on the third floor below roof T3 (Ground floor)
		Topside of ground floor = 2282 + 16 kN
		$P_{T3} = 2298$ kN
		$M_{T3} = 31.5$ kNm
		Note: (1) The above moment values are at the centre of the beams and columns. The actual design moments can be the reduced value at the soffit of the beams. If M_1 is the moment at the junction of beams and column and H its distance from the point of inflection of the column D the depth of the beam. We may use the design moment in the column as $$M_d = \frac{M_1(H - D/2)}{H}$$ (2) Design of each junction of column has to be checked for the maximum M values and the two values of P.

REFERENCES

1. Reynolds C.E. and Steedman J.C., *Examples of the Design of Building to CP 110 and Allied Codes*, Cement and Concrete Association, Slough, U.K., 1978.

2. Varghese P.C., *Limit State Design in Reinforced Concrete*, Prentice-Hall of India, New Delhi, 1994.

3. BS 8110, *British Standard Structural Use of Concrete*, British Standards Institution, London, 1985.

CHAPTER **9**

Analysis of Frames under Horizontal Loads

9.1 INTRODUCTION

The method by which a multi-storey building frames resist horizontal or lateral forces (due to wind, earthquakes, bomb blasts, etc.) depends on how the structure has been laid down or planned to bear these loads. In frames, properly braced lateral loads with braces such as shear walls, the lateral forces can be taken by these devices with only negligible forces transferred to the frames. On the other hand, open frames which have no special devices for lateral resistance, take the horizontal forces by deformation of the frame itself. They produce moments and shears in the columns and beams. In this chapter, the approximate methods of analysis of the unbraced frames for lateral forces as recommended in IS 456 Clause 22.4.3 is dealt with. Such frames can be used economically for building as high as 12 storeys. Frames with shear walls or special structural systems, like tubes and multiple tubes, used for tall buildings, should be analysed by more exact methods given in the specialised literature [1,2,3]. Similarly, unsymmetrical structure have also to be analysed by more complex methods.

9.2 EFFECT OF LATERAL LOADS

The calculation of forces, due to wind acting on structures, is dealt with in Chapter 7 and those due to earthquake in Chapter 18. For analysing the frame for the effects of these horizontal wind forces, these forces in each floor height from middle of upper floor to middle of lower floor are summed up and considered as concentrated at the corresponding frame joints, as shown in Fig. 9.1.

9.3 METHODS OF ANALYSIS

The methods commonly used for analysing a frame subjected to lateral loads are as under:

1. Elastic analysis of the whole frame by classical methods
2. Approximate analysis by the portal method
3. Approximate analysis by the cantilever mehod
4. Approximate analysis by the factor method [4]

148   ADVANCED REINFORCED CONCRETE DESIGN

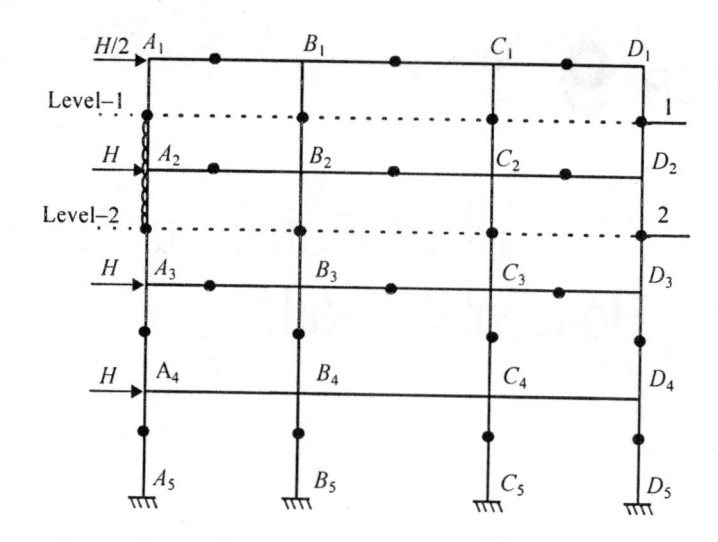

Fig. 9.1 Assumed points of loading and points of contraflexure for analysis of frames with lateral loads.

Exact analysis by elastic analysis of the whole structure cannot be easily done without the aid of a computer. In most moderately tall buildings up to 12 storeys, the approximate methods presented in this chapter give good results which can be safely used in practice. The portal method, because of its simplicity is more popular than the cantilever method. IS 456 Clause 22.4.3 allows such analysis for symmetrical building of moderate height. In this chapter, we shall deal with the portal and cantilever methods of analysis only.

9.4 PORTAL METHOD (METHOD OF PROPORTIONAL SHEAR)

The *portal method* is also known as the *method of proportional column shear forces*. The following assumptions are shown in Fig. 9.2.

1. Points of contraflexure of columns are at their mid-height.
2. Points of contraflexure of beams are at the mid-span. (These two assumptions make the structure determinate.)
3. Horizontal forces are concentrated at panel points, i.e. junctions of beams and columns.
4. The frame at each level of the contraflexure of the columns can be assumed to be made up of a series of single bay portals subjected to horizontal forces.
5. The shear force induced in the columns of the portals (at the level of the contraflexure of the columns) due to horizontal forces is proportional to the span of the individual portals.
6. The shear force in each portal is assumed to be equally distributed between the two columns legs Fig. 9.2.

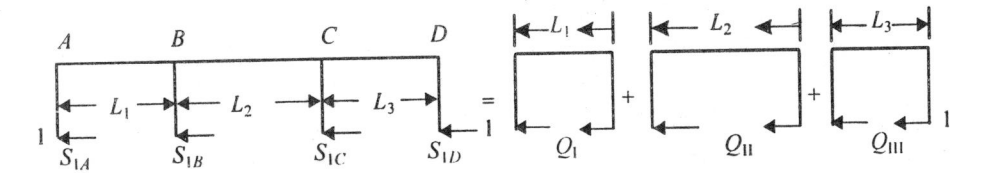

Fig. 9.2 Column shears by portal method.

9.4.1 Calculation Procedure

We start from roof downwards at level 1-1 shown in Fig. 9.1. From Fig. 9.2 the total shear in the legs of the first portal is given by:

$$Q_1 = \frac{H_1 L_1}{\Sigma L} \qquad (9.1)$$

where H_1 is the horizontal force due to wind, L_1 is the span of the first portal, and ΣL is the sum of the spans of all portals. Q_1 is distributed equally between the legs.

It should be noted that each interior column carries the sum of the shear from the adjacent portals formed when the frame is cut into a series of portals. Accordingly, the shear in the first interior leg will be the sum of the one-half column shears from the two adjacent portals. Taking L_1 and L_2 are the spans adjacent to the column being considered.

$$S_{1B} = \frac{H_1(L_1 + L_2)}{2\Sigma L} \qquad (9.2)$$

Similarly the shears in the other column in the first level can be calculated.

Next considering the second level from the top, the shear at the points of inflection just below it can be calculated. Taking successive storeys below, the shears in the columns at the various points of inflection can be calculated.

Having obtained the horizontal forces in the legs, the moments and shears in the columns and beams of the determinate structure can be directly obtained by simple statics as shown in Fig. 9.3. The procedure is illustrated by Example 9.1.

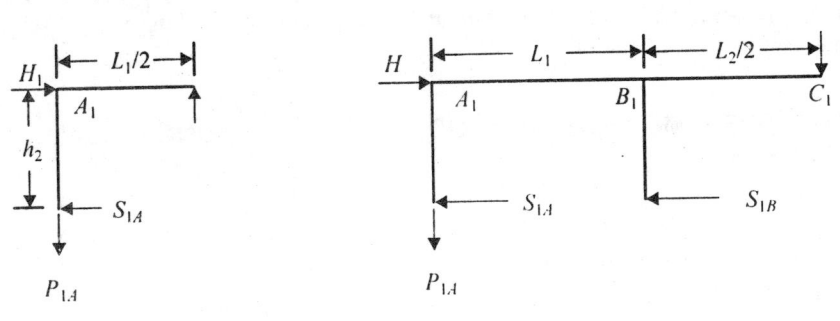

Fig. 9.3(a) Analysis of frames by portal method.

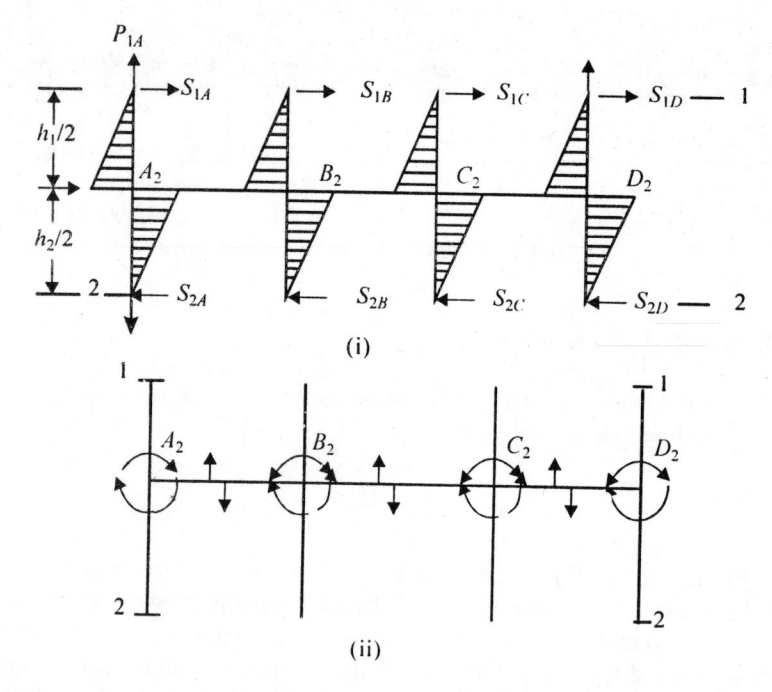

Fig. 9.3(b) Portal method: (i) column shears and moments and (ii) beam shears and moments.

The sequence of analysing the frame is to proceed from the left to the right of each floor starting from the top floor. The resultant moment is always drawn on the tension side of the member. This procedure avoids confusion of signs of moments. Another sign convention is to treat all clockwise moments produced by beams and columns on the joints as positive, so that, the anticlockwise moments at the end of individual members are positive.

The errors brought out in this analysis are mostly due to the incorrect assumption regarding the points of inflection of the columns at the mid-span. In an actual multi-storey frames the inflection points in its upper portion are slightly below the mid-heights, and in the lower portion they are slightly above the mid-heights. Only in the middle portion of the frame, the inflection points tend to be at the mid-points. However, at the foundation level the point of inflection depends on the fixity of the foundation. In the bottom storey the point of inflection lies between the base and one third the height of the column depending on the fixity exerted by foundation.

9.5 CANTILEVER METHOD (METHOD OF PROPORTIONAL AXIAL STRESSES)

The cantilever method is also known as the *method of proportional column axial stresses*. The following assumptions are made:

1. Points of contraflexure of columns are at their mid-heights.
2. Points of contraflexure of beams are at their mid-spans.
3. Horizontal forces act as concentrated load at points as in the portal method.

ANALYSIS OF FRAMES UNDER HORIZONTAL LOADS   151

4. Then we make the fourth assumption that the axial stress in the columns at any level is proportional to its distance from the 'centroid of the cross sections' of all columns at that level.

We made the first-three assumptions in the portal method as well. Knowing the cross-sectional area of the columns the axial forces in each column at each level can be determined from the fourth assumption. The equation for the columns with relevance to Fig. 9.4 for the

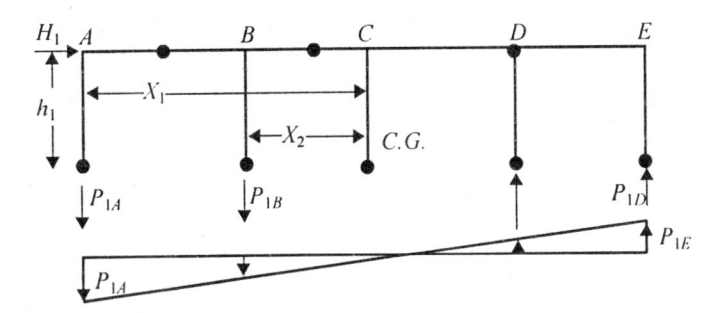

Fig. 9.4 Analysis of frames by cantilever method.

level below the top floor will be as follows:

$$\frac{P_{1A}}{X_1} = \frac{P_{1B}}{X_2} = R$$

$$P_{1A} = RX_1; \quad P_{1B} = RX_2$$

The signs of the forces on either side of the centre of gravity are opposite as the total vertical force is zero. R is obtained by taking the moment of all the vertical and horizontal forces about the point which is the intersection of the horizontal line along the points of inflection of the columns (at mid-height) and the vertical line through the C.G. of the columns, so that the horizontal shears S_1 vanishes:

$$\Sigma P_{1A} X_1 = \Sigma H_1 h_1$$

Substituting $P_{1A} = RX_1$, we get

$$R = \frac{\Sigma H_1 h_1}{\Sigma X_1^2}$$

Hence

$$P_{1A} = \frac{\Sigma H_1 h_1}{\Sigma X_1^2} X_1$$

$$P_{1B} = \frac{\Sigma H_1 h_1}{\Sigma X_1^2} X_2 = \frac{P_{1A} X_2}{X_1} \tag{9.3}$$

The value of the vertical axial forces P in all the columns at the level can thus be obtained. Proceeding on similar lines the forces in columns at other levels can also be obtained. Then the horizontal shears in the columns can be calculated by isolating parts of the determinate system proceeding from left to right by using simple statics as shown in Figs. 9.3 and 9.2. Any other procedure using simple statics can also be used for the analysis.

9.6 COMPARISON OF RESULTS OF ANALYSIS

It should be noted that the three methods of analysis (the exact method, the portal method and the cantilever method) yield similar results but different values. This may appear unsatisfactory. But as the principles of statics are satisfied in each method, the values obtained by any of the methods can be considered as acceptable values if one remembers that before a structure can fail a certain amount of redistribution of moments will take place in the structure. For this reason, further redistribution of moments is not admissible for the value of the moments obtained by the approximate methods of analysis such as portal or cantilever method. Redistribution of moment can be allowed only for the solution obtained by exact elastic analysis, where in addition to principles of statics compatibility conditions are also satisfied. The shears and moments are shown in Fig. 9.3(b).

The conclusions arrived at by a committee of the Structural Division of the American Society of Civil Engineers in 1940 were as follows [4].

1. The portal method is generally satisfactory for analysis of building with moderate height to width ratios.
2. The cantilever method is satisfactory for high narrow buildings.
3. In high frames, the change in length of the columns must also be considered in the analysis.

With the availability of modern computers, these methods nowadays serve also as a check on the results of computer analysis.

9.7 ANALYSIS OF RIGID FRAMES WITH TRANSFER GIRDERS

Building frames with transfer girders can be analysed by first solving the top part by an appropriate method and transferring the vertical and horizontal loads to the transfer girders which may then be assumed to transfer the load to the lower storeys through pin-jointed connections. The lower part is then again analysed by the same method.

9.8 DRIFT LIMITATION IN VERY TALL BUILDINGS

Horizontal loads produce double curvature of beams and columns by bending in unbraced frames and there is a horizontal movement of the frames. Similarly, axial tension in columns on one side of the centre of gravity of the columns with axial compression on the other side will also cause deformation and horizontal movement. The deflection at the top of the tall buildings due to these effects of horizontal forces is called the *drift*. A wind drift limit of 1/500 of the height of the building for each storey and also the total height is usually specified for reinforced concrete buildings [IS 456 (2000) Clause 20.5]. This criterion applies to tall buildings. The procedure for calculation of total drift is to determine the horizontal deflection for each storey starting from the ground level, due to double curvature of the

9.9 CLASSIFICATION OF STRUCTURAL SYSTEM FOR TALL BUILDING

From the variety of the structural system used for multi-storeyed buildings, four basic vertical loading systems can be identified:

1. The flat-wall system (load bearing walls and shear walls)
2. The framed system (special frames skeleton)
3. The core trunk system (by special cores in the building)
4. The tube or envelope system (load bearing structure along the exterior walls as tube)

By combining these four systems it is possible to have various types of structural systems for tall and very tall buildings. For moderately tall reinforced concrete buildings in low seismic areas, the most commonly used structural system is the rigid framed construction together with core walls and shear walls.

EXAMPLE 9.1 (Analysis of frames with lateral loads portal method)
A building frame is as shown in Fig. 9.5 in cross-section. It is subjected to horizontal loads at the joints as shown. Determine the bending and shear force diagram for the various beams and columns of the frame by the portal method.

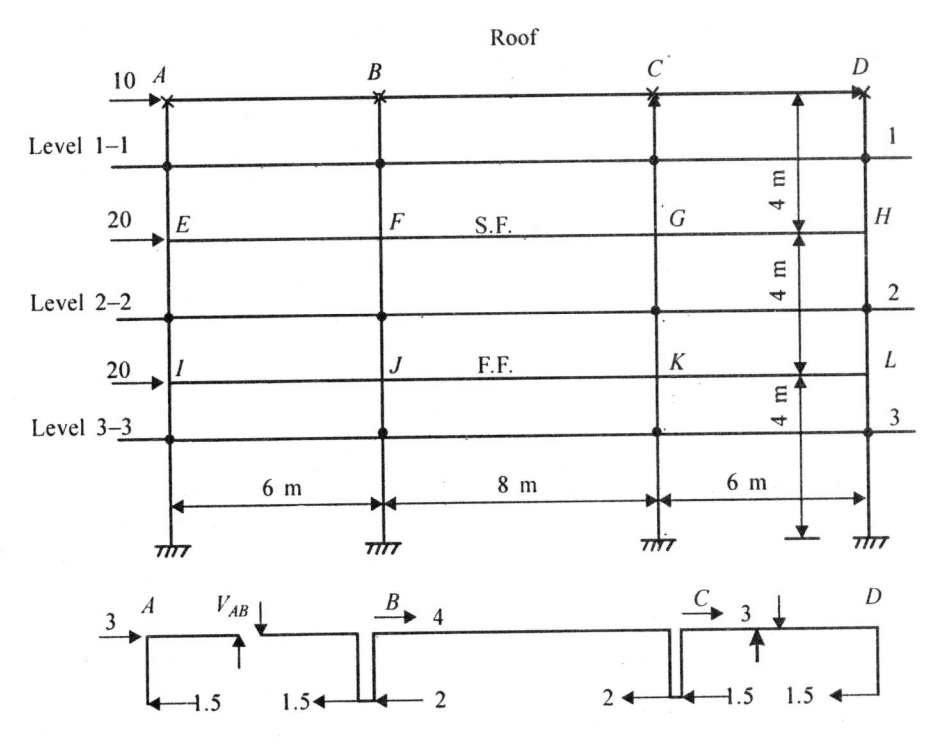

Fig. 9.5 Example 9.1 (contd.)

154 ADVANCED REINFORCED CONCRETE DESIGN

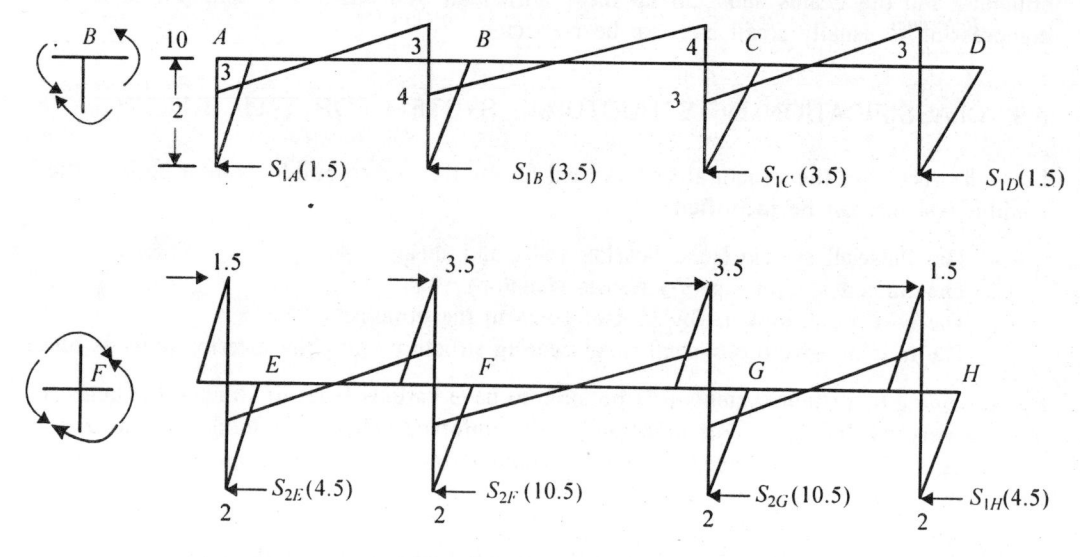

Fig. 9.5 Example 9.1.

Reference	Step	Calculations
	1	**Assumption** (a) Points of contraflexure of columns and beams are at their mid-points. Mark off levels 1-1, 2-2 through points of contraflexure at each level starting from the top. (b) Shear in columns is in proportion to the spans.
	2(a)	*Take level 1-1 and calculate the shear in each column* Assume each beam as forming part of a portal sub-frame and the horizontal shear in each column is proportional to the span of the portal. Each internal column will form part of two portals so that the shear will be the sum of the one-half shear from the adjacent portals.
Eq. (9.2)		$$S_{1A} = \frac{10(3+0)}{20} = 1.5 \text{ kN} = S_{1D}$$ $$S_{1B} = \frac{10(3+4)}{20} = 3.5 \text{ kN} = S_{1C}.$$
	2(b)	*Analyse joint-by-joint for moments* Starting from left to right analyse the beam and column moments. Joint A,

ANALYSIS OF FRAMES UNDER HORIZONTAL LOADS   155

Reference	Step	Calculations
Fig. 9.5		$M_{1A} = V \times$ (Storey height/2) $= 1.5 \times 2 = 3.0$ kNm (clockwise)
		As $\Sigma M = 0$, $M_{AB} = -3$ kNm (anticlockwise)
		Joint B,
		$M_{1B} = 3.5 \times 2 = 7.0$ kNm
		$\Sigma M = 0$, $M_{BA} = -3$ KnM
		Hence, $M_{BC} = 7 - 3 = -4$ kNm
		Joint C,
		$M_{1C} = 3.5 \times 2 = 7.0$ kNm
		$M_{CB} = -4$ and $\Sigma M = 0$
		Hence, $M_{CD} = -3$ kNm
		Joint D,
		$M_{1D} = +3$ and $M_{DC} = -3$ kNm
	2(c)	Calculate the beam shears from the moments
		$V_{AB} = \dfrac{-3}{3} = -1$ kN; $V_{BC} = \dfrac{-4}{4} = -1$ kN
		(*Note:* The beam shears at the same storey is the same as there are no vertical loads.)
	3.	*Consider level 2-2 along points of inflection and repeat the operation*
		Total horizontal force above level 2-2 = 30 kN
		Column shears are
		$S_{2E} = \dfrac{30(3+0)}{20} = 4.5$ kN $= S_{2H}$
		$S_{2F} = \dfrac{30(3+4)}{20} = 10.5$ kN $= S_{2G}$
	(a)	Analyse joint-by-joint for moment
		Joint E
		$M_{2E} = 4.5 \times 2 = 9$ kNm; $M_{1E} = 1.5 \times 2 = 3$ kNm
		$\Sigma M = 0$, hence $M_{EF} = -12$ kNm
	(b)	Joint F
		$M_{2F} = 10.5 \times 2 = +21$; $M_{1F} = +7.0$,
		$M_{FE} = -12$ as $\Sigma M = 0$, $M_{FG} = -16$

156   ADVANCED REINFORCED CONCRETE DESIGN

Reference	Step	Calculations
	(c)	*Calculate the shear in beams*
		The shear in beams is the same at each floor.
		$$\text{Shear} = \frac{M_{EF}}{\text{One-half length of } EF} = \frac{-12}{-3} = 4 \text{ kN}$$
	4	*Continue the operation for the next levels*
	5	*Determine the column loads*
		(1) Beam shears are same at all the levels. Hence loads in the interior columns are = 0
		(2) The outer column will have loads so that the moments due to column loads = External moment at each level.

EXAMPLE 9.2 (Cantilever method)
Analyse the frame by cantilever method.

Reference	Step	Calculations
	1	*Assumptions made*
		(a) Points of contraflexure of beams and columns are at their mid-points
		(b) Axial loads are in proportion to their distances from their centroid. On one side of C.G. it is upwards and on the other side downwards.
	2	*Find the C.G. of columns*
		In this example with symmetry, the C.G. passes through the centre line.
	3	*Taking level 1-1, calculate the axial loads in the columns*
		The axial loads are calculated by assuming that the moment of the lateral forces is in equilibrium with the column load moments, both taken about the C.G. of the columns. Hence the load in the column:
Eq. (9.3)		$$P_{1A} = \frac{x_1 \Sigma H_1 h_1}{\Sigma x_1^2}$$
		where x_1 = Distance of column: A from the C.G. of the columns, then
		$$P_{1A} = \frac{(10)(10 \times 2)}{2(10)^2 + 2(4)^2} = 0.86 t$$
		Hence $P_{1B} = \dfrac{0.86 \times 4}{10} = 0.35 \, t$

Reference	Step	Calculations
Ex. 9.1	4	*Analyse the determinate system*
		Shear in the column at its point of inflection (S_{1A}) is found by taking moments at mid-span of beam
Fig. 9.5		$2S_{1A} = 0.86 \times 3$
		$S_{1A} = 1.29$ kN (against 1.5 in Ex. 9.1)
		Similarly we can analyse S_{1A}, S_{1C} and S_{1D} and C;
Ex. 9.1	5	*Analyse the frame by simple statics*

REFERENCES

1. Smith B.S. and Coull, A., *Tall Building Structures: Analysis and Design*, Wiley, New York, 1991.

2. Kazim S.M.A. and Chandra R. *Analysis of Shear Walled Buildings*, Tor-Steel Research Foundation in India, 1970.

3. Fintel M. and Ghosh S. *Advanced Course of Design of High Rise Concrete Buildings for Wind and Earthquake Forces*, Engineering Staff College of India, Hyderabad, 1985.

4. Benjamin J.R. *Statically Indeterminate Structures: Approximate Analysis by Deflected Structures and Lateral Load Analysis*, Mc-Graw Hill, New York, 1959.

CHAPTER 10

Preliminary Design of Flat Slabs

10.1 INTRODUCTION

A two-way slab can be constructed in one of the following ways [Fig. 10.1]:

1. As solid slabs or ribbed slabs resting on walls or on beams and columns
2. As flat plates resting directly on straight columns
3. As flat slabs with or without drops supported on columns with enlargements. (column heads)

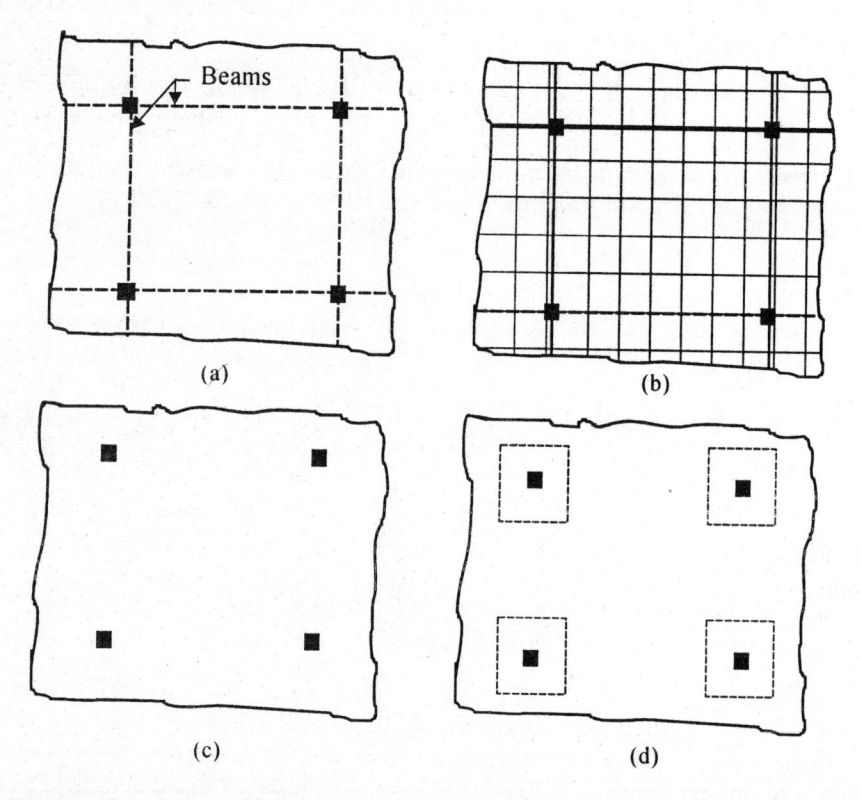

Fig. 10.1 Types of two-way slabs: (a) two-way slab on rigid or flexible beams (b) waffle slabs (c) flat plate (d) flat slab.

The details of the supports are as shown in Fig. 10.2. According to IS 456 and BS 8110 [1,2] two-way slabs on rigid beams are designed by the traditional method of coefficients. This procedure was also followed by ACI code till 1963. However, tests have shown that such designs give large factors of safety against ultimate failure, as much as 2.7 to 3.4, as compared

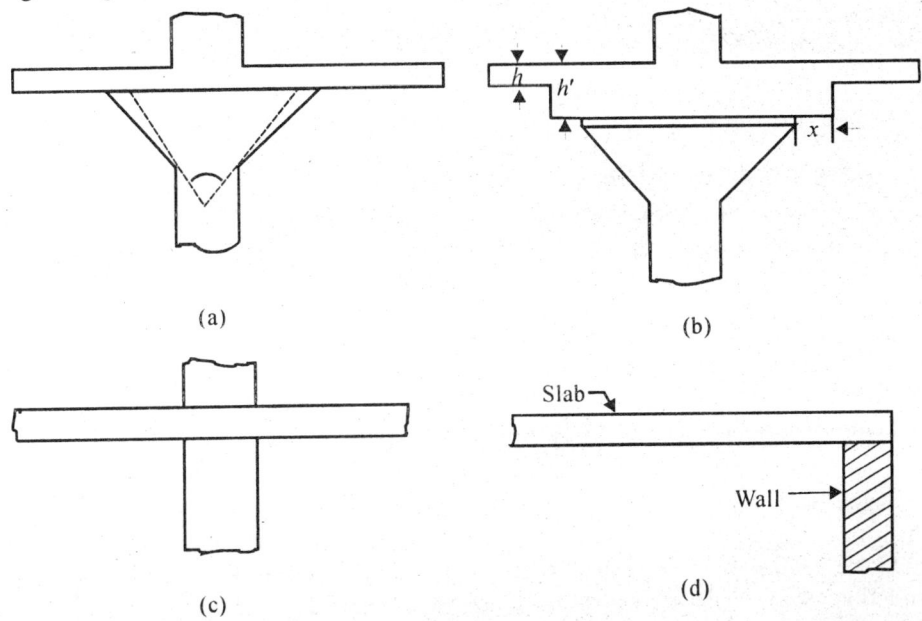

Fig. 10.2 Types of supports for flat slabs and flat plates: (a) flat slab with capital (b) flat slab with drops and column capital (c) flat plate (d) slab supported on wall.

to the method used for flat slabs in which it is of the order of 1.9 to 2.3. Hence, since 1963, ACI code adopted a common approach for design of all types of regular two-way slabs. The details of the method are described in Chapters 11 to 14. This chapter will mainly deal with the preliminary dimensions of the flat slabs and examine in detail the minimum thickness prescribed by ACI 318 for two-way slabs.

10.2 ADVANTAGES AND DISADVANTAGES OF FLAT SLABS

The main advantages of flat slabs and flat plates are the following:

1. Reduction in the total height required for each storey thus increasing the number of floors that can be built in a specified height.
2. Saving in materials of construction as illustrated in Table 10.1.
3. More uniform access to daylight and easier accommodation of the various ducts in the building.

The main disadvantage of flat slabs and flat plates is their lack of resistance to lateral loads, such as those due to high winds and earthquakes. Hence, special features like shear walls must be always provided if they are to be used in high-rise constructions or in earthquake regions.

TABLE 10.1 COMPARISON OF APPROXIMATE QUANTITIES PER SQ. METRE FOR TWO WAY SLABS DESIGNED BY VARIOUS METHODS
(6 m × 6 m Inner bay. Live load 10 kN/m²)

Case	Type of floor	Concrete (m²)	Form work (m²)	Steel (kg)
1.	(a) Flat slab ACI method	0.203	1.04	13.8
	(b) Flat slab by BS method	0.203	1.04	15.7
2.	(a) Flat plate with edge beam	0.213	1.02	15.6
	(b) Flat plate by BS method	0.213	1.02	17.0
3.	Two-way slab on main and secondary beams	0.185	1.63	23.6
4.	One-way slab on edge grinders only	0.204	1.29	31.0

10.3 HISTORICAL DEVELOPMENT

As in many other types of Civil Engineering Structures, construction of flat slabs preceded its theory of analysis and design. C.A.P. turner constructed flat slabs in U.S.A. as early as in 1906 mainly using intuitive and conceptual ideas. This was the start of these types of construction. Many slabs were load-tested between 1910–20 in U.S.A. It was only in 1914 that Nicholas proposed a method of analysis of these slabs based on simple statics. This method is used even today for the design of flat slabs and flat plates and is known as the direct design method and is explained in Chapter 11.

10.4 ACTION OF FLAT SLABS AND FLAT PLATES

The differences in the actions of flat slab and flat plates from that of conventional two-way slabs is shown in Fig. 10.3. In flat slabs and flat plates, the loads on the slab are directly

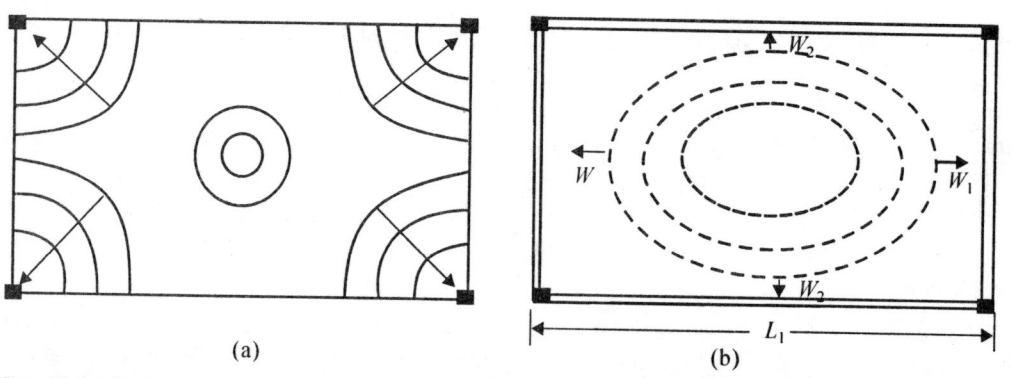

Fig. 10.3 Deflection contours of two-way slabs: (a) flat slabs and flat plates (b) two-way slab on beams.

transferred by plate action to the columns, whereas in conventional slabs with beams, the loads are first transferred from slabs to the beams and then from beams to the columns. A good appreciation of this load-transfer mechanisms (or load path) is necessary for a clear understanding of the analysis of these structures.

10.5 PRELIMINARY DESIGN OF FLAT SLAB

The following proportions are commonly used for the preliminary layout of the various parts of a flat slab [1]:

Columns. The *column* of a flat slab consists of the main column and the column head, which is the enlarged portion just below the slab. The size of the main part of the column of two-way slabs are usually made of about 1/16th the length of the larger span of the slab and of 1/8th to 1/9th the storey height of the building. In flat slabs, the *column head* which is also called *column capital* is usually made of 1/5th but not more than 1/4th the shorter span. Its height should not be less than 15 cm. It should be noted that for design calculations, only the portion of the column head which lies within the largest circular cone or pyramid of vertical angle 90° is considered as the active head of the column, (Fig. 10.2). It is preferable to have this cone lie inside the column head. The term 'equivalent square column' of a non-rectangular column is defined as a square column of the same area as the real column. (This term should not be confused with the equivalent column described in Section 13.3.)

Drop panel. In flat slab construction, the slab is sometimes thickened around the column. The standard practice is to thicken it to at least 1.25 times but it should not be more than 1.5 times the thickness of the main slab. This over-thickness reduces the deflection and assists in the resistance against punching shear. Such thickening is usually made rectangular in plan with its length measured from the centre line of the column to not less than 1/6th the smaller span of the surrounding panels. The total width of the enlargement from both sides of the centre line of the column is called the *drop width*. At the exterior column, this enlargement is only on one side of the column and will extend to the end of the slab. Drops with widths less than one-third the smaller span in internal columns are to be ignored in the design calculations.

There is also another requirement that when calculating the area of reinforcement for this region the thickness of the drop panel measured below the main slab should not be taken as greater than one-fourth the projection of the drop beyond the edge of the column capital. Referring to Fig. 10.2 the condition can be stated as $(h' - h) \not> x/4$.

Slab. The main horizontal member is the *slab* and the various rules have been specified to fix its minimum depth so that its deflection will lie within the required limits. These are dealt separately with in Section 10.7. The area of the slab enclosed by the centre lines of the columns at the four corners is called a *panel*.

10.6 BASIC ACTION OF TWO-WAY SLABS

Let us take the case of a conventional two-way slab on beams and columns with L_1 as the longer and L_2 as the shorter spans of the slab. Assume it is subjected to a uniformly distributed load w per m² which is divided into two loads w_1 (in the longer E-W direction of the slab) and w_2 (in the shorter N-S direction) so that $w = w_1 + w_2$. This is illustrated in Fig. 10.4. Let us first calculate the maximum bending moments in the E-W direction which will be composed of the B.M. in the slab and those in the beams spanning in that direction.

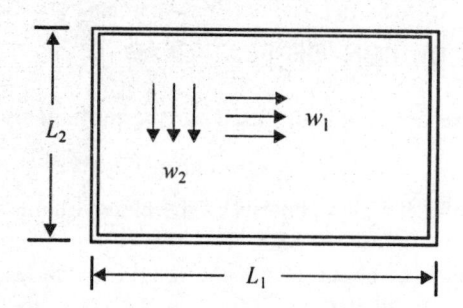

Fig. 10.4 Load transfer in two-way slab with beams.

$$\text{Total B.M. in the slab of width } L_2 \text{ and span } L_1 = \frac{w_1 L_1^2 L_2}{8}$$

$$\text{Total B.M. in the two-side beams of span } L_1 = \frac{w_2 L_2 L_1^2}{8}$$

Hence the total bending moment M_{o1} from the slab and the beams together in the E-W direction:

$$M_{o1} = \frac{1}{8}(w_1 + w_2)L_2 L_1^2 = \frac{1}{8}W(L_2 L_1) L_1 = \frac{1}{8}W L_1$$

where W is the total weight on the slab.

Similarly if we denote the total maximum bending moment M_{o2} in the N-S direction:

$$M_{o2} = \frac{1}{8}W L_2$$

The above relations demonstrate the important principle that even in the case of conventional design of two-way slabs we actually design it by taking the whole load both in the longitudinal X and the transverse Y directions. This forms the intuitive approach to the general method of designs of all two-way slabs including flat slabs and flat plates. After determining the above total static (bending) moment it remains to assign it in different proportions to the various parts of the slab depending on their respective rigidities. This method has been further refined from results of extensive tests especially in U.S.A. and the present clauses for design of such slabs in ACI 318 (89) have been evolved. The details of the method are explained in Chapters 11 to 14.

10.7 DETERMINATION OF MINIMUM THICKNESS OF SLAB

Reinforced concrete slabs should have sufficient thickness so that its long-term deflection under loads will be within the specified limits. Empirical methods are used in routine designs and in special cases we can use the principles given in Chapter 1 to estimate the deflections. The empirical method are usually based on span/depth ratios. In addition, ACI gives three empirical formulae for depth/span ratios which can be used for all cases of two-way slabs. These are dealt with in this Section.

10.7.1 Empirical Method for Minimum Depth of Slab

The minimum thickness recommended by ACI for different situation and when using Fe 415 steel as reinforcements are given in Table 10.2 [3]. It should be noted that it is preferable to have an *edge* beam (also called *marginal* beam or *spandrel* beam) along the outer edges of flat slabs to reduce deflections of the exterior slabs. This also helps in transferring some of the column moments as torsional moments to the beam. The depth of these edge beams should be at least 1.5 times the depth of the slab and α should not be less than 0.8 (see Section 10.7.2 for the definition of α). A second method to reduce the deflection in the exterior slabs is to cantilever the slab edges so that the edge panels will behave like interior panels. A third method to satisfy deflection requirements of exterior panel is to increase its depth of this portion of the slab to about 1.10 times the thickness of interior panels.

BS 8110 recommends the use of basic span effective depth ratio as given in IS 456 Clause 23.2.1 for deflection control of flat slabs and flat plates also. According to IS 456 Clause 31.2.1, the larger of the spans between the drops should be used as the controlling span for flat slabs. For slabs without drops an additional modification factor of 0.9 is also recommended. The modification factor for tensile reinforcement is based on the average value of M/bd^2 of the column and the middle strip (see Chapter 11 for definition of column and middle strips).

TABLE 10.2 MINIMUM THICKNESS OF TWO-WAY SLABS FOR DEFLECTION REQUIREMENTS (Fe 415 steel)

Case No.	Type of slab	Overall minimum thickness h	
		Based on L_n	h (mm)
1.	Flat plate without edge beam	$L_n/30$	125
2.	Flat plate with edge beam	$L_n/33$	100
3.	Flat slab (enlarged column) without edge beams	$L_n/33$	100
4.	Flat slab with edge beams	$L_n/36$	100
5.	Two-way slabs on rigid beams	$L_n/48$	90

Notes: 1. For the depth of flat slabs, provided with drop panels of dimensions, not less than those specified in Section 10.5, we can reduce the value got by 10%.
2. L_n is the clear span in the long direction.
3. Thickness also depends on required fire resistance.
4. The value of α for edge beams shall not be less than 0.8 (see Section 10.7.2 item 2).

The minimum thickness allowed by IS 456 (2000) Clause 31.2.1 for flat slab is 125 mm.

10.7.2 ACI Formulae for Span–Depth Ratio

For deflection control, the larger of the two spans L_1 and L_2 is designated as L_n. Taking h as the total depth of the slab three equations for h/L_n have been given in ACI code [3]. For a flat slab or flat plate, the larger value of the centre-to-centre distance between columns minus the side length of the column (or the equivalent square column in case of circular and non-rectangular columns) is taken as L_n. For a two-way slab on beams the larger span between the beam edges is taken as L_n. The parameters used are the following:

1. Aspect ratio = Ratio of larger effective span to the shorter effective span. That is,

$$\beta = \frac{L_n}{S_n}$$

2. Stiffness ratio of beams to slab = $\alpha(L_2/L_1)$, where α is the ratio of the cross-sectional flexural stiffness (EI) of the beam (along with the prescribed slab attached to it as shown in Fig. 10.5) to the flexural stiffness of the slab between the centre of the adjacent panels in the

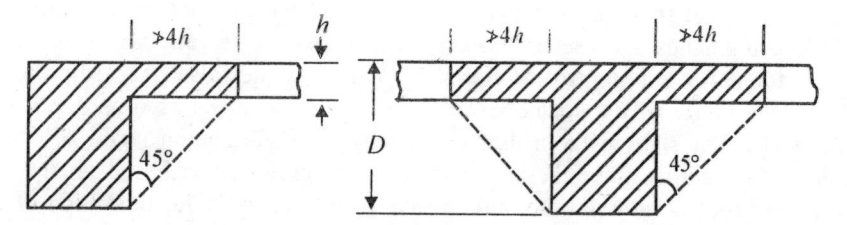

Fig. 10.5 Portion of slab acting with beams in two-way slabs with beams.

transverse cross-section. Thus we have:

$$\alpha = \frac{E_b I_b}{E_s I_s} = \frac{I_b}{I_s}$$

where

$$I_s = \frac{L_2 h^3}{12}$$

Then the stiffness ratio of the beam to the slab will be

$$\frac{I_b/L_1}{I_s/L_2} = \frac{\alpha L_2}{L_1}$$

$\alpha_1(L_2/L_1) = 1$, as the criterion for a beam to be considered as rigid.

3. Yield strength of steel = f_y (in N/mm^2)

Equations (10.1)–(10.3) (ACI 318 clause 9.5.3) given below refer to slabs whose discontinuous edges have a beam with α not less than 0.8 attached to it. Otherwise the thickness of slab as obtained from the equations should be increased by 10%. Similarly for flat slabs with drops not less than the minimum specified, the thickness obtained from the equations can be reduced by 10%.

$$\frac{h}{L_n} > \frac{(0.8 + f_y/1400)}{36 + 5\beta[\alpha_m - 0.12(1 + 1/\beta)]} \qquad (10.1)$$

where a_m is the average value of α on all the edges and L_n the clear span in the long direction.

$$\frac{h}{L_n} > \frac{0.8 + (f_y/1400)}{36 + 9\beta} \qquad (10.2)$$

$$\frac{h}{L_n} \ngtr \frac{0.8 + (f_y/1400)}{36} \qquad (10.3)$$

Substituting for $f_y = 415$ Eq. (10.3) reduces to $h/L_n \ngtr 0.03$. So we can see that h/L_n can be reduced by using steel of lower values of f_y but it is not an economic method to reduce deflections of slabs.

In addition to the above requirements ACI requires that the minimum thickness should be 155 mm (5 inches) when α_m is less than 2.0 and 88 mm (3.5 inches) when α_m is greater than 2.0.

10.8 MOMENT OF INERTIA OF FLANGED BEAMS

The exact procedure to find I values of flanged beams to determine α is given in Appendix A. However for most beams used in practical construction and for preliminary calculations, I values of T beams may be taken as 1.5 times and that of L beams 1.25 times the moment of inertia of the rectangular part of the respective beams.

EXAMPLE 10.1 (Preliminary design of flat slab)
Estimate the dimensions of a flat-slab system (with drops) for a four-storey building with 5 spans of 7.5 m in the longer direction, 5 spans of 6 m in the shorter directions and a storey height of 3 m.

Reference	Step	Calculations
	1	*Estimate the size of columns*
Text Sec. 10.5		Internal column = $\dfrac{L_1}{16}$ or $\dfrac{H}{8}$
		$\dfrac{L_1}{16} = \dfrac{7500}{16} = 469$; $\dfrac{H}{8} = \dfrac{3000}{8} = 375$
		Adopt 400-mm diameter columns.
	2	*Estimate the size of column capital*
		$D = \dfrac{L_2}{5} = \dfrac{6000}{5} = 1200$ mm; adopt 1500-mm diameter.
		Size of equivalent square column,

Reference	Step	Calculations
	3	$b = D\sqrt{\dfrac{\pi}{4}} = 1500 \times 0.886 = 1329$ mm $\not> \dfrac{6000}{4}$ *Estimate the length of drop* $\dfrac{1}{3}$ long span $= \dfrac{7500}{3} = 2500$ mm $\dfrac{1}{3}$ short span $= \dfrac{6000}{3} = 2000$ mm Adopt 2.5 × 2.0 m (this should be checked by step 5).
Text Sec. 10.7 and steps 1 and 2 IS 456 Clause 31.2.1	4	*Estimate the minimum slab thickness* $L_n = L_1 - b$ $h = \dfrac{L_n}{36} = \dfrac{7500 - 1330}{36} = 171$ mm Assuming drops, reduce h by 10%. Thickness = 171 − 17 = 154 mm Minimum thickness by IS code = 125 mm Adopt a thickness of 155 mm.
Sec. 10.5	5	*Check for other conditions as in Example 10.2. Estimate the thickness of drop H and check for limiting conditions.* $H = 1.25$ to $1.5h$ $1.25 \times 155 = 194$ and $1.5h = 233$ mm Adopt 225 mm as drop thickness. $H - h = 225 - 155 = 70$ mm Allowable maximum projection from edge of column capital to edge of drop $= 4(H - h) = 4 \times 70 = 280$ mm Allowable maximum size of drop = (column capital) + (2 × 280) $= 1500 + 560 = 2060$ mm adopt 2000 mm.
Sec. 10.5	6	*Estimate the size of exterior column to be used* Adopt a square column to match with the interior column and to accommodate the edge beam = 350 × 600 mm (say) (Edged beam height < $1.5h$) Adopt 400 mm square column.

PRELIMINARY DESIGN OF FLAT SLABS   167

Reference	Step	Calculations
	7	*Preliminary dimension adopted*
		Internal column diameter = 400 mm
		Size of capital = 1500 mm
		Size of drop panel = 2.0 mm
		Thickness of slab = 155 mm
		Thickness of drop = 220 mm
		Size of external column = 400 mm square
		Size of edge beam = 350 × 600 mm

EXAMPLE 10.2 (Minimum thickness of a flat plate)
Determine the minimum thickness of a flat plate having edge beams with 7.5 × 6 m panels on 500 mm square columns. (Fig. 10.6). Assume 415 grade steel.

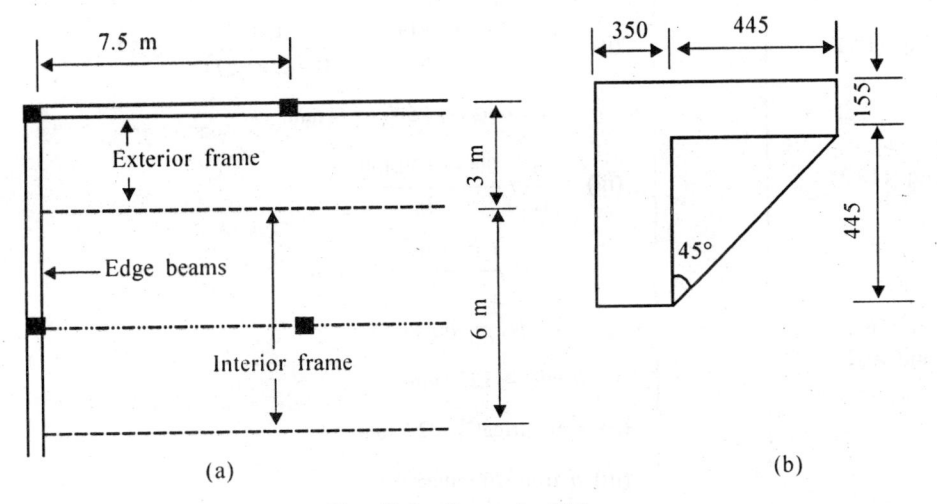

Fig. 10.6 Example 10.2.

Reference	Step	Calculations
Table 10.2	1	*Estimate by empirical method*
		$h = \dfrac{L_n}{33} = \dfrac{7500}{33} = 227$ mm; Minimum thickness = 125 mm
Sec. 10.7.1	2	*Determine the parameters affecting deflection*
		Assume interior spans

Reference	Step	Calculations
		(i) β = Ratio of clear spans = L_n/S_n $$= \frac{7500-500}{6000-500} = 1.27$$ (ii) $\alpha = 0$ (interior spans and no beams) (iii) $f_y/1400 = 415/1400 = 0.3$
	3	Estimate thickness by the three formulae
Eq. (10.1)		(i) $\dfrac{h}{L_n} > \dfrac{0.8+(f_y/1400)}{36-[0.6\beta(1+1/\beta)]}$ (with $\alpha = 0$) $$= \frac{0.8+0.3}{36-\left[(0.61 \times 1.27)\left(1+\dfrac{1}{27}\right)\right]} = 0.032$$ $h = 0.032 \times 7000 = 224$ mm
Eq. (10.2)		(ii) $\dfrac{h}{L_n} > \dfrac{0.8+(f_y/1400)}{36+9\beta} = \dfrac{1.1}{36+(9 \times 1.27)} = 0.0232$ $h = 0.0232 \times 7500 = 174$ mm
Eq. (10.3)		(iii) $\dfrac{h}{L_n} \not> \dfrac{0.8+(f_y/1400)}{36}$ $h \not> \dfrac{1.1 \times 7000}{36} = 214$ mm
IS 456 and ACI	4	Estimate design thickness (i) h min = 125 mm (ii) h empirical = 227 mm (iii) h from formulae not less than 224 mm and 174 mm but not more than 214 mm. Adopt a thickness of 210 mm for the design.

EXAMPLE 10.3 (Calculation of α values)

The corner panel of a flat slab 155 mm thick is 7.5 × 6 m (in X and Y directions with 350 × 600 mm edge beam around the corner. Calculate the α values of the long and short beams of the exterior frames into which the slab is to be divided for analysis.

(*Note:* In exterior span the slab does not stop at the centre line of columns but extends to the outer side of the beams.)

PRELIMINARY DESIGN OF FLAT SLABS

Reference	Step	Calculations
Fig. 10.6 Appendix A	1	Calculate I_b of the edge beams α = Stiffness of beams with slab attached/stiffness of slab between centres of adjacent panels Length of slab acting with beam = Depth of beam below slab or $4h$ $D - h = 600 - 155 = 445$ mm $4h = 4 \times 155 = 620$ mm Moment of inertia of beam $= \dfrac{K bH^3}{12}$; calculate K $\dfrac{B}{b} = \dfrac{350 + 445}{350} = 2.27 = x$ $\dfrac{h}{H} = \dfrac{155}{600} = 0.26 = y$ $K = \dfrac{1 + (x-1)[4 - 6y + 4y^2 + (x-1)y^3]y}{1 + (x-1)y} = 1.45$ $I_b = \dfrac{(1.45 \times 350)(600)^3}{12} = 9.13 \times 10^9$ mm^4
	2	Determine dimension of exterior frames in X direction (see Chapter 11, Fig. 11.2 for definition and formulae) $L_1 = 7.5$ $L_2 = 6.0$ (for computing L_2/L_1 values for tables and formulae Actual L_2 = Distance from slab edge to panel centre line) For calculating loads for M_o and for I_s $L_2 = 3000 + 175 = 3175$ mm $I_s = \dfrac{3175(155)^3}{12} = 9.85 \times 10^8$ mm^4 $\alpha_1 = \dfrac{I_b}{I_s} = \dfrac{91.3 \times 10^8}{9.8 \times 10^8} = 9.3$
	3	Determine dimension of exterior frame in Y direction Width of slab $= \dfrac{7500}{2} + 150 = 3900$ mm $I_s = \dfrac{3900 \times 155^3}{12} = 12.1 \times 10^8$ $\alpha_2 = \dfrac{91.3 \times 10^8}{12.1 \times 10^8} = 7.54$

Reference	Step	Calculations
	4	*Find α values in the interior frames* In X direction = $L_1 = 7.5$ and $L_2 = 6.0$ m In Y direction = $L_1 = 6.0$ and $L_2 = 7.5$ m α values can similarly be calculated for exterior frames

REFERENCES

1. IS 456 (2000), Code of Practice for Plain and Reinforced Concrete, Bureaus of Indian Standards, New Delhi, 1978.

2. BS 8110 (1985), *Structural Use of Concrete* (Parts 1 and 2), British Standards Institution, London, 1985.

3. ACI 318 (1989), Building Code Requirements of Reinforced Concrete, American Concrete Institute, Detroit, Michigan, 1989.

4. Wang, C.K. and Salmon, C.G., *Reinforced Concrete Design,* Harper and Row, Cambridge, 1965.

CHAPTER 11

Design of Two-way Slabs by Direct Design Method

11.1 INTRODUCTION

According to design procedure recommended by ACI code all types of two-way slabs including those supported on beams can be designed by one of the two methods: direct design method (DDM) and equivalent frame method (EFM). In both the methods, the two-way slab system is converted into series of rigid frames (one set spanning in the E-W direction and the other in the N-S direction) by means of vertical cuts through the slabs at section mid-way between the column lines as shown in Fig. 11.1.

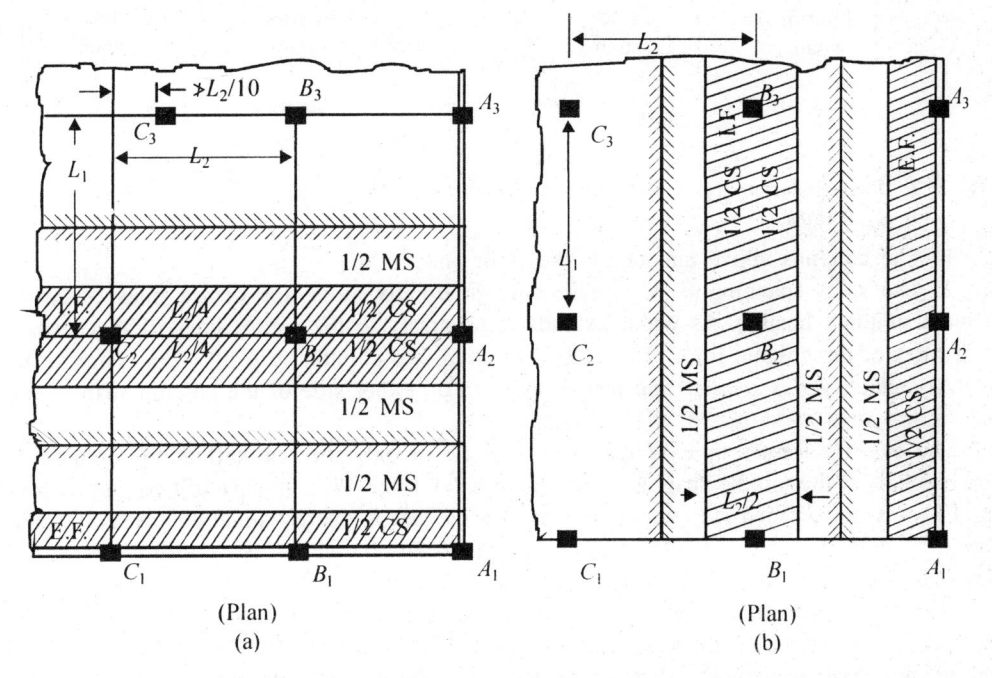

Fig. 11.1 Division of two-way slab into frames and strips: (a) frames in E-W direction—x frames, (b) frames in N-S direction—y frames. (C.S.—column strip; M.S.—middle strip; L_2—shortest span strip.

The resultant rigid frames in the E-W and N-S directions are analyzed separately for carrying all the loads in the span in each direction. As shown in Fig. 11.1 there will be 'interior frames' and 'exterior frames' to be analyzed separately. In each of these frames, we have 'interior spans' and 'exterior spans' to be dealt with. A typical interior frame in the E-W direction will consist of the following elements as shown in Fig. 11.2.

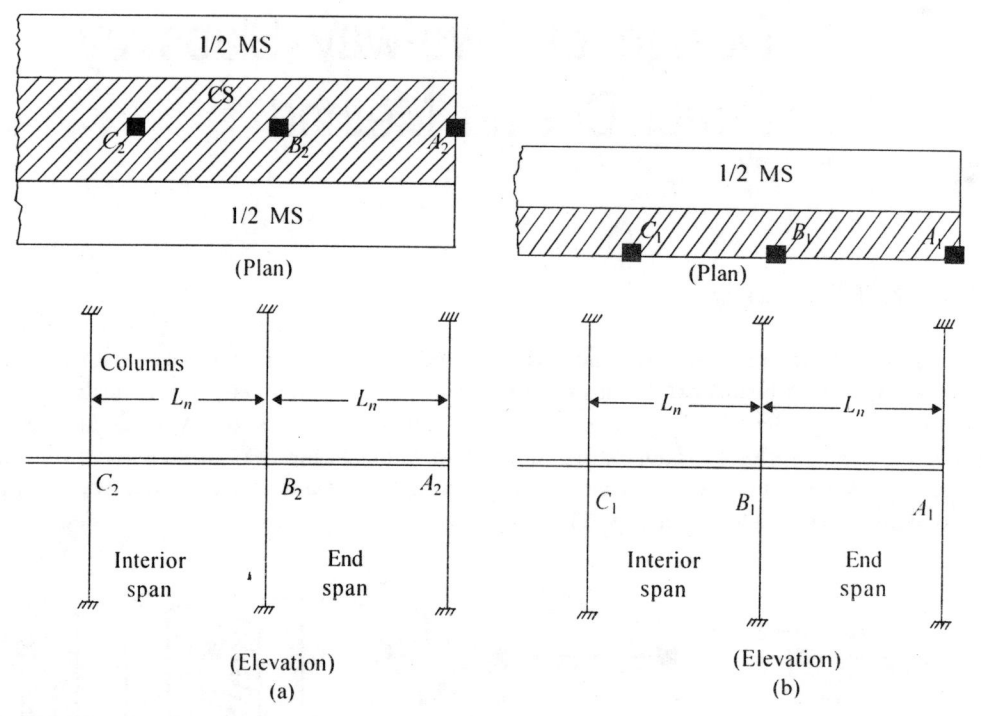

Fig. 11.2 Frames for analysis of flat slabs: (a) interior frame—x_1 frames (b) exterior frame—x_2 frames.

1. The columns above and below the floor and
2. The slabs spanning E-W direction longitudinally along the centre lines of columns (with or without beams). Its width extends transversely to the centre lines of the panels on both sides of the column lines. This slab is divided into column strip of *width equal to* $0.5L_1$ *or* $0.5L_2$ *whichever is smaller* and rest of the slab on either side of the column strip is called the *middle strip* as shown in the figure.

The exterior frames in E-W directions consist of only the one-half of the column strip and one-half of the middle strip as shown in Figs. 11.1 and 11.2. In case of two-way slab with rigid beams or stiffened ribs which acts as flexible beams there will also be beams running along the column centres. IS 456 however recommends the above type of analyses only for flat slabs and flat plates. The difference between the analysis by DDM and EFM is that in the 'direct design method' fixed coefficients are used for moments for the various parts of the slabs while in the 'equivalent frame method', the actual frames have to be analyzed by one of the classical methods of structural analysis, like moment distribution. Because of this procedure, DDM will only be applicable to more or less symmetrical layout of columns and slabs. These limitations are indicated in Section 11.2. Analysis by EFM can be used for any layout. It should also be clearly understood that the effect of lateral loads, unsymmetrical panels, etc., can be analyzed only by EFM. Similarly ACI 318 (89) commentary clause R.13.6.15 recommends that inverted raft foundations in which column loads are known and the soil reaction is assumed to be uniform, the analysis has to be carried out by the equivalent frame method. This chapter explains the design of two-way slabs by direct design method.

11.2 LIMITATIONS OF DIRECT DESIGN METHOD

Application of DDM is restricted to two-way slab that satisfies the following seven conditions:

1. There should be a minimum of three continuous spans in each direction.
2. The panels should be rectangular and the ratio of the longer to the shorter span, centre-to-centre of supports within the panel should not be greater than two.
3. The successive span lengths (centre-to-centre of supports) in each direction should not differ by more than one-third the longer span.
4. Columns should not be offset from either axes between the centre line of successive columns by more than 10% of the span in the direction of the offset (i.e. if the offset is in L_1 direction, maximum offset allowed is $L_1/10$).
5. Loads should be gravitational (vertical) only with the live load not more than 2 times dead load. They should also be uniformly distributed in the entire panel.
6. In two-way slabs with beams on all sides it should also satisfy the following additional condition. Denoting α_1 as the average of the two ratios of the flexural stiffness (EI) of the beams in the longer direction L_1 to that of the slab between centres of adjacent panels on each side of the beam, and α_1 as the above average ratio in the shorter directions L_2. The ratio of beam relative stiffness in the two directions is given by the expression $(\alpha_1/\alpha_2)(L_2/L_1)^2$ must lie between 0.2 and 5.0.
7. Redistribution of the resultant moments obtained by DDM is not permitted.

11.2.1 Explanation of the Limiting Conditions

The first condition ensures that the first interior support is neither fixed against rotation nor discontinuous. The second condition ensures two-way slab action. If the ratio exceeds 2 one-way slab action pre-dominates. In older codes, this ratio was 1.33 but it has now been increased to 2. The third and fourth conditions are related to the possibility of developing negative moments at or near mid-span. The fifth assumption ensures that there are no lateral loads and the limit on the LL to DL ratio of 2 obviates the necessity for checking for the effect of pattern loadings for symmetrical layouts. In the earlier codes this ratio was 3 and then it was necessary to check for pattern loads. For most practical dimensions of slabs and columns, the requirements for minimum α_c (see Section 11.7) is satisfied for LL/DL ratio up to 2. Hence effect of pattern loads can be neglected.

The sixth condition is applicable to slab with beams and it ensures that the beams are stiff enough relative to each other. Otherwise the distribution of moments will deviate very much from that assumed in the elastic analysis. From Section 11.8.1, it can be seen that the ratio specified is the relative stiffness of the beams in the two principal directions is shown below:

$$\frac{\alpha_1 L_2/L_1}{\alpha_2 L_1/L_2} = \frac{\alpha_1}{\alpha_2}\left(\frac{L_2}{L_1}\right)^2$$

This is specified to be between 1/5 and 5.

11.3 CALCULATION OF TOTAL STATIC MOMENTS FOR INTERIOR AND EXTERIOR FRAMES

The expression derived by Nicholas in 1914 for the total static moment M_0 for flat slabs is the basis of the calculation of the moment in the spans for the DDM. Nicholas derived his formula as in Fig. 11.3 that shows an isolated interior panel of a flat slab.

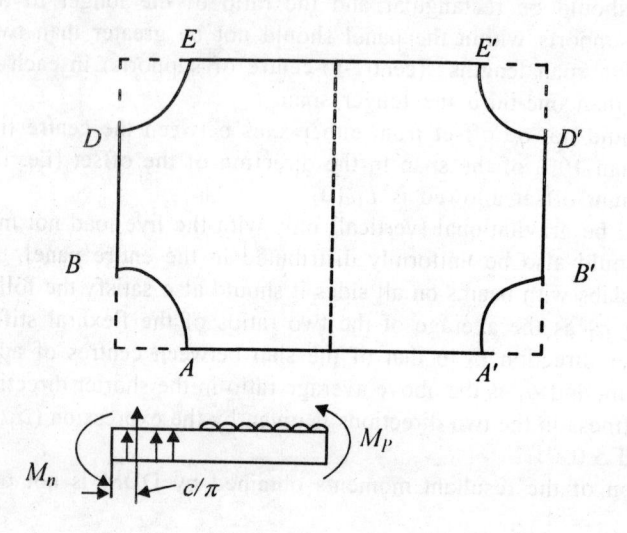

Fig. 11.3 Derivation of design static moment in flat plate by Nicholas.

The free body $ABDE\text{-}E'D'B'A'$ is acted on by a uniformly distributed gravity load w with end moments and shears. Because of continuity of spans, there is no shear along BD, EE', $D'B'$ and $A'A$ and the loads are carried by shears only along the column faces. The values of the total shear in the two corner support is given by the expression:

$$\text{shear } V = 2\left(\frac{wL_1L_2}{4} - \frac{w\pi c^2}{16}\right) = \frac{wL_1L_2}{2} - \frac{w\pi c^2}{8}$$

where c is the diameter of the column. Taking moments of the free body in the L_1 direction about the centre line of the columns along BD, we get the equilibrium equation

$$M_P + M_N - \left[\frac{wL_1L_2}{2}\frac{L_1}{4} - \frac{w\pi c^2}{8}\frac{2c}{3\pi}\right] + \left(\frac{wL_1L_2}{2} - \frac{w\pi c^2}{8}\right)\frac{c}{\pi} = 0$$

where

M_p = The positive moments at the centre line of the slab

M_N = The negative moment at the line of column support

$2c/3\pi$ = The centre of gravity of the circumference of a quadrant

c/π = The centre of gravity of a quadrant

In the above expression, the third term gives the moment due to the downward load and the last term the moment of support reaction. Simplifying the expression (taking proper signs into consideration) the total static moment M_0 is given by the formula.

$$M_0 = M_p + M_N$$

$$= \frac{1}{8} w L_2 L_1^2 \left(1 - \frac{4c}{\pi L_1} + \frac{c^3}{3 L_2 L_1^2}\right) = \frac{1}{8} w L_2 \left(L_1^2 - \frac{4cL_1}{3}\right)$$

$$= \frac{1}{8} w L_2 \left(L_1 - \frac{2c}{3}\right)^2 = \frac{w L_2 L_n^2}{8} \tag{11.1}$$

This formula can be looked upon as the bending moment of a simply-supported slab of effective span of $(L_1 - 2c/3)$ which is approximately equal to the clear span L_n. The corresponding moment in L_2 direction can also be written down in the similar way.

11.4 ACI 318 FORMULA FOR MOMENTS

In general analysis of two-way slabs nowadays we replace the above panel analysis by the frame analysis as described in Section 11.1. Vertical cuts though the midway between columns are made and the resultant frames are analyzed for moments. The expression for M_0 derived as above from simple statics was used in the early ACI codes for the values of moments to be resisted by flat slabs. However, since higher strengths, than that given by the above equation, were reported in actual tests the coefficient $1/8 = 0.125$ was at one time reduced to 0.09. Subsequent investigations by Westergard and Slater in 1921 have recommended the return to the theoretical value of 0.125. The present generally-accepted expression for design moment in L_1 direction is given in IS 456 (2000) Clause 31.4.2.2 as follows:

$$M_0 = \frac{w L_2 L_n^2}{8} = \frac{W L_n}{8} \tag{11.2}$$

where

M_0 = Total moment

w = Uniformly distributed load on slab

W = Design load on area $L_2 L_n$ (this is less than the load on $L_2 L_1$.)

L_n = Clear span extending from face to face of columns, capitals, brackets or walls but not less than $0.65 L_1$ (for the L_1 direction and similarly for L_2)

L_1 = Length of span in the direction of M_0

L_2 = Length of span transverse to L_1

Moment M_0 in the L_2 direction can also be found similarly. We should clearly note that L_n is the clear span extending from face to face of columns and W is the total load on the span $L_2 \times L_n$ only. (See also examples at the end of this chapter for a clearer understanding of the procedure.) The ACI moment will be lesser than the theoretical M_0, using L_1 the centre-to-centre distance as the span in the ratio $(L_n/L_1)^2$.

11.4.1 General Description of DDM

Each interior and exterior frame as shown in Fig. 11.1 into which the slab system has been divided in the E-W and N-S directions are analyzed separately for the value of M_0. Analysis of both frames are similar in this procedure except that the following conditions should be remembered when dealing with the exterior frames.

1. M_0 for exterior frames is calculated with the actual load on the panel of the slab and the clear span L_n. It should be noted that only one-half of the slab extends transversely on each side of the column line.

2. To calculate the moment of inertia of the slab I_s in the transverse direction for the exterior frame in α values, only one half of the span in the transverse direction is to be considered since I_s is calculated for the slab between the centres of adjacent panels (Fig. 11.4).

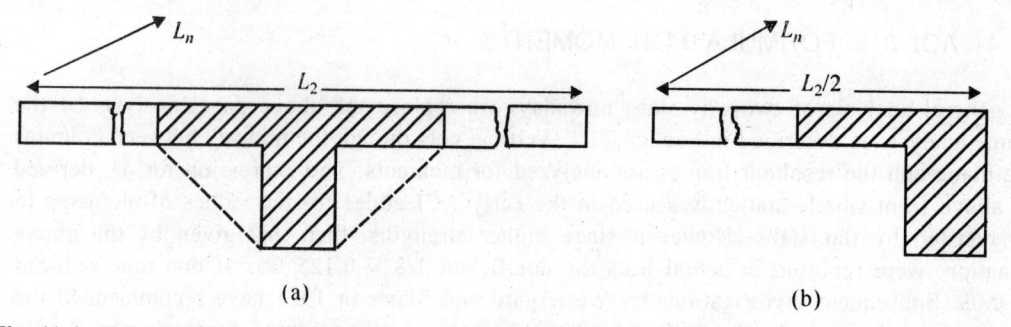

Fig. 11.4 Effective beam and slab sections for stiffness ratio: (a) interior frame (b) exterior frame.

3. To compute L_2/L_1 for transverse distribution to column and middle strips in the exterior frame, the full value of L_2 in the transverse direction is to be used in Eqs. (11.9)–(11.1) and in Table 11.1.

11.5 DISTRIBUTION OF MOMENTS M_0

The total static moment M_0 should be distributed firstly as positive and negative moments in the longitudinal direction and secondly these assigned positive and negative moments are again distributed transversely to the column and middle strips. The longitudinal distribution is dealt with in Section 11.6 and the transverse distribution in Section 11.8.

TABLE 11.1 TRANSVERSE DISTRIBUTION: ASSIGNMENT OF MOMENTS TO COLUMN STRIPS (PERCENTAGES) [Section 11.8]

Moment to be distributed	Type of beams present	$\dfrac{\alpha_1 L_2}{L_1}$	β_t	L_2/L_1 0.5	1.0	2.0
1. Positive moment in all spans [see Eq. (11.9)]	(a) No internal beam	0	Nil	60	60	60
	(b) With internal beam	≥ 1	Nil	90	75	45
2. Negative moment in interior spans [see Eq. (11.10)]	(a) No internal beam	0	Nil	75	75	75
	(b) With internal beam	≥ 1	Nil	90	75	45
3. Negative moment at exterior support [see Eq. (11.11)]	(a) No internal beam, No edge beam	0	0	100	100	100
	(b) No internal beam, With edge beam	0	≥ 2.5	75	75	75
	(c) With internal beam, No edge beam	≥ 1	0	100	100	100
	(d) With internal and edge beams	≥ 1	≥ 2.5	90	75	45

Note: For flat plates, $\alpha_1 L_2/L_1 = 0$.

11.6 LONGITUDINAL DISTRIBUTION OF M_0

The longitudinal distribution of the total moment M_0, into positive as well as negative moments, are made separately for the interior and exterior (or end) spans. This is shown in Table 11.2. In the interior spans, the distribution is simple as it corresponds to the 'fixed-end' beam cases. *Values of 0.65 at the supports and 0.35 at centre (negative and positive respectively) have been recommended.*

The distribution of the moment at the end or exterior spans depends on the type of exterior support and the ratio of exterior column to spandrel beams stiffness. The method used in the former ACI codes is described in Section 11.6.1. However, in the new code [from ACI 318 (1989) matters have been greatly simplified to make DDM a true Direct Design Method [1,2]. Fixed values as given in Table 11.2 have been prescribed for the various cases that can occur in practice and these coefficients can be directly read off for design. When using these values, we should be aware that *the interior negative and positive values are closer to the upper bound values and the exterior negative moment is close to the lower bound values.* This is not considered serious, as the exterior negative steel is governed by the minimum steel requirement in slabs for control of cracking. However, care should be taken not to use these values as a measure of moment to be transmitted to the external columns. The old method of modified stiffness as given in Section 11.6.2 is also allowed in ACI code

TABLE 11.2 MOMENT COEFFICIENTS FOR LONGITUDINAL DISTRIBUTION OF STATIC MOMENT IN TWO-WAY SLABS IN END-SPANS

Type of span	Description	Percentage distribution Interior / Exterior
Interior span	Interior span of all slabs	65 / 35 / 65
Exterior spans	Flat slab with edge beams (no internal beams)	70 / 50 / 30
Exterior spans	Flat slab without edge beams (no internal beams)	70 / 52 / 26
Exterior spans	Beams and slab (Beams between all columns)	70 / 57 / 16
Exterior spans	Flat slab on masonary wall (Exterior edge unrestrained)	75 / 63 / 0
Exterior spans	Flat slab on R.C. wall (exterior edge fully restrained)	65 / 35 / 65

as an alternate method. IS 456 uses the later method with modification to arrive at the distribution factors for the end spans.

[*Note:* As already stated in Section 10.7.1, ACI 318 (1989) Clause 9.5.3.5 recommends that all discontinuous edges should preferably be provided with an edge beam of stiffness ratio α not less than 0.80 to control deflection of end spans. It is good to follow this recommendation.]

The coefficients given in Table 11.2 for longitudinal distribution when using DDM are for the condition of the spans fully loaded with (DL + LL). The effect of different disposition of load (called *pattern loading*) can increase the positive moment and this has to be investigated as explained in Section 11.7. However, by limiting the LL/DL ratio to 2 (which is the usual case dealt with in practice in DDM) the necessity to check for the effect of pattern load has been dispensed with ACI 318 (1989).

11.6.1 Modified Stiffness Method for Longitudinal Distribution of M_0 in End-span (Alternate Method in ACI 318)

In earlier ACI codes, like ACI 318 (1977) and also in IS 456, the longitudinal distribution of M_0 into positive and negative moments in the end span were functions of the ratio of the stiffness of the equivalent column (this concept is explained in Chapter 13) above and below the slab to the combined flexural stiffness of the slab and beam (if any) at the junction taken in the direction of the span in which moments are considered. As it is used as an alternate method for design we shall examine it in more detail.

Let α_{ec} represents the ratio of stiffness of the equivalent column (as explained in Section 13.3.3) to the sum of the stiffness of the slab and the beam.

$$\alpha_{ec} = \frac{K_{ec}}{\Sigma(K_s + K_b)} \tag{11.3}$$

where

K_{ec} = Stiffness of equivalent column

K_s = Stiffness of the slabs = $4EI_s/L_1$, where $I_s = L_2 h^3/12$ (see Fig. 11.4)

K_b = Stiffness of beam if present (and = 0, in flat slabs) = $4EI_b/L_1$

Using this ratio the following moments as given in IS 456 Clause 30.4.3.3 are assigned for the positive and negative moments as fractions of M_0 for the end or exterior spans

$$\text{Interior negative moment} = M_0 \left[0.75 - \frac{0.10}{1 + (1/\alpha_{ce})} \right] \tag{11.4}$$

$$\text{Positive span moment} = M_0 \left[0.63 - \frac{0.28}{1 + (1/\alpha_{ce})} \right] \tag{11.5}$$

$$\text{Exterior negative moment} = \frac{0.65 M_0}{1 + (1/\alpha_{ce})} \tag{11.6}$$

In place of the modified stiffness method and the coefficient methods given by ACI, a third method is given for flat slabs in IS 456 Clause 31.4.3.3, [3]. Here the ratio stiffness α_{ec} has been replaced by α_c, where

$$\alpha_c = \frac{\Sigma K_c}{K_s + K_b} \qquad (11.6a)$$

where

ΣK_c = Sum of the flexural stiffness of column meeting at the joint

K_s = Flexural stiffness of slab expressed as moment per unit rotation

K_b = Flexural stiffness of beam (= 0, for flat slabs)

Thus in IS 456, the torsion member of the equivalent column is completely omitted. Of the three methods the coefficient method given in ACI 318 (1989) is the simplest and most popularly used in DDM. It has also been found to give satisfactory designs.

11.7 EFFECT OF PATTERN LOADING ON POSITIVE MOMENT (MINIMUM STIFFNESS OF COLUMNS)*

The positive moment in the continuous slab can be altered due to pattern loading. It can be shown that in symmetrical layouts obeying DDM limitations this can occur only when the LL/DL ratio is greater than 2. By limiting the DDM to cases where LL/DL is not more than 2, the necessity for checking for the effects of pattern loading has been eliminated in the DDM calculations. However a discussion of pattern loading is presented below for academic interest.

From the results of rigid frame analysis ('a study of its influence lines and maximum bending moment envelope' by Jersa and others) the following effects of pattern loading were noticed:

1. The larger the LL/DL ratio the larger will be the effect of pattern loading as pattern load effect is mainly related to the live load.

2. As the maximum positive moment in the span is more affected than the maximum negative moment by pattern loading, we need to concentrate our attention to the effects of pattern loading on the positive moment only.

3. The larger the ratio of the column stiffness to the slab (and beam) stiffness, the lesser will be effect of pattern loading on the positive moment. This becomes clear if we imagine that, with higher column stiffness the ends of the spans are closer to fixed conditions and loading from adjacent spans have less effect on the span moment.

From the study by Jersa, Sozen and Seiss in 1996, Table 11.3 was evolved. This table gives the values of the minimum column to beam stiffness that is required (with respect to the DL/LL ratio and the L_2/L_1 ratio) only below which the pattern loading will have any effect. Slabs which are identified as susceptible to the effects of pattern loading need to be checked for increased value of the positive span moment due to pattern loading. It should be noted that Table 11.3 is applicable to all types of slabs including two-way slabs on beams.

---

*This section may be omitted on first reading.

DESIGN OF TWO-WAY SLABS BY DIRECT DESIGN METHOD 181

TABLE 11.3 VALUE OF $A_{c(MIN)}$ FOR PATTERN LOADING
[Table 17 of IS 456 (2000)]

$\beta_a = \dfrac{DL}{LL}$	$\dfrac{LL}{DL}$	$\dfrac{L_2}{L_1}$	Value of $\alpha_{c(min)}$ for relative beam rigidity = α				
			0	0.5	1.0	2.0	4.0
2.0	0.5	0.50–2.0	0	0	0	0	0
		0.50	0.6	0	0	0	0
1.0	1.0	0.80	0.7	0	0	0	0
		1.00	0.7	0.1	0	0	0
		1.25	0.8	0.4	0	0	0
		2.00	1.2	0.5	0.2		
0.5	0.2	0.50	1.3	0.3	0	0	0
		0.80	1.5	0.5	0.2	0	0
		1.00	0.6	0.6	0.2		
		1.25	1.9	1.0	0.5	0	0
		2.00	4.9	1.6	0.8	0.3	0
0.33	3.0	0.50	1.8	0.5	0.1	0	0
		0.80	2.0	0.9	0.3	0	0
		1.00	2.3	0.9	0.4	0	0
		1.25	2.8	1.5	0.8	0.2	0
		2.00	13.0	2.6	1.2	0.5	0.3

[*Note:* For flat slabs and flat plates, $\alpha = 0$, for which IS values are given.]

The following are the definitions of the parameter listed in Table 11.3:

(i) $\alpha_c = \dfrac{\text{Sum of stiffness of upper and lower columns}}{\text{Sum of stiffness of beam and slab if any}} = \dfrac{K_{c1} + K_{c2}}{K_s + K_b}$

(ii) $\beta_a = \dfrac{DL}{LL}$ (unfactored)

(iii) $\alpha = \dfrac{I \text{ of beam with slab attached}}{I \text{ of slab between centres of adjacent panels}}$

(iv) $\dfrac{L_2}{L_1}$ = Aspect ratio

[*Note:* Usually the term stiffness or stiffness factor refer to EI/L values and term flexural rigidity refer to EI values.]

Table 11.3 gives the minimum value of α_c which will not produce an increase in positive moment to more than 30% of the moment obtained with uniform loading of live load. If α_c in a particular case is less than $\alpha_{c(min)}$ given in the table, we can either increase the column size or, alternately, the positive moment calculated should be multiplied by a coefficient $\delta_s > 1$ given by the following equation (see ACI 318 Eq. 13.5):

$$\delta_s = \left(1 + \frac{2 - \beta_a}{4 + \beta_a}\right)\left(1 - \frac{\alpha_c}{\alpha_{c(\min)}}\right) \tag{11.7}$$

The smallest possible value of β_a was 0.33, as in earlier ACI codes we were not allowed to use DDM with LL more than three times the DL. With this value, the value of δ_s theoretically can be as high as 1.39. Its usual value for buildings is only around 1.2. In the revised ACI code by specifying the direct design method restricted to cases where LL/DL ratio is not more than the more commonly occurring value of two, the effect of pattern loading can be omitted in the design calculations for most of the practical cases (*see* Section 11.2.1). Alternatively, we can use the above method to find $\alpha_{c(\min)}$, the minimum size of column for $\delta_s = 1$. By adopting such size of column there will be no effect due to pattern loading.

11.8 TRANSVERSE DISTRIBUTION OF MOMENTS AND FACTORS AFFECTING THE DISTRIBUTION

Having distributed the moment M_0 longitudinally as positive moment in the span and negative moments at the two ends, the next problem is the apportioning of these negative and positive moments to the column strips and middle strips of the respective sections. This is carried out using Table 11.1.

Taking the case of two-way slabs, in general, the interior column strips may have beams spanning between columns and the end column strips may have spandrel edge beams. The main factors that effect the transverse distribution are the following four quantities as given in Table 11.1:

1. The relation stiffness of the beam in the column strips to that of the slab $a_1 L_2/L_1$
2. The aspect ratio, i.e. ratio of width of slab to span of the slab L_2/L_1
3. The torsional resistance of the edge beam if the exterior span has a beam (β_t)
4. The type of wall in the case of the slab whose exterior end is supported on a wall (this is explained in Section 11.11).

Let us examine the first three factors in more detail.

1. The term $\alpha_1 L_2/L_1$ gives the relative stiffness of the beam in the column strip to that of the slab as shown below:

$$\frac{I_b/L_1}{I_s/L_2} = \frac{I_b L_2}{I_s L_1} = \alpha_1 \frac{L_2}{L_1}$$

When there is no beam in the column strip its value is zero. But if there is a beam in the column strip it will be more stiff and it will attract more of the moment to that section. We can also consider that if this factor is equal or greater than unity the beams can be considered as rigid. Hence $\alpha L_2/L_1 = 1$ is taken as a break point to define the nature of beams (rigid or flexible) in two-way slabs.

2. L_2/L_1 is aspect ratio. It is evident that larger the value of L_2 (width of slab) compared to length L_1 (span length) smaller will be the moment shared by the column strip.

3. β_t is torsional resistance of edge beam. If the exterior column has an edge beam (spandrel beam) running transversely to the span, then the negative moment at the end will

be distributed to the middle strip also whereas without such an edge beam the whole negative moment at this end will have to be taken by the column strip only. The magnitude of these effects depends on the value of β_t which can be defined as follows (*see* Fig. 11.5):

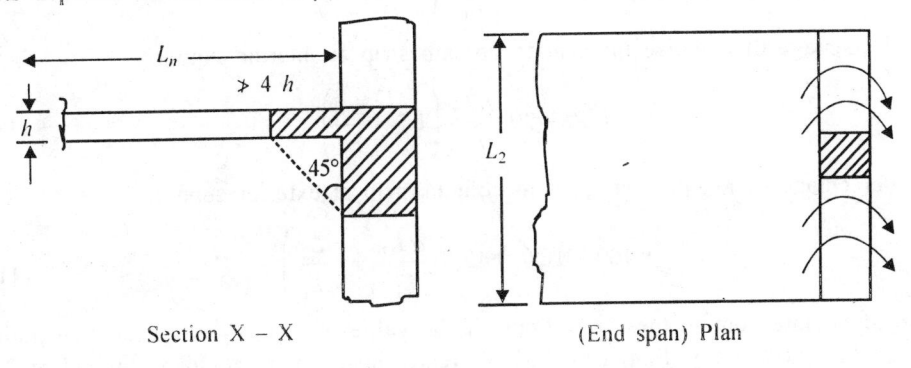

Fig. 11.5 Torsional rigidity of spandrel (or edge) beams.

$$\beta_t = \frac{\text{Torsional rigidity of edge beam}}{\text{Flexural rigidity of slab normal to the beam}} = \frac{GC}{E_c I_s} = \frac{C}{2I_s}$$

where

C = Torsion rigidity of the beam together with a length of the slab acting with it

$$= \Sigma\left[\left(1-0.63\frac{x}{y}\right)\frac{x^3 y}{3}\right], \text{ where } x<y \quad (11.8)$$

The largest volume obtained by splitting the beam into rectangles is taken as C, where

$$C = \frac{L_2 h^3}{12}$$

$$I_s = \frac{L_2 h^3}{12}$$

as L_2 is the length of slab acting with the beam, i.e. span length of the beam. It should be clearly noted that if $\beta_t \geq 2.5$ the edge beam is considered as rigid. In the formulae given below if $\beta_t \geq 2.5$, it should be taken as equal to 2.5 only. After obtaining the above three parameters, the lateral distribution factors can be read off from Table 11.1. [*Note:* The portion of the slab acting with T and L beams on its sides for calculation of β_t is given in ACI Clause 13.2.4 as four times the thickness of the slab or $(D - h)$ whichever is smaller (Fig. 10.3). However, for calculation of torsion capacity of beams, ACI Clause 11.6.1.1 recommends this value to be $3h$ or $D - h$ whichever is smaller.]

11.9 GENERAL EQUATIONS FOR TRANSVERSE DISTRIBUTION OF MOMENTS TO COLUMN STRIPS

Instead of using Table 11.3 the following formulae involving the three parameters described above can also be used for obtaining the transverse distribution factors.

1. Percentage of positive moment to column strip

$$= 60 + 30 \frac{\alpha_1 L_2}{L_1} \left(1.5 - \frac{L_2}{L_1}\right) \quad (11.9)$$

2. Percentage of negative moment to column strip at interior supports

$$= 75 + 30 \frac{\alpha_1 L_2}{L_1} \left(1.5 - \frac{L_2}{L_1}\right) \quad (11.10)$$

3. Percentage of negative moment to column strip at exterior support

$$= 100 - 10\beta_t + 12\beta_t \frac{\alpha L_2}{L_1} \left(1 - \frac{L_2}{L_1}\right) \quad (11.11)$$

[As already pointed out in these equations, if the value of $\alpha_1 L_2/L_1$ is greater than unity it should be limited to unity. Similarly if β_t is greater than 2.5 it should be limited to 2.5.].

[*Note*: In flat slabs with no beams the values in the above three equations will be 60%, 75% and 100% respectively as given in IS 456.]

11.10 ALLOCATION OF MOMENTS TO MIDDLE STRIPS

This procedure is fully explained in IS 456 Clause 31.5.5.4, some of them are discussed below:

1. At each section (at the middle and the ends of the slab) the balance of the moment, which remains after distribution, to the column strip as explained in Sections (11.8) and (11.9) is assigned to the half middle strips on either side of the column strips in proportion to their widths.

2. Each middle strip is to be designed to resist the sum of the moments assigned to its two half middle strips.

3. The full middle strip adjacent to an edge supported on a wall should be designed to resist twice the moment assigned to the first row of interior support as is explained in Section 11.11.

11.11 ANALYSIS OF EXTERIOR FRAME SUPPORTED ON A WALL

The following rules are to be used as guidelines when DDM is used for flat slabs whose exterior end is supported on walls along the column lines.

1. These walls are to be considered as very stiff beams.
2. The torsional resistance of such beams is to be taken as zero for masonry walls and as 2.5 (fully rigid) for R.C. walls.
3. With beams of large moment of inertia the column strip cannot deflect and no distribution of moments can occur to the middle strip. ACI 318 recommends that in such cases the half middle strip of exterior frames are to be designed for the same moment per unit length as the half middle strip corresponding to the adjacent interior frame (IS 456 Clause 31.5.5.4). Usually the exterior column strip will need only the minimum steel prescribed for slabs for temperature and shrinkage.

DESIGN OF TWO-WAY SLABS BY DIRECT DESIGN METHOD

11.12 TREATMENT OF UNEQUAL MOMENTS IN ADJACENT SPANS

Since redistribution of moments is allowed only in the case of theoretical analysis by equivalent frame method (EFM) with unequal negative moments occurring in adjacent panels when using DDM, the larger of the two moments should be used for design of that section (IS 456 Clause 31.4.3.5). This applies to the negative moments of the column and middle strips. However, with theoretical analysis (using EFM) the unbalanced moment is allowed to be distributed to the slab and column (preferably by using the equivalent column stiffness dealt with in Section 13.3.4).

11.13 DESIGN LOADS ON THE BEAMS

Beams in the column strip can be considered as rigid or flexible depending on the relative stiffness $\alpha_1 L_2/L_1$. If this value is unity, the beams can be considered as rigid and it takes care of 85% of the moment in the column strip. If its value is zero, no moment is transferred to beams. If $\alpha_1 L_2/L_1 < 1$ they are considered as flexible beams and the moment carried will be in proportion to its relative stiffness value as shown in Table 11.4.

TABLE 11.4 DISTRIBUTION OF MOMENTS TAKEN BY BEAMS IN TWO-WAY SLABS ON BEAM

$\alpha L_2/L_1$	Moment from column strip to beam (in %)
0	0
≥ 1	85

The same concept can be used to determine the shear of the beams. If $\alpha_1 L_2/L_1 \geq 1$ the beam can be considered as rigid and the load it will take is assigned by the 45 degree distribution as in the case of ordinary two-way slabs on beams. This load is carried by the beams and transmitted to the column. If $\alpha_1 L_2/L_1 = 0$, then all the loads are directly conveyed by the slab to the column by flat slab action. For intermediate values of $\alpha_1 L_2/L_1$, the load is partly carried by the beam to the columns and the balance is carried directly by the slab to the column. In addition to these loads from the slab, the beam has to carry all the loads that are directly placed on the beam such as walls built on the beam. The shear in the beams is calculated by using the above loads acting on the slab.

11.14 DESIGN OF REINFORCEMENTS

For design, first of all we have to consider the critical section for moments. This is dealt with in Section 13.6 of Chapter 13. In general they need not be considered along the centre line of the support but the critical sections are taken at some point towards the span as shown in Figs. 11.6 and Fig. 13.5.

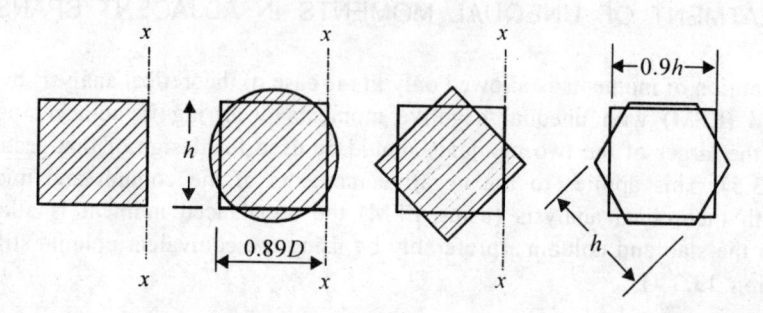

Fig. 11.6 Equivalent square sections for columns of non-rectangular sections.

11.14.1 Breadth and Effective Depth for Design of Steel in Column Strip

The usual practice for determining the width and depth of the column strips is as follows:

1. In flat slabs with drops, the British practice is to assume that the width of the column strip is equal to that of the drop panel. Drops are ignored if their smaller dimension is less than one-third the smaller dimension of the panel. In other types of slabs like flat plates, the width of column strip is taken as one-half the lesser span length.

2. The effective depth for design for negative moment at the drop should be smaller than the two following values:

   (i) The actual depth of the drop or slab.
   (ii) A depth $= h + (x/4)$, where h is the thickness of the slab and x is the distance from edge of drop to edge of column or column capital (see Section 10.5 and Fig. 10.2) in the case of slabs with drops.

3. As larger spans have larger moments, the steel in the larger direction should be placed below the steel in the smaller direction in the positive moment section. In the negative moment section at supports this should be suitably modified. Alternately, a mean effective depth can be used for design.

4. We should also take care of empirical requirements with respect to spacing and size of steel for crack control as in all reinforced concrete slabs.

[*Note:* In practice, the slabs are under reinforced and the area of steel required will be only about 0.38% Fe 415 steel. The minimum steel for shrinkage and temperature is 0.2% (ACI) and the spacing should not exceed twice the slab thickness IS 456 Clause 31.7.1.]

11.15 MOMENTS IN COLUMNS IN DDM

DDM ACI 318 Section 11.2 permits the following conservative and approximate method to be used to obtain the moments in the columns. As moment transfer is very critical for exterior and corner columns we should use conservative values for these moments as shown in Fig. 11.7.

DESIGN OF TWO-WAY SLABS BY DIRECT DESIGN METHOD 187

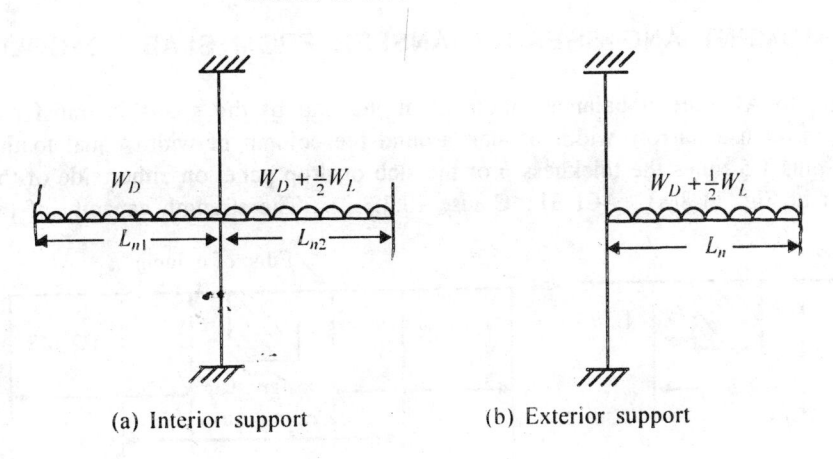

(a) Interior support   (b) Exterior support

Fig. 11.7 Moments in columns by direct design method.

11.15.1 Moments in Internal Columns

We assume that 65% of the static simply-supported bending moment obtained by loading the larger span with [DL + (LL/2)] and loading the shorter span by DL only is transmitted to the column junction as negative moments. We make a further assumption that 7/8th of this moment is transferred to the columns (top and bottom columns) and the rest to the slab through the column strip. Based on these assumptions the following formulae can be obtained:

$$\Delta M = \frac{0.65}{8}\left[(w_D + 0.5w_L)L_2 L_{n1}^2 - w_D L_2 L_{n2}^2\right] \tag{11.12}$$

$$M_C = \frac{7}{8}\Delta M = 0.07\left[(w_D + 0.5w_L)L_2 L_{n1}^2 - w_D L_2 L_{n2}^2\right]$$

IS 456 Clause 31.4.5.2 gives a formula for this moment with factor 0.8.

The value of M_c is shared by the columns above and below in proportion to their stiffness. When the spans are equal $L_{n1} = L_{n2} = L_n$, and if the columns are also of equal stiffness, we get the following moment in each column.

$$M_c' = \frac{0.035}{2} W_L L_2 L_n^2 \tag{11.13}$$

IS 456 Clause 31.4.5.2 gives a more refined method for column moments.

11.15.2 Moments in External Columns

Even though ACI does not give any guidance in this case we can get a conservative estimate by assuming that one of the spans does not exist to balance the moment of the other. Hence

$$\Delta M = 0.07\left(w_D + \frac{w_L}{2}\right)L_2 L_n^2 \tag{11.14}$$

11.16 MOMENT AND SHEAR TRANSFER FROM SLAB TO COLUMNS

According to ACI the unbalanced moment at the end of the spans is transferred to the columns through a narrow width of slab around the column of width equal to the column width c_2 plus 1.5 times the thickness h of the slab or drop panel on either side of the column as shown in Fig. 11.8(a). (ACI 318 Clause 13.3.3.2). The strength capacity of the slab in

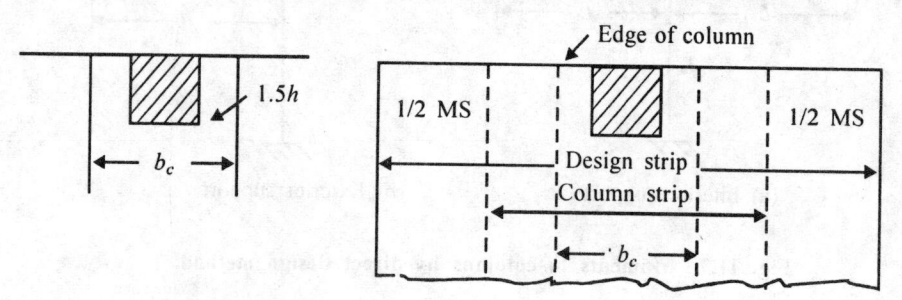

Fig. 11.8(a) Breadth of effective moment transfer strip involved in transfer of unbalanced moment [ACI 318].

this transfer zone should be checked and in many cases additional steel may be provided for the safe transfer. It should also be noted that when considering shear produced due to moment transfer (torsion) as explained in Section 12.7. The shear should be based on the actual moment capacity including this additional steel of the slab around the column.

11.16.1 Design of Edge Panels for Moment Transfer

BS 8110 Clause 3.7.4.2 [4] is very specific in its recommendations for design of moment transfer between slab and columns at edges and corners as the width of column strip available for this transfer at these locations can be considerably narrower and less confined than those available at the internal columns. Some of the typical cases are shown in Fig. 11.8(b) and in no case this

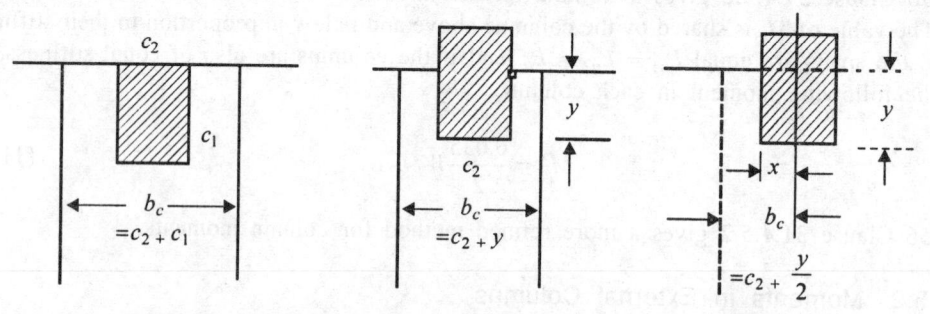

Fig. 11.8(b) Breadth of effective moment transfer strip [BS 8110].

transfer zone should be wider than the column strip. We check the capacity by the following formula.

$$M_{t(max)} = 0.15 f_{ck} b_c d^2$$

where

d = Effective depth of the top reinforcement in the slab

b_c = The width of the slab as shown in Fig. 11.8

Hence the reinforcements in the column strip will consist of the following:

(i) The uniformly distributed reinforcement in the strip for the negative moment in the strip.

(ii) Additional steel provided in the narrow strip specified for the transfer of unbalanced moment equal to kM_u given in Eq. 12.6 of Chapter 12. The moment capacity of this narrow strip in no case should be less than 50% of the total negative moment of the column strip. If it is less the structural arrangement should be suitably modified (BS Cl. 3.7.4.2).

11.16.2 Limitation of Moment Transfer

Normally the column moment should not be more than $M_{t(max)}$ as described in Section 11.16.1. If by exact analysis (EFM) the theoretical moment in external column is more than $M_{t(max)}$, we proceed as follows:

1. In slabs without edge beam we have to limit the moment in the column strip to $M_{t(max)}$ and transfer the balance as positive moment to the span.

2. We can provide an edge beam or extension of the slab beyond the centre line of columns to take care of additional moment.

3. We can specially reinforce the moment transfer strip by additional steel to transfer the larger moment to the column.

11.17 DETAILING OF STEEL IN FLAT SLABS

Detailing of flat slabs can be carried in two ways, viz., by using bent bars and by using straight bars.

The recommended detailing by straight bars is shown in Figs. 11.9 and 11.10. That with bent bars can be obtained from IS 456 (2000) Fig. 16. Considerable economy in placing of

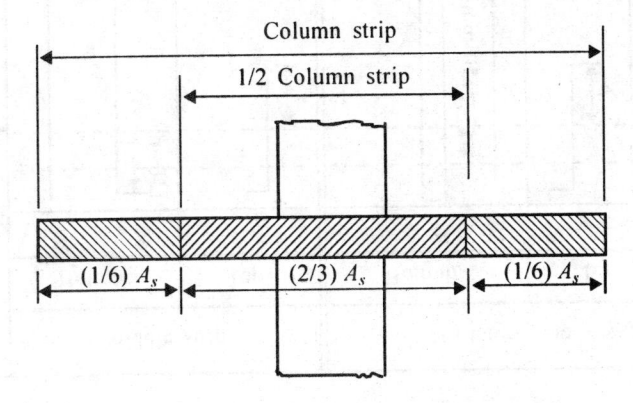

Fig. 11.9 Detailing of steel in column strip.

190 ADVANCED REINFORCED CONCRETE DESIGN

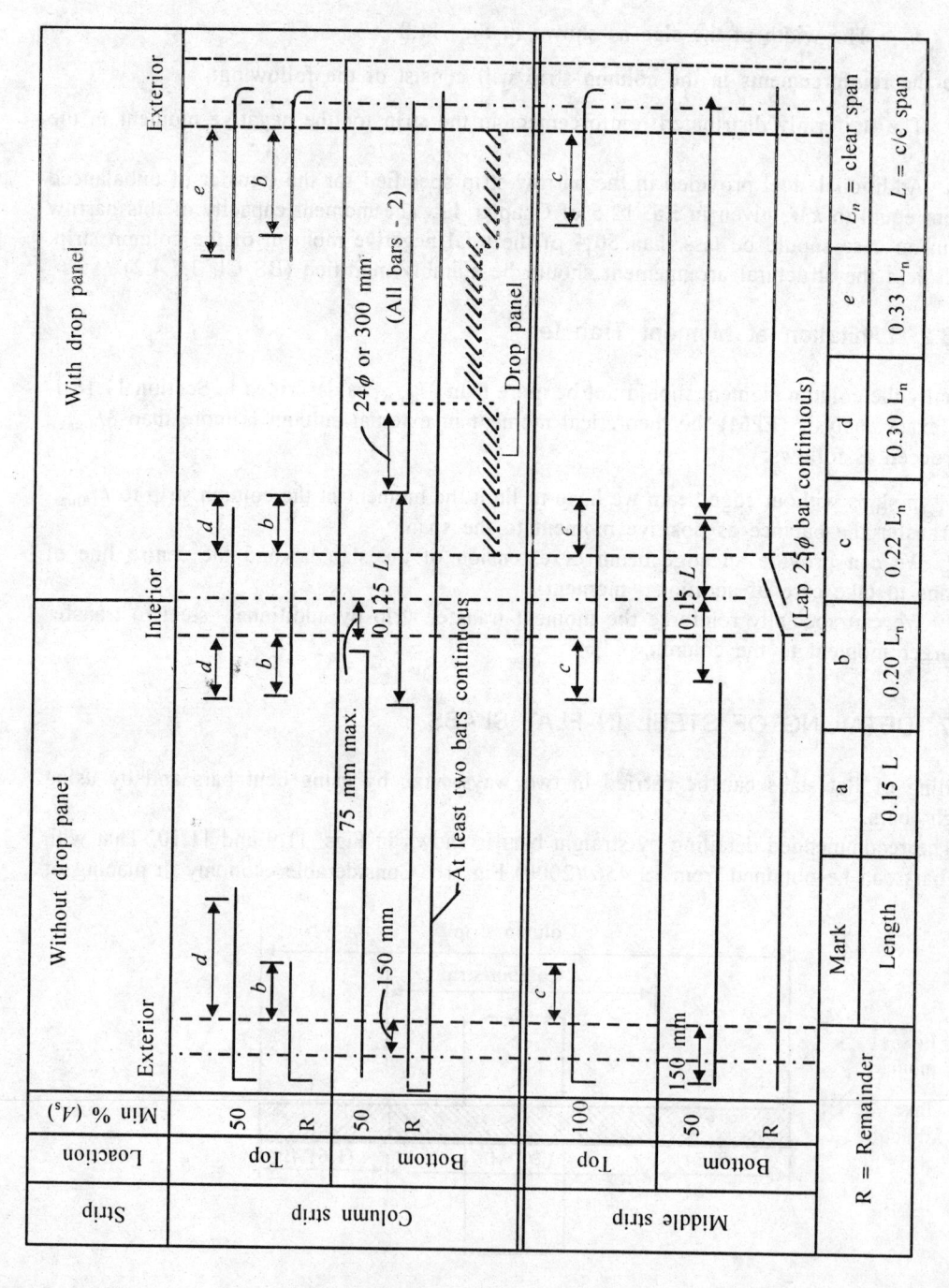

Fig. 11.10 Detailing of flat slabs and flat plates by straight rods.

steel can be achieved if no bent bars are used and the negative steel is placed as preformed mats [5]. It should be specially pointed out that, according to revised ACI 318 (1989) code Clause 13.4.8.5 it is mandatory to place at least two of the bottom bars in the column strip as continuous reinforcement properly spliced in each direction within the column core and properly anchored to the exterior supports. They are called 'Integrity Bars'. They give residual strength in punching shear failure. Punching shear produces cracking at the top of the slab where negative moment steel is also provided. This will leave only concrete at the bottom of the slab to resist shear. Hence this detailing is very important in flat slabs. We should also note that when adjacent spans are unequal the extension of negative steel beyond the face of support should be based on the longer span. The method of detailing of steel in the column strip is shown in Fig. 11.10. In the middle strip the steel can be distributed uniformly. The spacing of bars in flat slab shall not exceed two times the slab thickness (IS Cl. 31.7.1).

11.18 DESIGN OF TWO-WAY SLABS

The procedure for design of two-way slabs by DDM can be summarised in Table 11.5:

TABLE 11.5 STEPS FOR DESIGN OF TWO-WAY SLABS

Step	(A) *Slabs without internal beams*	(B) *Slabs on rigid beams or stiffened strips*
1	Check applicability of DDM.	As in A.
2	Select depth from (a) deflection (b) punching shear. Increase depth slightly for moment transfer.	Select depth from (a) deflection (b) one-way shear.
3	Divide slab into frames in the E-W and N-S directions.	As in A.
4	Take interior frames in E-W direction. Find $M_0 = wL_2L_n^2/8$ and L_n = Distance column to column faces.	As in A.
5	Find end conditions and assign positive and negative moments using Table 11.2.	As in A.
6	Distribute the moments in step 5, to column strips from Table 11.1 or Eqs. (11.9) to (11.11) and the rest moments to middle strips.	(i) As in A to column and middle strips (ii) Assign moments in the column strips to beam using Table 11.4 and the balance to the slab.
7	Take exterior frames in the E-W direction. Repeat steps 4 to 6. Assign also moments to edge beams.	As in A steps 4 to 6.
8	Take interior frames in the N-S direction. Repeat steps 4 to 6. Find $$M_0 = wL_1L_n^2/8$$	As in A steps 4 to 6.

Step	(A) *Slabs without internal beams*	(B) *Slabs on rigid beams or stiffened strips*
9	Take exterior frames in the N-S direction and repeat steps 4 to 6. Assign moments to edge beam.	As in A steps 4 to 6.
10	Design and detail slab for reinforcement.	Design beams for bending and shear. Design slab reinforcement for the slab moments.
11	Check the slab capacity for moment transfer by flexure and eccentric shear (Sections 11.15.3 and 11.16).	
12	Design edge beams for bending, shear and torsion.	
13	Irregular slabs to be designed by the equivalent frame method.	Irregular slabs can be designed by the coefficient methods IS 456 Clauses 23.2 to 23.4.

CONCLUSION

All types of regular two-way slabs (with rigid or flexible beams) which satisfy the limitations prescribed in Section 11.2 can be designed by DDM. Two-way slabs which are not regular, are designed by equivalent frame method (EFM). Two-way slabs which are not regular and which are on rigid beams can be conveniently analyzed by the coefficient method given is IS 456 (2000) Annexure D.

The axial loads in columns are calculated by conventional analysis. An approximate method giving conservative values for the moments in the columns is given in Section 11.15.

The moment transfer from slab to external columns and also the moment transfer due to pattern loading in internal columns takes place through the specified narrow strip of slab around the columns as described in Section 11.16. These zones should be properly designed for the good performance of flat slabs and flat plates.

EXAMPLE 11.1 (Direct design method of a flat slab)
A flat slab on a series of columns with column heads has the following dimensions:

Column spacings 7 m × 6 m in X and Y directions, respectively; thickness of main slab 180 mm; size of drop 3 m × 2.5 m in the interior; total thickness of drop 260 mm; interior column 750 m; column head diameter 1.53 m; exterior column 750 mm square with edge beam; edge beam 750 × 900 mm; storey height 3.3 m; factored dead and live loads—7 kN/m^2 and 5 kN/m^2, respectively.

Analyze an interior frame in X direction by DDM and determine the design moments by ACI and IS 456 methods (see Fig. 13.8).

DESIGN OF TWO-WAY SLABS BY DIRECT DESIGN METHOD   193

Reference	Step	Calculations
Ch. (10) Sec. 11.7	1	*Check preliminary dimensions* Equivalent square column head of 1.53 diameter $b = 0.89 \times 1.53 = 1.36$ m
IS 456 Clause 31.4	2	*Check applicability of DDM* Minimum three spans—O.K. Aspect ratio $\dfrac{L_1}{L_2} < 2.0$—O.K. Differences in spans are nil—O.K. Column offset—nil LL/DL ratio is less than 2
	3	*Divide the slab with frame in X and Y directions and obtain the dimensions of X and Y frames* **TABLE 1** Dimensions of X and Y Frames (in metres)

Fig. 11.1
Fig. 11.2

No.	Item	X frames X_1	X_2	Y frames Y_1	Y_2
1	c-to-c span	7.0	7.0	6.0	6.0
2	Clear span (L_n)	7 − 1.36 = 5.64	7 − 0.75 = 6.25	6 − 1.36 = 4.64	6 − 0.75 = 5.25
3	Width of span (L_2)	6.0	3 + (.75/2) = 3.375	7.0	3.5 + (.75/2) = 3.875
4	Width of column strip (BS)	2.5	1.25 + (.75/2) 1.625	3.0	1.75 + (.75/2) = 2.125
5	Width of mid strip	3.5	1.75	4.0	1.75
6	L_2/L_1	0.86	0.86	1.17	1.17

Sec. 11.14.1

Sec. 11.10

| | 4 | *Analyze the interior X frame ($L_1 = 7$ m and $L_2 = 6$ m)*
$$M_0 = \dfrac{wL_2 L_n^2}{8}$$
$$\dfrac{12 \times 6(5.64)^2}{8} = 286.3 \text{ kNm}$$
Note: We will distribute the moments first by ACI method and then by IS 456. |

Reference	Step	Calculations
	5	**Part 1 Determine distribution factors by ACI method**
Sec. 11.6	5.1	*Find Longitudinal distribution for flat slab with edge beam*
		1. *Interior spans*
		(Same as in IS 456)
		Support −ve = 0.65; Span +ve = 0.35
		2. *End spans (flat slab with edge beams)*
Table 11.2		Interior −ve = 0.70; Span +ve = 0.50; Exterior −ve = 0.30
		(IS 456 uses formulae to arrive at these values)
Section 11.8	5.2	*Find transverse distribution factors (ACI)*
		Here we use Table 11.1 or Eqs. (11.9) to (11.11)
		1. *Interior span*
		(a) Support moment (negative): 75% to column strip; 25% to mid-strip
		(b) Span moments (positive): 60% column strip; 40% mid-strip
		2. *End spans*
		End span has an edge beam, the distribution depends on the following factors as given in Table 11.1 or Eqs. (11.9) to (11.11).
Sec. 11.8		1. Stiffness of column strip $\alpha = 0$ for flat slab
		2. Aspect ratio $\dfrac{L_2}{L_1} = \dfrac{6}{7} = 0.86$
		3. β_t of edge beam $= \dfrac{C}{2I_s}$
		where $C = \sum \left(1 - \dfrac{0.63x}{y}\right) \dfrac{x^3 y}{3}$
		Edge beam 900 × 750 with slab 180 × 720 attached
		$x_1 = 0.75;\quad y_1 = 0.90$
		$x_2 = 0.18;\quad y_2 = 0.72$
		$C = 0.60\ \text{m}^4$
		$I_s = \dfrac{L_2 t^3}{12} = \dfrac{6 \times (0.18)^3}{12} = 0.0029\ \text{m}^4$
		$\beta_t = \dfrac{0.06}{2 \times 0.0029} = 10.3 > 2.5$
		Use value 2.5 only.

DESIGN OF TWO-WAY SLABS BY DIRECT DESIGN METHOD   195

Reference	Step	Calculations
Table 11.1		*Summarise transverse distribution factors of end-span*
		1. Interior support
		Negative moment: 75% to column strip; 25% to mid-strip
		Span positive moment 60% to column strip; 40% to mid-strip
		2. Exterior support (Negative moment) (with edge beam) 75% to column strip; 25% to mid-strip
	6	**Part 2 Determine the distribution factors by IS 456 method**
		We will first make the longitudinal and then the transverse distribution.
	6.1	*Find longitudinal distribution*
Sec. 11.6		*Interior span (same as in ACI)*
		Support (−ve) = 0.65;
		Span (+ ve) = 0.35
		End spans
		End-span distribution depends on α_c as given below.
		$\alpha_c = \dfrac{\sum K_c}{K_s}$ (omitting 4E)
		$K_c = \dfrac{I_c}{L_c} = \dfrac{750(750)^3}{12 \times 3300} = 7990 \text{ mm}^3$
		$\sum K_c$ that of column above and below
		Taking $I_s = \dfrac{L_2 t^3}{12}$
		$K_s = \dfrac{I_s}{L_1} = \dfrac{6000(18)^3}{12 \times 7000} = 416.6 \text{ mm}^3$
IS 456 Sec. 31.4.3.3		Ratio of flexural stiffness of exterior columns to flexural stiffness of slab in the direction of moments
		$\alpha_c = \dfrac{2 \times 7990}{416} = 38.4$
		Distribution factor = $\dfrac{1}{1+(1/\alpha)}$ = 0.975 = R
		Distribution factors are
		Interior (−ve) = 0.75 − 0.1R = 0.65

Reference	Step	Calculations
		Span (+ve = 06.3 − 0.28R = 0.36
		Exterior −ve = 0.65R = 0.63
	6.2	*Find transverse distribution factors is end span in flat slabs (IS 456)*
IS 456		Interior −ve: 75% to column strip; 25% to mid-strip
Clause 31.5.5		Span +ve: 60% to column strip; 40% to mid-strip
		Exterior −ve: 100% to column strip
		(The presence of edge beam as flexible beams between columns in interior spans can not be taken care of by information available in IS 456. Such designs can be made by ACI method.)
	7	*Determine final distribution of moments*
		$M_0 = 286.3$ kNm
Step 6.1		$0.65\ M_0 = 186.1$ kNm
		$0.35\ M_0 = 100.2$ kNm (similarly other values can be found)

TABLE 2. Longitudinal and Transverse Directions of Moments (Summary)

Types of moments	Longitudinal direction		Factors	Transverse direction	
	Factor	Moment		CS*	MS**
Negative	0.65	186.1	0.75 → 139.6		
			0.25		46.5
Positive	0.35	100.2	0.60 → 60.0		
			0.40		40.2

End span analysis (ACI)

Interior (−ve)	0.70	200.4	0.75 → 150.3		
			0.25		50.1
Span (+ve)	0.50	143.0	0.60 → 85.8		
			0.40		57.2

(Step 5.1(2) reference applies to End span analysis (ACI) section)

Reference	Step	Calculations				
		Types of moments	Longitudinal direction		Factors	Transverse direction
			Factor	Moment		CS* MS**
		Exterior (−ve)	0.30	85.9	→ 0.75 → → 0.25 →	64.4 21.5
		End span analysis (IS 456)				
		Interior (−ve)	0.65	186.1	→ 0.75 → → 0.25 →	139.6 46.5
		Span (+ve)	0.36	103.1	→ 0.60 → → 0.40 →	61.9 41.2
		Exterior (−ve)	0.63	180.4 → 100 →		180.4

*CS = Column strip

**MS = 2 nos half middle strips together

Step 8. *Estimate the design moment in the external columns*

Eq. (11.13)

$$M = 0.035\,(w_d + 0.5w_L)L_2 L_n^2$$
$$= 0.035(9.5)(6)(5.64)^2$$
$$= 63.5 \text{ kNm}$$

Step 9. *Estimate design moment in the internal column*

$$M'' = 0.035\,(0.5w_L)L_2 L_n^2$$
$$= 0.035(2.5)(6)(5.64)^2$$
$$= 16.7 \text{ kNm}$$

Note: Analysis of internal frames in the Y direction (Y_1 frames) can be made as shown above for X_1 frame. X_2 and Y_2 frames with beams along the span are analyzed by the method shown in Example 11.4. In the corner column, the effects of bending in X and Y directions should be considered.)

EXAMPLE 11.2 (Parameters in two-way slab design)

Define the following quantities and explain their significance in direct design method of two-way slabs including flat slabs.

$$\alpha_1;\ \alpha_1\,(L_2/L_1);\ (\alpha_1/\alpha_2)\,(L_2/L_1)^2;\ \beta_t \text{ and } \alpha_c$$

Item	Description
1	*Value of α_1 (beam-to-slab stiffness for L_1 beam)* Let us take a two-way slab with beams between columns. Consider bending along span L_1 (with the width of the slab L_2) The relative stiffness of the beams in the L_1 direction with that of the slab in the same direction will be given by the following equations. $I_b = KbD^2$ (with the attached slab) $I_s = \dfrac{L_2 t^3}{12}$ (width of slab being L_2) The ratios of their stiffness EI/L_1 will be given by $I_b/I_s = \alpha_1$
2 Eq. (11.8)	*Value of $\alpha_1(L_2/L_1)$ (for transverse distribution)* Let us consider the slab with beams. When we deal with the transverse distribution of moments between the column strip and the middle strip the column strip will take more moments if there is a beam in that strip. The value will depend on the relative stiffness of the beam in the L_1 direction and on slab in the L_2 direction. $\dfrac{I_b/L_1}{I_s/L_2} = \dfrac{\alpha_1 L_2}{L_1}$ where $I_s = L_2 t^3/12$ A value equal to unity for this relative stiffness is used as the break point value, so in Eqs. (11.9) to (11.11), if $\alpha L_2/L_1$ is greater than 1, it is taken as 1 only. Thus the factor $\alpha L_2/L_1$ gives the rigidity of the beam. If it is unity, 85% of the moment in the column strip is taken by the beam. When it is zero, the full moment is taken by the slab. For intermediate values proportionate distribution is assumed.
3	*Value of $(\alpha_1/\alpha_2)(L_2/L_1)^2$* $\dfrac{\alpha_1 L_2/L_1}{\alpha_2 L_1/L_2} = \dfrac{\alpha_1}{\alpha_2}\left(\dfrac{L_2}{L_1}\right)^2$ It is the ratio of the two values in L_1 and L_2 directions. For direct design of two-way slabs, its value should be between 0.5 and 2.0.
4	*Value of $\beta_t = C/2I_s$* This is the ratio of the torsion stiffness of spandrel beam to the bending stiffness of the slab. When we consider the end spans of frames which have edge beams, the torsional stiffness of the edge beam is a factor to be considered in the transverse (as well as longitudinal) distribution of moments. The torsional stiffness C includes the portion attached with the beam also [smaller of $(D - h)$ and $4t$ on each side]. Bending in the L_1 direction is the edge beam being in the L_2 direction. $I_s = \dfrac{L_2 t^3}{12}$ This value of β_t is substituted to give the distribution of moments in Eqs. (11.9) to (11.11) in ACI method for transverse distribution). As shown in Eq. (11.11), in a flat

DESIGN OF TWO-WAY SLABS BY DIRECT DESIGN METHOD 199

Item	Description
	slab the presence of an edge beam will reduce the column strip moment by a factor depending on β_t.
	The break point value of β_t is taken as 2.5. For any value larger than 2.5 we put 2.5 only in eq. (11.11).
5	Example α_{ec} and α_c
	When using ACI modified stiffness method the longitudinal distribution of moments in two-way slab in the end spans is calculated from equations containing the term
	$$\alpha_{ec} = \frac{\sum K_{ec}}{\sum (K_s + K_b)}$$
	involving stiffness of equivalent column, stiffness of slab and stiffness of beam. However in IS 456, the 'equivalent column' concept of ACI and the existence of the edge beam are neglected and the above expression becomes
	$$\alpha_c = \sum \frac{K_c}{K_s}$$
	It is to be noted that the present ACI code replaces the equations by Table 11.1. This makes the direct design method very simple and truly empirical.

EXAMPLE 11.3 [Design of two-way slabs with stiffened ribs (flexible beams)]

A floor consists of a series of spans with 165 mm thick slab on a grid of columns 375 × 375 mm² spaced 6 × 4.8 m in the X and Y directions. The slabs are stiffened by shallow beams (75 mm deep below the slab and 375 mm wide) between the columns. Analyze the slab by direct design method and also estimate the load that is to be carried by the beams. Assume a factored total design load of 10 kN/m².

Reference	Step	Calculations
	1	Calculate I of beams below slab
Fig. (10.4)		$b = 375$, $B = 375 + 2(75) = 525$ mm
		$h = 165$; $H = 165 + 75 = 240$ mm
Appendix (A)		$x = \frac{B}{b} = \frac{525}{375} = 1.4$, $y = \frac{h}{H} = \frac{165}{240} = 0.69$
		$k = 1 + (x-1)y^3 + \frac{3(1-y)^2(y)(x-1)}{1 + y(x-1)} = 1.193$
		$I_b = \frac{1.193 \times 375 \times 240^3}{12} = 5.15 \times 10^8$ mm units

Reference	Step	Calculations
Sec. (10.7.2)	2	Calculate $\alpha = I_b/I_s$ of the system For span 6 m, $I_s = \dfrac{L_2 t^3}{12} = \dfrac{4800(165)^3}{12} = 17.98 \times 10^8$ mm units $\alpha_1 = \dfrac{I_b}{I_s} = \dfrac{5.15}{17.98} = 0.29$
Fig. 11.4		For span 4.8 m, $I_s = \dfrac{6000(165)^3}{12} = 22.46 \times 10^8$ $\alpha_2 = \dfrac{5.15}{22.46} = 0.23$ Both values are less than 1 and hence the beams have to be considered as flexible.
	3	Calculate relative stiffness of beams $RS = \dfrac{\alpha_1}{\alpha_2}\left(\dfrac{L_2}{L_1}\right)^2 = \dfrac{0.29}{0.23}\left(\dfrac{4.8}{6.0}\right)^2 = 0.81$ This value is between 0.2 and 5.
Sec. (11.2)	4	Check applicability of DDM Minimum three spans; Aspect ratio $L_1/L_2 < 2.0$; Differences in spans, nil (O.K.); Column offset, nil; LL/DL ratio is less than 2 (assumed); For two-way slabs on beam, relative stiffness is between 0.2 and 5. Hence DDM is applicable.
Sec. 10.7.2	5	Take internal frame of span 6 m for analysis $\alpha_1 = 0.29$; $\dfrac{L_2}{L_1} = 4.8/6 = 0.8$; $\dfrac{\alpha_1 L_2}{L_1} = 0.23$ (for long beam) For span in the 4.8 m direction, $\alpha_2 = 0.23$ Corresponding $\dfrac{L_2}{L_1} = \dfrac{6}{4.8} = 1.25$ $\dfrac{\alpha_2 L_2}{L_1} = 0.29$ (for short beam)
	6	Determine the longitudinal distribution of moments Interior span Support (−ve) = 0.65, Span (+ ve) = 0.35 End span: As in Example 11.1.
	7	Calculate the transverse distribution of moments by formulae

DESIGN OF TWO-WAY SLABS BY DIRECT DESIGN METHOD

Reference	Step	Calculations
Eq. (11.10)	7.1	*Determine distribution of negative moment (Span L_1) to column strip* $= 75 + 30 \dfrac{\alpha_1 L_2}{L_1}\left(1.0 - \dfrac{L_2}{L_1}\right)$ $= 75 + 30(0.23)(0.2) = 76.3\%$ (say 76) To middle strip (2 half strips) $= 100 - 76 = 24\%$
Sec. (11.13)	7.2	*Determine the distribution of positive moment* To column strip, balance $= 60 + 30\alpha_1 \dfrac{L_2}{L_1}\left(1.5 - \dfrac{L_2}{L_1}\right)$ $= 60 + 30(0.23)(0.7) = 64.8\%$ (say 65%) To middle strip (balance) $= 35\%$
Sec. 11.13	7.3	*Determine the portion taken by beam from column strip moments* With $\alpha_1 \dfrac{L_2}{L_1} = 1$, portion taken by beam $= 85\%$ Portion taken when $\alpha_1 \dfrac{L_2}{L_1} = 0.23$ $= 0.85 \times 0.23 = 0.19$, i.e. 19% Balance taken by slab $= 81\%$
	8	*Calculate the simply-supported bending moment M_0* $M_0 = \dfrac{wL_2 L_n^2}{8}$ $L_n = 6 - 0.375 = 5.625\,m$ $M_0 = \dfrac{10 \times 4.8(5.625)^2}{8} = 190\,kNm$
	9	*Distribute moment as shown in Table 2* Negative moment (support) $= 0.65 \times 190 = 124\,kNm$ Positive moment (span) $= 0.35 \times 190 = 66\,kNm$

TABLE 1

Type of moment	Total moment	M.S.	Column strip Total	Slab	Beam
−ve at support	124 (0.65)	30 (0.24)	94 (0.76)	76 (0.81)	18 (0.19)
+ve in span	66 (0.35)	23 (0.35)	43 (0.65)	35 (0.81)	8 (0.19)

Reference	Step	Calculations
	10	*Determine the load taken by beam for shear design* Load taken by beam if the beams were rigid, is first calculated Since $k = \dfrac{6}{4.8} = 1.25$
Chapter 8 Eq. 8.2		% load on each short beam $= \dfrac{25}{k} = \dfrac{25}{1.25} = 20\%$ Load taken by rigid short beams of a slab of 6 m × 4.8 m $= 0.20 \times 10 \times 6 \times 4.8 = 57.6$ kN % load an each long beams = 30% Hence, the load taken by rigid long beams of slab $= 0.30 \times 10 \times 6 \times 4.8 = 86.4$ kN % load taken is 85%, if $\alpha_1 L_2/L_1 = 1$
Step 5		when $\alpha_1 L_2/L_1 = 0.23$ (long beam), the load taken $= 86.4 \times 0.85 \times 0.23 = 16.9$ kN when $\alpha_1 L_2/L_1 = 0.29$ (short beam), the load taken $= 57.6 \times 0.85 \times 0.29 = 14.2$ kN
	11	*Determine the width of column and middle strips for design* We should note that in flat plates and two-way slabs the width of column strip is equal to one-half the shorter span. One-half the shorter span $= \dfrac{4.8}{2} = 2.4$ m Width of column strip in X_1 and Y_1 frames = 2.4 m The balance of the width of slab is taken as the middle strip.

TABLE 2. The Properties of Frames (Example 11.3)

No.	Item	X_1 frames (in metre)	Y_1 frames (in metre)
1.	C.C. span	6.0	4.8
2.	Width of frame	4.8	6.0
3.	Width of column strip	2.4	2.4
4.	With of middle strip	2.4	3.6

DESIGN OF TWO-WAY SLABS BY DIRECT DESIGN METHOD   203

EXAMPLE 11.4 (Design of flat slab)

A flat slab floor has panels of 6.6 × 5.4 m in X and Y directions between centres of columns which are 450 × 450 mm in size. It has an edge beam all around the periphery of 250 × 500 mm which carries an exterior wall of weight 6 kN/m. The slab thickness is 150 mm and the characteristic live load it has to carry is 5.25 kN/m². The height of each storey is 3 m.

1. Analyze exterior frame in 6.6 m direction and determine the distribution of moments.
2. Determine the maximum torsional moment produced in the edge beam.
3. Check whether the moment can be transferred to column without transferring it on edge beam.

Reference	Step	Calculations
		Part 1 Analysis of exterior frame
Fig. 11.2 (b)	1	*Dimensions of exterior frame*
		$L_2 = \dfrac{5.4}{2} + \dfrac{0.45}{2} = 2.925$ for M_0 (to outer edge of beam)
		$L_2 = 5.4$ (for other calculations)
		$L_1 = 6.6$ (for interior span)
		$L_n = 6.6 - \dfrac{0.45}{2} = 6.375$ m
		Dead load = 0.15 × 25 = 3.75 kN/m²
		Live load = 5.25 kN/m²
		Total factored load w = 1.5(3.75 + 5.25) = 13.5 kN/m²
		$\dfrac{L_2}{L_1} = \dfrac{5.4}{6.6} = 0.82$
	2	*Determine total static moment to design by DDM which is applicable*
		$M_0 = \dfrac{wL_2 L_n^{\,2}}{8}$
		L_n = 6.375 m (between supports)
Step 1		$M_0 = \dfrac{13.5 \times 2.925\,(6.375)^2}{8}$
Eq. (11.2)		= 200.6 kNm
	3	*Calculate stiffness of edge beam (span in X direction)*
		α of edge beam I_b/I_s (assume K of beam for M.I. = 1.5)

Reference	Step	Calculations
		$I_b = \dfrac{1.5bd^3}{12} = \dfrac{1.5 \times 250 \times 500^2}{12} = 3.90 \times 10^{-3}$ m units
		$I_s = \dfrac{L_2 t^3}{12}$
		$= \dfrac{(2.925)(0.150)^3}{12}$
		$= 8.22 \times 10^{-4}$ m units
		$\alpha = \dfrac{I_b}{I_s} = \dfrac{3.90}{0.82} = 4.75$
	4	*Find the torsional resistance ratio of edge beam*
Eq. (11.8)		Torsional constant $C = \sum \dfrac{x^3 y}{3}\left(1 - 0.63 \dfrac{x}{y}\right)$
		$= \dfrac{(250)^3(500)}{3}\left(1 - 0.63 \times \dfrac{250}{500}\right)$
		$+ \dfrac{(150)^3(350)}{1}\left(1 - 0.63 \times \dfrac{150}{350}\right)$
		$= (1.78 + 0.29) \times 10^{-3} = 2.07 \times 10^{-3}$ m units
		$\beta_t = \dfrac{C}{2I_s} = \dfrac{2.07}{2 \times 0.82} = 1.26 \text{ (is } < 2.5)$
	5	*Determine unfactored DL/LL ratio*
		$\beta_a = \dfrac{DL}{LL} = 1 < 2$
		(Hence pattern loading need not be considered)
	6	*List lateral distribution factors* (for $L_1 = 6.6$ m)
		$\dfrac{L_2}{L_1} = \dfrac{5.4}{6.6} = 0.82 \text{ (use the full } L_2)$
		$\alpha \dfrac{L_2}{L_1} = 4.75 \dfrac{5.4}{6.6} = 3.89 > 1.0 \text{ (use 1.0)}$
		$\beta_t = 1.26 \text{ (less than 2.5)}$
	7	*Find longitudinal distribution of moments by Table 11.2*
		(see also Table in step 9)

Reference	Step	Calculations
	8	Calculate the transverse distribution factors to column strips by Table 11.1
Eqs. (11.9)–(11.11)		$$+\text{ve(spans)} = 60 + \frac{30\alpha L_2}{L_1}\left(1.5 - \frac{L_2}{L_1}\right)$$ $$= 60 + 30(1)(1.5 - 0.82) = 80.4\%$$ $$-\text{ve interior supports} = 75 + \frac{30\alpha L_2}{L_1}\left(1 - \frac{L_2}{L_1}\right)$$ $$= 75 + 30(1)(1 - 0.82) = 80.4\%$$ $$-\text{ve (exterior supports)} = 100 - 10\beta_t + \frac{12\beta_t \alpha L_2}{L_1}\left(1 - \frac{L_2}{L_1}\right)$$ $$= 100 - (10 \times 1.26) + (12 \times 1.26)(1.0)(1 - 0.82)$$ $$= 90.1 \text{ (check from Table 11.1)}$$
	9	Tabulate the distribution of moments in end frame (kNm) in long-span direction

TABLE 1 Distribution of Moments of Exterior Frame

Location	Longitudinal		Transverse strip		
	Factor	Moment	Factor %	1/2 column	1/2 middle
$M_0 = 200.6$ kNm					
Interior span					
−ve (supports)	0.65	130.4	80.4 → 104.8		
			19.6		25.6
+ve (span)	0.35	70.2	80.4 → 56.4		
			19.6		13.8
End span (flat slab with edge beams)					
−ve (interior support)	0.70	140.4	80.4 → 112.9		
			19.6		27.5
+ve (span)	0.50	100.3	80.4 → 80.6		
			19.6		19.7

Step 7

Table 11.1

ADVANCED REINFORCED CONCRETE DESIGN

Reference	Step	Calculations
		−ve (exterior support) 0.30 60.2 → 90.1 → 54.3 9.9 → 5.9
	10	Determine the moments to be carried by edge beam $\alpha L_2/L_1 > 1.0$ Hence edge beam taken 85% of moments The distribution can be tabulated as follows: **TABLE 2. Distribution of Column Strip Moments to Beams and Slabs** <table><tr><th rowspan="2">Location</th><th colspan="3">Distribution of moments (kNm)</th></tr><tr><th>Total</th><th>Beam</th><th>Slab</th></tr><tr><td>*Interior spans* −ve support (C)</td><td>104.8</td><td>89.0</td><td>15.8</td></tr><tr><td>+ve span</td><td>56.4</td><td>48.0</td><td>8.4</td></tr><tr><td>*End span* −ve interior supports (B)</td><td>112.9</td><td>96.0</td><td>6.9</td></tr><tr><td>+ve span</td><td>80.6</td><td>68.5</td><td>12.1</td></tr><tr><td>−ve exterior support (A)</td><td>54.3</td><td>46.1</td><td>8.2</td></tr></table>
Fig. 11.2	11	Find the moments in beam due to the wall load Design load = 6 kN/m Self weight = 0.25 × 0.5 × 25 = 3.2 kN/m Total factored load = 1.5 × 9.2 = 13.8 kN/m Use moment coefficients (a) End support $(A) = \dfrac{wl^2}{16} = \dfrac{13.8\,(6.6)^2}{16} = 37.6$ kN/m (−ve) (b) Middle of end span $= \dfrac{wl^2}{12} = 50.0$ kN/m (+ve)

Reference	Step	Calculations
		(c) Support next to end (B) $= \dfrac{wl^2}{10} = 60.1 \text{ kNm } (-\text{ve})$
		(d) Middle of interior span $= \dfrac{wl^2}{24} = 25.0 \text{ kNm } (+\text{ve})$
		(e) Interior supports $C = \dfrac{wl^2}{12} = 50.0 \text{ kNm } (-\text{ve})$
	12	Determine the total moments in the spandrel beam

TABLE 3. Table Moments is the Spandrel Beam

From wall load	From slab analysis	Total	Support [see Fig. 11.2(b)]
50.0	89.0	139.0	← C
25.0	48.0	73.0	
60.1	96.0	156.1	← B
50.0	68.5	118.5	
37.6	46.1	83.7	← A

Part 2 Determine maximum torsion in edge beam internal frame

Reference	Step	Calculations
Fig. 11.2(b)	1	Find static moment in interior frame (end span)
Fig. 11.2(a)		Maximum torsion will be felt by the end span of the interior frame with $L_n = 6.375$ m
		$M_0 = \dfrac{wL^2 L_n^2}{8} = \dfrac{(13.5 \times 5.4)(6.375)^2}{8} = 370$ kNm
Table 11.2 Case B1	2	Moment at exterior end
		30% of $M_0 = 370 \times 0.3 = 111$ kNm
	3	Transverse distribution
		$I_s = \dfrac{(5.4)(0.150)^3}{12} = 1.5 \times 10^{-3}$ m units
		$I_b = 3.9 \times 10^{-3}$
		$\alpha = \dfrac{3.9}{1.5} = 2.6$
		$\dfrac{\alpha_1 L_2}{L_1} = 2.6 \times 0.82 = 2.13 > 1.0$ (use 1.0)
Step 4		$\beta_t = \dfrac{C}{2I_s} = \dfrac{2.07}{2 \times 1.5} = 0.69 < 2.5$

Reference	Step	Calculations
Eq. (11.11)	4	*Find transverse distribution to column strip–exterior support* $= 100 - 10\beta_t + 12\beta_t\left(1 - \dfrac{L_2}{L_1}\right)$ $= 100 - 6.9 + 8.28(1 - 0.82) = 94.59$ (say 95%)
	5	*Calculate the moment in column strip at the exterior end* Total exterior moment = 111 kNm Column strip $M = 111 \times 0.95 = 105$ kNm
Page 218, Eq. (12.6)		In a square column, 60% of this moment is carried directly to column and 40% by torsion (and shear) through slab and beam. Torsional moment = $0.4 \times 105 = 42$ kNm This will be carried by two beams Torsion on each side $= \dfrac{42}{2} = 21.0$ kNm
	6	*Find the portion of moment carried by (1/2) middle strip on the side of the (1/2) column strip* $= \dfrac{111 - 105}{2} = 3.0$ kNm
	7	*Calculate total torsion from centre of edge beam to the junction of column in edge beam* Total Torsion = 21.0 + 3.0 = 24.0 kNm (*Note:* See also Example 14.5) for design of edge beam **Part 3 Moment transfer without edge beam**
	1	*Find moment to be transferred through column strip to column* 100% of end moment = 111 kNm
	2	*Check the value of column moment for approximate design* $M_0 = 0.035\left(w_D + \dfrac{w_L}{2}\right)L_2 L_n^2$ $= 0.035\,(1.5)\,(4.5 + 2.25)\,(5.4)\,(6.375)^2 = 77$ kNm which is less than 111 kNm. (Assume moment to be transferred is 111 kNm.)
	3	*Find dimensions of transfer zone* $h = 150$ mm; $d = 120$ mm (say) Breadth $b = 450 + 3h = 900$ mm

DESIGN OF TWO-WAY SLABS BY DIRECT DESIGN METHOD

Reference	Step	Calculations
BS 8110 Sec. 11.16.1	4	*Calculate moment capacity of b_0* $M_{t(max)} = 0.15 f_{ck}bd^2 = 0.15 \times 20 \times 900 \times (120)^2 = 39$ kNm (This is less than 50% of 111, hence not satisfactory.)
	5	*Determine moment to be transferred to column by bending* As column is square assuming the moment to be transferred is 60%. $111 \times 0.6 = 66.6$ kNm As this moment is larger than $M_{t(max)}$ we should change the structural arrangement as follows. (a) Transfer the moment in excess of the capacity of the slab $M_{t(max)}$, the edge of the slab should have an edge beam or an edge strip. (b) Alternately, determine the theoretical values of moments by exact analysis (*see* Chapter 13) and redistribute the moment, in excess of 39 kNm to the span.

Example 11.5 (Minimum size of columns for pattern loading)
Table 11.2 gives the value of $\Sigma K_c/\Sigma K_s$ for the pattern loading to be neglected in two-way slabs. Using Table 11.3, find the minimum size of internal square column for a flat slab of 6 m × 5.4 m panels with a LL/DL ratio of 1.67 so that the effects of pattern loading can be neglected. Assume thickness of slab is 200 mm and storey height is 3 m.

Reference	Step	Calculations
	1	*Determine α_{min} for $\beta_a = 1.67$* Consider an internal span in a frame with 5.4 m spans. $L_1 = 5.4$ m and $L_2 = 6$ m $\dfrac{L_2}{L_1} = \dfrac{6}{5.4} = 1.1$
Table 11.2		$\beta_a = 2.0$ for all L_2/L_1 and $\alpha = 0$, $\alpha_{c(min)} = 0$ $\beta_a = 1.0$ and L_2/L_1 and 1.1, $\alpha = 0$, $\alpha_{c(min)} = 0.74$ Hence $\beta_a = 1.67$ and L_2/L_1 and 1.1, $\alpha = 0$; find $\alpha_{c(min)}$
Sec. (11.7)		(Difference between 2.0 and 1.67 = 0.33) Taking K_c as column stiffness, above and below the slab. $\alpha_{c(min)} = (0.74)(0.33) = 0.25 = \dfrac{2K_c}{K_s}$

Reference	Step	Calculations
	2	Calculate K_c for the given K_s for the span in 6 m direction $$I_s = \frac{L_2 t^3}{12} = \frac{5.4(0.2)^3}{12} = 0.0036 \text{ m units}$$ $$K_s = \frac{4EI_s}{L_1} = \frac{4 \times 0.0036}{6} \text{ (omitting } E) = 0.0024 \text{ m}^3$$ Required $K_c = \dfrac{0.25 \times 0.0024}{2} = 3.0 \times 10^{-4}$ m units Assuming square column for ($L_c = 3$ m) $$K_c = \frac{d^4}{12 \times 3} = 3.0 \times 10^{-4} \text{ m units } (h = 3 \text{ m})$$ $d = 0.322$ m i.e. 322 mm (If α_c is less than $\alpha_{c(\min)}$ we can either increase the column size or increase positive moment by β_s.

REFERENCES

1. ACI 318 (1989), Building Code Requirements of Reinforced Concrete, ACI Committee, Detroit, Michigan.

2. Proposed Revisions to Building Code Requirements of Reinforced Concrete, ACI Committee, 318 Concrete International, December 1994.

3. IS 456 (2000), Code of Practice for Plain and Reinforced Concrete, Bureau of Indian Standards, New Delhi.

4. BS 8110 (1985), *Structural Use of Concrete*, British Standards Institution, London, 1985.

5. Mark Fintel (Ed.), *Handbook of Concrete Engineering*, Mark Fintel, Van Nostrand Reinhold, New York, 1974.

CHAPTER 12

Shear in Flat Slabs and Flat Plates

12.1 INTRODUCTION

It is very important to thoroughly check the safety of flat plates and flat slabs in shear. Failures in flat plate construction have been reported many times in engineering due to improper design for shear-transfer, especially at the exterior columns. This is due to inadequate appreciation of shear forces in flat slabs and the bending moments produced in the external columns. The following three types of shear action are to be checked in flat-slab and flat-plate design:

1. One-way shear of the slab (wide beam action)
2. Two-way shear around columns (punching shear action)
3. Shear caused by moment transfer (combined with torsional moment) around columns, especially the external columns, along with two-way shear.

Slabs should be safe in one-way shear without shear steel. In addition, if analyses show that punching shear strength along with the moment transfer is inadequate, the areas around the columns have to be reinforced to resist this shear by 'designed shear reinforcements' in the form of 'bar reinforcements' or 'shear-head reinforcements' (by means of rolled steel section). These problems are dealt with in this chapter. ACI 318 Clause 11.12 deals with design for shear in flat slabs and the same principles are also used in IS 456.

12.2 CHECKING FOR ONE-WAY (WIDE BEAM) SHEAR

The critical section for one-way shear is at a distance equal to the effective depth from the face of the column. The areas from which the load is to be transferred is known as the 'tributary areas'. These areas, for wide beam action, is shown in Fig. 12.1. The resultant shear is transferred by the full width of the section. The magnitude of the shear stress is given by the expression

$$v = \frac{V}{bd} \qquad (12.1)$$

where d is the effective depth. The maximum value permitted for one-way shear is the usual value allowed in codes for one-way slabs. Generally this type of failure is highly improbable in any type of practical construction for normal loads acting on the slab, but routine check is always carried out in design calculations. We may also apply the enhanced shear value for slabs given in IS 456 Clause 40.2.1.1.

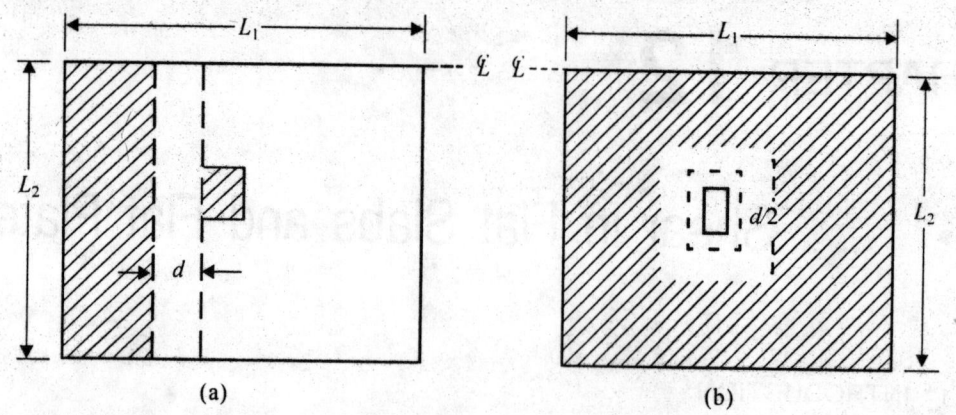

Fig. 12.1 Design shears at an interior column support in flat slabs: (a) one-way shear or beam shear (b) punching shear or two-way shear.

12.3 TWO-WAY (PUNCHING) SHEAR

It is easy for us to visualise that if a flat plate is gradually subjected to increase loading, the first crack will appear at the top of the slab around the column due to negative moment near the column. On further loading if the structure is not strong in shear, a truncated pyramid of concrete will be pushed out of the slab as shown in Fig. 12.2. Such a failure is called a *punching shear* also referred to as *two-way shear*. The critical sections of punching shear are assumed to be situated in footings at $d/2$ from the periphery of the column capital (drop panel) perpendicular to the plane of the slab. Where d is the effective depth of the section. The shape in the plan is geometrically similar to the support immediately below the slab as shown in Fig. 12.2. For flat slabs with drops, there can be two critical sections for checking

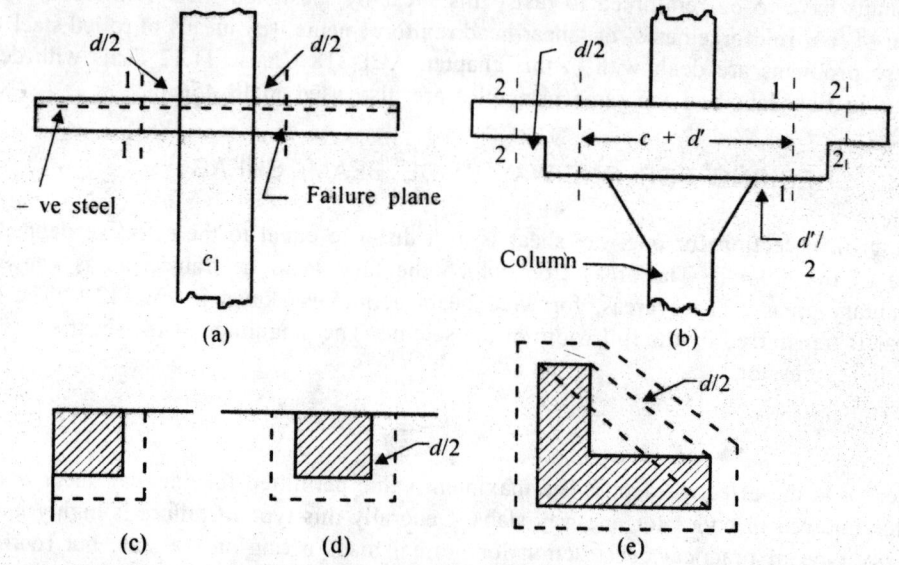

Fig. 12.2 Punching shear perimetres in flat slabs and flat plate: (a) flat plate (b) flat slab (c) corner column (d) external column (e) non-rectangular column.

punching shear, one at 1/2 (drop thickness) from face of column or capital and the other 1/2 (slab thickness) from the edge of the drop.

The formula for the critical perimeter b_0 for checking this punching shear can be different for different situations as shown in Fig. 12.2. These can be summarised as follows:

Case 1. For normal punching shear calculations without extra shear reinforcements with rectangular column size $c_1 \times c_2$ and effective depth of slabs d

$$b_0 = 2(c_1 + d + c_2 + d) = 4(c + d) \tag{12.2}$$

for a square column of size c

Case 2. When shear reinforcement in the form of steel bears are provided as shown in Fig. 12.7, for a length a beyond face of column for a square column of size e

$$b_0 = 4\left(c + \sqrt{2}a\right) \tag{12.2a}$$

Case 3. When fabricated shear head reinforcement as in Fig. 12.9 is provided for a square column

$$b_0 = 4\sqrt{2}\left[\frac{c}{2} + \frac{3}{4}\left(L_v - \frac{c}{2}\right)\right] \tag{12.2b}$$

The tributary area of the slab for punching shear will be the area outside the critical section of the panel being examined. (For location of critical perimeter of different types of column, refer IS 456 Figs. 13 to 15.)

The area resisting the shear = (Perimeter) × (Depth of slab)

The punching shear stress V_p is calculated as follows:

$$v_p = \frac{V_w}{b_0 d}$$

where

v_w = Factored shear from relevant contributory area

b_0 = Perimeter length of the critical section

d = Effective depth

12.4 PERMISSIBLE PUNCHING SHEAR

For a structure to be safe, the punching shear stress should be less than the safe value. The ultimate safe value of the punching shear of concrete is given by the least value of the following three expressions (IS 456 Clause 31.6.3)[1,2]:

$$\tau_p = 0.25\sqrt{f_{ck}} \tag{12.3}$$

$$\tau'_p = (0.5 + \beta_c)\tau_p \tag{12.4}$$

$$\tau''_p = \left(0.5 + \frac{\alpha_s d}{4b_0}\right)\tau_p \tag{12.5}$$

Equation (12.3) is the initial value suggested for a square column but research has shown that its magnitude is affected by the shape of the column, so that it can be expressed for a rectangle by Eq. (12.4), where β_c = ratio of short side to long side of column capital. [*Note:* β_c in ACI is the inverse of that in IS 456.] Further research has shown that the ratio of the effective depth d to critical perimeter b_0 also has an influence on τ_p as in Eq. (12.5)[2]. In this equation we take the following values for α_s:

$$\alpha_s = 40 \text{ for an interior column}$$
$$= 30 \text{ for an edge column}$$
$$= 20 \text{ for a corner column}$$

The value used is the smallest of the τ_p values got from Eqs. (12.3) to (12.5). When shear strength is inadequate we may resort to the following changes:

1. Use a thicker slab
2. Use a large column
3. Use higher strength concrete
4. Use additional shear reinforcement

12.5 SHEAR DUE TO UNBALANCED MOMENT (TORSIONAL MOMENTS)

The two types of shear considered so far are one-way shear and punching shear. We shall now deal with the third type of shear, which is due to moment transfer. Let us consider an internal column with column strips in E-W direction as shown in Fig. 12.3 with the punching shear

End column   Interior columns

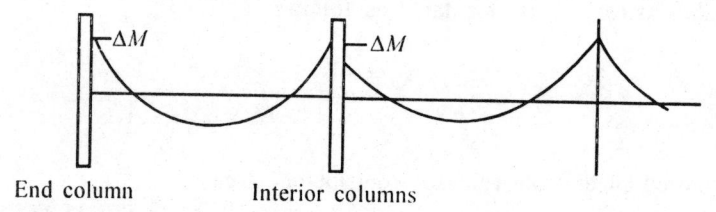

Fig. 12.3 Transfer of unbalanced moments between slabs without beams with columns at their junctions.

forces acting around it. Let M_1 and M_2 be the moments of the column strips on either sides of the column. If $M_1 = M_2$, there is no moment to be transferred from the slab to the column. If $M_1 > M_2$, then that part of the column strip that abuts directly on the column will be called upon to transfer some part of the unbalanced moment directly to the column, as bending moment, through side c_2 of the column strip. The other part of the unbalanced moment in the column strip will be transferred as torsion through the portion of the slab along the transverse direction adjacent to side c_1. This torsional moment produce additional shear which should be added to the punching shear. Thus, the unbalanced moment $(M_1 - M_2)$ will be transferred through the column, as bending moment in the E-W direction and as torsion in the N-S direction, of magnitude

$$(1 - k)(M_1 - M_2)$$

Following are the forces that act on the critical perimeter of the column:

1. The shear force V_u producing uniform shear stress along the periphery of the column.
2. A bending

$$M_b = M_1 - M_2$$

acting on face c_2 producing only tension and compression on face c_2 and no shears (Fig. 12.4).

3. A torsion

$$M_t = (1 - k)(M_1 - M_2) = \eta(M_1 - M_2)$$

along face c_1. This produces shear stresses around the critical section as shown in Fig. 12.4 on all faces. The shear stresses produced by (1) and (3) are added together and the resultant shear should be less than the specified safe value. (It should be noted that the simple addition of those stresses means that we are still using an elastic method of design for this purpose. Many methods for an ultimate load analysis have been proposed recently but have not been accepted in codes.)

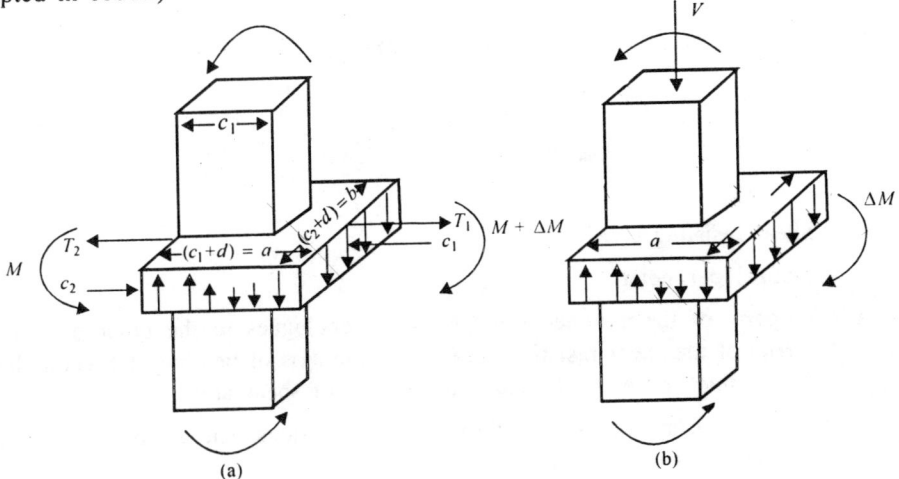

Fig. 12.4 Combined action of (a) punching shear and (b) shear due to moment transfer at column junctions.

Since the expressions for shears due to torsion around circular columns are complex, circular columns are first reduced to their equivalent square columns for simplicity (see Section 10.5). It is also assumed that the critical perimeter is the same as in the case of shear without moments. The overall lengths that takes place in shear transfer are $d/2$ away from the sides of the columns so that the moment transfer distances are:

$$a = c_1 + d \quad \text{(in longitudinal direction)}$$
$$b = c_2 + d \quad \text{(in the transverse direction)}$$

where d is the effective depth of the slab.

Tests by Hanson and Hanson (1968) have shown that in square columns when moments are transferred between a column and a slab, 60% of the moment is transferred by flexure and 40% by torsional moment M_t. Therefore to apply this result to rectangular shape, we make an assumption that the value of k is a function of the moment transfer distances a and b and can be given by the expression:

$$k = \frac{1}{1+(2/3)\sqrt{a/b}} \qquad (12.6)$$

so that $k = 0.6$ when $a = b$.

As we have already seen, $(1 - k) = \eta$ expresses the torsional component and it produces shear around the critical section. The magnitude of the shear can be expressed as:

$$v = \frac{Tx}{J} \qquad (12.7)$$

which is as derived in the following section.

12.5.1 Combined Effect of Shear and Torsional Moment

The shear stresses produced by the combined effect of shear force V_u and torsional moment M_t are taken as the sum of their effects. This can be conveniently expressed as:

$$v_{max} = \frac{V_u}{A_c} + \frac{\eta(M_1 - M_2)}{J} x_1$$

$$v_{min} = \frac{V_u}{A_c} - \frac{\eta(M_1 - M_2)}{J} x_2 \qquad (12.8)$$

where

A_c = Shear resisting area ($= b_0 d$)

b_0 = Critical perimeter

J = Property of the assumed critical section analogues to the polar moment of inertia of the shear resisting area, about the axis of bending perpendicular to the column section and located at the CG of shear area.

x_1, x_2 = Distances from the centre of twist to the sections where the maximum and minimum shear stress act.

The values of J, x_1, x_2 and A for various cases are given in Table 12.1 and Fig. 12.5.

TABLE 12.1 CALCULATION OF J VALUES FOR COLUMN JUNCTIONS

Fig. 12.5	J	x_1	x_2	A
Case 1	$\dfrac{2ad^3}{12} + \dfrac{2da^3}{12} + 2bdx_1^2$	$\dfrac{a}{2}$	$\dfrac{a}{2}$	$2(a+b)d$
Case 2	$\dfrac{2ad^3}{12} + \dfrac{2da^3}{12} + 2bdx_1^2 + 2ad\left(\dfrac{a}{2} - x_1\right)^2$	$\dfrac{a^2}{2a+b}$	$a - x_1$	$2(2a+b)d$
Case 3	$\dfrac{ad^3}{12} + \dfrac{da^3}{12} + 2bd\,x_1^2$	$\dfrac{a}{2}$	$\dfrac{a}{2}$	$2(b+a)d$
Case 4	$\dfrac{ad^3}{12} + \dfrac{da^3}{12} + bdx_1^2 + ad\left(\dfrac{a}{2} - x_1\right)^2$	$\dfrac{a^2}{2(a+b)}$	$\dfrac{a(a+2b)}{2(a+b)}$	$(a+b)d$

SHEAR IN FLAT SLABS AND FLAT PLATES 217

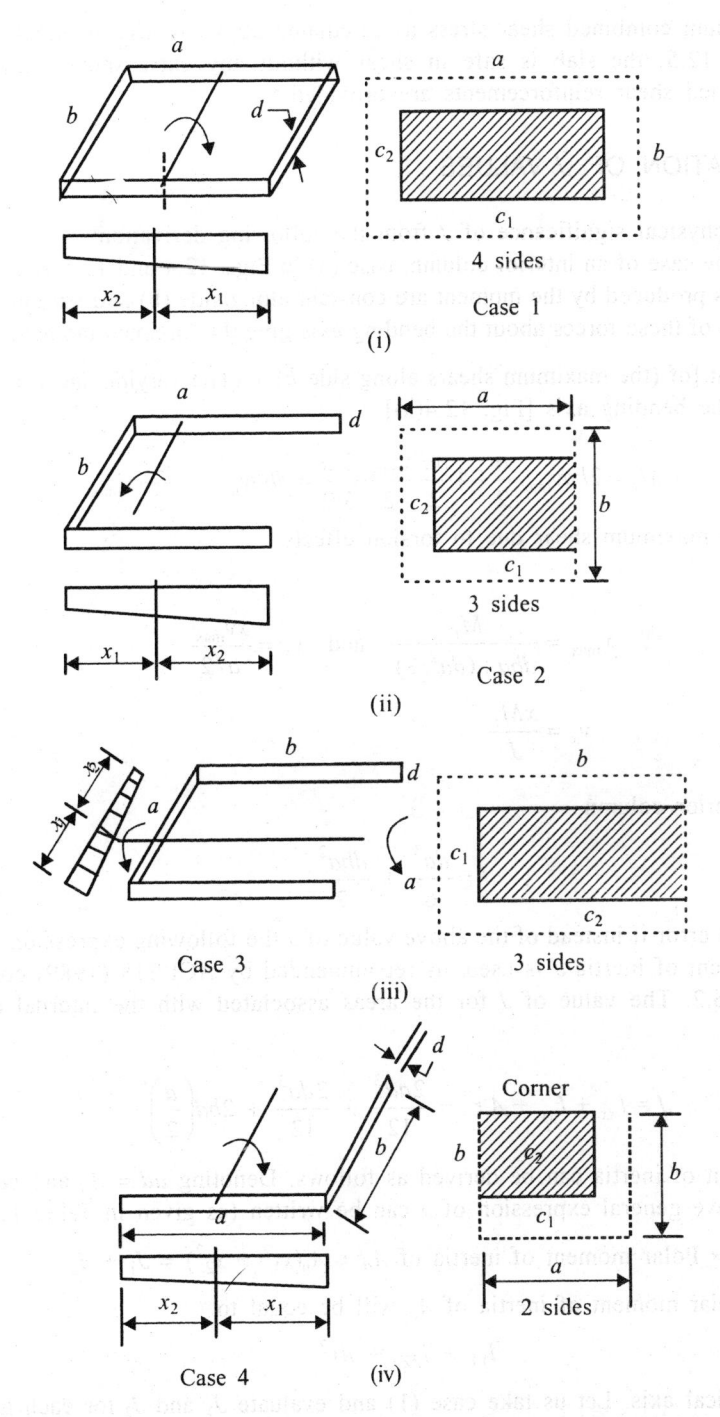

Fig. 12.5 Column junctions: (i) interior column (ii) edge column (bending perpendicular to edge) (iii) edge column (bending parallel to edge) (iv) corner column.

If the maximum combined shear stress as calculated above is less or equal to τ_p given in Eqs. 12.3 to 12.5, the slab is safe in shear without any extra shear reinforcements. Otherwise, designed shear reinforcements are provided.*

12.6 CALCULATION OF *J* VALUES

We can get the physical significance of *J* from the following derivation:

Let us take the case of an interior column, case (a) in Figs. 12.4 and 12.5 and Table 12.1. The shear stresses produced by the moment are constant along side (b) and varying along side (a). The moments of these forces about the bending axis give the torsional moment M_t. Hence,

M_t = Moment [of (the maximum shears along side b) + (The varying shear alongside a)] about the bending axis [Fig. 12.4(a)].

$$M_t = 2bdv_{max}\frac{a}{2} + 4\frac{ad}{2}\frac{v_{max}}{2}\frac{2}{3}\frac{a}{2} = dbav_{max} + \frac{da^2}{3}v_{max}$$

where v_{max} is the maximum shear due to torsion effects.

Therefore

$$v_{max} = \frac{M_t"}{dba + (da^2/3)} \quad \text{and} \quad v_x = \frac{xv_{max}}{a/2}$$

$$v_x = \frac{xM_t}{J}$$

Hence for an interior column,

$$J = \frac{da^3}{6} + \frac{dba^2}{2}$$

There is not much error if instead of the above value of *J* the following expression analogous to the polar moment of inertia *J* is used, as recommended by ACI 318 (1989) commentary Clause R.11. 12.6.2. The value of *J* for the areas associated with the internal column is given by

$$J = I_{xx} + I_{yy} + Ar^2 = \frac{2ad^3}{12} + \frac{2da^3}{12} + 2bd\left(\frac{a}{2}\right)^2$$

[The polar moment of inertia can be derived as follows: Denoting $ad = A_a$ and $bd = A_b$, in Fig. 12.5, the above general expression of *J* can be written (as given in Table 12.1) as

$$J = \text{Polar moment of inertia of } A_a + A_b(x_1^2 + x_2^2) = J_1 + J_2 \quad (12.9)$$

In general, the polar moment of inertia of A_a will be equal to

$$I_{YY} + I_{ZZ} + Ar^2$$

I_{ZZ} being the vertical axis. Let us take case (1) and evaluate J_1 and J_2 for each area.

---

*See Section 12.8.

For each side ($a \times d$),

$$J_1 = \frac{ad^3}{12} + \frac{da^3}{12} + da\left(\frac{a}{2} - x\right)^2$$

When $x = a/2$ as in cases 1 and 3, the third term becomes equal to zero.

To evaluate J_2, we take the moment of inertia of A_b with two (b) sides, as in case (1), about the centre of gravity. Therefore,

$$J_2 = A_b (x_1^2 + x_2^2)$$

In the other three cases as shown in Table 12.1 and Fig. 12.5 only one (b) side will be present so that $J_2 = A_b x_1^2$. Adding up J_1 and J_2 we get J. The values of J are given in Table 12.1 and Fig. 12.5 for various cases and can easily be written down by inspection.]

12.7 STRENGTHENING OF COLUMN AREAS FOR MOMENT TRANSFER BY TORSION WHICH PRODUCES SHEAR

We have already seen that the forces that act on the faces of the column due to load and unbalanced moments are:

1. B.M. and S.F. along c_2 edges
2. Torsion and S.F. along c_1 edges

It is therefore advisable, as already pointed out in Section 11.15.3, to strengthen the portion that transmits these forces. Accordingly, the unbalanced moment at column junctions is considered to be transferred to the critical section ($c_2 + d$) through a slab width equal to the (column width + 3h) where h is the total thickness of the slab. Closer spacing of designed steel or provision of additional steel in this region for the additional unbalanced moment is recommended when detailing the reinforcements. As already pointed out in sections 11.5 and 11.16, this moment transfer is especially important for external and corner columns of flat slabs.

When end-moments are determined by DDM using approximate moment coefficients, Clauses 13.6.3.3 and 13.6.3.6 of ACI code require that the fraction of moment transferred by torsion producing shear must be based on the full strength capacity of the whole column strip. Taking also into account the special reinforcements including the additional steel available in this region, the theoretical column strip moment is calculated from loads. This avoids shear failure under all circumstances.*

12.8 SHEAR REINFORCEMENT DESIGN

Generally, flat slabs (with drops) can be made safe in punching shear without extra shear steel by increasing the thickness of the slab near the columns as drop panels. But this increases the cost of shuttering and, in addition, drops are also not very pleasing to the eyes. In flat plates, however, punching shear becomes critical and extra reinforcements to cater for punching shear will be found necessary. These shear reinforcements can be in the form of (i) bar reinforcements and (ii) shear head reinforcements fabricated from rolled-steel sections. These are shown in Figs. 12.7 and 12.9. The summary of recommendations for shear design of flat slabs and flat plates is given in Table 12.2.

* This is explained in the Example 12.6.

TABLE 12.2 DESIGN FOR PUNCHING SHEAR (ACI)

Value of punching shear, v_p	Type of shear reinforcement
$v_p \leq \tau_p$	No special steel needed
$v_p \geq \tau_p$ but $\leq 1.5\tau_p$	Provide designed-bar reinforcements
$v_p \leq \tau_p$ but $\leq 1.75\tau_p$	Provide designed fabricated-shear head reinforcement
$v_p \leq 1.75\tau_p$	Redesign the slab

Note: When providing steel for shear assume concrete takes only $0.5\tau_p$.

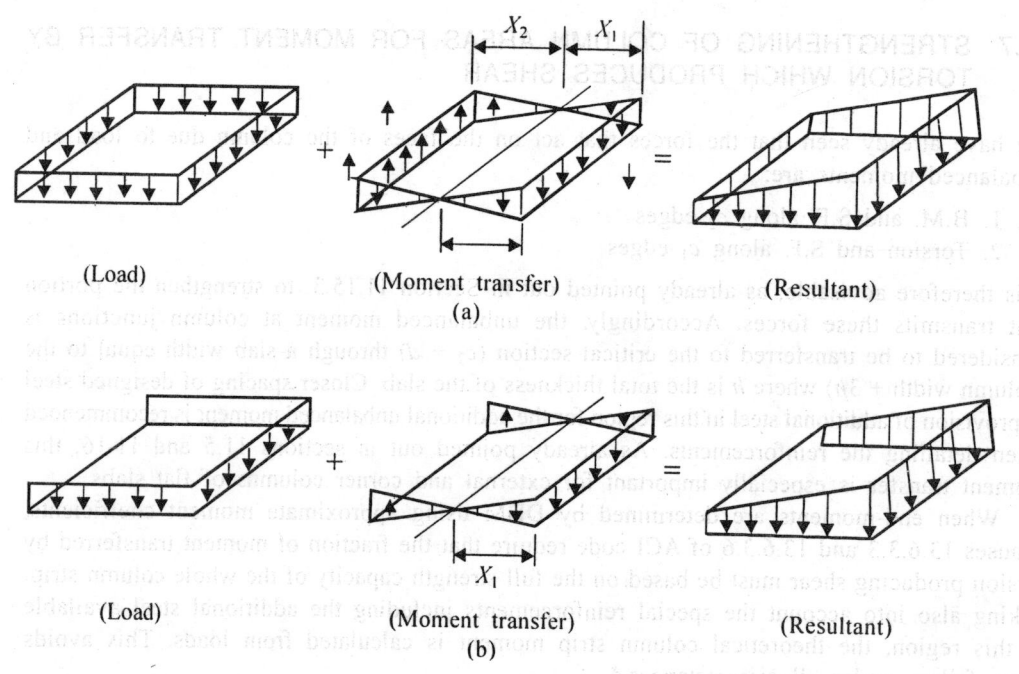

Fig. 12.6 Combined action of punching shear and shear due to moment transfer at column junction: (a) internal column (b) external column.

12.8.1 Design of Bar Reinforcements for Punching Shear

We know that the principal difficulty in using bars as shear reinforcements in slabs is to find the necessary anchorage for the bars. Bar reinforcements fabricated in the form of a hat, called *shear hats*, were once used in flat plates construction, but they have not been found as effective as was imagined. Single U stirrups or double U stirrups as shown in Figs. 12.7 and 12.8 are commonly used nowadays. They should be provided in the region extending from the column face to a distance away from the critical perimeter, where the shear stress does not exceed one-half of the allowable design shear strength value of τ_p.

12.8.2 Maximum Shear Stress Allowed in Punching Shear with Steel Bars

When designing these shear reinforcements with steel bars we should note that the ACI code allows these to be used only in situations where the maximum value of the calculated punching shear at conventional section is equal to or less than $1.5\tau_p$ (Table 12.2).

$$v_p \leq 1.5\tau_r \tag{12.10}$$

According to ACI 318 (89) Clause 11.12.3.1, when designing the shear steel we should assume that concrete can be allowed to take $0.5\tau_p$ of the shear and the balance should be taken by steel. The reinforcements should extend to a distance beyond which it is not required as shown in Fig. 12.7. In addition, the new critical perimeter is given by Eq. (12.2b).

Anchorage of stirrups may be difficult in slabs less than 250 mm thick. In thinner slabs stirrups should be used only if they are closed and enclose at each corner a longitudinal bar as shown in Figs. 12.6–12.8. [ACI 318 Clause R11–12.3]. The total shear capacity of concrete and steel in this case is given by the equation:

$$V_u = V_c + V_s = 0.5\tau_p b_0 d + \frac{A_{sv} f_s d}{s} \tag{12.11}$$

where

A_{sv} = Total area of stirrups in the critical shear perimeter

Single U stirrups give $A_{sv} = 2$(Area of bar)

Double U stirrups give $A_{sv} = 4$(Area of bar)

s = Spacing of stirrup = $d/2$

d = Effective depth of slab

b_0 = Critical shear perimeter

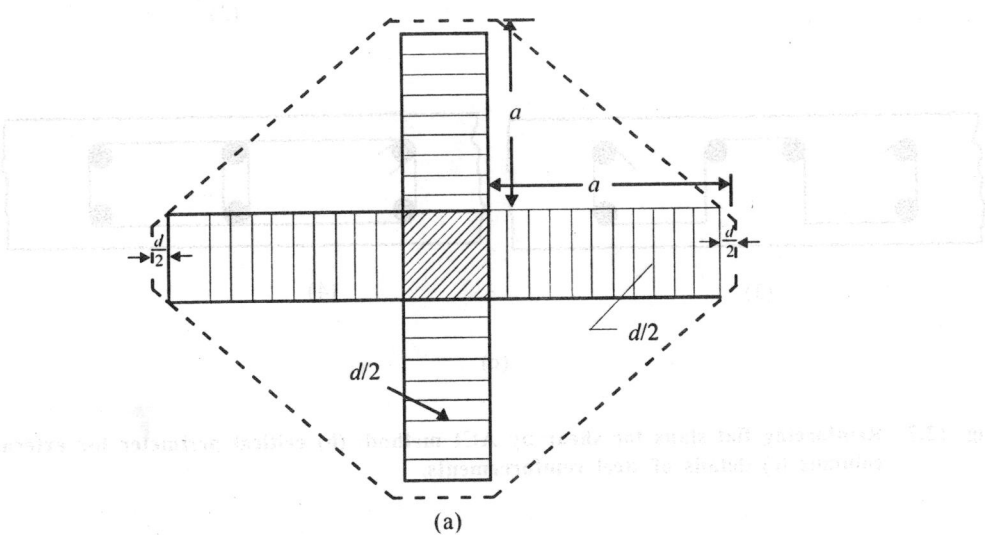

Fig. 12.7 Reinforcing flat slabs for shear by ACI method: (a) critical perimeter for checking shear for internal columns (*contd.*).

(b)

(1) (2)

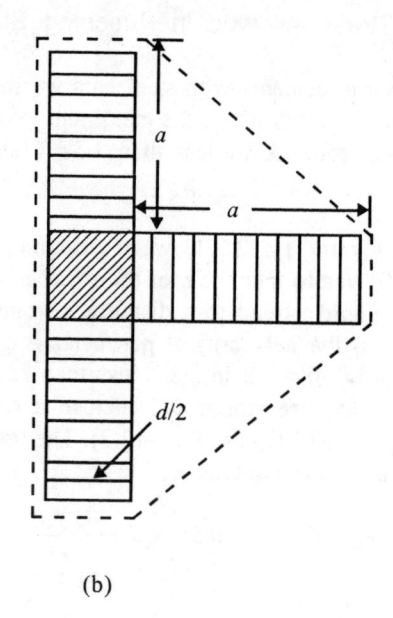

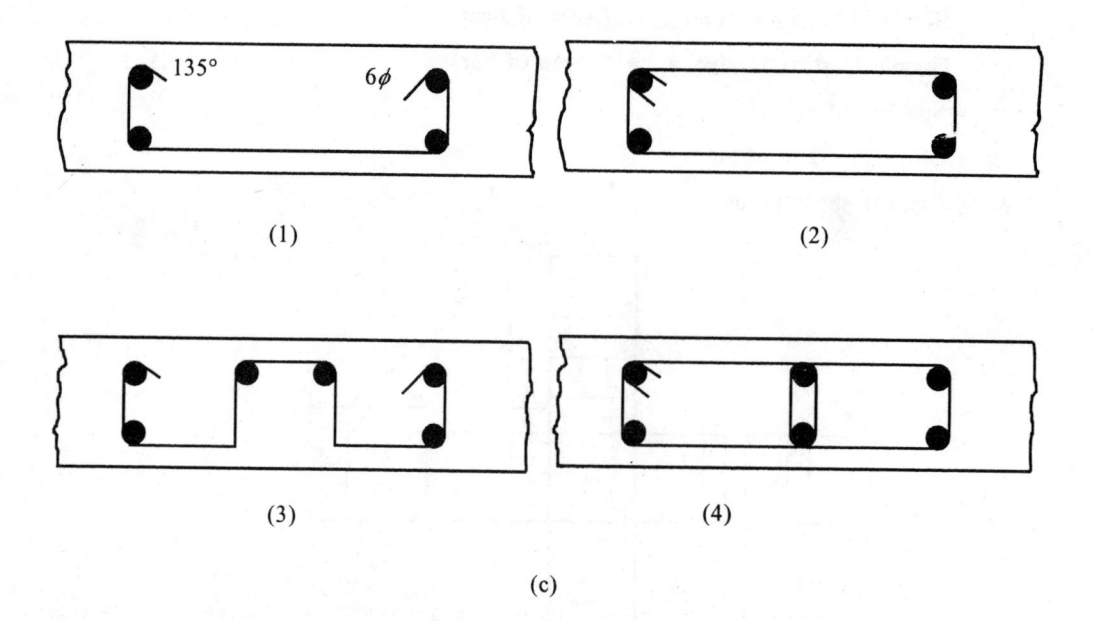

(3) (4)

(c)

Fig. 12.7 Reinforcing flat slabs for shear by ACI method: (b) critical perimeter for external columns (c) details of steel reinforcements.

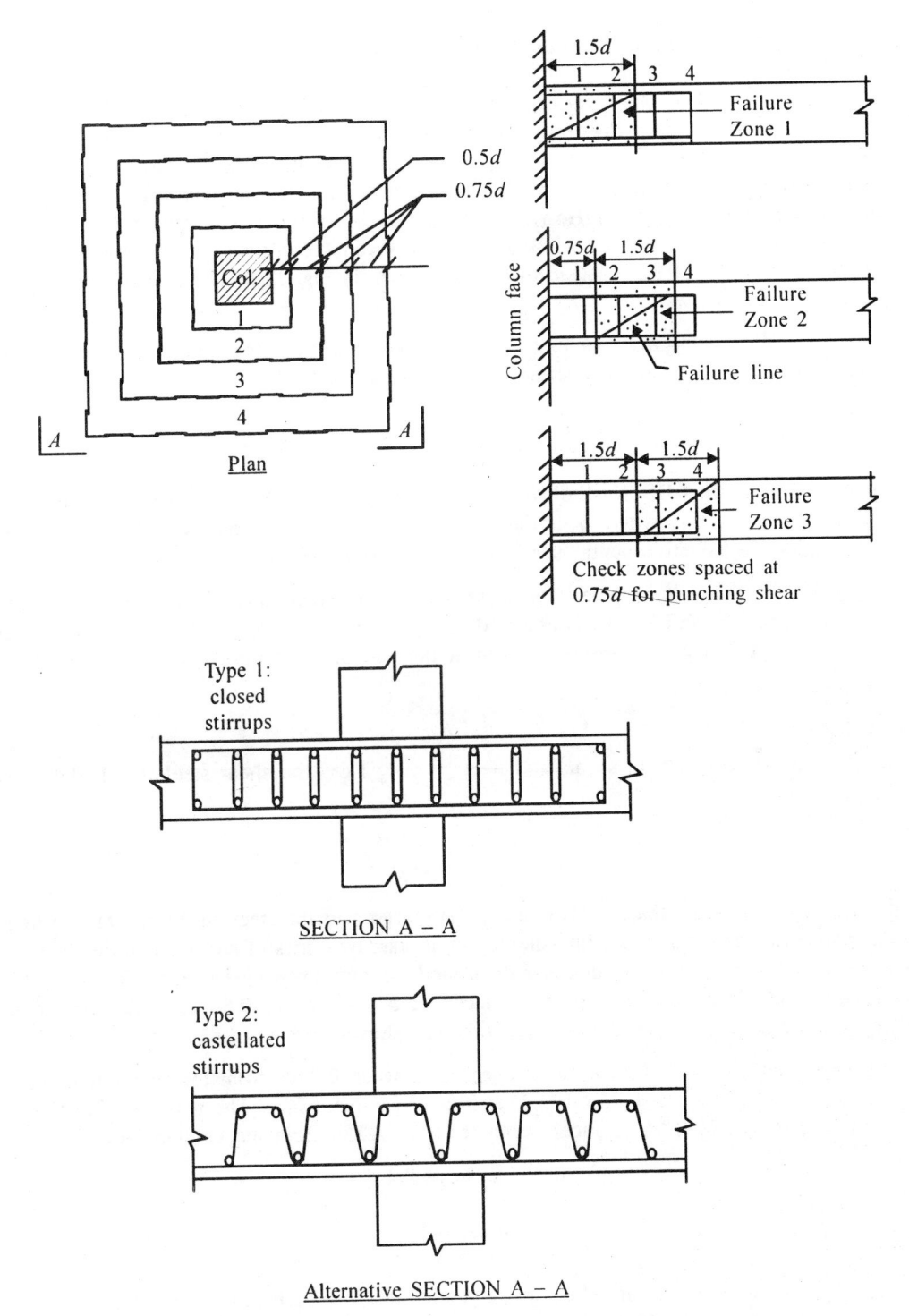

Fig. 12.8 Reinforcing flat slab for shear by BS 8110 method.

12.8.3 Design of Shear Reinforcements by BS 8110

BS Clause 3.7.7.5 allows the shear capacities in slabs, over 200 mm in depth, to be increased by reinforcing it with stirrups arranged in a specified way. The allowable punching shear value in BS 8110 is the same as in one way (beam) shear (Table 19 of IS 456), the percentage of steel to be considered for shear value being the mean value of tension steels in the X and Y directions near the column [3]. However, to compensate for this lesser value of the allowable beam shear compared to the IS and ACI punching shear, the critical perimeters are taken in BS 8110 at distances of 1.5d instead of 0.5d as shown in Fig. 12.8. The design procedure to be adopted is as follows:

Step 1: Calculate V_0 the punching shear at the column face. Its value should be less than the maximum allowable one-way shear (Table 20 of IS 456)

$$v_0 = \frac{V_0}{2(c_1 + c_2)} < \tau_{(max)} \text{ or } 5 \text{ N/mm}^2 \text{ or } 0.8\sqrt{f_{ck}}$$

It should be noted that BS 8110 allows an enhanced value of shear for sections taken near to the support as $2d\tau_c/a_v$ where a_v is the distance of concentrated load from the face of the support and d the effective depth (see also IS 456 Clause 40.5).

Step 2: The first failure zone is taken as the zone between column face and 1.5 d from the column face. The failure line is assumed as shown in Fig. 12.8 and the critical perimeter $b_1 = 2(c_1 + c_2 + 3d)$ is the perimeter taken at the end of the failure line.

$$v_1 = \frac{V_1}{b_1 d}$$

If $v_1 < \tau_c$ the slab is safe in shear, and if $v_1 > \tau_c < 1.6\tau_c$ the shear steel is calculated as follows:

$$A_{sv} = (v_1 - \tau_c)\frac{b_1 d}{0.87 f_y}$$

with $(v_1 - \tau_c)$ is not less than 0.4 N/mm². A_{sv} is the area of the steel cut by the failure line. As it is specified that the failure line should cut at least two lines of stirrups the area of steel obtained above should be distributed evenly around the zone in two perimeters. To satisfy this requirement, the first perimeter steel is placed at a distance of 0.5d and the subsequent perimeters are spaced at 0.75d from each other as shown in Fig. 12.8.

Step 3: The second failure zone is taken as between 0.75d from the first failure zone and extending for 1.5d (i.e. 0.75d to 2.25d) from the column face. The potential failure line is again shown in Fig. 12.8. Critical perimeter is at 2.25d from the column face.

$$b_2 = c_1 + c_2 + 9d$$

$$v_1 = \frac{V_2}{b_2 d}$$

A_{sv} is calculated as in step 2 and the corresponding steel will include the second row of steel already considered by the first failure zone and the third perimeter row.

Step 4: Similar successive zones are checked as shown in Fig. 12.8 until a zone is reached which does not require reinforcement.

Step 5: Detailing of steel. As already stated, the first perimeter of stirrups should be at $0.5d$ from the column face and the subsequent perimeters are spaced at $0.75d$. Either closed stirrups sor castellated stirrups, as shown in Fig. 12.8, can be used. They must pass around steel rods of normal size running perpendicular to the stirrups at each face as shown in the figure.

12.8.4 Design of Fabricated Shear Heads

Another method for providing shear is to use fabricated shear heads. The shear heads fabricated from rolled steel sections or two channel sections as shown in Fig. 12.9 are found to be very

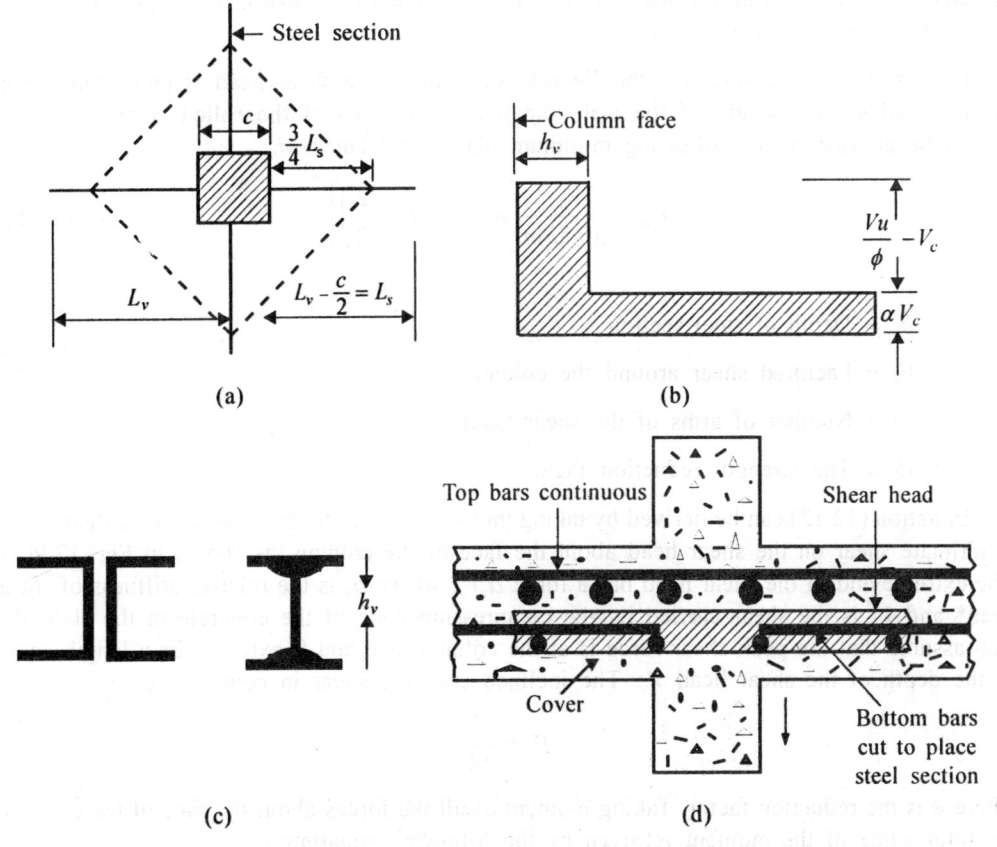

Fig. 12.9 Design of fabricated shear heads (a), (b), (c) and (d).

efficient against failure with large punching shear values (Table 12.2). They are made from rolled sections welded to form four identical arms at right angles to each other for placing over the internal columns. For exterior columns there can be only three arms and in corner columns only two arms. (Eventhough, it is not specified in the codes, it may be advisable in the case of external columns, especially the corner column to extend the shear head into the column also by an arm at right angles to the plane of the shear head reinforcements.) The shear head construction permits the use of a thinner slab above inner columns when the shear controls slab thickness. According to the design, recommended by ACI 318 Clause 11.12.4 shearheads should satisfy the following nine conditions:

Condition 1: We have seen in Section 12.8.1 that the maximum punching shear calculated at $d/2$ from face of column for use of bar reinforcement is $1.5\tau_p$. With shear head, the maximum punching shear at the conventional section can be more but should not exceed $1.75\tau_p$, where τ_p is given by Eqs. 12.3 to 12.5.

Condition 2: The length of the arms of the shear head should be such that the punching shear stress calculated on the new critical perimeter should not exceed $\tau_p = 0.25 f_{ck}$. This new critical perimeter for punching shear calculation with shear heads is taken at 0.75 times the distance of the shear head arm from the face of the column as shown in Fig. 12.9. Shear stress is calculated by using Eq. (12.2b).

Condition 3: To ensure that the flexural strength of the shear head is much more than the required shear strength of the slab, the plastic modulus of the rolled section selected should be at least of the following minimum plastic resistance M_p:

$$M_p = \frac{V_u}{2(0.85)n}\left[h_v + \alpha_v\left(L_v - \frac{c}{2}\right)\right] \qquad (12.12)$$

where

h_v = Depth of steel section

V_v = Factored shear around the column

n = Number of arms of the shear head

0.85 = The strength reduction factor

Equation (12.12) can be derived by taking moments of the forces assumed to be distributed at ultimate shear on the shear head about the face of the column, as shown in Fig. 12.9(b). The extreme end of the shear head has a force $\alpha_v V_c$, where α_v is the relative stiffness of shear head* and V_c is the shear carried by the compression zone of the concrete in the slab. We also assume that the peak shear force is at the column face and it extends for a length equal to the depth of the shear head V_v. The inclined cracking shear in concrete is

$$V_c = \frac{V_u}{2\phi}$$

where ϕ is the reduction factor. Taking moment of all the forces about the face of the column, the total value of the moment is given by the following equation:

$$M_p = \frac{V_u}{2\phi}h_v + \frac{\alpha_v V_u}{2\phi}\left(L_v - \frac{c}{2}\right) \qquad (12.13)$$

With four arms (n = 4) each of them should have an M_p given by Eq. (12.12).

---

*See condition 6.

$$M_p = \frac{V_u}{(0.85)(8)}\left[h_v + \alpha_v\left(L_v - \frac{c}{2}\right)\right] \tag{12.13a}$$

Condition 4: The overall depth of the steel section must not exceed 70 times the web width, i.e.

$$h_v > 70t_w$$

Condition 5: The bottom flange (compression flange) of the steel section must be located within a distance of $0.3d$ of the bottom of the slab.

Condition 6: The value α_v can be defined as the ratio of the stiffness of each shear head arm to the stiffness of the surrounding composite cracked slab having a width equal to the width of the column *plus* the effective depth c (i.e. $c + d$). This should not be less than 0.15. Taking $E_s/E_c = m$ and I_{xx} = moment of inertia of the arm, we get

$$\alpha_v = \frac{mI_{xx}}{I(\text{of composite section})} \geq 0.15 \tag{12.14}$$

The composite cracked section will have a width $(c + d)$ with the concrete and steel sections and the negative steel placed on top side of the slab. The bottom side of the slab will be in compression in this zone.

Condition 7: The shear head can be assumed to add to the column strip negative moment resistance by a value M' given by the equation below:

$$M' = \frac{V_u}{2n}\alpha_v\left(L_v - \frac{c}{2}\right) \tag{12.15}$$

However this value should not be more than 30% of the column strip negative moment or the change in the column strip moment over the length L_v.

Condition 8: The steel shapes must be fabricated by welding with a full penetration weld into identical arms at right angles.

Condition 9: The ends of the steel joints must be square cut or cut at an angle not less than 30° with the horizontal, provided the plastic moment strength of the remaining tapered section is equal or greater than M' given in equation (12.15).

(*Note:* These steel sections are placed under the top mat of negative steel at supports which should be uninterrupted, and by cutting some of the bottom steels in the slab at the top of the columns. This is shown in Fig. 12.9(d).

12.9 EFFECT OF OPENINGS IN FLAT SLABS

There are many situations in which openings cannot be avoided in flat slab construction. These are dealt with in detail in IS 456 Clause 31.8. Codes classify openings into the following two types:

1. Those which do not need special analysis
2. Those which need special analysis

The main recommendations for dealing with these holes are as follows:

Openings which do not need special analysis. They include the following:

(i) Openings of any size completely within the intersecting middle strips (within the middle half of the spans in each direction).

(ii) Openings of one-quarter width of each strip in the common areas of a column strip and a middle strip of limited size (viz., one-eighth of the span length in each direction). If the reinforcements interrupted in these openings are put back on all sides of the openings, these slabs are found to behave well.

Situations which need special analysis for shear. Openings of 1/8th width of the strip in areas common to two column strips are found to be generally safe. The presence of other openings near the column of flat slabs is not desirable especially when they are situated on the more stressed part of the perimeter. One of the methods to take care of these openings is to check its safety by reducing the effective perimeter in design calculation for shear as indicated in IS 456 Clause 31.6.1.2. According to ACI 318 (1989) Clause 11.12.5.2., for slabs with shear head reinforcements, the ineffective portion of the perimeter is taken as one-half the theoretical value calculated for flat slabs without shear heads.

12.10 RECENT REVISIONS IN ACI 318

The following revisions have been proposed in the ACI 318 (1993) code Clause 13.5.3.3 to simplify design [3]. The unbalanced moment about the axis parallel to the edge may be assumed to be carried more by moment transfer (with less by torsion and shear transfer) provided that the V_u calculated for the edge support is only 75% or less of its shear capacity in the case of edge columns and 50% of its shear capacity at corner columns. Under these cases, the tedious calculations for shear due to shear transfer by torsion need not be carried out. If a member size falls short of the theoretical requirements, its safety can be checked by adjusting the level of torsion without revising the member sizes.

12.11 SHEAR IN TWO-WAY SLABS WITH BEAMS

In two-way slabs with rigid beams the load is transferred from slabs to the beams from the contributory areas, assumed to be formed by the 45° lines at the corners. The beams carry the load to the columns. The shear in the beams and the design moments in the external and corner columns are based on this load transfer mechanism [4,5]. If the value of $\alpha L_2/L_1 < 1$, the beam cannot be considered as rigid. In such case the load in the slab is transferred to the column partly through the beams and the rest of the load goes from the slab directly to the column as already explained in Section 11.13 and Table 11.4. Design for shear of the slab and beam must be made using this principle.

EXAMPLE 12.1 (Checking for shear)
A flat plate with 7.5 × 6 m panels on 500 × 500 mm columns has a slab thickness of 185 mm, designed for a total characteristic load (DL + LL) of 9.3 kN/m². Check the safety

of the slab in shear if grade 25 concrete and grade 415 steel are used for its construction. How can we increase the shear capacity of the slab?

Reference	Step	Calculations
	1	*Design Data* Effective depth assuming 25 mm cover and use of 10 mm steel $d = 185 - 25 - 5 = 155$ mm Factored load $= 1.5 \times 9.3 = 13.95$ kN/m^2
	2	*Check for one-way shear for a central column* Assume no shear along cuts through the mid-points between slabs Take section at distance $d = 155$ mm from face of the column Contributory area for shear $$= 6\left(\frac{7.5}{2} - 0.25 - 0.155\right) = 20.07 \text{ m}^2$$ V_u for one-way shear $20.07 \times 13.95 = 280$ kN Length resisting shear $b = 6000$ mm $$v = \frac{V_u}{bd} = \frac{280 \times 10^3}{6000 \times 155} = 0.30 \text{ N/mm}^2$$
IS 456 Table 19		Assume $p = 0.2\%$; τ_c for $M_{25} = 0.33$ Hence, safe in one-way shear.
	3	*Check for punching shear* Distribution of critical perimeter from column face $= d/2 = 77.5$ mm Area inside critical perimeter $= (c + d)^2$ $= (500 + 155)^2 = (655)^2$ mm^2 Load on critical perimeter $V = 13.95 \, [(7.5 \times 6) - (0.655)^2]$ $= 621.76$ kN Critical perimeter $b_0 = 4(c + d) = 4 \times 655 = 2620$ mm Punching shear $= \dfrac{V}{b_0 d} = \dfrac{622 \times 10^3}{2620 \times 155} = 1.53$ N/mm^2
Eqs. (12.3)– (12.5)	4	*Check allowable punching shear* $\tau_p = 0.25\sqrt{f_{ck}} = 1.25$ N/mm^2 $\tau'_p = (0.5 + \beta_c)\tau_p = 1.5\tau_p$

Reference	Step	Calculations
Eqs. (12.3) to (12.5)		$\tau_p'' = \left(0.5 + \dfrac{\alpha_s d}{4b_0}\right)\tau_p$ with $\alpha_s = 40$ $= 0.5 + \dfrac{40 \times 155}{4 \times 2620} = 1.09\tau_p$ Allowable = 1.25 N/mm². Hence the slab is not safe in punching shear.
	5	*Method of increasing shear capacity* (a) Increase the strength of concrete (b) Provide a drop and if necessary provide a capital also (c) Provide shear reinforcement in the form of bar reinforcement or fabricated-shear head reinforcement
	6	*Determine necessary strength of concrete to withstand punching shear* $0.25\sqrt{f_{ck}} = 153$; $f_{ck} = \left(\dfrac{1.53}{0.25}\right)^2 = 37.5$ N/mm² Use 40 grade concrete [*Note:* Example 12.2 shows the design of stirrups for the shear.]

EXAMPLE 12.2 (Design of reinforcement for shear)

Design the necessary stirrups for reinforcing the slab in Example 12.1 against the shear assuming $f_x = 25$ and $f_y = 415$ N/mm² as in Fig. 12.7.

Reference	Step	Calculations
	1	*Design data (As in Example 12.1)*
Ex. 12.1		$V_p = 1.53$ N/mm²; $\tau_p = 1.25$ N/mm²
Step 3		$b_0 = 2620$ mm; $V_u = 622$ kN
	2	*Check suitability for shear reinforcement* Allowable maximum shear stress with reinforcement
Eq. (12.10)		$v_p = 1.5\tau_p = 1.5 \times 1.25 = 1.875$ N/mm² Hence shear steel can be safely used. As slab is thinner than 250 mm, closed stirrups with one bar at each corner is to be provided.
	3	*Calculate shear that resists at the critical section with shear steel*
Step 4, Example 12.1		τ_p' allowable with shear steel $= \dfrac{\tau_p}{2} = \dfrac{1.25}{2} = 0.625$ N/mm²

Reference	Step	Calculations
Steps 3		Shear that can carry concrete = $V_c = \tau'_c b_0$ = 0.625 × 2620 × 155 = 254 kN Shear to be carried by steel = V_s = 622 − 254 = 368 kN
Eq. (12.11)	4	Design of stirrups as shown in Fig. 12.7 $\dfrac{V_s}{d} = \dfrac{A_{sv}(0.87 f_y)}{s_v}$ For internal columns shear steel is provided as four strips of two-legged closed stirrups of maximum spacing $s_v = d/2 = 77$ mm Adopt spacing of 60 mm $A_{sv} = \dfrac{V_s}{d} \dfrac{s_v}{0.87 f_y} = \dfrac{268 \times 10^3 \times 60}{155 \times 0.87 \times 415} = 287 \text{ mm}^2$ for all the four sides. Steel for each side = $\dfrac{287}{4} = 71.75 \text{ mm}^2$ For each leg = $\dfrac{71.75}{2} = 36 \text{ mm}^2$ Provide 8 mm bar giving an area of 50.3 mm². (Alternately, design by SP.16 Table 62.)
	5	Determine distance to which stirrups should extend Shear steel can be stopped at $d/2$ (i.e. 78 mm) for the perimeter where shear in concrete is $\tau_p/2$.) Let the distance of this perimeter be (a) from face of column $\lambda'_0 = 4(\text{column size} + a\sqrt{2}) = 4(500 + 1.41a)$ $\dfrac{1}{2}\tau_p = 0.625 \text{ N/mm}^2$ Assume $V_u = 622$ kN (it will be much less) 0.625 × 4(500 + 1.41a) × 155 = 622000 a = 784 mm from edge of column $a - \dfrac{d}{2} = 784 - 78 = 706$ Provide closed loops of width 500 mm and depth 135 mm with cover of 25 mm with 10 mm rods at each of the corners of the column and extend it to 700 mm from the face of the column.

EXAMPLE 12.3 (Shear in flat plate and flat slabs)

Indicate how the flat plate in Example 12.1 can be converted into a flat slab so that it is safe against shear (see Fig. 12.10).

232 ADVANCED REINFORCED CONCRETE DESIGN

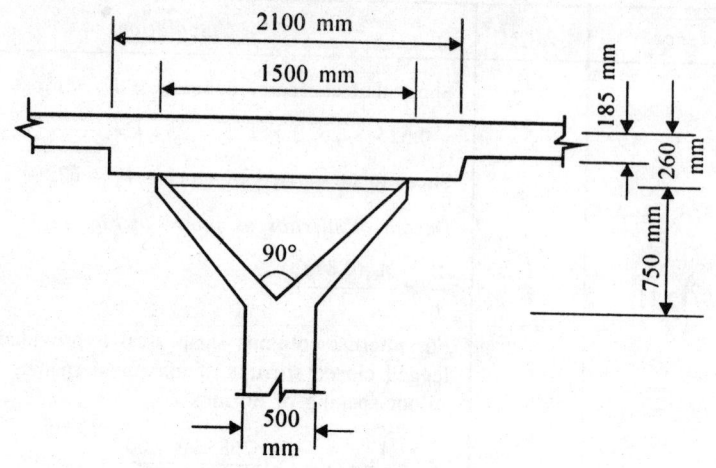

Fig. 12.10 Example 12.3.

Reference	Step	Calculations
	1	*Provide a drop and column head for layout*
		Thickness of drop = 1.25 to 1.5h (assume 1.4h)
		h' = 1.4 × 185 = 259 mm (say 260 mm)
		h' − h = 260 − 185 = 75 mm
		Projection of drop = 4 × 75 = 300 mm from end of capital
		Approximate size of column capital = $L_1/5$ = 7.5/5 = 1.5 m
		Size of drop = 1500 + 600 = 2100
		Final dimensions
		Size of column = 500 × 500 mm
		Size of column capital = 1500 × 1500 mm
		Size of drop panel = 2100 × 2100 mm
		The capital should be so dimensioned that the 90° line lies inside the capital (Fig. 12.10)
	2	*Take first critical section for punching shear and calculate shear*
		Critical section at d/2 from end of column capital
		d' = 260 − 30 = 230 mm
		Size of critical perimeter area = 1500 + 230 square
		Area inside perimeter = 1.73 × 1.73 m²
		Load from contributory area for punching shear
		= 13.95 [(7.5 × 6) − (1.73)²] = 586 kN

SHEAR IN FLAT SLABS AND FLAT PLATES 233

Reference	Step	Calculations
Ex. 12.1 Step 4	3	Critical perimeter = 4 × 1730 mm = 6920 mm Punching shear = $\dfrac{586 \times 10^3}{6920 \times 230} = 0.37 \text{ N/mm}^2$ This is less than the allowable value. *Take second critical section for punching shear* Critical section at $d/2$ from the end of the drop Size of critical section = 2100 + 155 square Punching shear = $\dfrac{13.95[(7.5 \times 6) - (2.255)^2]}{4 \times 2255} = 0.062 \text{ N/mm}^2$ This section is also safe in punching shear. Hence the flat slab is safe in shear without any reinforcements.

EXAMPLE 12.4 (Design of shear head reinforcement)

Design a fabricated shear head for an internal column for the slab in Example 12.4 (*see* Fig. 12.11).

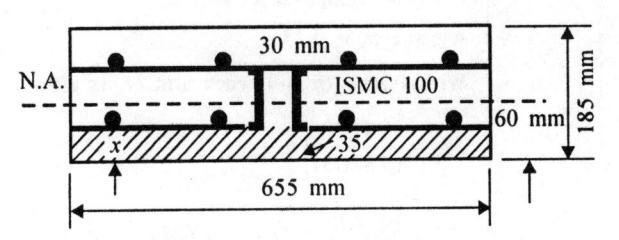

Fig. 12.11 Example 12.4.

Reference	Step	Calculations
Section 12.8.4	1	*Check whether shear in slab is less than the allowable one* For shear head, punching shear should not exceed $1.75 \tau_p$. Allowed $\tau_p = 1.25 \text{ N/mm}^2$ $1.75 \times 1.25 = 2.18 \text{ N/mm}^2$ v_p of slab = $1.53 < 2.18$ Shear head can be provided to resist shear.
	2	*Determine the required critical perimeter for safety in shear* $V_u = 622 \text{ kN} = b_0 \times \text{Depth} \times \tau_p$ $b_0 \times 155 \times 1.25 = 622 \times 10^3$

234 ADVANCED REINFORCED CONCRETE DESIGN

Reference	Step	Calculations
Eq. (12.2b)	3	$b_0 = 3210$ mm The shear head layout should give this value of b_0. Find L_v to give required b_0 $b_0 = 4\sqrt{2}\left[0.5c_1 + 0.75\left(L_v - \dfrac{c_1}{2}\right)\right]$ $4\sqrt{2}\,[250 + 0.75(L_v - 250)] = 3210$ $L_v = 675$ mm
Eq. (12.12)	4	Determine the required plastic moment of shear head to prevent premature bending failure $M_p = \dfrac{V_u}{8(0.85)}\left[h_v + \alpha_v\left(L_v - \dfrac{c}{2}\right)\right]$ Assume for slab thickness of 185 mm, the depth of steel section $h_v = 100$ m.
Eq. (12.14)		$\alpha_v = \dfrac{m\,I_{xx}}{I(\text{of composite section})} \geq 0.15$ Assume $\alpha_v = 0.25$
Eq. (12.13a)		With 4 arms ($n = 4$) each arm M_p is given by $M_p = \dfrac{V_w}{(0.85)(8)}\left[h_v + \alpha_v\left(L_v - \dfrac{c}{2}\right)\right]$ $= \dfrac{622 \times 10^3}{0.85 \times 8}\,[100 + 0.25\,(675 - 250)]$ $= 189 \times 10^5$ Nmm $= 189 \times 10^3$ kg-cm $Z_p = \dfrac{M_p}{f_y} = \dfrac{189 \times 10^3}{2500} = 75.6\text{ cm}^3$ From steel tables choose 2 channels to this plastic modulus. Alternately, $Z_p = 1.15$ (elastic Z) $Z_{xx} = \dfrac{75.6}{1.15} = 65.7\text{ cm}^3$ Two channels ISMC 100 of $Z_{xx} = 37.3\text{ cm}^3$ Total $Z_{xx} = 2 \times 37.3 = 74.6\text{ cm}^3$ Area $= 2 \times 11.7 = 23.4\text{ cm}^3$ $I_{xx} = 2 \times 186.7 = 373.4\text{ cm}^4$

Reference	Step	Calculations
	5	Shear head consists of two channels extending to 675 mm from centre of column on all the four sides of the column (bottom cover 35 mm). CG of steel section = 35 + 50 = 85 mm from bottom. *Check the assumed relative stiffness of section (α_v)* Moment of inertia of composite section Width of section = $c + d$ = 500 + 155 = 655 Depth of section = Depth of slab = 185 mm Assume 4 nos 16 mm bars at top of section (800 mm^2). Cover to steel = 35 mm Two channels I_{xx} = 373.4 × 10^4 mm^4, Area = 2340 mm^2 $m = \dfrac{280}{f_{ck}}$ (assume m = 9) We have to find NA of section and I of equivalent section.
	5.1	*Find NA (let it be x from the bottom of section)* Equivalent areas of steel is mm^2 in section Top steel rods 9 × 800 = 7200 Steel section 9 × 2340 = 210.6 × 10^2 CG of combined section (from bottom) $\dfrac{(655)x^2}{2} = 210.6 \times 10^2(85-x) + 7200(150-x)$ x = 60 mm Concrete is in compression at bottom at supports.
	5.2	*Find I_c of composite section about CG* = That of (steel section + of steel rods + concrete) $= 9(373.4)10^4 + 210.6 \times 10^2(85-60)^2 + 7200(90)^2 + \dfrac{655(60)^3}{3}$ $= 152.25 \times 10^6$ mm^4
	5.3	*Find relative stiffness* $\alpha_v = \dfrac{mI_{xx} \text{ of steel}}{I_c} = \dfrac{(9)(373.4)10^4}{(152.25)10^6} = 0.221$ This is more than 0.15. Section adopted is permissible.

Reference	Step	Calculations
Eq. (12.15)	6	*Estimate the contribution of shear head towards negative moment* $$M' = \frac{V_u}{2n}\alpha_v\left(L_v - \frac{c}{2}\right) = \frac{622}{8}(0.221)(675-250) = 7302 \text{ kNm}$$ M' should not be taken more than 30% of column strip negative moment or change in column strip moment over a length equal to L_v.

EXAMPLE 12.5 (**Critical perimeter of non-rectangular columns**)
Determine the ultimate punching shear strength of a slab of effective depth 150 mm supported on a L shaped column ABCDE as shown in Fig. 12.12. Assume $f_{ck} = 25$ N/mm^2 AB = 600, BC = 1200 and AF = 200 mm.

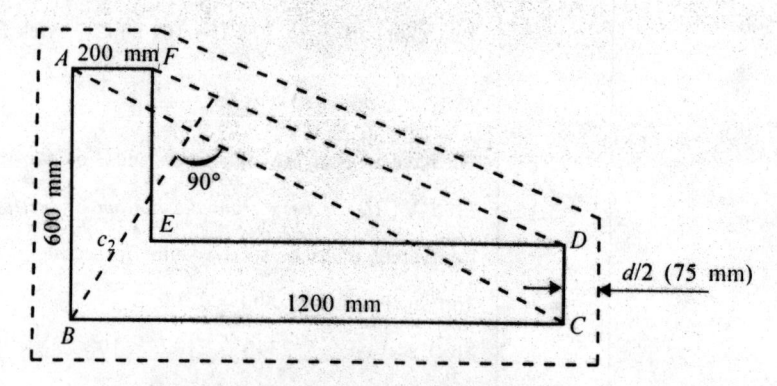

Fig. 12.12 Example 12.5.

Reference	Step	Calculations
	1	*Effective loaded area ABCDFA* Determine the area by measurement of the loaded area c_1 and c_2. Large dimension = $AC = c_1 = 1340$ mm Short dimension = perpendicular from B to AC $c_2 = 537$ mm $\beta_c = c_2/c_1 = 537/1340 = 0.40$
Eqs. (12.3)–(12.6)	2	*Determine the punching shear value from the three equations* $\tau_p = 0.25\sqrt{f_{ck}} = 1.25$ N/mm^2 $\tau'_p = (0.5 + 0.40)\tau_p = 1.125$ N/mm^2

Reference	Step	Calculations
		$\tau_p'' = \left(0.5 + \dfrac{\alpha_s d}{4b_0}\right)\tau_p$
		b_0 = Perimeter at a distance of $d/2$ from the column face = 4200 mm
		α_s = 40(for interior column)
		$\tau_p'' = \left[0.5 + \dfrac{40 \times 150}{4 \times 4200}\right]\tau_p = 1.07$ N/mm^2
		Adopt the least among the three = 1.07 N/mm^2
	3	Estimate the ultimate shear strength
		$V = b_0 d \tau_p = 4200 \times 150 \times 1.07 = 674$ kN
		Punching shear strength = 674 kN

EXAMPLE 12.6 (Shear due to moment transfer in columns)
A flat plate 185 mm thick (effective depth 150 mm) has 7 m × 6 m panels supported on 500 mm square columns. The total factored design load is 12 kN/m^2. Assuming that the column strip negative moment $M = 65$ kNm:

1. Design the strip around an *external column* for the moment transfer and check for safety in punching shear at the external columns where $f_{ck} = 25$ N/mm^2 and $f_y = 415$ N/mm^2. (*Note:* Transfer of moment from slab to column is very important in flat slab design. First, this moment transfer may be assumed to take place through a narrow width of slab (1.5d) around the column. This portion should be specially designed for this moment transfer. Secondly, shear capacity of the columns for withstanding the moment transfer should not be based not on the theoretical value obtained by coefficients but it should also be based on the ultimate moment capacity of the slab including the special reinforcements provided. So that shear failure (which is sudden and brittle) can never take place at these places. (*see* Sections 11.15.3 and 11.15.4).

Reference	Step	Calculations
	1	Design column strip for negative moment
		$M = 65$ kNm, $d = 150$ mm, $b = 6000/2 = 3000$ mm
		$\dfrac{M}{bd^2} = 0.96$ Hence $p = 0.282$
		Area of steel = $\dfrac{0.282 \times 3000 \times 150}{100} = 1269$ mm^2
		Provide 12 bars of 12 mm (1352 mm^2)
	2	*Determine moment transfer components (bending moment part + torsion part)*

Reference	Step	Calculations
Eq. (12.6)		Unbalanced moment at edge = 65 kNm = M Portion transferred as bending = kM Portion transferred as torsion = $(1-k)M$ $k = \dfrac{1}{1+(2/3)\sqrt{a/b}}$ $a = c_1 + \dfrac{d}{2} = 500 + 75 = 575$ mm (external column) $b = c_2 + d = 500 + 150 = 650$ mm $k = \dfrac{1}{1+(2/3)\sqrt{575/650}} = 0.615$ *Moment transferred as bending* $M_1 = 0.615 \times 65 = 40$ kNm Only 25 kNm is in torsion Torsion transfer coefficient = $1 - k = 1 - 0.615 = 0.385$
	3	*Design additional steel for bending moment part* Area within the effective slab width of $2(1.5d)$ around the column has to be specially reinforced for transfer of M_1. Width of transfer = $c_2 + 2(1.5d) = 500 + (3 \times 150) = 950$ mm
SP 16 Table 3		$\dfrac{M}{bd^2} = \dfrac{40 \times 10^6}{950(150)^2} = 1.87$ $p = 0.574$ $A_s = \dfrac{0.574 \times 950 \times 150}{100} = 818$ mm^2 Put 8 nos. of 12 mm (900 mm^2) in 950 mm width.
Steps 1 and 3		Total number of bars in column strip = 12 + 8 = 20 bar of 12 mm Area = 2262 mm^2 Percentage = $\dfrac{2262 \times 100}{3000 \times 150} = 0.50$
SP 16 Table 3		Corresponding $M/bd^2 = 1.65$ $M = 1.65 \times 3000 (150)^2 = 111$ kNm

SHEAR IN FLAT SLABS AND FLAT PLATES 239

Reference	Step	Calculations
See the 'Note', Example 12.6	4	This is about 1.7 times the moment (67 kNm) obtained by coefficients. According to ACI code, the torsion transfer for design of shear in column should be based on this increased value. Determine the torsion moment capacity of column strip $1 - k = \eta = 0.385$; $T_m = 0.387 \times 111 = 42.7$ kNm (*Note:* In step 2, torsion transfer of 65 kNm was only 25 kNm.)
	5	*Calculate shear in column* Shear produced by torsional moment plus direct load from slab.
	5.1	*Sectional properties for shear in torsion*
Step 2		Section $d/2$ around column $a = 500 + 75 = 575$ mm $b = 500 + 150 = 650$ mm Critical perimeter $= 2a + b = 1800$ mm
Table 12.2 Case 2 External column		$x_1 = \dfrac{a^2}{2a+b} = 183.7$ $x_2 = 575 - 183.7 = 391.3$ $A_c = (2a+b)d = 2.70 \times 10^5$ mm^2 $J = \dfrac{575(155)^3}{6} + \dfrac{155(573)^3}{6} + 575 \times 155 \left(\dfrac{a}{2} - x_1\right)^2 + bdx_1^2$ $= 9449 \times 10^6$ mm^4 $\dfrac{J}{x_1} = 51.4 \times 10^6$ $\dfrac{J}{x_2} = 24.15 \times 10^6$
	5.2	*Calculate shear force for punching shear* $V = $ (Load)(Contributory area) $= \dfrac{12 \times 7 \times 6}{2} = 252$ kN Perimeter area resisting shear $= d \times b_0$ $A_c = 150 \times 1800 = 2.70 \times 10^5$ mm^2
	5.3	*Determine maximum and minimum punching shear* $u = \dfrac{V}{A_c} \pm \dfrac{M_t}{J} x$

Reference	Step	Calculations
		Maximum shear $= \dfrac{252 \times 10^3}{270 \times 10^3} + \dfrac{42.7 \times 10^6}{51.4 \times 10^6} = 1.76 \text{ N/mm}^2$
		Minimum shear $= \dfrac{252 \times 10^3}{270 \times 10^3} - \dfrac{42.7 \times 10^6}{24.15 \times 10^6} = -0.83 \text{ N/mm}$
		The negative sign indicates that the shear due to torsion is more than that due to direct shear.
	6	*Calculate permissible value by the three formulae*
Eq. (12.3) to (12.5)		$\tau_p = 0.5\sqrt{f_{ck}} = 1.25 \text{ N/mm}^2$ $\tau'_p = (0.5 + 1)\tau_p = 1.5\tau_p$ $\tau''_p = 0.5 + \dfrac{30 \times 150}{4 \times 1800}\tau_p = 1.13\tau_p$
		Allowable shear = 1.25 N/mm² 1.76
		Shear resistance can be increased by designed shear reinforcements as shown in Example 12.2.

REFERENCES

1. IS 456 (2000), Code of Practice for Plain and Reinforced Concrete, Bureau of Indian Standards, New Delhi.

2. ACI 318 (1989), Building Code Requirements for Reinforced Concrete, American Concrete Institute, U.S.A.

3. Proposed Revisions to Building Code Requirements for Reinforced Concrete, ACI 318 (1989), Concrete International, December 1994.

4. Ferguson, P.M., Breen, J.E. and Jirsa, J.O., *Reinforced Concrete Fundamentals*. John Wiley and Sons, New York, 1988.

5. Purushothaman, P., *Reinforced Concrete Structural Elements*, Tata McGraw-Hill, New Delhi, 1984.

CHAPTER 13

Equivalent Frame Analysis of Flat Slabs

13.1 INTRODUCTION

Two-way slab systems that do not satisfy the limiting conditions for the direct design method (DDM) should be analyzed by the equivalent frame method (EFM). The equivalent frame method is very similar to DDM but it uses classical methods of analysis, instead of coefficients, to give the positive and negative moments in the longitudinal direction. They should be then distributed transversely in the same way as already described in the DDM in Section 11.9. Thus, the difference between the DDM and EFM analysis for gravity loads lies only in the procedure of getting the magnitude of the longitudinal negative and positive moments.

As already pointed out, when dealing with effects of lateral loads due to wind or earthquake on the flat slab frames, direct designs procedure is not permitted. Analysis of all such frames, irrespective of whether they obey the conditions for DDM or not, have to be made by the equivalent frame analysis or other theoretical methods. Similarly, raft slabs whose column loads are fixed, if designed as two-way slabs, should be also analyzed by the equivalent frame method. ACI Clause 13.3.1.2 states that lateral load analysis of unbraced frames shall take into account the effects of cracking and reinforcement on reducing the stiffness of frame members.

This chapter gives a brief outline of the equivalent frame analysis for two-way slabs according to ACI code. It should be noted that the method given in IS 456 uses approximation of the ACI method. A comparison of the results obtained by the two methods is given in *Reinforced Concrete Structural Elements, Behvior Analysis and Design* [1].

13.2 HISTORICAL DEVELOPMENT OF THE CONCEPT OF EQUIVALENT FRAME

The idea of equivalent frame was first proposed in its elementary form in the California Building Code as early as in 1933. A modified version of the same was introduced in ACI code in 1941. Since then this method has appeared with various improvements in all the subsequent ACI codes. In the earlier version of ACI codes, the stiffness of the slab which is referred commonly as the *slab beam* and the columns were to be calculated as in ordinary frame analysis form basic principles. In 1961, Corely and others suggested that the column stiffness should be modified by taking into account the torsional stiffness of slabs on the sides of the column too. It was also pointed out by Sozen, Seiss, et al., on the controlled test results in the laboratory that the observed span moments were more and 'support moments' were less than those obtained by the conventional analysis of that time. To get calculated values nearer to the test values it was suggested that the stiffness of the column should be reduced by introducing the concept of an additional torsional member. Thus the concept of 'equivalent column'[1] was introduced. In 1970, Corely and Jirsa developed an empirical formula for K_t the torsional stiffness of the transverse member, to reduce the support moment and the calculation to agree with test results. Such a modified analysis was later (1977) recommended in the ACI code. However, it was shown in 1981 by Vanderbilt that modification can as well be made in the beam stiffness and analysis can be made using *equivalent beam method*. Accordingly in revised ACI 318 (1983) code the provisions of ACI (1977) code have been revised and presented in a more general form permitting any acceptable method for design. It may however be noted that the equivalent column analysis for which readymade tables are available is more often used than the equivalent beam method.[2]

13.2.1 Definition of Equivalent Frame

As we have already seen in Chapter 11, the equivalent frame is a simple substitute of a two-dimensional model for a three-dimensional frame consisting of slabs and columns. Thus, the building frame is cut vertically from top to bottom first in the longitudinal (E-W) direction and then in the transverse (N-S) direction along the centre lines of adjacent panels. In general the frames used for analysis in EFM methods will consist of three members as shown in Fig. 13.1:

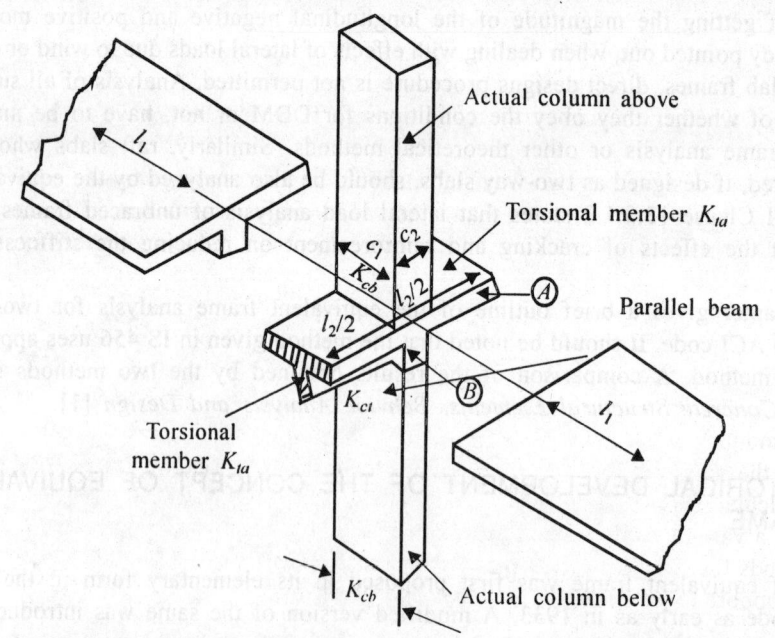

Fig. 13.1 Equivalent column (column and torsional member).

---
[1] See Section 13.3.4.
[2] See Section 13.7.2.

1. A series of slabs (or slab beams) in the longitudinal direction
2. Columns extending above and below the slabs
3. The transverse moment transfer elements (torsional members) consisting of the slab and beams (if any) at the columns in the transverse direction.

For the exterior row of column the slab beam will consists of only one-half the interior panel. It should be noted that in EFM analysis the spans are considered from centre-to-centre of slabs, but the design moments are reduced[1] to those at the face of the supports. These resulting frames are then reduced to two-dimensional model with columns as vertical members and the slabs compressed to equivalent beams as horizontal members. This two-dimensional models are the equivalent frames and they should be analyzed by methods of elastic analysis such as moment distribution, slope deflection or matrix displacement methods. Each floor is analyzed separately with the columns assumed fixed at the floors above and below for the worst live load condition (pattern load) on the 'slab-beams'.[2] As this method is considered to be an exact method, redistribution of moments is also allowed. The analysis is made in both the E-W and N-S frames for the factored dead and live loads.

13.3 BACKGROUND OF ACI (1977) METHOD—EQUIVALENT COLUMN METHOD

An equivalent frame is a two-dimensional representation of the three-dimensional structures. For the sake of structural analysis, it is obvious that appropriate moment of inertia and expressions for rigidity should be assigned to the beams and columns. The values to be assigned were determined from results of extensive tests so that the results obtained by the theoretical analysis agree very well with the results of the laboratory tests.

13.3.1 Moment of Inertia of Slab-beams

The first problem is to assign suitable rigidity value for the slab-beam. ACI 318 (1989) in Clause 13.7.3 gives the rules regarding assumptions to be made for calculating the moment of inertia of these slab-beams. In flat plates where the moment of inertia of the slab is constant over the whole span it is given by the simple equation:

$$I_s = \frac{L_2 h^3}{12} \tag{13.1}$$

where L_2 is the transverse span and h the thickness of the slab.

The moment of inertia of the plates inside the columns is corrected but it is indeterminate. However the omission for this correction does not make much difference in the overall results of the calculations for flat plates. In flat slabs with drop panels and column capitals the slab will have a varying cross-section as shown in Fig. 13.2 and corrections have to be made. ACI recommends the moment of inertia of such slab beams from the face of support to the centre line of support to be assumed as equal and given by.

$$I'_s = \frac{I_{sd}}{[1-(c_2/L_2)]^2} \tag{13.2}$$

---

[1]See Section 13.6.
[2]See Section 13.5.

where

I_{sd} = Moment of inertia of the slab-beam at the face of column-capital or bracket.

c_2 = Size of the column (or equivalent square section column as described in Section 10.5) in the transverse direction

L_2 = Span in the transverse direction.

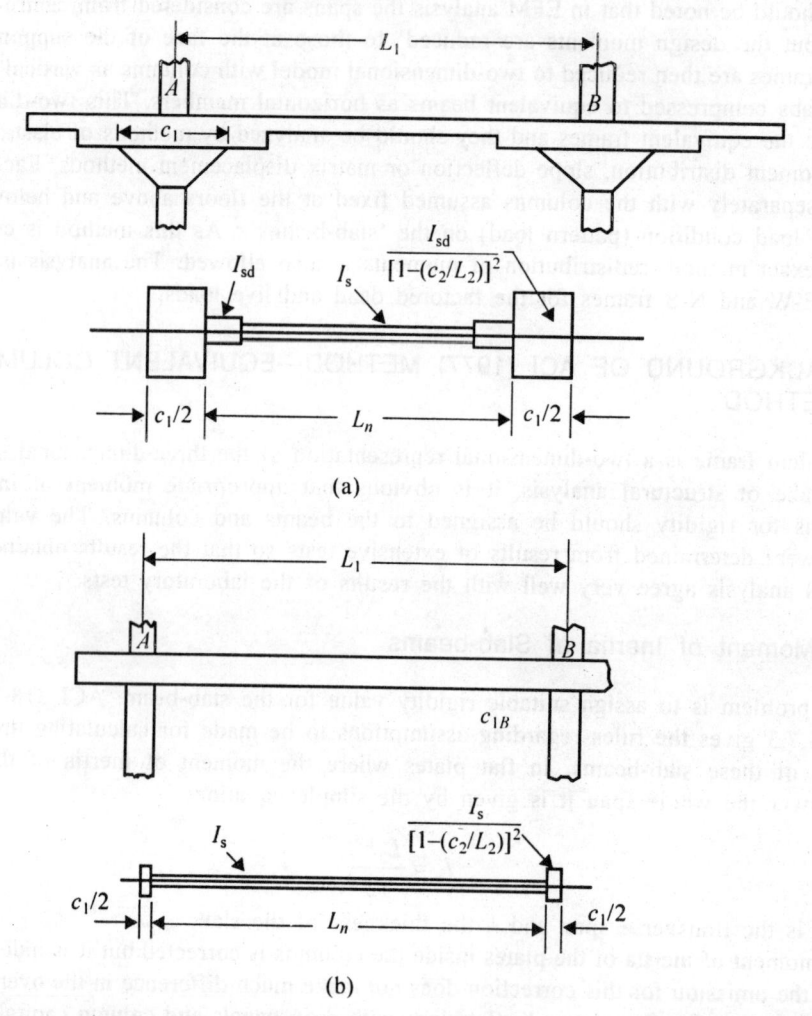

Fig. 13.2 Idealisation of slab elements for slab-beam stiffness: (a) slab with drops and capital (b) simple slabs.

For analyzing the vertical gravity loads, the moment of inertias are usually calculated on the gross area of concrete without taking into account the effects of cracking and amount of reinforcement but for analysis of horizontal load the effect of cracking and consequent decrease of I has also to be taken into account.[1]

---

[1] *Refer* Section 13.8.

13.3.2 Theoretical Column Stiffnesses

The theoretical column heights L_c is measured between the centre lines of the slabs, as shown in Fig. 13.3. The basic column stiffness is given by the expression:

$$K_c = \frac{4E_c I_c}{L_c} \tag{13.3}$$

where I_c is the moment of inertia of the column.

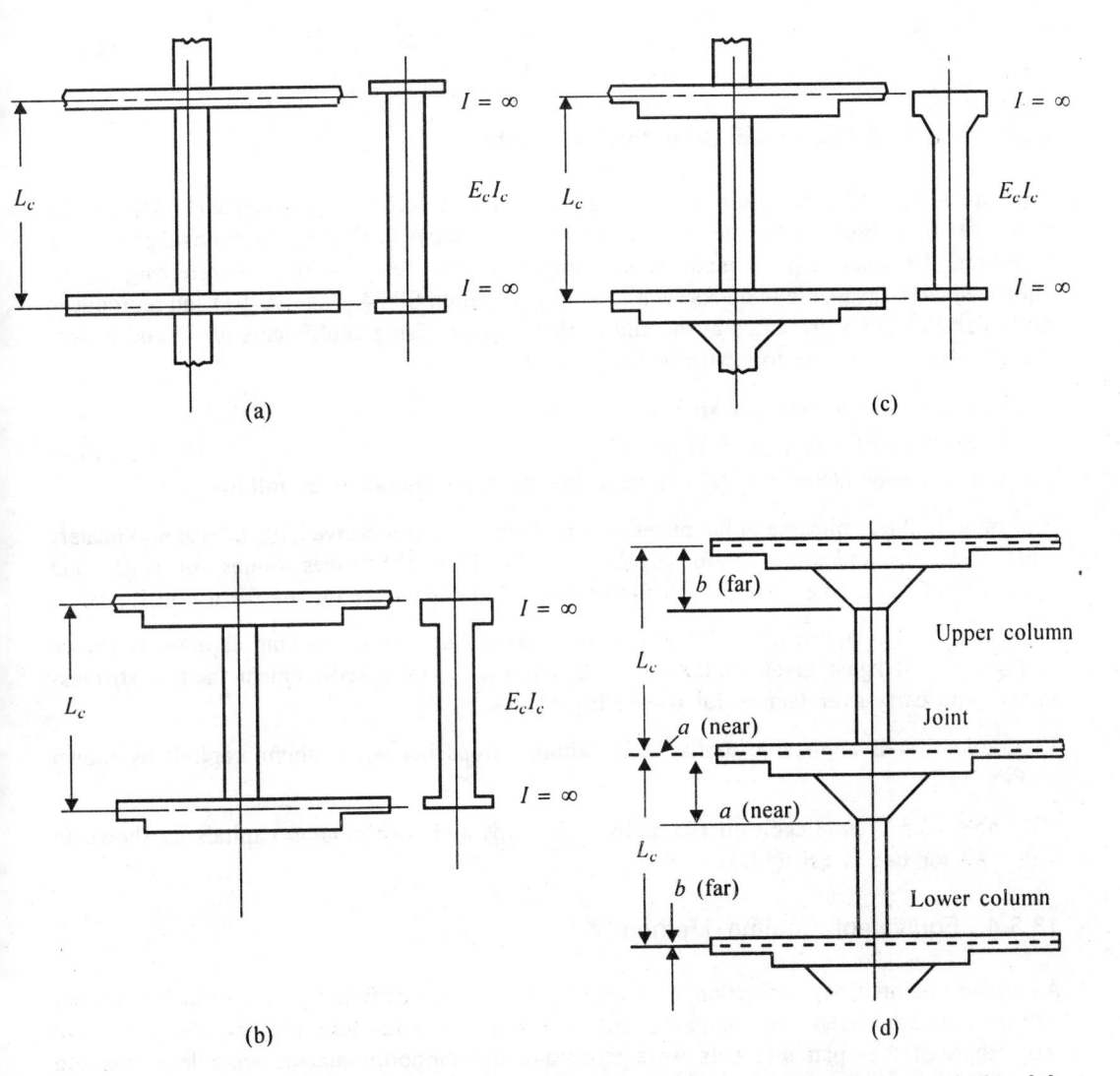

Fig. 13.3 Sections for calculating column stiffness: (a) slab without drop panel (b) slab with drop panel (c) slab with drop panel and capital (d) idealised slab column joint coefficients.

Calculation of I_c for a straight column is easy but when drop panels and column heads are present some sort of empirical procedure has to be introduced for simplicity. Accordingly, the stiffness of the column from the centre line of the upper slab to the bottom of the drop panel is taken as infinity. It is then taken as varying linearly to the base of the column capital after which it remains constant till the top of the bottom slab is reached. The moment of inertia inside the bottom slab is again taken as infinity as in Fig. 13.3. With infinite value of I, (the reciprocal of stiffness), which we call as flexibility, becomes zero. With values of a and b as shown in Fig. 13.3(d), Table 13.1 gives values of coefficients to calculate the stiffness of column for various values of a/L_c and b/L_c as

$$K_c = \frac{k_c E_c I_c}{L_c} \tag{13.3a}$$

13.3.3 Use of Published Data for Flat Slabs

The calculation of factors for stiffness, carry-over and fixed-end moments (FEM) for the beams can be worked out for various situations from the above fundamentals sectional properties. However, the work can be made much simpler in the design office by referring to the published tables available in books such as in IS publication SP 24 (Page 88–91) and reproduced as Tables 13.2–13.5 presented at the end of this chapter. Using coefficients m, K_s and c from these tables, we get the following values:

1. Fixed-end moment $(FEM) = mwL_2L_1^2$
2. Stiffness of slab $K_s = k_s E L_2 h^3/(12L_1)$ (13.4)
3. Carry-over factor = (COF) is read directly from the tables as follows:

Table 13.2 is applicable to flat plates shown in Fig. 13.7. It is derived, c_{1A}/L_1 is approximately equal to c_{2A}/L_2, and similarly for $c_{1B}/L_1 = c_{2B}/L_2$. This table gives values for c_{1A}/L_1 and c_{1B}/L_1, the fixed moment factors, stiffness factors and carry-over factors for use in Eq. 13.4.

Table 13.3 is applicable to flat slabs with drops but without column capitals as shown in Fig. 13.7. It again gives values for c_{1A}/L_1 and c_{1B}/L_1, the fixed-moment factors, stiffness factors and carry-over factors for use in Eq. (13.4).

Table 13.4 is applicable to flat slabs without drops but with column capitals as shown in Fig. 13.7.

Table 13.5 is applicable to flat slabs with drops and with column capitals as shown in Fig. 13.7 for use in Eq. (13.4).

13.3.4 Equivalent Column Method

As already pointed out in Section 13.2, when the equivalent frames with the simple column stiffness were analyzed and compared with test values, it was found (around 1963) that the test values of the span moments were more and the support moments were less than the theoretical values. This showed that the column sides were not as rigid as imagined. This can now be physically explained by the fact that under the conditions prescribed for the early theoretical model the column and the slab-beam are considered as fixed. However, in reality it is not so fixed. Beyond the immediate vicinity of the columns, slab tends to rotate freely as it is restrained only by the torsional stiffness of the slab which is quite small.

As shown in Fig. 13.4, there will be considerable 'leakage of moments' at supports. This naturally tends to increase the positive bending moment in the span and decrease the negative bending moment at the supports. In other words, the actual rotational restraint of the columns on the slab is less than that indicated by the 'simple equivalent frame'. This is due to lack of torsional stiffness of the slab that is supposed to fix the supports. The fixity offered therefore should not be only a function of the bending stiffness of the column but also the torsional stiffness of the slab and that of any transverse beam if present. If there are no beams to fix, the slabs along the column line, the torsional stiffness of the strip of slab of width equal to c_1 (i.e. the width of the column in the longitudinal direction) and length equal to the transverse span (L_2) should be taken into account, as shown in Fig. 13.1. The real column with the torsional members is called the equivalent column. It should be noted that IS 456 simplification has not introduced this aspect in its recommendations).

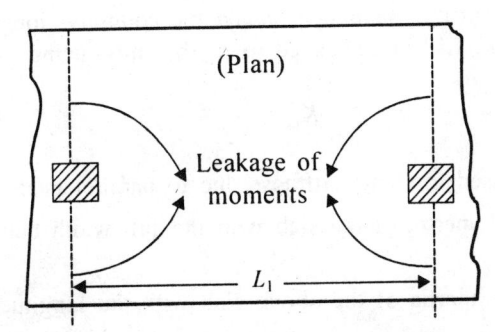

Fig. 13.4 Leakage of moment from supports to span.

For physical concept, it may be imagined that the end-moments are first transferred to the transverse beam or strip and then by torsion to the column. Thus the flexibility (inverse of stiffness) of an equivalent column is taken as the sum of the flexibility of the actual column above and below the 'slab beam' plus the flexibility of the 'attached torsional members' on each side of the column. This relationship can be expressed by the equation:

$$\frac{1}{K_{ec}} = \frac{1}{\Sigma K_c} + \frac{1}{K_t}$$

$$K_{ec} = \frac{\Sigma K_c \, \Sigma K_t}{\Sigma K_c + \Sigma K_t} \qquad (13.5)$$

where

K_{ec} = Flexural stiffness of equivalent column (moment per unit rotation)

ΣK_c = Sum of flexural stiffness of the actual column above and below the joint

K_t = The torsional stiffness of each of the transverse slab or the slab and beam (if any), that acts on the face of the column (attached torsion member).

Thus K_{ec} will always be smaller than K_c, i.e., the equivalent column is more flexible than the actual one. When $K_t = 0$ then $(1/K_t)$ is infinity and the value of K_{ec} will also be small.

Equation (13.5) was stated in the ACI 1977 code. In 1970, Corely and Jirsa had developed an empirical formula for the torsion stiffness of the 'attached torsional member' mentioned above so that the results of the analysis of equivalent frame with the 'equivalent column' member described above yield results close to those obtained in the laboratory. The expression derived for the torsional stiffness of the attached torsional member is as follows:

$$K_t = \frac{(\sum 9E_c C)}{L_2[1-(c_2/L_2)]^3} \tag{13.6}$$

where

C = Torsional constant of transverse torsional members given by Eq. (13.7)

E_c = Modules of elasticity of concrete

L_2 = Span of member subjected to torsion on either side of column

c_2 = The length of the side of the column in the transverse direction in line with L_2.

When there is a parallel beam spanning between the columns along L_1 as in two-way slab on beams, the value of K_t will be enhanced to K_{ta} by introducing a factor I_{sb}/I_s. Therefore,

$$K_{ta} = \frac{K_t I_{sb}}{I_s} \tag{13.7}$$

where

K_{ta} = The increased torsional stiffness due to parallel beam B in Fig. 13.1.

I_s = Moment of inertia of the slab with the full width between panel centre lines
    = $L_2 h^3/12$

I_{sb} = Moment of inertia of the above slab with the longitudinal beam

(*Note:* In Eq. (13.6) the coefficient 9 is based more on tests with gravity loads and its value may change for horizontal loads with more test data with such loads.)

The torsional constant C for the transverse member is to be evaluated for the cross-section as usually done in R.C. design by dividing it into separate rectangular parts and carrying out the following summation:

$$C = \sum \left(1 - 0.63\frac{x}{y}\right)\left(\frac{x^3 y}{3}\right) \tag{13.8}$$

where x is less than y in the rectangle (refer also Section 14.6).

This method of analysis is called the 'equivalent column method' recommended for design in ACI (1973) commentary and it should be distinguished from the alternate 'equivalent beam method'.[1] The availability of the alternate equivalent beam methods, which have since been developed the present ACI code, has omitted the equivalen column methods from the main body of the code.

13.4 SUMMARY OF PROVISIONS IN ACI 318

In the analysis recommended by ACI code, any two-way slab system can be considered as equivalent frames on column lines both in the longitudinal and transverse directions. The

---
[1] For equivalent beam method, *see* Section 13.7.2.

whole load is assumed to be carried on both directions as in the direct design method in these frames. The moment of inertia of slab-beams and the equivalent moment of inertia of the columns will be determined as described above. In the absence of a beam in the L_2 direction, the torsional member 'attached' to the column will be a slab of width c_1 equal to that of the column in the longitudinal direction L_1 and length equal to one half of the transverse span L_2 on each side of the column (as in Fig. 13.1). If there are any beam in the L_1 direction its effect should be considered in detail in Eq. (13.7). Beam in the longitudinal direction increases from K_t to K_{ta} as in Eq. (13.7). As already indicated, ACI 318 (1977) recommended this equivalent column method for analysis of the effects of all types of loads. Substitute frame as described in Chapter 8 may also be used for the simplification of the analysis for gravity loads. Redistribution of moments are also allowed.

13.5 ARRANGEMENT OF LIVE LOAD

Analysis should also be made for effect of pattern loading. However if the live load does not exceed 75% of the dead load or if all the spans are likely to be loaded simultaneously ACI and IS 456 Clause 31.5.2.2 allow the frame to be analyzed with factored live and dead load on all the spans. It may be noted that according to ACI, when using DDM in the case of frames of equal spans, effect of pattern load can be neglected when live load is even up to twice the dead load. For two-way slabs with beams and loads acting directly on the beams, an additional frame analysis has to be made for the effects of such loads.

13.6 REDUCTION IN NEGATIVE MOMENTS

As the frame analysis is made with centre line distance as spans (*see* IS 456 Clause 31.5.3.1), codes allow the negative moment so obtained to be reduced for design purposes. The moment reduces on the face of the supports at interior (square or equivalent square) column and to one-half of the distance from the face of support to centre line for exterior columns as well as to specified Sections for rectangular columns as shown in Fig. 13.5. However, the numerical

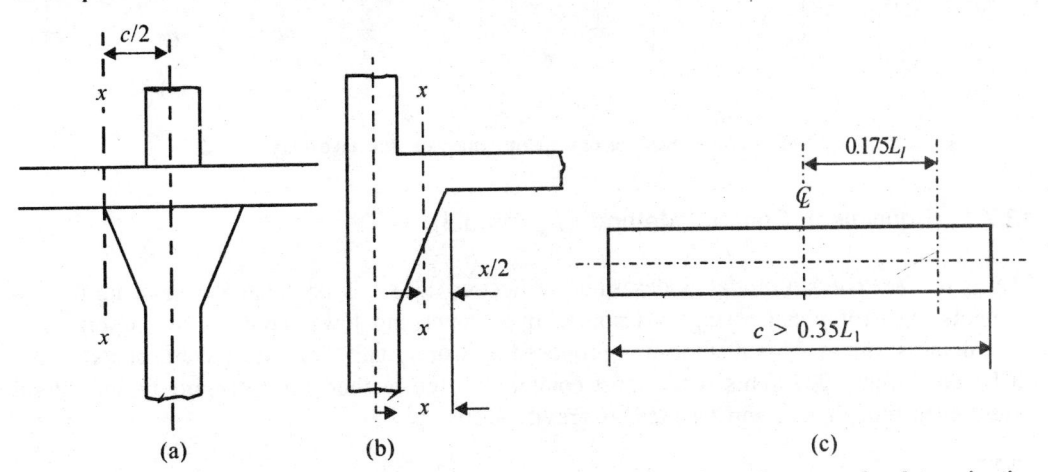

Fig. 13.5 Correction of moments computed from centre lines to face of support for determination of design moments.

value of the sum of the negative and positive moments should not be less than M_0 in Eq. (11.2) as stated in the following equation.

$$\frac{M_1 + M_2}{2} + M_3 = \frac{1}{8}\omega L_2 L_n^2$$

where M_1, M_2 are the negative moments and M_3 the positive moment.

13.7 GENERALISED SPACE FRAME MODEL

ACI 318 (1983) Clause 13.3.1 permits the analysis to be carried out by "any procedures satisfying conditions of equilibrium and geometric compatibility" thus recognising the findings of Venderbilt [3,4,5]. The basic principle is, a true model of the space structure should take into account the effect of constraints along the column lines. Accordingly an analytical model shown in Fig. 13.6 has been proposed by Venderbilt to analyse the two-way slab systems especially for lateral loads. The beam having a moment of inertia equal to that the entire cross-section across the full width of the equivalent rigid frame is located along $A'B'C'$, offset from the plane defined by the columns. Beam ABC located on the column lines will have no stiffness but the beam $A'B'C'$ will be connected to the column by torsion elements AA', BB'. This model is called the *space frame model*. The word 'space' being used only to differentiate it from the 'plane model'. This 'space model' can further be simplified for easy analysis by manipulating it into equivalent plane models which can be analyzed by ordinary methods. Two of the important plane models that can be evolved from the space model are the commonly used equivalent column method and the equivalent beam method.

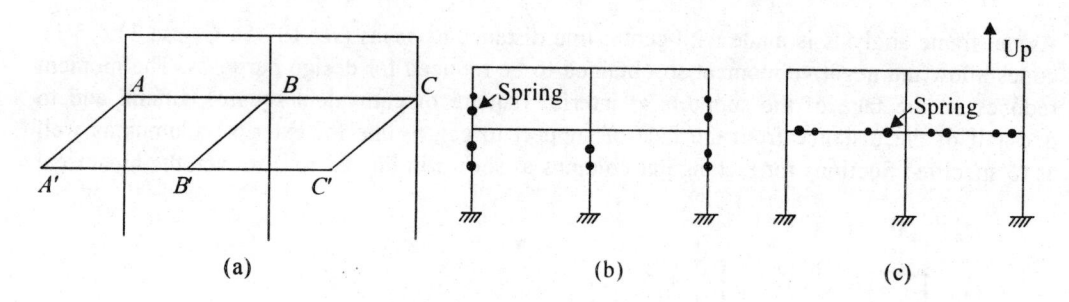

Fig. 13.6 Equivalent beam analysis for two-way slabs.

13.7.1 Equivalent Column Method (K_{ec} Model)

The *equivalent column model* as described in Section 13.3.4 is obtained by replacing torsion elements from rotational springs distributed over upper and lower columns in proportion to the column stiffness. This method was proposed by Corely and Jirsa and recommended in the ACI Codes from 1977. This is the most commonly used method for both gravity and lateral loads, even though it is most suited to gravity loads.

---

[1]More details of this method can be found in references given at the end of this chapter.

13.7.2 Equivalent Beam Method (K_{ec} Model)

The model got by replacing the torsion element by rotational springs at the slab beam ends is called the *equivalent beam model*. This was first suggested by Venderbilt in 1981. It is generally believed that this model gives better results for analysis of frames for lateral loads like wind and earthquake loads. This model has not yet been recommended in IS 456.

13.7.3 Other Modifications

Other modifications of the space models have also been proposed. For example, a space model, in which the moment of inertia of the column strip is placed on *ABC* in Fig. 13.6 and the moment of inertia of the middle strip is placed on *A'B'C'*, has been proposed by Wang and Salmon for the analysis of both gravity and lateral loads [6].

13.8 LATERAL LOADS AND TWO-WAY SLAB SYSTEMS

Two-way slab systems without beams, like flat slabs and flat plates, are inherently not very strong under lateral loads. In tall buildings they should preferably be braced by devices such as shear walls, so that the frames have to withstand only gravity loads. In moderately high buildings they should be analyzed by conservative methods for lateral load analysis and designed with sufficient margin of safety.

There is considerable difference of opinion as what should be the stiffness of the 'beam slab' for the analysis of the equivalent frame for horizontal loads. BS 8110 Clause 3.7.23 recommends that 'for horizontal loading' it is more appropriate to take one-half the width that is taken for vertical load analysis. This reduction is mainly due to cracking of the slab especially near the columns. Hence, it stands to reason that this factor will be higher for flat slabs with drops surrounding the columns. The ACI code Clause 13.3.12 only states, for lateral load analysis the effects of cracking and reinforcement on stiffness must be considered. This stiffness depends on parameters such as L_2/L_1, c_1/L_1, c_1/c_2 depths of drop and concentration of reinforcements provided for unbalanced moment transfer by flexure. Values reported from tests for reduction vary from 0.25 to 0.5. From these discussions it is clear that the BS 8110 recommendation of using 0.5 times width of the slab as effective width for stiffness calculation can be considered as a very acceptable method for analysis of horizontal loads like wind loads.

13.9 DESIGN PROCEDURE

The following procedure is usually recommended for orderly analysis of flat slab by equivalent frame method:

Step 1. Calculate from the geometry c_1/L_1, c_2/L_2

Step 2. Determine the stiffness of column from the expression $K_c = K_c I_c/L_c$, with k_c obtained from Table 13.1 and Eq. 13.3a

Step 3. Calculate the torsional stiffness of the attached member by Eq. 13.6

252   ADVANCED REINFORCED CONCRETE DESIGN

Step 4. Estimate the stiffness of the equivalent column from results of step 2 and 3 by Eq. 13.5

Step 5. Read off the B.M. coefficient carryover factor and coefficient for stiffness of slab from readymade tables. Determine also the stiffness of the slab (Tables 13.3–13.5)

Step 6. From stiffness of slab and column calculate the distribution factors

Step 7. Consolidate steps 1–6 in a table

Step 8. Analyze the equivalent frame by moment distribution method

Step 9. Make correction for design moments as in Section 13.6 and compare them with the direct design method. (Also if the slabs can be analyzed by DDM the total B.M. need not exceed M_0

Step 10. Having obtained the negative moments at supports and the positive moments in the span distribute them to the column strip and middle strips as explained already in Chapter 11 using the same coefficients as in DDM

Step 11. The column moments can be taken as those obtained from the analysis instead of using the empirical method given in Section 11.15

As steps 9 to 11 are self-explanatory only steps 1 to 7 are worked out in Example 13.1.

TABLE 13.1   COLUMN STIFFNESS COEFFICIENTS k_c
[See Eq. (13.3a), Fig. 13.3]

a/L_c	\multicolumn{12}{c}{b/L_c}
	0.00	0.02	0.04	0.06	0.08	0.10	0.12	0.14	0.16	0.18	0.20	0.22	0.24
0.00	4.000	4.082	4.167	4.255	4.348	4.444	4.545	4.851	4.762	4.878	5.000	5.128	5.263
0.02	4.337	4.433	4.533	4.638	4.747	4.862	4.983	5.110	5.244	5.384	5.533	5.690	5.856
0.04	4.709	4.882	4.940	5.063	5.193	5.330	5.475	5.627	5.787	5.958	6.138	6.329	6.533
0.06	5.122	5.252	5.393	5.539	5.693	5.855	6.027	6.209	6.403	6.608	6.827	7.060	7.310
0.08	5.581	5.735	5.898	6.070	6.252	6.445	6.650	6.868	7.100	7.348	7.613	7.897	8.203
0.10	6.091	6.271	6.462	6.665	6.880	7.109	7.353	7.614	7.893	8.192	8.513	8.859	9.233
0.12	6.659	6.870	7.094	7.333	7.587	7.859	8.150	8.461	8.796	9.157	9.546	9.967	10.430
0.14	7.292	7.540	7.803	8.084	8.385	8.708	9.054	9.426	9.829	10.260	10.740	11.250	11.810
0.16	8.001	8.291	8.600	8.931	9.297	9.670	10.080	10.530	11.010	11.540	12.110	12.740	13.420
0.18	8.796	9.134	9.498	9.888	10.310	10.760	11.260	11.790	12.370	13.010	13.700	14.470	15.310
0.20	9.687	10.080	10.510	10.970	11.470	12.010	12.000	13.240	13.940	14.710	15.560	16.490	17.530
0.22	10.690	11.160	11.660	12.200	12.800	13.440	14.140	14.910	15.700	16.690	17.210	18.870	20.150
0.20	11.820	12.370	12.960	13.610	14.310	15.080	15.920	16.840	17.870	19.000	20.260	21.650	23.260

TABLE 13.2 ACI MOMENT DISTRIBUTION CONSTANTS FOR FLAT PLATE
[See Eq. (13.4), Fig. 13.7] (SP 24 Table E.4)

Column dimension		Uniform load		Stiffness factor		Carryover factor	
c_{1A}/L_1	c_{1B}/L_1	m_{AB}	m_{BA}	k_{AB}	k_{BA}	COF_{AB}	COF_{BA}
0.00							
	0.00	0.083	0.083	4.00	4.00	0.500	0.500
	0.05	0.083	0.084	4.01	0.04	0.504	0.500
	0.10	0.082	0.086	4.03	4.15	0.513	0.499
	0.15	0.081	0.089	4.07	4.32	0.528	0.498
	0.20	0.079	0.093	4.12	4.56	0.548	0.495
	0.25	0.077	0.097	4.18	4.88	0.573	0.491
	0.30	0.075	0.102	4.25	5.28	0.603	0.485
	0.35	0.073	0.107	4.33	5.78	0.638	0.478
0.05							
	0.05	0.084	0.084	4.05	4.05	0.503	0.503
	0.10	0.083	0.086	4.07	4.15	0.513	0.503
	0.15	0.081	0.089	4.11	4.33	0.528	0.501
	0.20	0.080	0.092	4.16	4.58	0.548	0.499
	0.25	0.078	0.096	4.22	4.89	0.573	0.494
	0.30	0.076	0.101	4.29	5.30	0.603	0.489
	0.35	0.074	0.107	4.37	5.80	0.638	0.481
0.10							
	0.10	0.085	0.085	4.18	4.18	0.513	0.513
	0.15	0.083	0.088	4.22	4.36	0.528	0.511
	0.20	0.082	0.091	4.27	4.61	0.548	0.508
	0.25	0.080	0.095	4.34	4.93	0.573	0.504
	0.30	0.078	0.100	4.41	5.34	0.602	0.498
	0.35	0.075	0.105	4.50	5.85	0.637	0.491
0.15							
	0.15	0.086	0.086	4.40	4.40	0.526	0.526
	0.20	0.084	0.090	4.46	4.65	0.546	0.523
	0.25	0.083	0.094	4.53	4.98	0.571	0.519
	0.30	0.080	0.099	4.61	5.40	0.601	0.513
	0.35	0.078	0.104	4.70	5.92	0.635	0.505
0.20							
	0.20	0.088	0.088	4.72	4.72	0.543	0.543
	0.25	0.086	0.092	4.79	5.05	0.568	0.539
	0.30	0.083	0.097	4.88	5.48	0.597	0.532
	0.35	0.081	0.102	4.99	6.01	0.632	0.524
0.25							
	0.25	0.090	0.090	5.14	5.14	0.563	0.563
	0.30	0.088	0.095	5.24	5.58	0.592	0.556
	0.35	0.085	0.100	5.36	6.12	0.626	0.548
0.30							
	0.30	0.092	0.092	5.69	5.69	0.585	0.585
	0.35	0.090	0.097	5.83	6.26	0.619	0.576
0.35							
	0.35	0.095	0.095	6.42	6.42	0.609	0.609

254 ADVANCED REINFORCED CONCRETE DESIGN

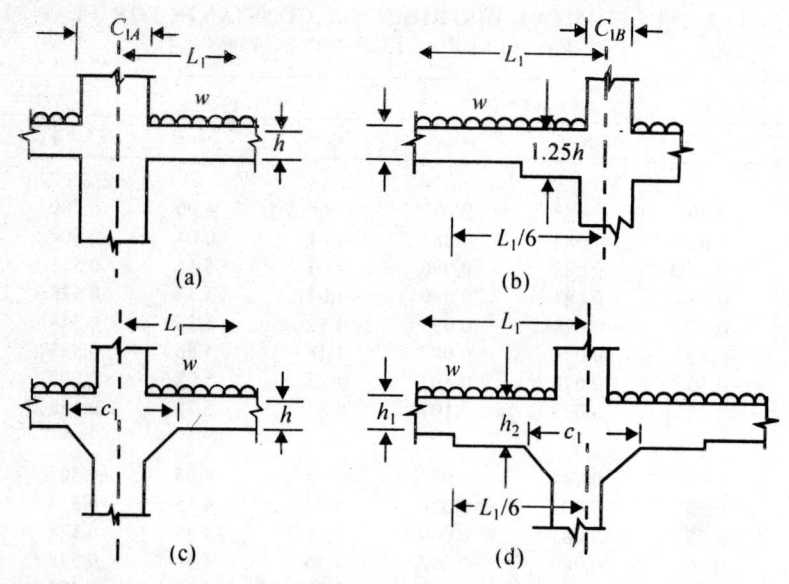

Fig. 13.7 Diagram for tables for design of flat slabs by the equivalent column method: (a) flat plate (Table 13.2) (b) flat slab with drop panel (Table 13.3) (c) flat slab with column capitals only (Table 13.4) (d) flat slab with capitals and drops (Table 13.5).

TABLE 13.3 ACI MOMENT DISTRIBUTION CONSTANTS FOR FLAT SLABS WITH DROP PANEL [See Eq. (13.4), Fig. 13.7] (SP 24 Table E.5)

Column dimension		Uniform load		Stiffness factor		Carry-over factor	
c_{1A}/L_1	c_{1B}/L_1	m_{AB}	m_{BA}	k_{AB}	k_{BA}	COF_{AB}	COF_{BA}
0.00							
	0.00	0.088	0.088	4.78	4.78	0.541	0.541
	0.05	0.087	0.089	4.80	4.82	0.545	0.541
	0.10	0.087	0.090	4.83	4.94	0.553	0.541
	0.15	0.085	0.093	4.87	5.12	0.567	0.540
	0.20	0.084	0.096	4.93	5.36	0.585	0.537
	0.25	0.082	0.100	5.00	5.68	0.606	0.534
	0.30	0.080	0.105	5.09	6.07	0.631	0.529
0.05							
	0.05	0.088	0.088	4.84	4.84	0.545	0.545
	0.10	0.087	0.090	4.87	4.95	0.553	0.544
	0.15	0.085	0.093	4.91	5.13	0.567	0.543
	0.20	0.084	0.096	4.97	5.38	0.584	0.541
	0.25	0.082	0.100	5.05	5.70	0.606	0.537
	0.30	0.080	0.104	5.13	6.09	0.632	0.532
0.10							
	0.10	0.089	0.089	4.98	4.98	0.553	0.533
	0.15	0.088	0.092	5.03	5.16	0.566	0.551
	0.20	0.086	0.094	5.09	5.42	0.584	0.549
	0.25	0.084	0.099	5.17	5.74	0.606	0.548
	0.30	0.082	0.103	5.26	6.13	0.631	0.541

TABLE 13.3 Contd.

Column dimension		Uniform load		Stiffness factor		Carry-over factor	
c_{1A}/L_1	c_{1B}/L_1	m_{AB}	m_{BA}	k_{AB}	k_{BA}	COF_{AB}	COF_{BA}
0.15							
	0.15	0.090	0.090	5.22	5.22	0.565	0.565
	0.20	0.089	0.094	5.28	5.47	0.583	0.563
	0.25	0.087	0.097	5.37	5.80	0.604	0.559
	0.30	0.085	0.102	5.48	6.21	0.630	0.554
0.20							
	0.20	0.092	0.092	5.55	5.55	0.580	0.580
	0.25	0.090	0.096	5.64	5.88	0.602	0.602
	0.30	0.088	0.100	5.74	6.30	0.627	0.571
0.25							
	0.25	0.094	0.094	5.98	5.98	0.598	0.598
	0.30	0.091	0.098	6.10	6.41	0.622	0.593
0.30							
	0.30	0.095	0.095	6.54	6.54	0.617	0.617

TABLE 13.4 ACI MOMENT DISTRIBUTION CONSTANTS FOR SLAB-BEAM MEMBERS WITH COLUMN CAPITALS
[See Eq. (13.4) Fig. 13.7] (SP 24 Table E.6)

c_1/L_1	c_2/L_2	m	k_s	COF	c_1/L_1	c_2/L_2	m	k_s	COF
0.00					0.30				
	0.00	0.083	4.000	0.500		0.00	0.083	4.000	0.500
	0.10	0.083	4.000	0.500		0.10	0.086	4.488	0.527
	0.20	0.083	4.000	0.500		0.20	0.089	5.050	0.556
	0.30	0.083	4.000	0.500		0.30	0.092	5.6	0.585
	0.40	0.083	4.000	0.500		0.40	0.095	6.41	0.614
	0.50	0.083	4.000	0.500		0.50	0.098	7.205	0.642
0.05					0.35				
	0.00	0.083	4.000	0.500		0.00	0.083	4.000	0.500
	0.10	0.084	4.093	0.507		0.10	0.087	4.551	0.529
	0.20	0.085	4.222	0.516		0.20	0.090	5.204	0.560
	0.30	0.085	4.261	0.518		0.30	0.093	5.979	0.593
	0.40	0.086	4.368	0.526		0.40	0.096	6.888	0.626
	0.50	0.086	4.398	0.528		0.50	0.099	7.935	0.658
0.10					0.40				
	0.00	0.083	4.000	0.500		0.00	0.083	4.000	0.500
	0.10	0.085	4.272	0.513		0.10	0.087	4.607	0.530
	0.20	0.086	4.362	0.524		0.20	0.090	5.348	0.563
	0.30	0.087	4.535	0.535		0.30	0.094	6.255	0.598
	0.40	0.088	4.698	0.545		0.40	0.097	7.365	0.635
	0.50	0.089	4.848	0.554		0.50	0.100	8.710	0.672

TABLE 13.4 Contd.

c_1/L_1	c_2/L_2	m	k_s	COF	c_1/L_1	c_2/L_2	m	k_s	COF
0.15					0.45				
	0.00	0.083	4.000	0.500		0.45	0.083	4.000	0.500
	0.10	0.085	4.267	0.517		0.10	0.087	4.658	0.530
	0.20	0.087	4.541	0.534		0.20	0.090	5.480	0.564
	0.30	0.089	4.818	0.550		0.30	0.094	6.517	0.602
	0.40	0.090	5.090	0.565		0.40	0.098	7.836	0.642
	0.50	0.092	5.349	0.579		0.50	0.101	9.514	0.683
0.20					0.50				
	0.00	0.083	4.000	0.500		0.00	0.083	4.000	0.500
	0.10	0.086	4.346	0.522		0.10	0.087	4.703	0.530
	0.20	0.088	4.717	0.543		0.20	0.090	5.599	0.564
	0.30	0.090	5.108	0.564		0.30	0.094	6.600	0.603
	0.40	0.092	5.509	0.584		0.40	0.098	8.289	0.645
	0.50	0.034	5.908	0.602		0.50	0.102	10.329	0.690
0.25									
	0.00	0.083	4.000	0.500					
	0.10	0.086	4.420	0.525					
	0.20	0.089	4.887	0.550					
	0.30	0.091	5.401	0.576					
	0.40	0.094	5.952	0.600					
	0.50	0.096	6.527	0.623					

TABLE 13.5 ACI MOMENT DISTRIBUTION CONSTANTS FOR SLAB-BEAMS WITH CAPITALS AND DROP PANELS

[See Eq. (13.4), Fig. 13.7] (SP 24 Table E.7)

c_1/L_1	c_2/L_2	Constants for $h_2 = 1.25h_1$			Constants for $h_2 = 1.5h_1$		
		m	k_s	COF	m	k_s	COF
0.00							
	0.00	0.088	4.795	0.542	0.093	5.837	0.589
	0.05	0.088	4.795	0.542	0.092	5.837	0.589
	0.10	0.088	4.795	0.542	0.093	5.837	0.589
	0.15	0.088	4.795	0.542	0.093	5.837	0.589
	0.20	0.088	4.795	0.542	0.093	5.837	0.589
	0.25	0.088	4.795	0.542	0.093	5.837	0.589
	0.30	0.088	4.797	0.542	0.093	5.837	0.589
0.05							
	0.00	0.088	4.795	0.542	0.093	5.837	0.589
	0.05	0.088	4.846	0.545	0.093	5.890	0.591
	0.10	0.089	4.896	0.548	0.093	5.942	0.594
	0.15	0.089	4.944	0.551	0.093	5.993	0.596
	0.20	0.089	4.990	0.553	0.094	6.041	0.598
	0.25	0.089	5.035	0.556	0.094	6.087	0.600
	0.30	0.090	5.077	0.558	0.094	6.131	0.602

TABLE 13.5 Contd.

c_1/L_1	c_2/L_2	\multicolumn{3}{c}{Constants for $h_2 = 1.25h_1$}	\multicolumn{3}{c}{Constants for $h_2 = 1.5h_1$}
		m	k_s	COF	m	k_s	COF
0.10							
	0.00	0.088	4.795	0.542	0.093	5.837	0.589
	0.05	0.088	4.894	0.548	0.093	5.940	0.593
	0.10	0.089	4.992	0.553	0.094	6.042	0.598
	0.15	0.090	5.039	0.559	0.094	6.142	0.602
	0.20	0.090	5.184	0.564	0.094	6.240	0.607
	0.25	0.091	5.278	0.569	0.095	6.335	0.611
	0.30	0.091	5.368	0.573	0.095	6.427	0.615
0.15							
	0.00	0.088	4.795	0.542	0.093	5.837	0.589
	0.05	0.089	4.938	0.550	0.093	5.986	0.595
	0.10	0.090	5.082	0.558	0.094	6.135	0.602
	0.15	0.090	5.228	0.565	0.095	6.284	0.608
	0.20	0.091	5.374	0.573	0.095	6.342	0.614
	0.25	0.092	5.520	0.580	0.096	6.579	0.620
	0.30	0.092	5.665	0.587	0.096	6.723	0.626
0.20							
	0.00	0.088	4.795	0.542	0.093	5.387	0.589
	0.05	0.089	4.978	0.552	0.093	6.027	0.597
	0.10	0.090	5.167	0.562	0.094	6.221	0.605
	0.15	0.091	5.361	0.571	0.095	6.418	0.613
	0.20	0.092	5.558	0.581	0.096	6.616	0.621
	0.25	0.093	5.760	0.590	0.096	6.816	0.628
	0.30	0.094	5.962	0.590	0.097	7.015	0.635
0.25							
	0.00	0.088	4.795	0.542	0.093	5.837	0.589
	0.05	0.089	5.015	0.553	0.094	6.605	0.598
	0.10	0.090	5.245	0.565	0.094	6.300	0.608
	0.15	0.091	5.485	0.576	0.095	6.543	0.617
	0.20	0.092	5.735	0.587	0.096	6.790	0.626
	0.25	0.094	5.944	0.598	0.097	7.043	0.635
	0.30	0.095	6.261	0.600	0.098	7.298	0.644
0.30							
	0.00	0.088	4.795	0.542	0.093	5.837	0.589
	0.05	0.089	5.048	0.554	0.094	6.099	0.599
	0.10	0.090	5.317	0.567	0.095	6.372	0.610
	0.15	0.092	5.601	0.580	0.096	6.657	0.620
	0.20	0.093	5.902	0.593	0.097	6.953	0.631
	0.25	0.094	6.219	0.605	0.098	7.258	0.641
	0.30	0.095	6.550	0.618	0.099	7.571	0.651

EXAMPLE 13.1 [Design by equivalent frame method (EFM)]

Using published tables (Tables 13.1–1 5) determine the coefficients for analysis of the equivalent frame of the following flat slab by moment distribution method (Fig. 13.8):

1. Spacing of columns: 7 m in x-direction and 6 m in y-direction
2. Internal columns: 750 mm diameter with 1.53 m circular column head of 0.9 m depth from top of the slab to bottom of the column capital
3. Peripheral columns: 750 mm square with 1.53 × 1.53/2 rectangular column heads
4. Thickness of main slab: 180 mm
5. Drops: 2.4 × 3 m size and 280 mm thick ($h_2 = 1.5h$ approximately)
6. Peripheral beams: 750 × 900
7. Storey height 3.3 m from centre-to-centre of main slabs

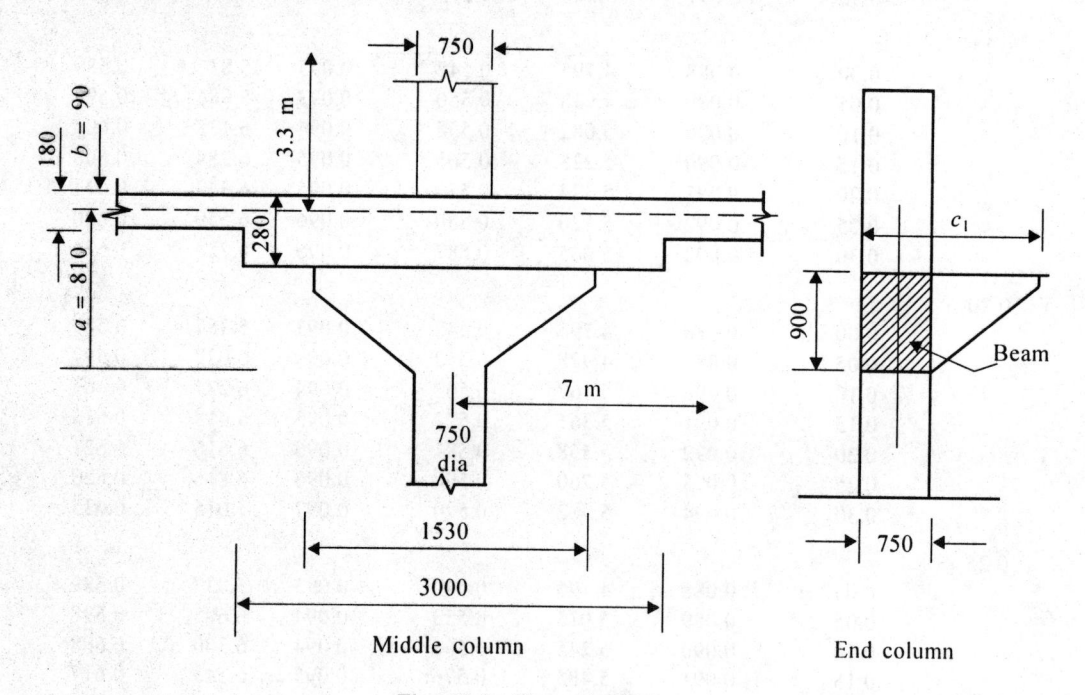

Fig. 13.8 Example 13.1.

Analysis of frame in XX direction (7 m span)

Step 1. Calculate from the geometry c_1/L_1, c_2/L_2 of the flat slab

Distance of centre of slab to top of slab = 0.180/2 = 0.09 m

Distance of centre of slab to bottom of capital = 0.9 − 0.09 = 0.81 m

Refer Fig. 13.3.

EQUIVALENT FRAME ANALYSIS OF FLAT SLABS

Column	External column	Internal column
Upper column $\dfrac{a}{L}$	$\dfrac{0.09}{3.3}=0.027$	0.027
Upper column $\dfrac{b}{L}$	$\dfrac{0.81}{3.3}=0.245$	0.245
Lower column $\dfrac{a}{L}$	$\dfrac{0.81}{3.3}=0.245$	0.245
Lower column $\dfrac{b}{L}$	$\dfrac{0.09}{3.3}=0.027$	0.027

Column heads	External span	Internal span
	$c_1 = \dfrac{1.53}{2}+\dfrac{0.75}{2}=1.14$	$c_1 = 0.89 \times 1.53 = 1.36$ (Equivalent square)
	$c_2 = 1.53$	$c_2 = c_1$
Left $\dfrac{c_1}{L_1}$	$\dfrac{1.14}{7.0}=0.163$	$\dfrac{1.36}{7.0}=0.194$
Left $\dfrac{c_2}{L_2}$	$\dfrac{1.53}{6.0}=0.255$	$\dfrac{1.36}{6.0}=0.227$
Right $\dfrac{c_1}{L_1}$	$\dfrac{1.36}{7.0}=0.194$	$\dfrac{1.36}{7.0}=0.194$
Right $\dfrac{c_2}{L_2}$	$\dfrac{1.36}{6.0}=0.227$	$\dfrac{1.36}{6.0}=0.227$

Step 2. Calculate the stiffness of the columns

External upper column	Internal upper column
$I = \dfrac{(0.75)^4}{12} = 0.02637 \text{ m}^3$	$I = \dfrac{\pi(.75)^4}{4} = 0.0155$
$\dfrac{a}{L_c} = 0.0227;\ \dfrac{b}{L_c} = 0.245$	$\dfrac{a}{L_c} = 0.027;\ \dfrac{b}{L_c} = 0.245$

ADVANCED REINFORCED CONCRETE DESIGN

External upper column	Internal upper column
Column stiffness coefficients from Table 13.1	
$k_c = 6.025$	$k_c = 6.025$
$K_c = \dfrac{k_c I_c}{L_c} = \dfrac{(6.025)(0.02637)}{3.3} = 0.048$	$K_c = \dfrac{(6.025)(0.0155)}{3.3}$
	$K = 0.0283$

External lower column	Internal lower column
$\dfrac{a}{L_c} = 0.2450;\ \dfrac{b}{L_c} = 0.027$	$\dfrac{a}{L_c} = 0.245;\ \dfrac{b}{L_c} = 0.027$
$k_c = 12.52$	$k_c = 12.52$
$K_c = \dfrac{(12.52)(0.02637)}{3.3} = 0.10$	$K_c = \dfrac{(12.52)(0.0155)}{3.3} = 0.0588$
$\Sigma K_c = 0.148$	$\Sigma K_c = 0.087$

Step 3. Calculate the torsional stiffness of the attached member

$$C = \Sigma\left(1 - 0.63\dfrac{x}{y}\right)\left(\dfrac{x^3 y}{3}\right) \qquad \text{Torsional stiffness } K_t = \dfrac{(\text{Two sides})(9EC)}{L_2(1 - c_2/L_2)^3}$$

External span	Internal span
The attached member is a L beam 750 × 900 mm with attached slab 620 × 280 mm	The attached member is a slab of 0.28 m depth on face of column ($c_1 = 1.36$ m)
$x_1 = 0.75;\ y_1 = 0.90$	$C = \left[\dfrac{1 - 0.63(0.28)}{1.36}\right]\left[\dfrac{(0.28)^3(1.36)}{3}\right]$
$x_2 = 0.28;\ y_2 = 0.62$	$= 0.0086$
$C = 0.0633$	
$K_t = \dfrac{2 \times 9 \times 0.0633}{6(1 - 0.255)^3} = 0.459$ (with $E = 1$)	$K_t = \dfrac{2 \times 9 \times 0.0086}{6(1 - 0.227)^3} = 0.0558$

Step 4. Calculate the stiffness of equivalent column

$$\text{Column stiffness } \dfrac{1}{K_{ec}} = \dfrac{1}{\Sigma K_c} + \dfrac{1}{K_t} \quad \text{[from Eq. (13.5)]}$$

$\dfrac{1}{K_{ec}} = \dfrac{1}{0.148} + \dfrac{1}{0.459} = 8.94$	$\dfrac{1}{K_{ec}} = \dfrac{1}{0.0872} + \dfrac{1}{0.0556} = 29.45$
$K_{ec} = 0.11$	$K_{ec} = 0.034$

(*Note:* The stiffness of column has been reduced due to the attached torsion member.

Step 5. Read off B.M. Coefficient carry-over factors and slab stiffness from Table 13.5

$$\text{Stiffness of slab } K_s = \frac{k_s h^3 L_2}{12 L_1}$$

Using Table 13.5, Eq. 13.5

End column	Interior column
$\dfrac{c_1}{L_1} = 0.162; \quad \dfrac{c_2}{L_2} = 0.255$	$\dfrac{c_1}{L_1} = 0.194; \quad \dfrac{c_2}{L_2} = 0.226$
(1) FEM Coefficient $m = 0.096$	(1) FEM Coefficient $= 0.096$
(2) Carry-over factor $= 0.623$	(2) Carry-over factor $= 0.624$
(3) $k_s = 6.65$ (Table 13.5)	(3) $k_s = 6.716$
$K_s = \dfrac{6.65 \times (0.18)^3 \times 6}{12 \times 7} = 0.0028 \text{ (approx)}$	$K_s = \dfrac{6.716 \times (0.18)^3 \times 6}{12 \times 7} = 0.0028 \text{ (approx.)}$

Step 6. Calculate the distribution factors

Exterior end	First interior column
Column stiffness $= 0.11$	Column stiffness $= 0.0340$
Slab stiffness $= 0.0028$	Stiffness of slab right of column $= 0.0028$
Total $= 0.1128$	Stiffness of slab left of column $= 0.0028$
	Total $= 0.0396$
Distribution to column $= \dfrac{0.110}{0.1128} = 0.975$	Distribution to column $= \dfrac{0.0340}{0.0396} = 0.86$
Slab $= 0.025$	To each slabs $\dfrac{100 - 86}{2} = 7\%$

Step 7. Consolidate steps 1–6 in a table

TABLE 13.6 DATA FOR ANALYSIS OF EQUIVALENT FRAME IN XX DIRECTION (SPAN 7 M)

Item	Description	External column	First internal column	Second internal column
1. Geometry	Column above			
	a/L_c	0.027	0.027	0.027
	b/L_c	0.245	0.245	0.245
	Slab			
	c_1/L_1 left	0.162	0.194	0.194
	c_2/L_2 left	0.255	0.227	0.227
	Column below			
	a/L_c	0.027	0.027	0.027
	b/L_c	0.245	0.245	0.245

TABLE 13.6 Contd.

Item	Description	External column	First internal column	Second internal column
2. Stiffness	K_c above	0.0480	0.0283	0.0283
	K_c below	0.1000	0.0589	0.0589
	ΣK_c	0.1480	0.0872	0.0872
	Torsion member			
	C	0.0633	0.0086	0.0086
	K_t	0.4590	0.0556	0.0556
	K_{ec}	0.1100	0.040	0.0340
	K (slab) left	0.0000	0.0028	0.0028
	K (slab) right	0.0028	0.0028	0.0028
	ΣK = slab + col.	0.1128	0.0396	0.0396
3. Factors	B.M. Coefficient	0.096	0.096; 0.096	0.096; 0.096
	COF	0.623	0.623; 0.624	0.624; 0.624
4. Distribution	Slab	0 | 2.5	7 | 7	7 | 7
	Column	97.5	86	86

REFERENCES

1. Purushothaman, P., *Reinforced Concrete Structural Elements—Behavior Analysis and Design*, Tata McGraw-Hill, New Delhi, 1984.

2. SP 24, Explanatory Hand Book on IS 456 (1978) Bureau of Indian Standards, 1984.

3. Venderbilt, M.D., *Equivalent Frame Analysis for Lateral Loads*, ASCE Journal Struct. Div. ST-10, October 1979.

4. Venderbilt, M.D. and Corley, W.G., *Frame Analysis of Concrete Buildings*, Concrete International, January 1983.

5. Venderbilt, M.D., *Equivalent Frame Analysis of Unbraced Reinforced Concrete Buildings For Static Lateral Loads*, Structural Research Report No. 36, Civil Engineering Department, Colorada State University, Collins, June 1981.

6. Wang, C.K. and Salmon, C.G., *Reinforced Concrete Design*, Harper and Row, Cambridge, U.S.A., 1985.

CHAPTER 14

Design of Spandrel (or Edge) Beams

14.1 INTRODUCTION

Spandrel beams or edge beams are provided to slabs in many situations as shown in Fig. 14.1. In conventional two-way slab system with beams, these edge beams give fixity to the end slab thus reducing deflection as well as span moments. In the case of flat slabs they also help to redistribute the negative moments at the end-spans to the middle strips.

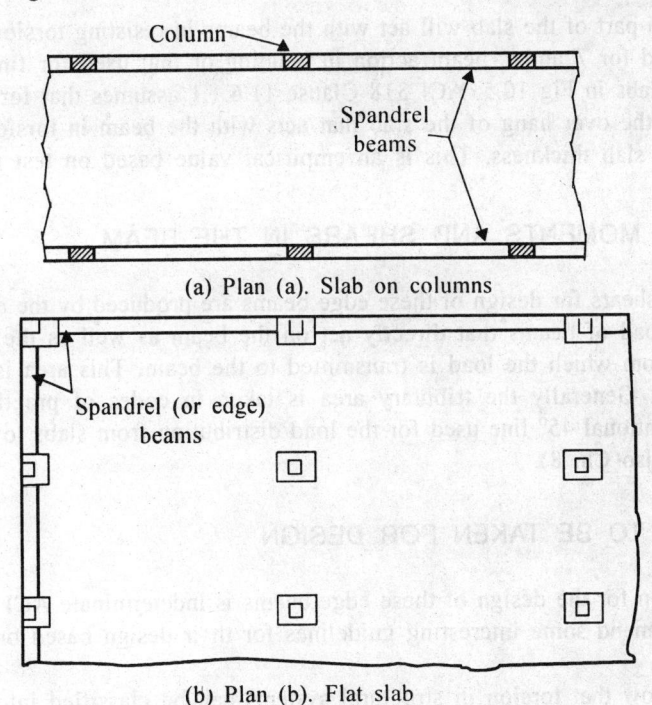

(a) Plan (a). Slab on columns

(b) Plan (b). Flat slab

Fig. 14.1 Action of spandrel beam: (a) a walk-way on columns (b) edge beams in flat slabs.

We have seen in Chapter 11 that without edge beams the entire exterior negative moment in the end-span of flat slabs have to be taken by the column strip itself. Being edge members these beams are subjected to torsion in addition to bending and shear. A large volume of research in the behaviour and design of spandrel beams have been conducted between 1968 and 1974, in U.S.A. [1]. The results of which have been incorporated in ACI 1977 and later codes. This chapter deals with the design of these beams in the light of ACI 318 (1989). *Design of Concrete Beams in Torsion* [2] can also be used as a good reference and aid for design of these beams. These aspects have not yet been incorporated in IS 456.

14.2 DESIGN PRINCIPLES

It is obvious that edge beams in slabs are designed for bending, shear and torsion. The routine empirical procedure used in the design of ordinary beam and slab system (as given in IS 456 Clause 21.5.2) assumes that the end-fixing moment is $wl^2/24$ for the slab. This is also taken as torsion acting on the beam. In many cases they are not separately designed for torsion but reinforcements for shear in these members are provided more liberally than in other situations so that these reinforcements are assumed to take care of the torsion as well. A much more realistic procedure would be to quantify the torsion that can come on the beam and design the edge beam for the combined moment, shear and torsion.

14.3 SIZE OF BEAM TO BE CONSIDERED

It is obvious that a part of the slab will act with the beams in resisting torsion. It is different from that part used for T and L beam action in bending or that used for finding the α and β_1 values in flat slabs in Fig 10.5. ACI 318 Clause 11.6.1.1 assumes that for calculating the torsional capacity the over hang of the slab that acts with the beam in torsion can be taken as three times the slab thickness. This is an empirical value based on test results.

14.4 BENDING MOMENTS AND SHEARS IN THE BEAM

The moments and shears for design of these edge beams are produced by the direct loads like wall loads, dead load of beams that directly act on the beam as well as the load from that area of the slab from which the load is transmitted to the beam. This area is usually called the *tributary area*. Generally the tributary area is taken in codes of practice as the area between the conventional 45° line used for the load distribution from slabs to beams (IS 456 Clause 24.5; see also Ch. 8).

14.5 TORSION TO BE TAKEN FOR DESIGN

The value of torsion for the design of these edge beams is indeterminate ACI code and PCA publications recommend some interesting guidelines for their design based on the following principles.

We already know that torsion in structural systems can be classified into the following two types [3,4]:

1. Statically determinate or equilibrium torsion
2. Statically indeterminate or compatibility torsion

As spandrel beams come under the second category, the amount of ultimate torsion that the member is required to resist is based not only on the requirements of statics but also on the relative stiffness of the member. This stiffness varies very much as the member cracks in torsion. Tests indicate that due to such cracking the torsional stiffness can come down to as low as 5 to 10% of the uncracked value, compared to about 50% in bending [4]. In the ACI publication SP 35 (1973), *Analysis of Structural System for Torsion* [5], Lampert proposed an expression for the torsional rigidity of the cracked reinforced concrete section. However, in the same publication, Collins and Lampert came to an interesting conclusion that "in case of beams in compatibility torsion, analysis of the structure assumes zero torsional stiffness. This stiffness can also result in a design which is as good as that of the torsional stiffness of the uncracked section". This principle can be explained as follows:

If we call the torsional moment that causes the cracking of a section as 'cracking torsional moment T_c', experiments show that adding extra steel for resisting torsional moments beyond the cracking moment has some effect on the twist or the rotation of the member. But this effect is very little on the increase in torsional moment capacity as shown in Fig. 14.2. This observation has been used as the 'key' to a simple method of design for torsion of reinforced concrete members in compatibility torsion. Accordingly, two methods of design of members subjected to compatibility torsion have been proposed:

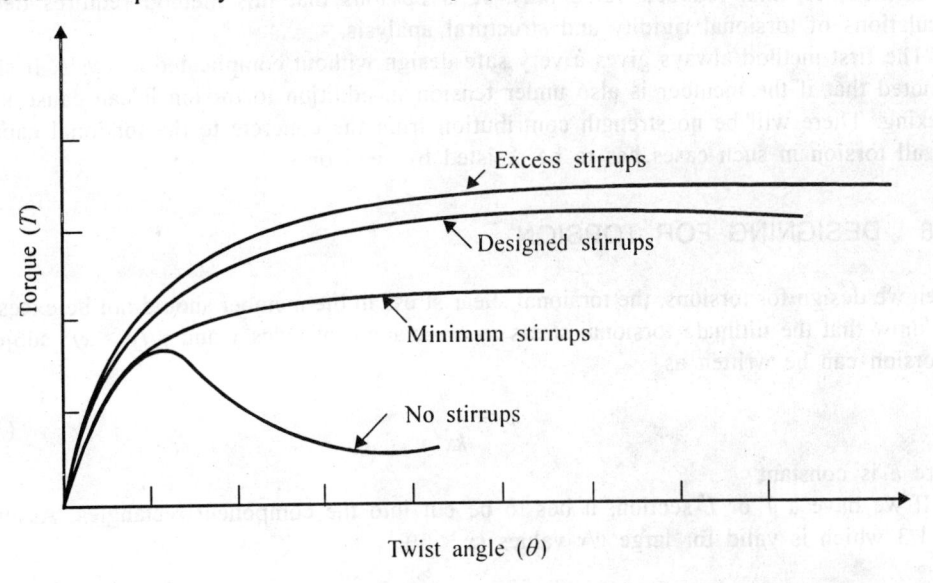

Fig. 14.2 Effect of amount of stirrups in beams under torsion.

Method 1. This simple and conservative method is given by ACI 318 (1989) Clause 11.6.3. For members in a statically indeterminate structure where redistribution of internal forces can occur, the maximum factored torsional moment of the member can be assumed as the cracking torsion. The value of this torsion T_c is given in pound units by the following expression in ACI 318 (1989) (commentary Clause 11.6.3.).

$$T_c = \frac{4}{3}\phi\sqrt{f_c'}\Sigma x^2 y \qquad (14.1)$$

where

ϕ = Reduction factor (=0.85)

f'_c = Cylinder strength in p.s.i.

If we assume that beyond this cracking moment, the member rotates and redistributes the moments along alternate load paths, there is little use in designing members beyond this torsional moment. The procedure to be adopted in this method is to analyze the system without taking into account the torsional stiffness of the member, and design the member to resist the cracking torsion, the bending moment and shear obtained by the above analysis.

Accordingly a maximum factored torsional moment, equal to the cracking torsion, is assumed to occur at the critical section. This torsional moment is also used to calculate new or adjusted values of shears and moments in the adjoining members by simple analysis that are designed for the stress resultants. This method is illustrated by Example 14.1.

Method 2. The second method is used for members in which the torsion may be much less than the cracking torsion described in Method 1. To find out this torsion, exact theoretical analysis of the structural system is made using the torsional rigidity of the uncracked section. If the magnitude is less than the value of compatibility cracking torsion, we need to design the member for that reduced value only. It is obvious that this method requires detailed calculations of torsional rigidity and structural analysis.

The first method always gives a very safe design without complicated analysis. It should be noted that if the member is also under tension in addition to torsion it can cause severe cracking. There will be no strength contribution from the concrete to the torsional capacity, and all torsion in such cases has to be resisted by steel only.

14.6 DESIGNING FOR TORSION

When we design for torsions, the torsional shear stress in the member should not be excessive. We know that the ultimate torsional stress in a rectangle of sides x and y $(y > x)$ subjected to torsion can be written as

$$v_t = \frac{T}{kx^2 y} \qquad (14.2)$$

where k is constant

If we have a T or L section, it has to be cut into the component rectangles. Assuming $k = 1/3$ which is valid for large y/x values $(x < y)$.

$$v_t = \frac{T}{\Sigma x^2 y / 3} \quad \text{or} \quad T = \frac{1}{3} v_t \Sigma x^2 y$$

14.6.1 Permissible Design Values for Compatibility Torsion

From extensive testing of T and L sections, the following values of torsion have been incorporated in the commentary of Clause 11.6.5 of the ACI 318 (1989) code for members in compatibility torsion.

(a) A torque can be considered as nominal (i.e. the torque effects can be neglected) when the factored torsional stresses obtained from Eq. (14.2) are of the order of $1.5\sqrt{f'_c}$ (p.s.i. units.)

(This is 0.375 times the cracking torsion given in (b) below. Thus, the magnitude of nominal torsion is given by the expression in pound inch units as [ACI 318 Clause 11.6.1]:

$$T_n = 0.85\left(1.5\sqrt{f_c'}\right)\sum \frac{x^2 y}{3} \qquad (14.3)$$

(*Note:* (1) This is also expressed as $T_n = 0.85\left(0.5\sqrt{f_c'}\right)\left(\sum x^2 y\right)$

(2) The above expressions are dimensionally incorrect unless we take the value of $1.5\sqrt{f_c'}$ as the allowable stress.

(3) Let us remember that T_n will work out to $0.375 T_c$ given in Eq. (14.4). We may take T_c as the base.

(b) When the torsional stress in concrete is $4\sqrt{f_c'}$ the section reaches cracking stage in compatibility torsion. Hence the cracking or compatibility torsion can be expressed by the following equation as given in ACI 318 Clause 11.6.3.

$$T_c = 0.85\left(4\sqrt{f_c'}\right)\frac{\sum x^2 y}{3} \quad \text{p.s.i. units}$$

Converting the above equation into S.I. units by assuming $f_{ck} = 0.8\, f_c'$ and including the reduction factor 0.85 we can write the formula for cracking torsion as follows:

$$T_c = 0.25\sqrt{f_{ck}}\left(\frac{\sum x^2 y}{3}\right) \qquad (14.4)$$

approximately.

(c) When the factored torsional moment is between the above nominal and cracking torsion the elastic analysis based on uncracked section can be made. When torsion is more than cracking torsion redistribution of moments can be assumed in indeterminate structures which are in compatibility torsion.

(d) The absolute maximum torsion allowed in a section in torsion (by ACI 318) clause 11.6.9.4) is $T_{max} = 4T_c$. If torsion is more as calculated from Eq. (14.2) we have to redesign the member so that the stresses are reduced to allowable values [6].

CONCLUSION

Based on theory and results of laboratory tests the following procedures of design of members in compatibility torsion can be recommended:

Procedure 1 (The Simple Procedure). For analyzing the indeterminate structural system under compatibility torsion we first assume that in any case the torsion in the member will not exceed the value of cracking torsion given by Eq. (14.4). The bending moment and the shear force values are obtained by a simple analysis without taking into consideration its torsional stiffness of the beam. The beam under torsion is designed for cracking torsion as assumed in the analysis. This method is simple and does not need elaborate structural analysis. This is illustrated by Example 14.1. In this case if the actual torsion is lower than the cracking torsion it will result only in an overdesign of the members.

Procedure 2. Where the torsion is expected to be very low, but more than the nominal value, we have the option to overdesign for cracking torsion as in procedure 1 or to determine the actual torsion by an exact elastic analysis of the structure, This determination is possible by using bending and torsional stiffness of uncracked section and provide steel only for the corresponding elastic torsion.

EXAMPLE 14.1 (R.C. beams in torsion)

The plan of a building shown in Fig 14.3 consists of a grid 10 m × 4 m in the x and y directions. Along the exterior edge in the y direction one of the columns is omitted on all the floors. The edge beam is 350 × 500 and the floor slab 125 mm thick. Determine:

1. The torsional moment for which the spandrel beam ACB should be designed
2. Find the bending moments in beam CE
3. Determine the design bending moment shears in ACB

Assume the total factored load on the beam such as $AD = 30$ kN/m, $f_{ck} = 25$, $f_y = 415$, the exterior columns are 300 × 600 mm, the interior columns are 600 × 600, and the height between floors is 3.3 m.

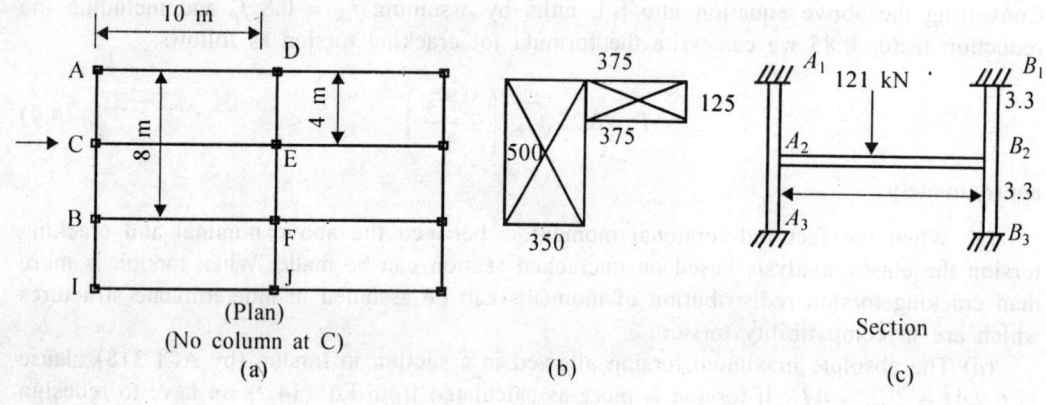

Fig. 14.3 Example 14.1.

Reference	Step	Calculations
	1	Find torsional capacity of edge beam in compatibility torsion (AC and CB) in y direction
		Size of beam acting is as shown in Fig. 14.3
Fig. 14.3(b)		$T_c = 0.25\sqrt{f_{ck}}\dfrac{\sum yx^2}{3}$
		$= (0.25)(5)\left(\dfrac{1}{3}\right)[(500)(350)^2 + 375(125)^2]$
		$= 28$ kNm
		Torsional capacity at C = Torsional capacity of CA and CB
		$= 2 \times 28 = 56$ kNm

DESIGN OF SPANDREL (OR EDGE) BEAMS 269

Reference	Step	Calculations
	2	*Analyze beam CE in x direction* Using moment distribution method assuming fixity at E
	2.1	*Calculate fixing moment at C* Fixing moment at $C = \dfrac{wl^2}{12} = \dfrac{30 \times 10^2}{12} = 250$ kNm However at the end C, the fixing moment cannot be more than the torsional capacity (=56 kNm)
	2.2	*Distribute the excess moment at C* |	C	E | |---|---|---| |	−250	+ 250 | |	+194 ⟶	+ 97 | | Total	−56	347 | These values give the bending moment distribution in beam CE.
	3	*Calculate reaction at C* $R_c = \dfrac{wL}{2} - \dfrac{M_A - M_B}{L}$ $= \dfrac{30 \times 10}{2} - \dfrac{347 - 56}{10} = 121$ kN
Fig. 14.3(c)	4	*Determine bending in ACB* Analyze the frame ACB with column at A and B with concentrated load 121 kN by subframe analysis. The resultant B.M., shear and torsional moment are to be used to design beam ACB.

EXAMPLE 14.2 (Design of spandrel beams)

A walkway consists of a slab 5.4 m between edges supported on spandrel beams 200 × 600 in size, which in turn is carried on 300 × 200 columns spaced at 7 m centres (Fig. 14.4). Assuming that the total factored load on the walkway is 6 kN/m² and the slab thickness is 150 mm, determine the design torsional moment in the spandrel (edge) beams and the walkway slab.

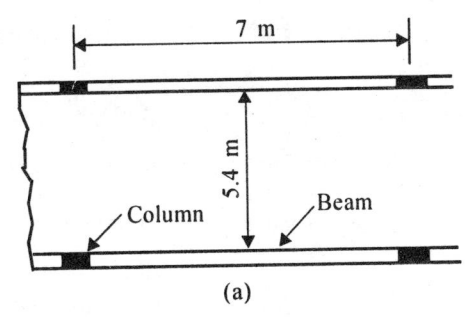

Fig. 14.4(a) Example 14.2.

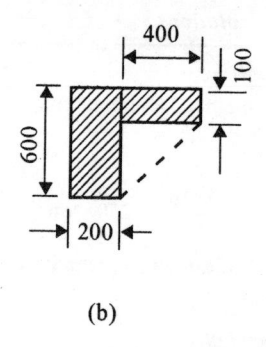

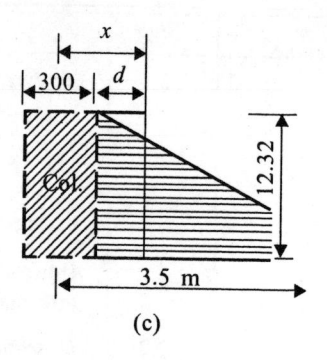

Figs. 14.4(b) and (c)   Example 14.2.

Reference	Step	Calculations
	1	*Determine torsional capacity of edge beams*
		Width of the slab attached to beam is lesser of
		$3h = 3 \times 150 = 450$ mm
		and
		$D - h = 600 - 150 = 450$ mm
		Adopt 450 mm
Text Sec. 14.6.1		$\dfrac{\sum yx^2}{3} = \dfrac{1}{3}[600(200)^2 + 450(150)^2] = 11.4 \times 10^6$ mm units
		$T_c = 0.25\sqrt{f_{ck}}\,(11.4 \times 10^6)$ N-mm (where $f_{ck} = 20$) $= 12.7$ kNm
	2	*Find fixing moment at junction of slab and edge beam*
		T_c is the maximum value of torsion the edge beam can take at support with zero at the centre of the beam.
		Assume, maximum T_c in the beam is taken at a distance from the centre line of column = (Depth of beam) + $\left(\dfrac{1}{2} \times \text{Size of column}\right)$
		$= 550 + \dfrac{300}{2} = 700$ mm
		Distance from the centre line of beam to the point of maximum torsional moment = $\dfrac{1}{2}(\text{Span}) - 0.7\,\text{m} = 3.5 - 0.7 = 2.8$ m
		The torsional moment per metre length = $\dfrac{12.7}{2.8} = 4.54$ kN/m

Reference	Step	Calculations
	3	This torsional moment is the fixing moment per metre length of the slab. *Calculate the span moment for the slab* $$M_0 = \frac{wl^2}{8} = \frac{6 \times (5.4)^2}{8} = 21.87 \text{ kNm}$$ Fixing moment = 4.54 kN Span moment = 21.87 − 4.54 = 17.33 kN/m

EXAMPLE 14.3 (Spandrel beams in flat slabs)

The peripheral spandrel beam of a flat slab is 250 × 475 mm in size and the depth of the main slab is 175 mm. The transverse distribution of moments at the end-span result in a negative moment of 5.8 kNm for the half column strip and 1.6 kNm for the half middle strip. The total negative moment being 7.4 kNm on each side of the column. Determine the torsional moment for which the spandrel beam should be designed. Assume $f_{ck} = 20$.

Reference	Step	Calculations
Eq. (14.4)	1	*Calculate cracking or compatibility torsion* $$T_c = \frac{0.25}{3}\sqrt{f_{ck}} \sum yx^2 = 0.373 \sum yx^2 \text{ for } f_c = 20$$ Length of the slab acting with beam for calculating torsional capacity = 3h = 3 × 175 = 525 D − h = 475 − 175 = 300 < 525 Adopt 300 mm $T_c = 0.373 [475 \times 250^2 + 300 \times 175^2]$ 14.5 kNm Negative moment from flat slab analysis is (5.8 + 1.6) = 7.4 kNm
Sec. 14.6.1a(3)	2	*Check whether the actual torsion is less than nominal torsion capacity of edge beam* Nominal torsional capacity = 0.375 T_c = 0.375 × 14.5 = 5.44 kNm As the actual torsion is more than the nominal torsional capacity torsion cannot be neglected.
	3	*However this torsion is less than the compatibility torsion* Hence we can proceed as follows:

Reference	Step	Calculations
		1. Design only for the actual torsion
		2. Alternately to be on the safer side design for compatibility torsion capacity.
		Note: If the actual torsion is larger than the compatibility torsion we proceed as in Example 14.4.

EXAMPLE 14.4 (Redistribution of moments in end-span of a flat plate)

The exterior span of the flat plate has the following dimensions: $L_n = 7.1$ m, $L_2 = 6$ m, and factored load $w = 10.4$ kN/m² (Fig. 14.5). The arbitrarily assumed coefficients of longitudinal distribution of moments are 0.26 at external support C and 0.75 at internal support A with 0.52 in the span at B. There is a spandrel beam at the exterior end and its torsional capacity is 23 kNm. Redistribute the moments in the slab so that the torsional capacity of the spandrel beam is not exceeded.

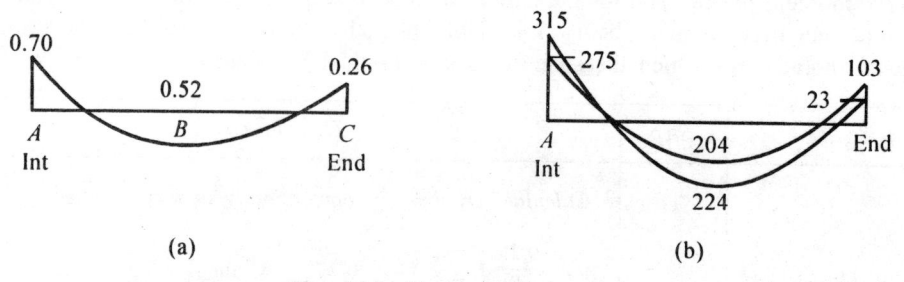

Fig. 14.5 Example 14.4.

Reference	Step	Calculations
	1	Value of M_0 and its distribution
		$M_0 = wL_2L_n^2/8 = 10.4 \times 6 \times (7.1)^2/8 = 393$ kNm
		$M_A = 393 \times 0.70 = 275$ kNm
		$M_B = 393 \times 0.52 = 204$ kNm
		$M_C = 393 \times 0.26 = 103$ kNm
		The spandrel beam has a capacity of 23 kNm but $M_c = 103$ kNm
	2	*Redistribute excess of moment at C*
		As the capacity of beam at C is only 23 kNm
		Excess of moment $103 - 23 = 80$ kNm
		has to be distributed. One-half of 80 (=40) will be carried over to A and the value of B will be increased by $80/4 = 20$.

DESIGN OF SPANDREL (OR EDGE) BEAMS 273

Reference	Step	Calculations
	3	Final moments for design will be $M_A = 275 + 40 = 375$ kNm $M_B = 204 + 20 = 224$ kNm $M_C = 103 - 80 = 23$ kNm *Carry out the statical check* $0.5(315 + 23) + 224 = 393 = M_0$

EXAMPLE 14.5 (Design of spandrel beam)

Design the edge beam of size 250 × 500 mm with slab thickness of 150 mm of a flat slab if the analysis shows that it is subjected to a maximum torsional moment of 37.3 kNm. Assume $f_{ck} = 25$ N/mm².

Reference	Step	Calculations
Example 11.4 Text Sec. 14.3	1	*Data* Torsion in beam $T = 37.3$ kNm Beam is 250 × 500 mm; Depth of slab 150 mm Portion acting with beam = $3h = 3 \times 150 = 450$ mm
Sec. 14.6.1	2	*Calculate cracking torsion* $\dfrac{\Sigma x^2 y}{3} = \dfrac{1}{3}=[(250)^2(500)+(150)^2(450)]=13.80\times 10^6$ Cracking torsion $T_c = \dfrac{0.25}{3}\sqrt{f_{ck}}\,\Sigma x^2 y$ $= 0.25 \times 5 \times 13.8 = 17.25$ kNm $T > T_c$
Sec. 14.6.1	3	*Calculate maximum torsion allowed* Maximum torsion allowed = $4T_c = 4 \times 16.25 = 65$ kNm $T < T$ maximum (allowed) Member can be designed for torsion as follows.
	4	*Method of design to be adopted* As the member is in compatibility torsion, we design the beam for cracking torsion combined with the shear and moments due to loads. We also distribute the balance of the torque to the slab as in Example 14.4. Considering each side excess of moment over cracking torsion

Reference	Step	Calculations
Example 14.4	5	$= 37.3 - 17.25 = 20.1$ kNm Excess from both sides of the column $= 2 \times 20.1 = 40.2$ kNm *Distribute excess moment to the span and to the adjacent interior support as follows* Let the increase in moments $= \Delta M$ ΔM at support $= \dfrac{40.2}{2} = 20.1$ kNm ΔM in span $= \dfrac{40.2}{4} = 10.05$ kNm

REFERENCES

1. Hsu, T.T.C. and Burton, K.T., *Design of Reinforced Concrete Spandrel Beams*, Proceedings ASCE, S.T.1., January 1974.

2. Skobie, *Design of Concrete Beams in Torsion*, Portland Cement Association, Illinois, 1983.

3. Varghese, P.C., *Limit State Design of Reinforced Concrete*, Prentice-Hall of India, New Delhi, 1994.

4. Ferguson, P.M., Breen, J.E. and Jersa, J.O., *Reinforced Concrete Fundamentals*, John Wiley and Sons, NY, 1988.

5. Collins, M.P. and Lampert, P., *Redistribution of Moments at Cracking—The Key to Simpler Torsion Design—Analysis of Structural Systems for Torsion*, SP 35, A.C.I., Detroit, 1973.

6. Ghosh, S.K. and Rabbat, B.G., (Eds.), *Notes on ACI 318 (1989) with Design Applications*, Portland Cement Association, Illinois, 1990.

CHAPTER 15

Provision of Ties in Reinforced Concrete Slab–Frame System

15.1 INTRODUCTION

The Indian code IS 456 does not give any specific recommendations for checking the stability of reinforced concrete slab frames for abnormal loads that may accidentally occur in a building. BS 8110 Clause 3.12.3 gives a few general recommendations for such design and these can be followed as a guide for design of slab and frame buildings [1,2]. Observations of failures of structures (especially those made from precast elements) due to explosion of gas cylinders and other accidental causes have shown that structures should be detailed to act as a whole, under all possible conditions of loadings. For this purpose, it is usual to check whether the superstructure meets the following three requirements:

1. Suitable protective works like earth work or other types of barriers are provided for columns and other vertical load bearing members at the road level where vehicular traffic can be expected either by design or by accident.

2. In regions subjected to heavy winds or earthquakes, the structure has overall stability to withstand the specified ultimate horizontal (wind and earthquake) loads for the region under consideration. In all cases, a load not less than 1.5% of the total characteristic vertical dead load above the level should be considered as horizontal load for such analysis.

3. Even in ordinary buildings, interaction between elements should be provided by tying the structure together using the following type of ties (Fig. 15.1):

    (i) Peripheral ties
    (ii) Internal ties
    (iii) Horizontal column and wall ties
    (iv) Vertical ties

This tying up helps the structure to stand up and prevent progressive collapse, by providing alternate load paths in case of any local failures. BS 8110 rules regarding provision of these ties are dealt with in this chapter. Progressive collapse can otherwise occur when due to a local failure debris of one floor falls on the next below which starts a train of successive failure of floors below.

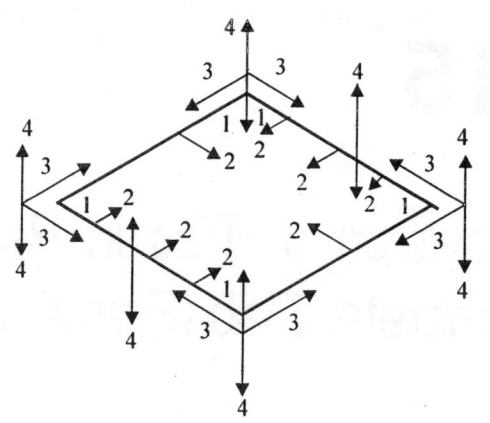

Fig. 15.1 Provision of horizontal and vertical ties in buildings.

15.2 DESIGN FOR OVERALL STABILITY (ROBUSTNESS)

The vertical loads to be considered along with horizontal loads for stability against lateral loads under item 2 above as given in IS 456 Clause 36.4 are the following:

$$1.2(DL + LL \pm HL)$$
$$1.5 \text{ or } 0.9DL \pm 1.5HL$$

See Section 18.10 and Table 18.6 in Chapter 18. HL indicates horizontal load which may be due to wind or earthquake. The minimum horizontal load should not be less than 1.5% of the total dead load above that level. [*Note:* The factor 1.5 for D.L. is to be taken when dead load does not add to the stability and the factor 0.9 when D.L. is beneficial to stability.)

15.3 DESIGN PROCEDURE FOR TIES

For the third requirement given in Section 15.1 we proceed as follows. Eventhough standard detailing ensures these provisions automatically, it is a good practice to check for the existence of the four types of ties to withstand the specified forces as detailed below. The purpose of these ties is to sustain the load of a floor and prevent the damaged part (or debris) from falling on to the lower floors. The usual procedure adopted is to first design the building for the specified loads and then carry out the final check for the tie forces. When designing for the provision of different types of ties, no forces other than those specified for the particular tie analysis are assumed to act. The reinforcements, already provided in slabs, beams and columns, can be assumed to act also for this stability considerations. It is always important to check for continuity of these steels in the detailing of the structure.[1]

15.3.1 Design of Peripheral Ties

Peripheral ties are type 1 ties shown in Fig. 15.1. At each floor and roof level, continuous peripheral ties should be provided in each direction. The horizontal force F_t to be resisted by

---

[1]Indicated in Section 15.4.

these ties at each level depends on the number of storeys above the level considered, including basement. Value of F_t is to be lesser in the following two expressions:

1. $F_t = 20 + 4n$ kN; (where n = number of storeys including ground floor
2. $F_t = 60$ kN

Thus $F_t = 24$ kN for single storey and $F_t = 60$ kN for a ten storeys and more.
With high yield reinforcement of grade 415 N/mm², an area of 2.5 mm² of steel is enough to be provided per kilo neuton of force. Thus for the maximum force of 60 kN only 150 mm² steel or 2 nos of 10 mm or 1 no of 16 mm bar is necessary for such ties. This steel can be easily provided in the perimeter beams, slabs or perimeter walls. BS 8110 also stipulates that this should be located within a distance of 1.2 m from the edge of the building or perimeter walls.

15.3.2 Design of Internal Ties

Internal ties are type 2 ties as shown in Fig. 15.1. These ties are provided in the slab or beams at each floor or roof level in X and Y directions approximately at right angles each other. They should fulfil the following conditions:

1. They should be continuous throughout the length of the slab or beam.
2. They may be spread evenly in the slab or grouped in beams, walls, etc. Their spacings should not be greater than $1.5L$, where L is the greater of the distances in metres between columns or vertical supports. These supports any two adjacent floors in the direction of the tie under consideration.
3. They should be anchored to the peripheral ties, described in Section 15.3.1, at their two ends or continue as horizontal ties to columns beams or walls as shown in Fig. 15.2.
4. When anchored in walls they should be placed within 0.5 m of the top or bottom of floor slabs.

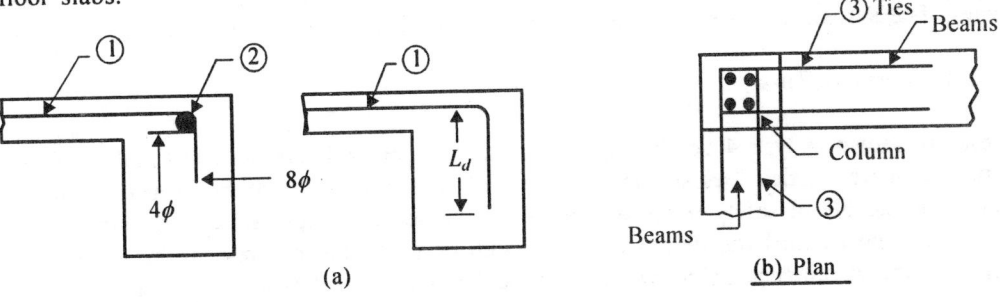

Fig. 15.2 Details of anchoring of ties: (a) internal ties (b) column ties.

For the above purposes, when detailing continuous slabs on beams, some of the bottom or the top bars in them are given continuity at supports. It should be noted that bars which are laid at bottom and bent to the top are not considered as 'un-interrupted tie bars'. These tie bars should be straight from end to end with continuity bars provided for laps. The tensile forces for which the internal ties are designed is greater of the following (expressed in kN/m² per metre width):

(i) F_t as in Section 15.3.1 above

(ii) $\dfrac{g_k + q_k}{7.5} \dfrac{L}{5} F_t = 0.0267(g_k + q_k)LF_t$

where

g_k = Characteristic dead load (kN/m^2) on slab

q_k = Characteristic live load (kN/m^2) on slab

L = Largest centre line distance between supports

Generally two 10-mm bars per metre width of slab will be ample for this purpose. They can be evenly distributed or also grouped at positions of beams. They are to be anchored in the peripheral beams by one of the methods as shown in Fig. 15.2 and as per requirements for anchorage.[1] The importance of providing integrity bars in two-way slabs has already been emphasized in Section 11.13.

15.3.3 Horizontal Ties from External Columns and Walls Back to the Floor

These are shown as type 3 ties in Fig. 15.1. All external load bearing vertical members such as external columns and walls should be tied back into the floor (roof) structure horizontally at each floor level by means of proper ties. In case of external walls every metre length of the wall should be anchored into the floor.

The force to be considered is greater of the following two.

(i) The lesser of $2F_t$ or $0.4 L_0 F_t$ in kN, where L_0 is the floor-to-ceiling height in metres and F_t is in kN.

(ii) Three per cent of the total ultimate vertical load in the column or wall at that floor level.

For corner columns this force is to be provided in each of the two directions at right angles. The system used for a column-to-floor beam anchorage is shown in Fig. 15.2(b).

15.3.4 Vertical Ties

These are shown as type 4 ties in Fig. 15.1. Each column and each wall of building of five storeys or more, vertical ties should be continued from foundation to roof level. These ties should be capable of safely resisting a tensile force equal to the maximum design ultimate dead and imposed load transferred to the column or wall from any one storey or the roof. Normally the reinforcements provided in the column will meet this requirement.

15.4 CONTINUITY AND ANCHORING OF TIES

A tie is considered as continuous if it is properly lapped, welded or mechanically joined at the points of discontinuity. When the ties to be anchored are at right angles. These are considered as properly anchored to the other if it satisfies the following requirements:

---

[1]See Section 15.5.

1. It should extend 12 times its diameters or equivalent anchorage beyond the other bars.
2. An effective anchorage length (based on the force in the bars) beyond the centre line of the other bars is provided.

CONCLUSION

It is mandatory that overall structural stability should be provided in all structures by proper ties. Engineers at the construction sites should also pay special attention to provision of these ties and their continuity in the various members. As stated in Section 15.2, the design procedure is to check whether the above requirement of ties is satisfied by the reinforcements already provided in the structure for the various other design requirements. Separate reinforcements are not normally needed for this purpose.

EXAMPLE 15.1 (Design of ties in buildings)
The plan of a six-storeyed framed building G + 5 designed as slab on beams supported on columns is shown in Fig. 15.3. The average characteristic dead load on the slab is 7 kN/m^2 and the imposed load 3 kN/m^2. The storey height is 3.5 m with floor-to-ceiling height 3 m. Design the various types of ties prescribed in BS 8110 for ensuring robustness of the building.

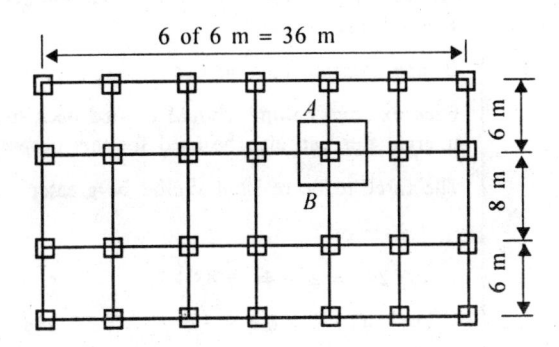

Fig. 15.3 Example 5.1.

Reference	Step	Calculations
Sec. 15.3.1	1	*Design peripheral ties* F_t is lesser of the following two quantities: (a) $20 + 4n_0 = 20 + (4 \times 6) = 44$ kN (b) 60 kN Lesser of the two = 44 kN. $A_{st} = \dfrac{44 \times 10^3}{0.87 \times 415} = 122$ mm^2 Provide two 10-mm bars (157 mm^2)

Reference	Step	Calculations
		Note: The maximum area for peripheral ties $$= \frac{60 \times 10^3}{0.87 \times 415} = 166 \text{ mm}^2$$ 2T10 will provide this steel as above.
Sec. 15.3.2	2	*Design internal ties in X and Y directions* $g_k = 7$ kN/m²; $q_k = 3$ kN/m², $L = 8$ m Tensile force is greater of the following two quantities: (i) $F_t = 44$ kN (ii) $0.0267(g_k + q_k) LF_t = 0.0267 \times 10 \times 8 \times F_t = 2.136 F_t$ $= 2.136 \times 44 = 94$ kN > 44 kN $$A_{st} = \frac{94 \times 10^3}{0.87 \times 415} = 260 \text{ mm}^2$$ Provide 2T12 bars (339 mm²) in the slab per metre width in X and Y directions.
Sec. 15.3.3	3	*Design external column ties* Each external column should be tied back to the floor or roof. The internal ties can also be used for this purpose. The force to be resisted should be greater of the following: (i) Lesser of (a) $2F_t = 2 \times 44 = 88$ kN (b) $0.4 L_0 F_t = 0.4 \times 3 \times F_t = 1.2 F_t$ and The floor-to-ceiling height $L_0 = 3$ m. $= 1.2 \times 44 = 52.8$ kN (ii) To find 3% of total ultimate load in the external column at that floor level. Contributory area $A = 6 \times 3 = 18$ m² Ultimate load from one floor to the external column $= 1.5 (10) (18) = 270$ kN per floor (total load $= 10$ kN/m²) Load on the first floor level $= 270 \times 5 = 1350$ kN 3% of 1350 $= 40.5$ kN Force to be resisted-adopt $= 52.8$ kN At corners provide tie backs in both directions. $$A_{st} = \frac{52.8 \times 10^3}{0.87 \times 415} = 146 \text{ mm}^2$$

Reference	Step	Calculations
Sec. 15.3.4 Fig. 15.3	4	*Design vertical ties (Building higher than 5 storeys)* Design force = Maximum vertical design load from floor on the most loaded column = 1.5(10)(42) = 630 kN (area = 7 × 6 = 42 m^2) Steel necessary in column $= \dfrac{630 \times 10^3}{0.87 \times 415} = 1745 \text{ mm}^2$ 4 of 25 mm^2 = 1963 mm^2

REFERENCES

1. BS 8110 (1985), *Structural Use of Concrete*, British Standards Institution, London.
2. Ray, S.S., *Reinforced Concrete Analysis and Design*. Blackwell Science, London, 1995.

CHAPTER 16

Design of Reinforced Concrete Members for Fire Resistance

16.1 INTRODUCTION

The Indian Standards IS 1641 (1988) and IS 1642 (1989) deal briefly with fire resistance of buildings. But details of design are not given in this code. BS 8110 Part 2 and CEB-FIP model code 'Design of concrete structures for fire resistance' can be used as the basis of design of concrete members against damage by fire [1,2]. The latter recommends the following three assessment procedures for assessing fire resistance of reinforced concrete members:

1. Assessment method I is based on tests under the standard heating conditions, as formulated by ISO 834. The design criteria is the fire resistance time of the member against the ISO 834 fire[1] should be equal or exceed that specified by the building regulations.

2. Assessment method II is the direct application of a direct test on structures or elements based on the concept of the equivalent of the fire exposure which tries to relate the effects of an arbitrary given fire to those of the ISO 834 standard fire. Thus it is the equivalent time concept.

3. Assessment method III is by means of engineering calculations based on experimental data. It is an analytical structural design of a member for a non-standard fire exposure using heat and mass balance equations for temperature. The assessment takes into account the conditions of fire load, ventilation and thermal properties of the structural element used.

However, at present, the most commonly used method is method I. The other methods are still in the developmental stages. Accordingly, this chapter describes method I (which is generally found in modern codes) and gives a resume of the other two methods. The latter require specialized knowledge for field application.[2]

16.2 ISO 834 STANDARD HEATING CONDITIONS [2]

Most of the present-day knowledge of fire resistance of reinforced concrete members have been obtained from tests on members which were experimentally exposed in a fire chamber

---

[1]*See* further explanation in Section 16.2.
[2]The model code for fire design of concrete structures 'Bulletin D Information No. 174 of Committee Euro-International Du Beton' may be consulted for further details regarding the assessment methods I and III. IS 456 (2000) clause 21 and Table 16 deal with fire resistance of reinforced concrete.

to standard fire. A standard fire can be defined as a standardised time-dependent temperature development, according to the standard heating conditions stated in ISO 834 as shown in Fig. 16.1. Thus, when applying method I, one should remember that all references are with respect to test conditions and not to real life situations.

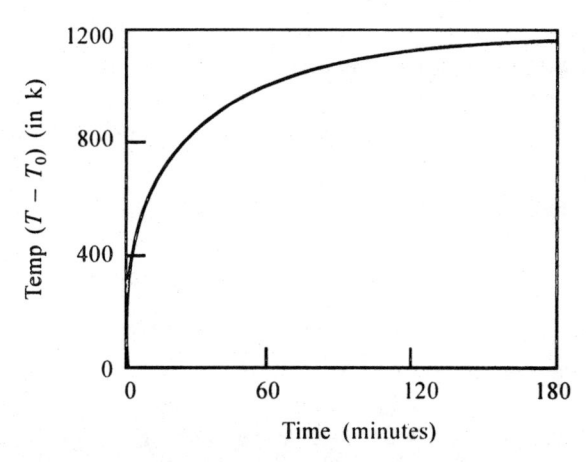

Fig. 16.1 Temperature–time relationship in ISO standard fire.

ISO 834 specifies a test method which provides the determination of fire resistance of a structural element on the basis of the length of time for which specimen satisfies the test criteria. The specimen is exposed to a furnace temperature rise which obeys the following time-tempreture relation:

$$T - T_0 = 345 \log_{10} (8t + 1)$$

where

t = Time in minutes

T = Furnace temperature in Kelvin at time t

T_0 = Furnace temperature in Kelvin at time $t = 0$.

16.3 GRADING OR CLASSIFICATIONS

The fire resistance of a structural member is expressed in two ways:

1. Directly by a fire resisitance time in hour or minutes, for example, 0.5 hour or 30 min.
2. Indirectly by fire resistance classes such as F30, F60, F90, F240 which respect the minimum fire resistance time in minutes.

16.3.1 Evaluation of Fire Resistance to ISO 834 Fire

As already mentioned in Section 16.1, there are three procedures for design and evaluation of fire resistance with respect to the standard fire:

Procedure 1. By use of tabulated data.

Procedure 2. By actual fire testing of full scale model in specialised laboratories and issue of certificates.

Procedure 3. By theoretical fire engineering calculations of ultimate strength of the members.

Of these, the first method is most commonly used in routine design. The second method is expensive and restricted in its applications. For the third method, at present data on limited type of members, like beams only, are available. It is not applicable to columns and walls.

16.4 EFFECT OF HIGH TEMPERATURE ON STEEL AND CONCRETE

Steel and concrete both undergo changes in properties at elevated temperatures, as described below [1,3].

16.4.1 Effect of Fire on Steel

The strength and deformation properties of steel undergo changes with increase in temperature. The magnitude of the change depends on the composition and the manufacturing process. The increased strength of cold-drawn and cold-twisted steel rods and prestressing steel is due to dislocation and distortion of microstructure of steel. At high temperature, this hardening effect is neutralised. These changes are shown graphically in Fig. 16.2. For most of the steels the temperature at which the recrystallisation of the microstructure starts on heating is about 300°C.

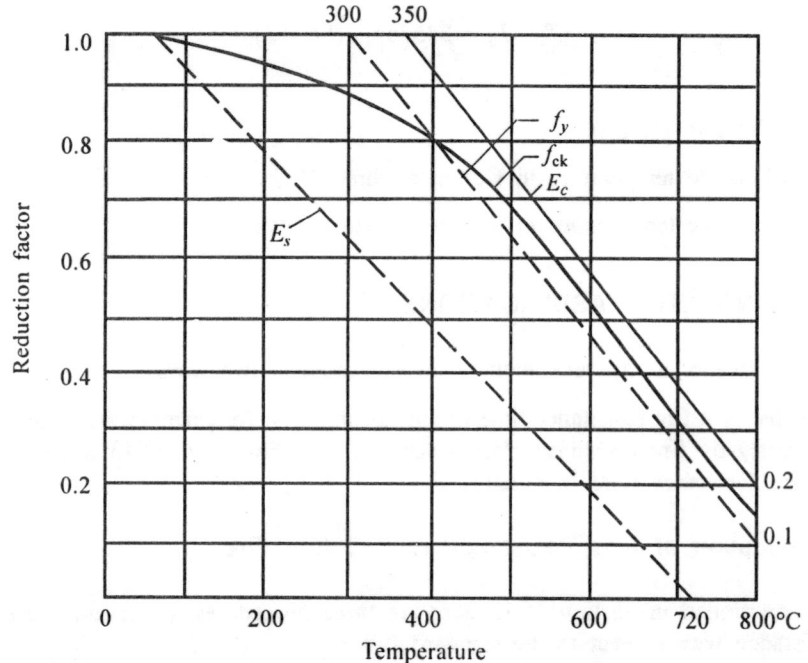

Fig. 16.2 Decrease in strengths and elastic moduli of concrete and steel with temperature.

16.4.2 Effect of Fire on Concrete

In concrete also, the strength decreases and deformation increases with rise in temperature. It is also seen that the critical concrete temperature, beyond which the change in properties is noticeable, is stress dependent, i.e. higher stresses result in lower critical temperature. The trend of decrease in strength of dense concrete with temperature is also shown in Fig. 16.2.

16.5 EFFECT OF HIGH TEMPERATURES ON DIFFERENT TYPES OF STRUCTURAL MEMBERS

Beams, slabs and columns are effected differently by fire. Some of these effects are as follows:

16.5.1 Effect of Fire on Simple Beams and Slabs

Members, like beams and slabs, are heated from below and the tensile zones are directly exposed to fire. In such cases the ultimate load is reached when the tensile reinforcement reaches the critical temperature at which the yielding starts, under the actual dead load, stresses. In most cases, the compressive zones (except webs of thin members) do not reach the compression limit state and collapse is mainly by yielding of tensile steel.

16.5.2 Effect of Fire on Continuous Beams and Slabs

In dealing with continuous beams, the effects under service condition in ordinary state should be superimposed with the effects of heating from below (temperature restraint) and the failure mechanism may be as shown in Fig. 16.3. During a fire the strength reduction at supports

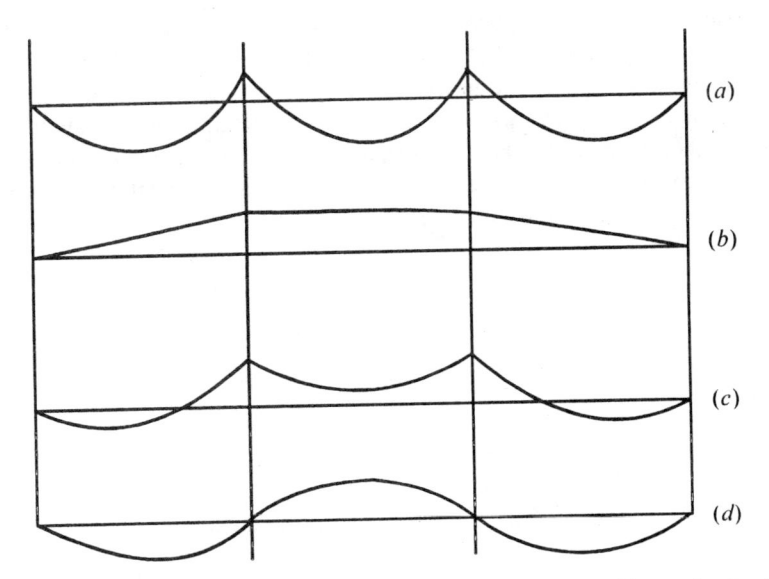

Fig. 16.3 Mode of failure of continuous beams under fire from below: (a) service load moments (b) temperature restraint moments (c) resultant moments (d) failure mode.

of continuous beams will be small and hence a high design moment at support helps greater fire endurance of continuous beams.

16.5.3 Effect of Fire on Columns

Under fire, the load carrying capacity of the steel in the outer parts of the cross-section of a column decreases rapidly. This results in large amount of load to be transferred to the concrete and this can lead to an earlier failure of the column. Slender columns tend to fail earlier than shorter columns mainly due to the further increases in slenderness. This slenderness is caused by the damage of the outer parts of the concrete breaking on spalling. Columns are generally monolithically connected at top and bottom to beams. Because of the larger mass at junctions, there will be slower temperature rise at these junctions. Thus the loss of stiffness at the ends will be much less and this is a positive effect on the final load-bearing capacity by maintaining the end-fixity conditions.

16.5.4 Effect of Fire on Tension Members

In design of tension members, the concrete is considered to act only as an insulation and the limit state is reached when the steel reaches the critical temperatures and yields under the load.

16.5.5 Spalling of Concrete with High Temperatures

Two types of spalling occur in concrete at high temperatures. The first is aggregate spalling which is of minor importance as it is confined to the outermost areas of the concrete mass. It occurs due to physio-chemical transformation of the aggregate structure. The second is the explosive spalling, which is dangerous and occurs at early stages of heating. The main reason of this spalling is the water-vapour movements and the temperature gradient in the cross-section. The resulting stresses along with the stresses due to loading cause the explosive spalling. Small sections are prone to explosive spalling than large cross-sections. Similarly, rapid changes in cross-sections promote explosive spalling. Explosive spalling also depends on the type of aggregates used. Dense aggregates from granite are more prone to spalling than light-weight aggregate or limestone. Figure 16.4 shows the relation between breadth and

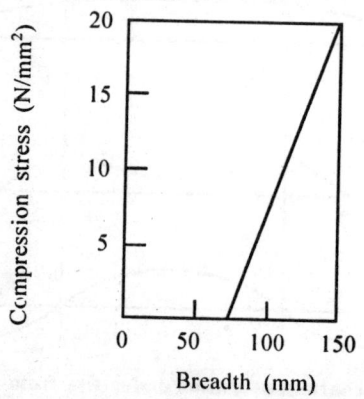

Fig. 16.4   Minimum breadth of R.C. members to avoid spalling.

compressive stress under service condition. These conditions are to be satisfied to prevent spalling in dense aggregates.

16.5.6 Protection against Spalling

The measures that can be used to avoid spalling are:

   (i) the application of special plaster as protective coat to avoid spalling
  (ii) the provision of false ceiling as a fire barrier
 (iii) the use of light-weight aggregates
 (iv) the use of sacrificial tensile steel in the design, where protective coatings are not used.

The insulation provided by various applications can be considered in terms of the following equivalent thickness of concrete:

   (i) Plaster or sprayed fibre = (1 × thickness) of concrete
  (ii) Vermiculite slabs = (1 × thickness) of concrete
 (iii) Mortar of gypsum plaster = (0.6 × thickness) of concrete

It may also be noted that the above equivalent finishes can also be used to remedy deficiencies in cover thickness for reconstruction after a fire damage.

16.6 FIRE RESISTANCE BY STRUCTURAL DETAILING FROM TABULATED DATA

Analysis of a large number of fire tests on concrete members under ISO 834 fire has established design rules principally with respect to the following:

1. cover to steel
2. size of members (minimum thickness for a given fire-rating)
3. other factors like detailing practice.

The values for these factors are given in the Indian codes for design [6]. The recommended values are conservative and safe for use under normal conditions. Most of these recommendations assume that the critical temperature at which collapse of a simply-supported and unrestrained member under its design service loads occurs when steel reaches a temperature of 500°C. It should however be remembered that this temperature may be lower with prestressing steel. On the other hand, hyperstatic structure may not collapse at the above critical temperature due to moment redistribution. Fire protection requirement in the latter can be slightly lower than that specified in tables which are meant for simply-supported structures.

16.6.1 Definition of Average Cover

Cover usually refers to the distance between the nearest heated face of the concrete and the surface of the main reinforcement in the case of single layer of steel. Sometimes, for many layers of steel, the term average cover C_{av} is also used and is determined below:

$$C_{av} = \frac{\Sigma AC}{\Sigma A}$$

where A is the area of tensile reinforcement and C the distance between the exposed surface and main reinforcement. In design for fire, 'cover' generally means the clear cover to the main reinforcement. In structural design, however, cover is expressed as the nominal cover which covers all reinforcements including stirrups. For convenience of accommodating 8 mm to 12 mm stirrups in beams and transverse steel in columns tabulated values sometimes give 'nominal cover' assuming that 10 mm diameter stirrup is used in the construction. Thus,

$$\text{Nominal cover} = (\text{Cover to main steel}) - (10 \text{ mm})$$

16.6.2 Tabulated Values of Cover and Minimum Size of Section

Based on the analysis of fire test results, tables such as Table 16.1 giving minimum requirements of sizes of sections and concrete cover to steel (heat protection to steel) have been published. It should be noted that the maximum clear cover allowed in reinforced concrete is 40 mm, and with larger covers additional reinforcement has to be provided in order to prevent the fall off of the concrete cover. The following comments refer to covers specified for fire protection:

1. *Cover for beams*: The usual tables apply to beams affected by fire from three sides. These tables fulfil the requirements of the minimum width b of beams and cover for the different fire resistance periods F30 to F240. These also give the minimum rib thickness required for T beams.

2. *Cover for columns*: The values of the minimum thickness of cover are given as it is fully exposed, or 50% exposed or only one face is exposed.

3. *Cover for continuous beams*: As already stated, continuous beams behave better under fire than simply-supported beams. Because of this favourable action the values of cover can be reduced somewhat in these beams. To ensure proper redistribution of moments at least 20% of the top steel should continue over all the spans, and the negative steel should extend to $0.15L$ over each support beyond the theoretical cut off point (where L is the larger span corresponding to the support).

4. *Cover for floor slabs*: The minimum thickness of floors and cover to be given are usually specified.

5. *Cover for walls*: The minimum thickness specified to be provided for walls depend on the amount of steel provided in the walls.

16.7 ANALYTICAL DETERMINATION OF THE ULTIMATE BENDING MOMENT CAPACITY OF REINFORCED CONCRETE BEAMS UNDER FIRE [5,6]

At the present state of development, only the analytical method for determination of ultimate bending capacity of beams under effects of fire is fairly well known. The types of failure due to loss of strength in shear, bond, anchorage, etc., are still in the experimental stage. The method of calculating the bending capacity is based on the three following assumptions first proposed in the Swedish manual *Fire Engineering Design of Concrete Structures*:

1. The temperature distribution (isotherms) in typical sections of R.C. members obtained from tests or calculations can be adopted for estimates of the temperature in steel and concrete.

TABLE 16.1 MINIMUM REQUIREMENTS FOR FIRE RESISTANCE OF MEMBERS IN DENSE CONCRETE [1,2]

[IS 456 (2000) Clause 21 and Table 16](Cover specified is nominal cover)

Type of member	\multicolumn{6}{c}{Minimum dimensions in (mm) excluding combustible finish for a fire rating of}
	F30	F60	F90	F120	F180	F240
1. Simply-supported beams						
Minimum width	200	200	200	200	240	280
Cover	020	020	020	040	060	070
2. Continuous beams						
Minimum width (not specified)	125	125	125	150	200	240
Cover	020	020	020	030	040	050
3. Floor slabs						
Minimum thickness	75	95	110	125	150	170
Cover (S.S. Slabs)	20	20	25	35	45	55
Cover (continuous slabs)	20	20	20	25	35	45

(*Note:* For two-way slabs, the cover may be reduced by 5 mm)

	F30	F60	F90	F120	F180	F240
4. Fully-exposed column						
Minimum width	150	200	250	300	400	450
Cover	40	40	40	40	40	40

(Data is also available for 50% and one side exposure.) See IS 456 (2000) Clause 21.2.

	F30	F60	F90	F120	F180	F240
5. Walls						
Minimum thickness						
(a) With less than 0.4% steel	150	150	175	–	–	–
(b) With 0.4 to 1.0% steel:	100	120	140	160	200	240
(c) With > 1.0% steel	100	100	100	100	150	180
Cover (recommended)	25	25	25	25	25	25
6. Waffle slab ribs						
Minimum rib width	125	125	125	125	150	175
Cover (simply supported)	20	20	35	45	55	65
Cover (continuous)	20	20	20	35	45	55

Notes: (1) Cover above 40 mm in beams and 35 mm in slabs or slab ribs require attention to prevent spalling.

(2) Both cover and minimum dimensions should be satisfied for fire resistance.

(3) Nominal cover + 10 mm = Cover to main steel.

For example, the isotherms of a T section for a standard fire is given in Fig. 16.5. Similar figures for rectangles, square (columns) are also available in the literature. It can be calculated also by theory of heat transfer. Readymade tables like Tables 16.2 and 16.3 are also available [5].

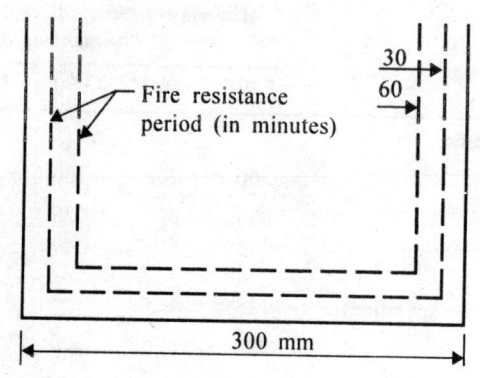

Fig. 16.5 Temperature zone (500°C) in R.C. beams exposed to ISO fire after 30 and 60 minutes.

TABLE 16.2 TEMPERATURE DISTRIBUTION IN SLABS UNDER A STANDARD FIRE

Time (in min)	Distance from exposed face (in mm)										
	0	10	20	30	40	50	60	70	80	90	100
30	700	480	345	255	200	145	120	100	90	75	65
60	840	665	520	400	320	255	200	165	140	130	100
90	–	790	655	535	445	365	310	255	210	180	155
120	–	–	730	605	510	430	380	320	300	230	200
180	–	–	840	700	600	520	465	405	360	320	275
240	–	–	–	765	680	600	535	475	435	400	365

TABLE 16.3 APPROXIMATE TEMPERATURE DISTRIBUTION IN THE BEAMS UNDER A STANDARD FIRE (Fire from below the beams)

Time (in min)	Distance from exposed face (in mm)										
	0	10	20	30	40	50	60	80	100	120	140	150
60	860	700	550	420	350	300	250	180	100	80	60	40
90	920	720	630	520	450	380	320	250	200	150	115	100
120	1000	840	700	580	520	470	400	320	260	220	200	180
180	1100	850	750	650	600	530	480	400	340	300	270	260
240	1140	900	800	700	650	600	550	480	400	370	310	300

2. Concrete exposed to more than 500°C can be assumed as incapable of contributing to axial load, bending amount and their combinations. The concrete below that temperature can be assumed to maintain their full grade strength that has been designed at ordinary room temperature. Otherwise a reduction in strength (as shown in Fig. 16.2) can be assumed.

3. The strength of steel at a given temperature can be estimated from Fig. 16.2. Even though tests show that the strength in tension depends on the $(A_s/(bdf_{ck})$ ratio and the strength of steel in compression corresponding to the 0.5% proof stress these requirements are not necessary for routine calculations. The assessment of strength of a beam in bending can be made as follows:

(a) Determine the temperature profile in the beam for the specified fire exposure time t by any available method.

(b) We may assume that the strength of concrete in the new beam is contributed by the full strength of concrete whose temperature is below 500°C, i.e. excluding the concrete that lie outside the 500°C isotherm. *Alternately*, calculate the mean temperature of concrete and assume its strength from Fig. 16.2.

(c) The strength of the individual steel bars are taken from the temperature profiles for the various sections. The reduced strength of these steel bars in tension and compression can be taken from Fig. 16.2.

(d) Calculate the ultimate load bearing capacity of the reduced cross-section with the above data.

(e) BS 8110 recommends γ_m for steel as 1 and for concrete 1.3 with γ_f values of DL as 1.05 and LL as 1 for these calculations.

16.8 OTHER CONSIDERATIONS

In addition to actual structural design as indicated in the above sections, non-structural measures against fire risk can also be considered during the design stage. By incorporating these factors at the design stage itself we can reduce the classification required for a particular structure. Some of these measures are:

1. Choice of proper materials of construction
2. Provision of fire escapes for safety of occupants
3. Installation of smoke and heat detection systems for fire warning
4. Provision of sprinklers and availability of fire fighting squads and equipment.

We should also envisage that in the event of a fire in a building it may not collapse but only be damaged to some extent. In many cases the damage may be moderate so that it can be reconstructed for original use. If the damage is too severe it may be demolished. One of the methods of ensuring repairability and reusability of structure in the future event of fire is to use an increased fire resistance grading requirements during the initial design stage. Though the effects of such a procedure cannot be quantified it will make the repairability of the structure more probable and easy. The assessment of the strength of the structure damaged by fire and the methods that can be used to repair them is a specialised field of study. The best sources of information for repair and rehabilitation are the case studies published in literature.

292 ADVANCED REINFORCED CONCRETE DESIGN

EXAMPLE 16.1 (Loss of strength by fire)
Estimate the loss in bending strength of a beam with the following dimensions (Fig. 16.6) and subjected to a standard fire (as given in Tables 16.2 and 16.3). The beam is 600 × 350 mm in size with 6 nos. of 16 mm rods placed in one row with a side and bottom clear cover to main steel of 40 mm. Assume $f_{ck} = 20$ and the steel is of grade 415.

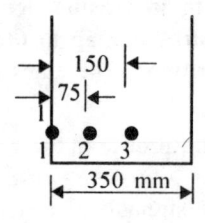

Fig. 16.6 Example 16.1.

Reference	Step	Calculations
Sec. 16.6.1	1	*Find distances of steel from face of concrete*
		Cover to outer surface of steel = clear cover = 40 mm
		Assume distance from exposed surface of all steel bars = 40 mm
		(If distances are different, each reinforcement should be considered separately.)
	2	*Estimate temperature and strength of steel with time*
		Distance of steel from face of concrete = 40 mm
Table 16.3 Fig. 16.2		<table><tr><td>Time</td><td>Temperature</td><td>Reduction factor</td><td>Strength (N/mm²)</td></tr><tr><td>0</td><td>Normal</td><td>1.0</td><td>415</td></tr><tr><td>60 min</td><td>350</td><td>0.90</td><td>373</td></tr><tr><td>90 min</td><td>450</td><td>0.70</td><td>290</td></tr><tr><td>120 min</td><td>520</td><td>0.60</td><td>250</td></tr><tr><td>180 min</td><td>600</td><td>0.44</td><td>138</td></tr><tr><td>240 min</td><td>650</td><td>0.35</td><td>145</td></tr></table>
	3	*Find temperature and strength of concrete during fire with time*
		T_1 = Temperature of face
		T_2 = Temperature of 1/4 width = say 80 mm from face
		T_3 = Temperature at 1/2 width = 175 mm from face (estimated)
		$T = (2T_1 + 2T_2 + T_3)/5$ = weighted mean temperature
		(T_3 is estimated)

DESIGN OF REINFORCED CONCRETE MEMBERS FOR FIRE RESISTANCE

Reference	Step	Calculations						
		Time	T_1 at face	T_2 (@ 80 mm)	T_3 (@ 175 mm)	T	Reduction factor for T	Strength (N/mm^2)
		\multicolumn{7}{c}{Estimate}
		0	–	–	–	–	–	20
Table 16.3		60	860	180	30	422	0.9	18
		90	920	250	90	486	0.83	16
Fig. 16.2		120	1000	320	170	562	0.63	13
		180	1100	400	250	650	0.45	9
		240	1140	680	280	784	0.25	5

Step 4: *Strength of beam by first method*

(Steel tension) + (Steel compression + Concrete compression) = 0

Moment of resistance = moment of forces

M = M(concrete) + M(compression steel) – M(tension steel)

BS recommends γ_m of 1.0 for steel and 1.2 for concrete

For concrete max stress is given by

$$\frac{0.67 f_{ck}}{\gamma_m} \quad [\text{with } \gamma_m = 1.2 \text{ (instead of usual 1.5)}]$$

x = Depth of neutral axis

Assume BS rectangular stress block of depth $0.9x$

Maximum concrete stress $= \dfrac{0.67 f_{ck}}{1.2} = 0.56 f_{ck}$

Compression block = $(0.56 f_{ck})(b)(0.9x) = (0.5 f_{ck}) bx$

A_{st} = 6 nos. of 16 mm = 1206 mm^2

Table steps 2 and 3: For 60 min fire (f_y = 373 and f_{ck} = 18)

$0.5 \times 18 \times 350 x = 1206(373)$

$x = \dfrac{1206 \times 373}{0.5 \times 18 \times 350} = 143 \text{ mm}$

$T = \dfrac{1206 \times 373}{1000} = 450 \text{ kN}$

Step 5: *Find M by taking moments about the top of beam*

d = 600 – 40 – 8 = 552 mm

$M = 450\left(552 - \dfrac{x}{2}\right)$ for the rectangular block

= (450)(552 – 72) = 216 kNm

Reference	Step	Calculations
	5	*Note:* This value should not be compared to its design strength where we assume $\gamma_m = 1.2$ for steel, 1.5 for concrete. We should only check whether the above value is ample to withstand the bending moment due to actual loads on the beam. *Determine the strength of beam by the second method* Assume, all concrete whose temperature is above 500°C will not contribute to the strength. With 60 minutes fire 20 mm all around reach about 500°C (New beam is of height = 600–40 = 560 mm and breadth = 350–40 = 310 mm. Its dimension is 560 × 310 mm, Therefore $(0.5)(20)(310)x = (1206)(373)$ $x = 145$ mm $d = 560 - 20 - 8 = 532$ mm $M = 450(532 - 72.5) = 207$ kNm In a similar way, the strength for 90, 100, 180 and 240 minutes of fire can be worked out and a curve showing M_t/M_0 on *y*-axis and temperature on *x*-axis can be drawn to represent the loss of strength with time.

REFERENCES

1. BS 8110 part 2, British Standards Institution, London, 1985.

2. FIP/CEB Model Code for Fire Design of Concrete Structures, FIP/CEB, London, 1969.

3. Australian Standard 3600 (1988), Concrete Structures, Standards Association of Australia, Sydney.

4. Anchor, R.D., Malhotra, H.L. and Purkess, J.A., *Design of Structures against Fire*, Elsevier Applied Science, London, 1986.

5. Martin, L.H., Croxton, P.C.L. and Purkess, J.A., *Structural Design in Concrete to BS 8110*, Edward Arnold, London, 1989.

6. IS 456 (2000) Indian Standard Plain and Reinforced Concrete–Code of Practice–Bureau of Indian Standards, New Delhi.

CHAPTER 17

Design of Plain Concrete Walls

17.1 INTRODUCTION

A vertical load bearing member whose breadth is more than four times its thickness is called a *wall*. The minimum thickness allowed for a concrete wall is 100 mm. A wall is called a *reinforced concrete wall* if the percentage of total compression steel in it is not less than 0.4% of the gross area of concrete so that the strength of the wall will include the strength of steel as well. If the total compression steel is less than 0.4%, it cannot be assumed to contribute to the strength of the wall and such walls are designated as *plain walls*. In actual practice, all walls with less than 1% compression steel are considered as plain walls from fire resistance consideration. Walls can be braced or unbraced.[1]

In addition, walls are also classified as *short* (stocky) or *slender* depending on their slenderness ratio.[2] According to BS 8110, walls are 'stocky' when this ratio does not exceed 15 for a braced wall and 10 for unbraced wall. In IS 456, the dividing ratio is 12. If it exceeds the above values the walls are considered as slender walls and are to be designed taking the additional eccentricity due to slenderness also into consideration. While reinforced concrete walls are designed by the theory of reinforced concrete columns taking unit width of the wall as a column, plain walls are designed by the theory of masonry walls.

Various types of loads that can act on a wall are shown in Fig. 17.1. They can be in-plane forces or out of plane force. The design procedure used to design these walls is shown in

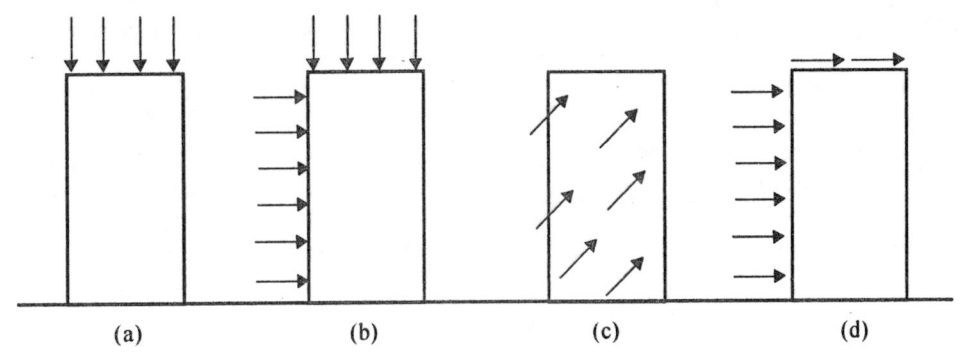

Fig. 17.1 Types of forces acting on walls.

---

[1]For braced and unbraced walls, *see* Section 17.2.
[2]For slenderness ratio of wall, *see* Section 17.3.

296   ADVANCED REINFORCED CONCRETE DESIGN

Table 17.1. Walls that carry in-plane vertical loads can be designed as reinforced concrete walls when the loads are heavy or eccentricity of load is large so that more than 0.4% of

TABLE 17.1 DESIGN OF CONCRETE WALLS

Type of load on wall	Design procedure
1. In-plane vertical forces in wall	(a) When loads are large and eccentricity is large; design as R.C. walls.
	(b) When loads and moments are moderate: design as plain walls.
2. In-plane vertical and horizontal forces in plain walls	(a) When section is wholly in compression; design separately for axial load and shear; neglect bending.
	(b) When section is partly in tension; design for axial load with flexure by IS 13920 [7] and also design separately for shear.
3. In-plane vertical forces and horizontal forces perpendicular to the plane of the wall	(a) If axial load does not exceed $0.04 f_{ck} \times A_g$; design as slab (height to thickness not to exceed 50); Otherwise design as wall.
4. Forces acting on a wall which forms part of a framed structure.	As R.C. wall for forces obtained by elastic analysis.

compression steel has to be used. They are designed as plain walls when the loads are moderate and steel is necessary only for cracking, temperature effects, etc. Walls subjected to in-plane horizontal loads are designed for bending and shear; walls with forces that act perpendicular to the plane of the walls (out-of-plane forces as in retaining walls) are designed as slabs. Panel walls which are constructed as infilling in structural framework are non-load-bearing walls but they should be sufficiently strong to resist the wind pressure acting at right angles to it as slabs. For this purpose, the panels should be given enough bearing by setting them in rebates in the members of the frame or by providing suitable steel dowels connecting them with the framework.

IS 456 (2000) clause 32 in its fourth revision has incorporated the design of braced plain concrete walls subjected to in-plane vertical and horizontal loads. It follows the practice in the Australian Code AS 3600 (1988) [3]. Design of plain and reinforced concrete walls is treated in detail also in BS 8110 [4] Clause 3.9. In this chapter, we shall deal only with the basic theory of design of plain and reinforced concrete walls for in-plane forces. Since the study of advanced topics such as layout and design of shear-walls (which are also designed on similar principles), design of retaining walls, basement walls, silo walls, etc., will require discussions on many other related topics that are not dealt with in this chapter.

17.2 BRACED AND UNBRACED WALLS

Braced wall is one in which the lateral stability against horizontal forces of the entire structure at right angles to the plane of the wall being considered is provided by other walls or by other means. Walls can be assumed as braced, if all the following conditions are satisfied:

1. Walls or other bracing elements are provided in two directions in the structure so as to provide lateral stability to the structure.
2. The lateral forces are resisted by shear in the planes of these elements.
3. The roof and floor systems are designed to transfer lateral forces.
4. The connections between walls and the lateral supports are designed to resist horizontal forces due to simple static reactions at the level of the internal support together with 2.5% of the vertical load that the wall is designed to carry at that level. This resistance should not be less than 2 kN per metre length of the wall.

Walls which do not comply with these requirements are considered as unbraced. Multistoreyed buildings should not be planned to depend only on unbraced walls for its overall stability. Incorporation of various types of ties discussed in Chapter 15 and provision of shear walls are necessary for its stability.

17.3 SLENDERNESS OF WALLS

The carrying capacity of concrete walls, as in the case of reinforced concrete columns, depends also on its slenderness ratio. Reinforced concrete walls are designed in the same way as reinforced concrete columns and their slenderness ratio is also calculated similarly. However, plain concrete walls design is similar to design of masonry walls. Its slenderness is to be taken as in masonry walls and is lesser of the following two ratios:

(a) Ratio of effective height along vertical direction and thickness = H_e/t
(b) Ratio of effective length along the horizontal direction and thickness = L_e/t

where H_e is the effective height and t the thickness. Effective length of plain walls will be taken as specified in Tables 17.2 and 17.3 and the maximum slenderness ratios allowed for various types of walls is given in Table 17.4. According to IS 456, when the slenderness ratio is equal to or more than 12 walls, these are considered slender.

TABLE 17.2 EFFECTIVE HEIGHT OF BRACED PLAIN CONCRETE WALLS (IS 456)

Restrained by	Restrained against rotation at both ends	Not restrained against rotation at both ends
1. Floors	$0.75H_0$	H_0
2. Intersecting walls	$0.75L_0$	L_0

where H_0 is the unsupported height and L_0 is the horizontal distance between centres of lateral restraints in Table 17.2. In ACI 318, restrained against rotation implies attachment to a member having flexural stiffness (EI/L) at least as large as that of the wall.

TABLE 17.3 EFFECTIVE HEIGHT OF UNBRACED PLAIN CONCRETE WALLS (BS 8110)

Nature of wall	Unbraced
1. With a roof or floor spanning at right angles on top of the wall	$1.5L_0$
2. With no roof or floor on top of the wall	$2L_0$

where L_0 is clear height distance of wall between supports.

TABLE 17.4 SLENDERNESS LIMIT OF CONCRETE WALLS (BS 8110)

Steel (percentage)	Type of wall	Braced or unbraced	L_e/H
0.4 or less	Plain	(a) Braced	30
		(b) Unbraced	30
0.4 to 1.0	Reinforced	(a) Braced	40
		(b) Unbraced	30
1 to 4	Reinforced	(a) Braced	45
		(b) Unbraced	30

Note: The IS requirement for minimum slenderness ratio of masonry walls [5] is as follows. For load bearing walls, cement-mortar 1:6 or cement-lime mortar ratio 1:2 the slenderness ratio should not exceed 18, except for dwellings of not more than 2 storeys where it shall not exceed 24. If only lime mortar is used the corresponding ratio shall not exceed 12 and 18, respectively.

17.4 ECCENTRICITIES OF VERTICAL LOADS AT RIGHT ANGLES TO WALL

Concrete walls are designed for vertical loads along with their eccentricities. The following guidelines are used to calculate the eccentricity of loads at right angles to the wall:

1. The vertical loads on a wall due to a discontinuous concrete floor or roof acting only on one side of the wall produce eccentricity of loading. The load is assumed to act at one-third the depth of the bearing area measured from the span face of the wall (Fig. 17.2).

2. When a roof or floor is continuous on both sides of the wall, the load may be assumed to act on the centre of the wall.

3. For a braced wall only the eccentricities of individual walls need to be considered. The eccentricity of the other vertical load on the braced wall at any level between horizontal lateral supports can be assumed to be zero.

4. For an unbraced wall however the resultant eccentricity of the total vertical load is calculated by taking into account the eccentricity of all the individual vertical loads and moments acting above that level.

5. In any case, the minimum eccentricity of not less than $t/20$ (or 20 mm according to BS 8110) should be assumed in design of all concrete walls.

DESIGN OF PLAIN CONCRETE WALLS 299

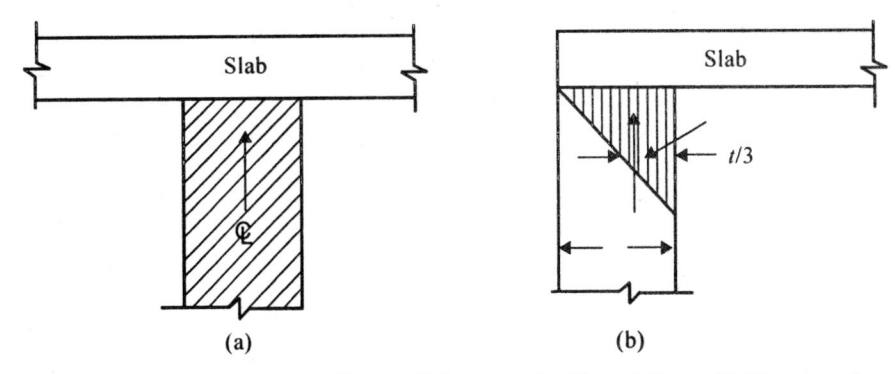

Fig. 17.2 Eccentricity of loads on walls: (a) slab on both sides of the wall (b) slab only on one side of the wall.

6. As per BS 8110 [4], concentrated loads on walls due to beams or column bases can be assumed to be distributed through the wall provided, the local stress does not exceed $0.6 f_{ck}$ for concrete of grade 25 and above, and $0.5 f_{ck}$ for lower grade concrete. This allowable pressure according to IS 456 Clause 34.4 is $0.45 f_{ck}$. A horizontal length of the wall is considered effective in carrying concentrated loads in walls. This should not exceed the bearing of the load plus four times the wall thickness or the centre-to-centre distance between the loads.

17.5 EMPIRICAL DESIGN METHOD FOR PLANE CONCRETE WALLS CARRYING AXIAL LOAD

Plain walls can be 'short-braced', 'slender-braced' or 'unbraced'. The empirical or simplified method of design of these plane concrete walls under axial load is derived from the theory of masonry walls [5]. The mode of failure of such walls is shown in Fig. 17.3. The effect of eccentricity of vertical load as well as the additional eccentricity for slenderness can simply

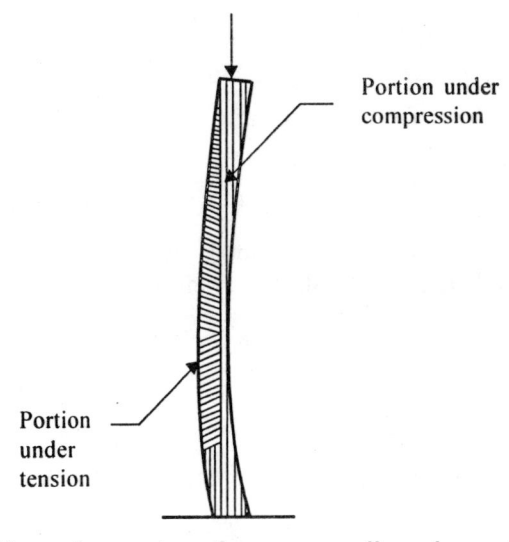

Fig. 17.3 Failure of concrete and masonry walls under vertical compressive load.

be taken as a reduction in the effective thickness of the wall. The formula derived is based on the fact that the presence of small quantities of steel, such as the minimum specified for these walls, do not increase its strength above that of the plain concrete wall as shown in Fig. 17.4. Thus the strength of the wall is only due to the strength of concrete. The steel does

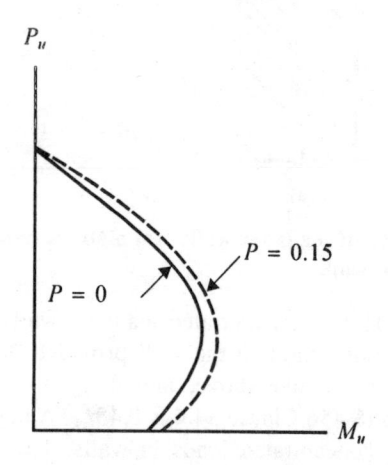

Fig. 17.4 Interaction curve for walls with low percentage of compression steel.

not contribute to the strength of the wall. Based on these assumptions, the strength of plain walls for unit length of the wall is given by the following formula (BS 8110 Clause 3.9.4.15):

(1) *Short-braced plain walls:*

$$P_W = (t - 2e_x)\alpha f_{ck} \qquad (17.1)$$

where

t = Thickness of the wall

e_x = The eccentricity of load at right angles to the wall, the minimum value being $t/20$

α = Stress reduction factor for f_{ck} which varies with f_{ck} and H/L ratio, but its average value is taken as 0.3

It may be noted that when $e_x = t/2$ the wall will not carry any load.

(2) *Slender braced plain walls:* The corresponding expression for the strength per unit length of a slender braced wall is given by the following formula:

$$P_W = 0.3 f_{ck}(t - 1.2e_x - 2e_a) \qquad (17.2)$$

where

$e_a = (H_e/t)^2(t/2500)$ = The addition eccentricity due to slenderness $\qquad (17.3)$

H_e = The effective height of the wall and e_x and α are the same as in Eq. (17.1)

(*Note*: In Eq. 17.3 for reinforced concrete column, we use 2000 instead of 2500.)

(3) *Unbraced plain walls*: For unbraced plain walls the effective length, as shown in Table 17.3, is larger than the actual length. The ultimate load of short- and long-unbraced walls can be taken as the lesser of the following expressions (BS 8110 Clause 3.9.4.17). (Here the factors of e_x and e_a are different.)

$$P_w = \alpha f_{ck}(t - 2e_{x1}) \tag{17.4}$$

$$P_w = \alpha f_{ck}(t - 2e_{x2} - e_a) \tag{17.4a}$$

where e_{x1} and e_{x2} are the resultant eccentricities of the load at the top and the bottom of the wall. It being not less than $t/20$ (BS 8110 Clause 3.9.4.17). e_a is the additional eccentricity. And $\alpha = 0.3$.

17.5.1 Limitation of Empirical Method

The expression for the strength in slender braced walls in Eq. 17.2. is the one recommended in the fourth revision of IS 456 and AS 3600 [3]. ACI 318 Clause 14.5 [6] also recommends an expression similar to the above for the design of plain walls by the empirical method which is recommended to be used only when the actual eccentricity of the vertical load is not greater than $t/6$, i.e. the resultant load acts within the middle third of the thickness of the wall. The empirical method is allowed to be used for rectangular cross sections only. Load bearing walls of non-rectangular section, such as ribbed wall panels, must be designed as reinforced concrete compression members.

17.6 DESIGN OF WALLS FOR IN-PLANE HORIZONTAL FORCES

In this section we examine the method of design for shear and flexure due to in-plane lateral forces on plane concrete walls.

17.6 1 Design for Shear Due to In-plane Forces

IS 456 (2000) clause 32.4 gives design procedures for checking concrete walls for shear due to in-plane lateral forces. As shown in Figs. 17.5–17.7, the deformations of walls under lateral forces depend on the ratio of height to width (H/L) of the wall. Accordingly, walls are divided in to the following two classes:

(a) High walls $H/L > 1.0$
(b) Low walls $H/L \leq 1.0$

where H and L are overall height and length of the wall. In ACI, the classification is based on H/L ratio = 2. For a gable wall the overall height is the average height. This division is made due to the fact that cracking due to shear in the above two cases is different as shown in Fig. 17.7. For resisting shear forces, vertical steel is more effective in low walls and the horizontal steel is more effective in high walls. The following rules are used for design of concrete walls for shear:

1. The effective depth d of the wall is taken as equal to $0.8L$ for shear considerations.
2. The critical section for shear is taken at a distance from the base equal to $0.5L$ or $0.5H$, whichever is less.

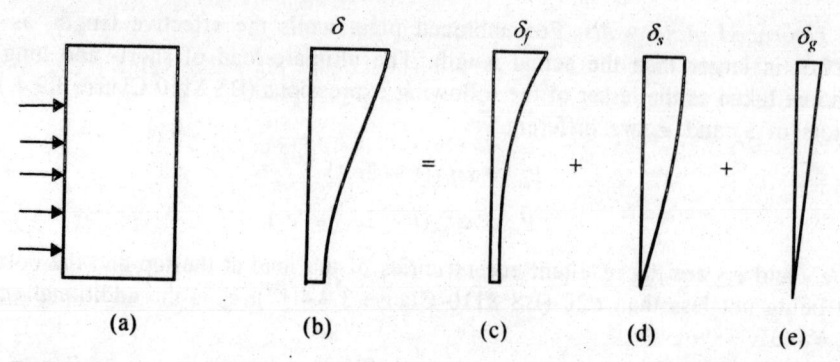

Fig. 17.5 Deflection of walls under horizontal loads: (a) wall (b) total deflection (c) defection by flexure (d) deflection due to shear (e) deflection due to foundation rotation.

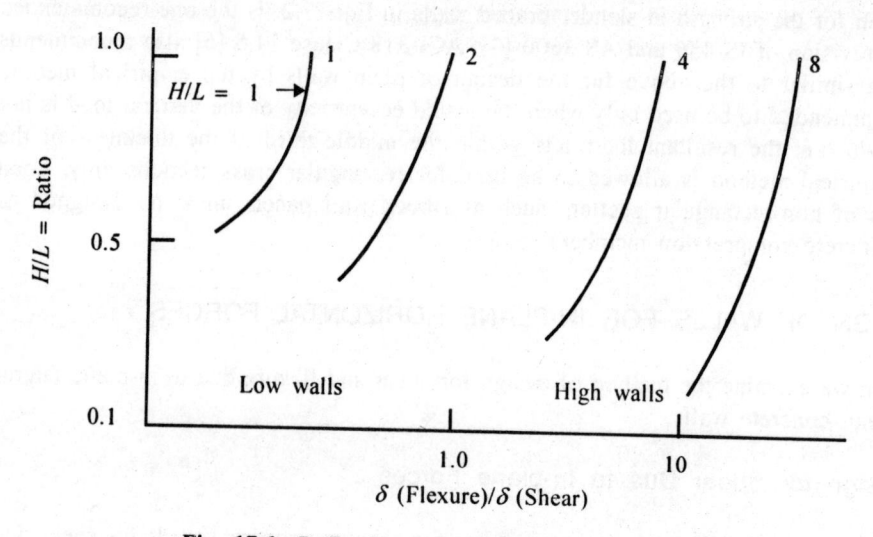

Fig. 17.6 Deflection characteristics of low and high walls.

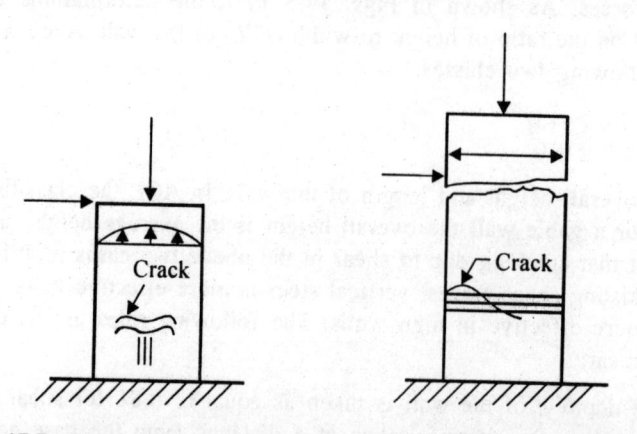

Fig. 17.7 Cracking due to shear in walls: (a) low walls (b) high walls.

DESIGN OF PLAIN CONCRETE WALLS   303

3. The nominal shear should not exceed the maximum allowable value (τ_{max}) as given below (IS 456 (2000) Clause 32.4.2.1

The nominal value of shear with nominal steel is v. Therefore,

$$v = \frac{V}{td} \tag{17.5}$$

where v must be less than τ_{max}.

$$\tau_{max} = 0.17 f_{ck} \tag{17.6}$$

IS 456 (2000) Table 20 for τ_{max} values are slightly lower than the values got from the above formula. But Table 20 may also be used for routine design.

4. In designing for shear we first check whether the minimum steel specified in these walls be enough to resist the given shear. For this purpose, we first calculate the nominal shear v from Eq. (17.5) and check whether the nominal shear value is less or greater than the allowable value as given in Eqs. (17.7) and (17.8). If it is less, we need to provide only the nominal vertical and horizontal steel as given in Section 17.1.1 for the wall. If it is more than the allowable value, the wall should be checked for safety in shear. The allowable value of shear strength recommended by Australian code AS 3600 Clause 11.5.4 for walls without reinforcement steel is given by the following formula [3]:

(a) For low walls, $H/L \leq 1$

$$\tau_c = 0.66\sqrt{f_c} - 0.21\frac{H}{L}\sqrt{f_c}$$

where f_c = cylinder strength. Taking $f_c = 0.8 f_{ck}$, we get (as in IS 456 Clause 32.4.3)

$$\tau_c = 0.6\sqrt{f_{ck}} - 0.2\frac{H}{L}\sqrt{f_{ck}} \tag{17.7}$$

for $H/L = 1$,
$$\tau_c = 0.40\sqrt{f_{ck}} \tag{17.7a}$$

(b) For high walls $H/L > 1$. The lesser values got by the following equations is the allowable shear:

$$\tau_c = 0.40\sqrt{f_{ck}} \tag{17.8}$$

$$\tau_c = 0.045\sqrt{f_{ck}}\,\frac{(H/L)+1}{(H/L)-1} \tag{17.8a}$$

But in any case

$$\tau_c \geq 0.15\sqrt{f_{ck}} \tag{17.8b}$$

5. If v is greater than t_c, we have to calculate the steel areas required in the wall. For this purpose IS 13920 (1993) Clause 9.2.2 [7] recommends the conventional design procedure for shear steel in walls. The shear stress that the concrete can resist is taken as those given

by the values in Table 19 of IS 456 (2000). We use the following procedure for design of this shear reinforcements.

As the effective depth is taken as $0.8L$, we have

$$V_c = 0.8\tau_c tL \qquad (17.9)$$

$$V_s = V - V_c \qquad (17.9a)$$

where

V_c = Shear taken by concrete

V_s = Shear taken by steel

V_u = Factored design shear

Since V_u is greater than V_c, we design the steel for the balance of the shear to be carried by steel by the conventional formula:

$$V_s = 0.87 f_y d \frac{A_{sv}}{S_v} \qquad (17.10)$$

where A_{sv} is the area of steel being considered within a distance S_v (the spacing)

(i) For design of vertical steel we take a cross-section of the wall, and the necessary steel is calculated as

$$\frac{A_{sv}}{S_v} = \rho_v t \qquad (17.11)$$

where ρ_v is the vertical steel ratio (not percentage of steel).

(ii) For design of horizontal steel we consider the cross-sectional area of the wall for unit vertical height, and the steel required is calculated as

$$\frac{A_{sv}}{S_v} = \rho_h t$$

where ρ_h is the horizontal steel ratio.

6. For low walls $H/L \leq 1$, where vertical steel is more effective, design is first made for vertical steel. The final area recommended should not be less than that of the minimum specified for walls. If the required steel is more than the specified minimum horizontal steel, the same steel as in the vertical direction is to be provided also in the horizontal direction. The minimum horizontal steel specified in walls is usually more or at least equal to the minimum vertical steel. The minimum amount of steels to be provided is given in Section 17.7.2.

7. For high walls $H/L > 1$, as horizontal steel is more effective than vertical steel, design is made for horizontal steel, which should not be less than the specified minimum. We also provide the minimum specified vertical steel. We must ensure that in all walls, shear strength of the wall should be made much greater than its bending strength so as to avoid any sudden and brittle failure.

17.6.2 Design of Concrete Walls for Flexure (In-plane Bending)

When walls are subjected to in-plane flexure due to horizontal forces in addition to vertical loads, design is carried out as follows (No. 2 of Table 17.1):

(i) When the whole section is in compression only, the walls are designed for axial load and then checked for shear separately as explained in Section 17.6.1. The in-plane bending can be neglected.

(ii) When part of the wall section is in tension, the wall is designed for flexure by the theory of bending of R.C. members under combined bending and axial load and then checked for the shear resistance by Section 17.6.1. The theory of bending of concrete walls for axial load and flexure and construction of interaction diagram is usually treated under design of shear walls. Details of design procedures are given in Appendix A of IS 13920 (1993).[1]

17.7 RULES FOR DETAILING OF STEEL IN CONCRETE WALLS

17.7.1 Design of Transverse Steel in Concrete Walls

In plain walls, transverse reinforcement need not be designed to support vertical bars against buckling as in columns as the vertical bars are not called upon to carry the load. However, in R.C. walls where the vertical steel bears the loads, the transverse horizontal steel should be properly designed as in columns to restrain the vertical steel against buckling.

17.7.2 Rules for Detailing of Longitudinal Steel in Plain Concrete Walls

The minimum vertical and horizontal steel provided in walls is expressed in terms of the gross area of concrete. The provisions are as follows (Table 17.5 can be used for easy reference):

1. Indian Standards [IS 456 (2000) Clause 32.4]

   (a) *Vertical steel*: The minimum vertical steel for plain walls should be 0.12% for high yield bars and welded fabric and 0.15% for mild steel bars. Also the maximum spacing should be 450 mm or three times the wall thickness, whichever is less.

   (b) *Horizontal steel*: The minimum horizontal steel for all types of walls (plain or reinforced) should be 0.20% for high yield bars with diameter not larger than 16mm and 0.25% for mild steel bars. The maximum spacing of bars should be 450 mm or three times the wall thickness, whichever is less.

   (c) For walls, more than 200 mm thick, the vertical and horizontal reinforcements should be provided in two grids, one near each face of the wall with proper cover.

(*Note*: According to IS 456 clause 31.4, the minimum horizontal steel is more than the minimum vertical steel. According to BS (1985) Clause 3.9.4.18, the minimum percentage of steel in each direction should be 0.25 for Fe 460 steel.)

---

[1]*See* further details in Chapter 19.

TABLE 17.5 MINIMUM REINFORCEMENT IN WALLS

Wall thickness (mm)	Steel spacings for given percentage of steel in two layers		
	0.20	0.25	0.4
100	6 mm at 280 mm	6 mm at 200 mm	8 mm at 250 mm
125	6 mm at 220 mm	6 mm at 175 mm	8 mm at 200 mm
150	6 mm at 175 mm	6 mm at 150 mm	10 mm at 250 mm
175	6 mm at 150 mm	8 mm at 225 mm	10 mm at 200 mm
200	6 mm at 140 mm	8 mm at 200 mm	12 mm at 275 mm
225	6 mm at 125 mm	10 mm at 275 mm	12 mm at 250 mm
250	8 mm at 200 mm	10 mm at 250 mm	12 mm at 225 mm
275	10 mm at 275 mm	10 mm at 225 mm	12 mm at 200 mm
300	10 mm at 250 mm	10 mm at 200 mm	12 mm at 175 mm

Note: Steel of Table 17.5 should be provided on both faces at the given spacing. If a single layer is used, adopt half the spacing.

EXAMPLE 17.1 (Design of a plane concrete wall)

A plain braced concrete wall of dimensions 8 m high, 5 m long and 200 mm thick is restrained against rotation at its base and unrestrained at the ends. If it has to carry a factored total gravity load of 180 kN and a factored horizontal load of 8.45 kN at the top, check the safety of the wall. Assume $f_{ck} = 20$ and $f_y = 415$.

Reference	Step	Calculations
ACI code	1	Check whether the wall will be wholly in compression under loading.
		Maximum B.M. in wall $M = 8.45 \times 8 = 67.6$ kNm (factored)
		Maximum load $P = 180$ kN
		$e = \dfrac{M}{P} = \dfrac{67.6}{180} = 0.375$ m
		$\dfrac{L}{6} = \dfrac{5}{6} = 0.83$ m
ACI 318 Cl. 14.5		As $e < L/6$, all the portions of the wall are under compression. Empirical design is applicable.
	2	*Determine the slenderness of the wall*
Table 17.2		$H_e = 0.75 \times 8000 = 6000$
		Maximum slenderness ratio $= \dfrac{H_e}{t} = \dfrac{6000}{200} = 30$
Table 17.4		If slenderness ratio > 12, the wall is slender. Here the maximum allowed ratio = 30. Hence it is slender within limits.
	3	*Find minimum eccentricity*

DESIGN OF PLAIN CONCRETE WALLS

Reference	Step	Calculations
		$e_{min} = \dfrac{t}{20} = \dfrac{200}{20} = 10$ mm
	4	*Find additional eccentricity*
Eq. (17.3)		$e_a = \left(\dfrac{H_e}{t}\right)^2 \dfrac{t}{2500} = (30)^2 \dfrac{200}{2500} = 72$ mm
	5	*Calculate reduction in thickness of braced wall*
Eq. (17.2)		$1.2\, e_{min} + 2e_a = (1.2 \times 10) + 2(72) = 156$ mm
		Balance thickness = 200 − 156 = 44 mm
	6	*Determine ultimate load carrying capacity per unit length of wall*
		$P_w = 0.3 f_{ck}\,(t - 1.2 e_{min} - 2e_a)$
		$= 0.3 \times 20\,(200 - 156) = 264$ N/mm
		Total capacity of wall for 5 m length
		$P = 264 \times 5000/1000$
		$= 1320$ kN > 180 kN (required)
Sec. 17.7.2	7	*Determine minimum steel to be placed in the wall*
		p_t = Horizontal steel percentage (=0.20)
		p_v = Vertical steel percentage = 0.15
		$A_h = 0.002 \times 200 \times 1000$ (per metre) = 400 mm^2
		As the thickness is more than 150 mm, the steel has to be placed in two layers of 200 mm^2/m.
		Provide T8 at 250 (201 mm^2/m).
		A_v should not be less than 0.15%. Provide the same steel in both the directions.
	8	*Check for shear*
		$\dfrac{H}{L} = \dfrac{8.0}{4.5} = 1.77$ (high wall)
		$d = 0.8L = 0.8 \times 4500 = 3600$ mm
		$t = 200$ mm
		Critical section 0.5L or 0.5H from base
Sec. 17.6.1		Shear stress $v = \dfrac{V}{td} = \dfrac{8.45 \times 10^3}{200 \times 3600} = 0.012$ N/mm^2
		Allowable τ_{max}

Reference	Step	Calculations
Eq. (17.6)		$\tau_{max} = 0.17 f_{ck} = 3.4$ N/mm^2 > 0.12 (or se IS 456 Table 20)
		Shear taken by concrete wall without steel for a high wall ($H/L > 1$) is lesser of the two value of τ_c:
Eq. (17.8)		$\tau_c = 0.40\sqrt{f_{ck}} = 0.40\sqrt{20} = 1.79$ N/mm^2
		$\tau_c = 0.045\sqrt{f_{ck}}\dfrac{(H/L)+1}{(H/L)-1} = 0.045\sqrt{20}\,\dfrac{2.77}{0.77} = 0.72$ N/mm^2
		but not less than
		$\tau_c = 0.15\sqrt{f_{ck}} = 0.15\sqrt{20} = 0.67$ N/mm^2
		v is much less than τ_c. Hence the wall is safe in shear with the minimum steel provided.
	9	Assume the shear is high, we proceed as follows:
		As an example, let $V = 864$ kN
Step 8		$v = \dfrac{864 \times 10^3}{200 \times 3600} = 1.2$ N/mm^2 < 3.4 N/mm^2
	10	Design of steel for shear
		As the wall is high and horizontal steel is more effective, we will check the shear that can be taken by minimum horizontal steel.
		$V_c = 0.8 L t \tau_c = 3600 \times 200 \times 0.72 = 518.4$ kN
Eq. (17.9a)		Shear to be taken by steel = 864 − 518 = 346 kN
		Minimum horizontal steel per metres height at 0.2% (with s_v = 1000 mm)
Eq. (17.11)		$\dfrac{A_{sv}}{S_v} = \rho_h t = 0.0020 \times 200 = 0.4$ mm^2
Eq. (17.10)		$V_s = 0.87 f_y d \dfrac{A_{sv}}{S_v}$
		$= 0.87 \times 415 \times 0.4 \times 3600$ N
		$= 520$ kN > 346 kN (required)
		Hence safe with minimum steel.

REFERENCES

1. Varghese, P.C., *Limit State Design of Reinforced Concrete*, Ch. 17, Prentice-Hall of India, New Delhi, 1994.

2. IS 456 (2000) Code of Practice for Plain and Reinforced Concrete, Bureau of Indian Standards, New Delhi, 2000.

3. AS 3600 (1988), Concrete Structures, Standards Association of Australia, Sydney.

4. BS 8110 (1985), Part 2, Structural Use of Concrete, British Standards Institution, London.

5. Explanatory Handbook on Masonry Code SP 20, Bureau of Indian Standards, 1981.

6. ACI 318 (1989), Building Code Requirements for Reinforced Concrete, American Concrete Institute, Detroit.

7. IS 13920 (1993), Ductile Detailing of Reinforced Concrete Structures Subjected to Seismic Forces, Bureau of Indian Standards, New Delhi.

CHAPTER 18

Earthquake Forces and Structural Response

18.1 INTRODUCTION

Earthquakes produce earthquake forces on structures. The resultant loads are called earthquake loads denoted by E or EL. When planning a building against natural hazards like earthquakes or cyclones we can design it to behave in one of the following three limit states:

1. *Serviceability limit state:* In this case, the structure will undergo little or no structural damage. Important buildings like hospitals, atomic power stations; etc., should be designed for elastic behaviour under expected earthquake forces.

2. *Damage controlled* (*damageability*) *limit state:* In such cases, if an earthquake or cyclone occurs there is some damage to the structure that can be repaired after the event and the structure can again be put into use. Most of the permanent buildings should come under this category. For this purpose, the structure is to be designed for limited ductile response only.

3. *Survival limit state:* In this case, the structure may be allowed to be damaged extensively in the event of an earthquake or cyclone but the supports should stand and be able to carry the permanent loads fully so that in all cases there will be no caving in of the structure and no loss of life. Earthquakes occur without warning and such behaviour is achieved by positively avoiding brittle failure of bending members and also ensuring that supporting elements like columns do not fail. Earthquake energy should also be dissipated by 'full ductile' response of the structures.

We can design structures for the first two limit states by elastic or restricted ductile response of the structure using conventional methods of design and incorporating ductile detailing. Full elastic design is costly. Limited ductile response is cheaper and full ductile response is the cheapest. However, full ductile design is carried out by the theory of plastic hinge formation, and careful detailing for ductility. Eventhough such structures are cheaper, they undergo large-scale damage under severe earthquakes, and therefore reconstruction may not be possible. The current practice is to design structures for one of the first two limit states as the subject of the full plastic design is still in the development stages only.

Seismic design of structures is a vast specialised subject and this chapter attempts to explain only some of the basic concepts of such design and also the provision of the present

Indian code to calculate the earthquake loads on buildings. Some of the features that are incorporated in the proposed revision of the code are also included in this chapter.

18.2 BUREAU OF INDIAN STANDARDS FOR EARTHQUAKE DESIGNS

The Bureau of Indian Standards has published the following documents about earthquake design of structures:

1. IS 4326 (1976) code of practice for earthquake resistant design and construction of buildings.
2. IS 1893 (1984) Criteria for earthquake design of structures (fourth revision). This code is currently in use in India, for design against earthquake forces. The code was first published in 1962 and was revised in 1966, 1970, 1975 and 1984. Now it is under its fifth revision which is expected to be published shortly. This code discusses the following problems:

    (i) Calculation of the horizontal seismic coefficient by (a) the seismic coefficient and (b) using the response spectrum method.

    (ii) Calculation of base shear by (a) pseudo static method (b) modal superposition technique and the response spectrum method.

3. IS 13920 (1993) code of practice for ductile detailing of reinforced concrete structures subjected to seismic forces. This code was written to supplement IS 4326.
4. SP 22 Explanatory Handbook on Codes for Earthquake Engineering. This handbook was published in 1983 to give explanation and examples for IS 1893 (1975) and IS 4326 (1976) and hence needs revision.

18.3 EARTHQUAKE MAGNITUDE AND INTENSITY

The terminologies used in earthquake are given in IS 1893 Section 2. The magnitude of earthquake is a measure of the energy released by an earthquake and is popularly measured by the Richter scale M1 to M8. The intensity of an earthquake is an estimate of the perceived local effects, which depends on the peak acceleration velocity and the duration. The scale used for the intensity is modified as Mercalli scale. The earthquake is reported in newspapers by the Richter scale and it is related to the maximum trace deformation of the seismogram wave recorded by a standard seismograph at 100 km from the epicentre.

18.3.1 Magnitudes of Some of the Past Earthquakes in India as Expressed in Richter Scale

The following are some of the major earthquakes that took place in India during the last hundred years.

1. The Assam earthquake in 1897 (M8.7)
2. The Bihar-Nepal earthquake in 1938 (M8.4)
3. The Quetta earthquake in 1935 (M7.6)

4. The Assam–Tibet earthquake in 1950 (M8.7)
5. The Koyna earthquake in 1967 (M6.5). (The Koyna dam designed with seismic coefficient of 0.05 withstood the earthquake well.*)
6. The Bihar–Nepal earthquake in 1988 (M6.6)
7. The Uttar Kashi (Uttar Pradesh) earthquake in 1991 (M6.6)
8. The Latur earthquake in 1993 (M6.4**)
9. Earth tremor at Kottayam, Idukki (Kerala) in 2001 (M4.8[†])
10. The Jabalpur earthquake in 1997 (M6.0[†])
11. The Chamoli (Uttar Pradesh) earthquake in 1999 (M6.8)
12. The Bhuj (Gujrat) earthquake in 2001 (M7.9[††])

Earthquakes of Richter scale less than M5 rarely cause any damage. Those of M5 to M6 cause damage to the structures located close to the epicentre. M6 and above release large amount of energy and can be dangerous to structures. The 1950 Assam earthquake was of magnitude 8.7 and is considered as the strongest ever that happened in India. (The approximate lengths of faultslips associated with M4, M5, M6, M7, M8 and M8.5 earthquakes are approximately 1, 3, 8, 40, 300 and 850 km respectively.)

18.3.2 Effect of Earthquakes

Eventhough earthquake motion involves vertical, horizontal and torsional oscillations, only horizontal motion is considered of importance in structural design. Vertical forces are only a fraction of gravity loads and these are assumed to be taken care of the factors of safety provided for dead and live loads. However, in certain cases, they may also have to be taken into consideration when dealing with stability analysis [1,2].

18.4 HISTORICAL DEVELOPMENT

It was realised long ago that the major effect of earthquake is the horizontal rocking force. A very conservative value of 0.1 times gravity for acceleration acting on all parts of the structure was used in early days in U.S.A. as a method of design for buildings in active earthquake zones. Other zones may have lesser values. The coefficient assigned to a given zone was known as the seismic coefficient. In 1950 Assam earthquake (Richter scale 8.7), it was found that structures designed with a basic seismic coefficient of 0.08 were able to withstand the earthquake effects well. This coefficient was hence accepted since long ago in India as the maximum basic horizontal seismic coefficient to be used for zone V, the highest zone in India. The lower zones then were assigned seismic coefficients as shown in Table 18.1.

---

*Reservoir induced earthquake.
**Latur is placed in Zone I of IS 1893 (1984) but still had a major earthquake.
[†]Jabalpur and Kerala are both in Zone III of IS 1893 (1984). Yet Jabalpur had a major earthquake in 1997—Earth tremors occur frequently in many places in India. (An earthquake with loss of human lives took place in Coimbatore, Tamil Nadu in 1900. The Himalayas and the Western Ghats of India are reported as the two most vulnerable zones in India for earthquakes.)
[††]Ahmedabad placed in Zone III suffered severe damages.

EARTHQUAKE FORCES AND STRUCTURAL RESPONSE

TABLE 18.1 CLASSIFICATION OF SOME TOWNS IN THE SEISMIC MAP OF INDIA
[IS 1893 (1984) APPENDIX E]

Place	Zone	Place	Zone
Agra	3	Hydarabad	1
Ahmedabad	3	Imphal	5
Allahabad	2	Jabalpur	3
Amritsar	4	Jamshedpur	2
Asansol	3	Jorhat	5
Bangalore	1	Kanpur	3
Bhilai	1	Kathmandu	5
Bikaner	3	Kochi	3
Mumbai	3	Lucknow	3
Calcutta	3	Nagpur	2
Calicut	3	Nellore	2
Chandigarh	4	Patna	4
Channai	2	Pondicherry	2
Coimbatore	3	Simla	4
Cuttack	3	Srinagar	5
Dehra Dun	4	Thiruvananthapuram	3
Delhi	4	Vijayawada	3
Guwahati	5	Visakhapatnam	2

(*Note:* In the next revision, Madras and Nellore will be in Zone 3. Ahmedabad also needs revision.)

The acceleration concept was used first in the following formula published in 1959 by the Seismological Committee of the Structural Association of California.

$$V_h = (\text{Mass})(\text{Acceleration}) = Ma \tag{18.1}$$

where

V_h = Horizontal shear force

M = Mass of all the permanent loads at that point

a = Acceleration due to earthquake

In the modern methods of analysis of buildings, we use a more theoretical approach which takes into account the natural frequency, and damping factor of the building. The structure is treated as a discrete system having lumped masses at each floor level. The loads covered in each floor, are the permanent loads consisting of all the dead loads on each floor and one-half of the columns above and below the floor, as well as an appropriate portion of the live load that always act on the structure. This live load is specified as one-fourth of live load for classes of loading upto 3 kN/m^2 and one-half of the live load for those carrying 4 kN/m^2 and above as given in IS 1893 (1984) Clause 4.1.1. This lumped mass system is then analyzed by simple methods or more complex methods.

18.5 BASIC SEISMIC COEFFICIENTS AND SEISMIC ZONE FACTORS

As already pointed out, the approach to earthquake design in a large country like India should be to divide the country into different zones, depending on the intensity of forces that can be expected to occur in that zone which to some extent depends on its geology. This division will not give any idea of the likely frequency of its occurance, which is very difficult to predict. Thus, at present, India is divided into five zones, zones I to V. The zone V gives the highest intensity. Earthquake forces are then to be calculated by coefficients called 'design horizontal seismic coefficient' which is a measure of the horizontal ground acceleration that can occur in the given zone. Each version of the Indian code has its own division of India into zones. By 1996, we had seven zones. This was reduced to five zones in 1970 Fig. 18.1. In the next revision of the code, it is proposed to combine zones 1 and 2 into a single zone thus reducing the number of zones to four (Fig. 18.2).

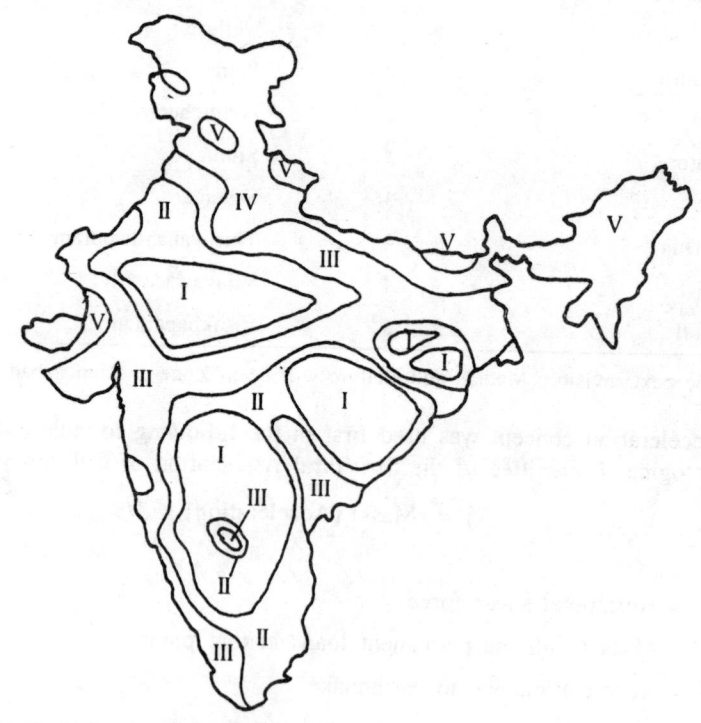

Fig. 18.1 Seismic zones in India—IS 1893 (1984).

Two methods have been recommended in IS 1893 (1984) to determine the horizontal seismic coefficient. These are the *traditional seismic coefficient* method and the *modern response spectrum* method. The second method is more refined and general than the first one as it is related to a response spectrum in which the earthquake forces also depend on the

EARTHQUAKE FORCES AND STRUCTURAL RESPONSE 315

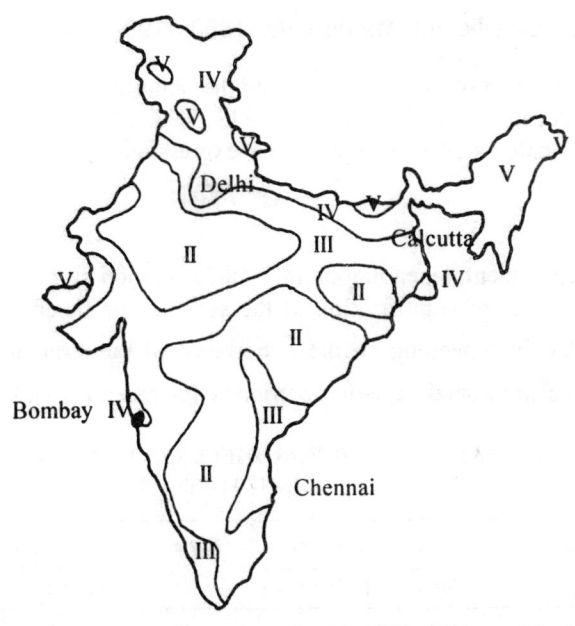

Fig. 18.2 Proposed seismic zones in IS 1893 (fifth revision).

dynamic characteristics, like the period of oscillation and the damping factor in addition to the ground motion. In order that both methods should give comparable results the coefficients used in the two cases are different. The traditional method uses the seismic coefficient α and the response spectrum method uses the seismic zone factor F_0. The value of F_0 is 5 times the value of α as can be seen in Table 18.2. The horizontal seismic coefficients are then calculated as shown in the following sections. It will be seen later in this chapter that values obtained by seismic coefficients method are the maximum values as obtained by the response spectrum method for 5% damping and period of oscillation $T = 0.2$ second.

TABLE 18.2 VALUE OF BASIC SEISMIC COEFFICIENT F_0 AND SEISMIC ZONE FACTORS Z FOR DIFFERENT ZONES IN INDIA (IS 1893 Table 2)

Zone	Seismic coefficient α_0	Seismic zone factor F_0	Z
V	0.08	0.40	0.36
IV	0.05	0.25	0.24
III	0.04	0.20	0.16
II	0.02	0.10	0.10
I	0.01	0.05	0.10

Notes:
  (i) F_0 Values are 5 times α_0 values
  (ii) Z is the zone factor in the proposed revision of IS 1893
  (iii) For underground structures, the values are to be modified as given in the note in Table 2 of IS 1893.

18.5.1 Seismic Coefficient Method [IS 1893 (1984) Clause 3.4.2.3] [Method 1]

In order to take the effect of foundation soil and the importance of the building, two coefficients β and I are introduced in the 1984-revision of IS 1893. The horizontal seismic coefficient α_h in this method is calculated from the simple expression:

$$\alpha_h = \beta I \alpha_0 \tag{18.2}$$

where

β = A coefficient depending on the soil-foundation system (Table 18.3). It represents the effect of amplification of the seismic coefficient

I = A factor depending on the importance of the structure (Table 18.4)

α_0 = Basic horizontal seismic coefficient for the region (Table 18.2)

TABLE 18.3 VALUES OF B FOR DIFFERENT FOUNDATION SYSTEMS
[IS 1893-Table 3]

Foundation Soil	Piles resting on Rock	Piles resting on Other soils	Rafts	Footings With ties	Footings Without ties	Walls
Hard soils (Rock)	1.0	—	1.0	1.0	1.0	1.0
Medium soil	1.0	1.0	1.0	1.0	1.2	1.2
Soft soils	1.0	1.2	1.0	1.2	1.5	1.5

[*Note:* It should be noted that the factors given in IS 1893 (1984) take only limited consideration of the foundation conditions. There are mainly four aspects of foundation that influence seismic forces:

1. Effect of types of soil under and around the foundation (soft soils amplify the ground motion more than that of stiff soils)
2. Soil structure interaction (this leads to higher damping and higher natural period of vibration and hence lesser seismic forces)
3. Effect of soil along the path of vibration from the centre of earthquake (soft soil amplify the ground motion)
4. Differential settlement (buildings are designed for larger lateral load if there is larger differential settlement).

Thus buildings on isolated footings on soft soils without ties are designed for 50% higher seismic forces than the buildings on raft or piles.]

TABLE 18.4 VALUE OF IMPORTANCE FACTOR* I
[IS 1893 (1984) Table 4]

Structure	Value of I
Dams	3.0
Containers of poisonous gases	2.0
Important community structures	1.5
Others	1.0

*It represents the seriousness of the failure of the structure.

18.5.2 Response Spectrum Method and Seismic Zone Factor for α_h [IS 1893 (1984) Clause 3.4 2.3(B)] (Method 2)

The maximum horizontal acceleration of an oscillating system depends very much on the fundamental period of oscillation of the structure itself and its relation to the period of the exciting force. Thus, it has been observed that during some earthquakes, in the same locality buildings of a certain height underwent more damages than those which were more or less than that height. The expected response of the structure can also be considered as elastic or non-elastic according to the limit state described in Section 18.1.

The elastic response spectrum is a plot of the peak response (velocity, acceleration, displacement, etc.) of an elastically behaving structure with a single degree freedom of oscillation of period T and a specified damping ratio when subjected to a ground acceleration [Fig. 18.3(a)].

The response spectra are computed for a range of periods and for different values of damping for standard conditions as shown in Fig. 18.3(b). The diagram that is used in practice is that of the average acceleration spectrum obtained by Prof. G.W. Housner from four of the very strong earthquakes that occurred in U.S.A. To use it for other zones of different degree of seismicity, the ordinates of the average spectrum are to be multiplied by a factor F_0. [Figure 18.3(b) is recommended by IS.]

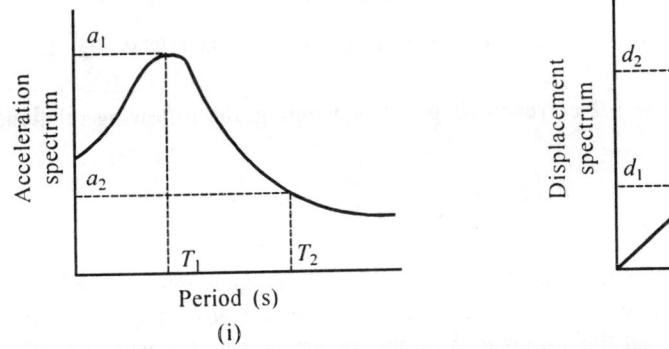

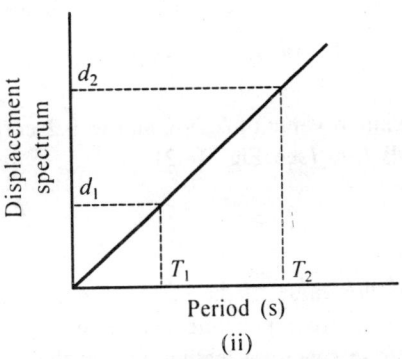

Fig. 18.3(a) Influence of the fundamental frequency of a structure on effects of an earthquake: (i) on design acceleration spectrum (ii) on design displacement spectrum.

From the response spectrum diagram, the value of S_a/g, called the 'spectral acceleration coefficient', corresponding to the natural frequency of vibration of the building, with the specified damping factor, can be read off. The value of the design seismic coefficient is given in Eqn. (18.3).

$$\alpha_h = \beta I F_0 \frac{S_a}{g} \qquad (18.3)$$

where β and I are same as before. F_0 is the seismic zone factor (also denoted by the symbol Z in the uniform building code of U.S.A.) given in Table 18.1 and equal to $5\alpha_0$ [Eq. (18.2)]. S_a/g is the average acceleration coefficient as given in Fig. 18.3b.

It can be seen that for 5% damping the maximum value of S_a/g is of the order of 0.2 or 1/5. This maximum value of critical damping occurs for an approximate value of the natural period of vibration T between 0.2 and 0.3. As the value of F_0 given in Table 18.1 is $5\alpha_0$ for

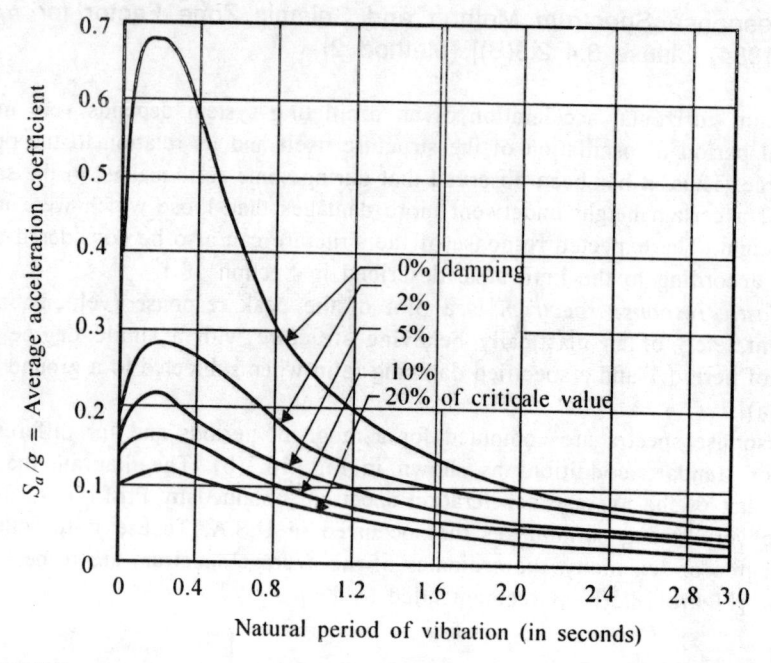

Fig. 18.3(b) Average acceleration spectra prescribed by IS 1893 (1984) Fig. 2.

maximum value of S_a/g equal to 0.2 corresponding to 5% damping, the following relationship holds true (see Fig. 18.2).

$$\frac{F_0 S_a}{g} = \frac{1}{5} 5\alpha_0 = \alpha_0$$

Thus the value of the seismic coefficient α_0 given in IS code is sometimes looked upon as the maximum possible acceleration for 5% damping. Hence the advantage of using the response spectrum method is that the values of S_a/g and α_0 for values of T less or more than the critical damping and also for damping factors greater than 5% can be calculated by this method. The forces can be much less than the value calculated by the seismic coefficient method which becomes a particular case of the general case. The response spectrum method is also more scientific than the old concept of seismic coefficient method. It can also accommodate inelastic design as shown below. However we should not compare the two methods built on different concepts.

18.5.3 Inelastic Response Spectrum

As already explained in Section 18.1, structures can also be designed for inelastic behaviour. For this purpose, we use an inelastic spectrum as shown in Fig. 18.4. The presently proposed inelastic spectrum is related to ductility factor μ which is defined as the ratio of total imposed displacement Δi at any instant to that at the onset of the yield displacement Δy. Therefore,

---

[1]*See* Chapter 22.

$$\mu = \frac{\Delta i}{\Delta y} > 1$$

Inelastic response should be used with ductile detailing. The forces involved are shown in Fig. 18.4. The strength requirement decreases with increasing ductility.

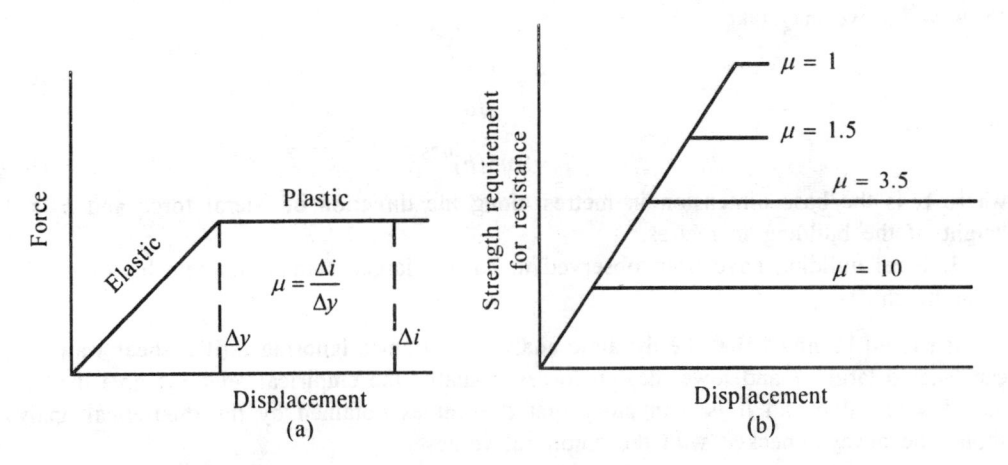

Fig. 18.4 Effect of ductility on strength requirement for resistance: (a) elastic-plastic response (b) reduction in strength requirement with ductility.

18.5.4 DAMPING

Vibrations can be damped or undamped. The damping factor depends on the material of the building and it has an important bearing on the seismic forces. IS 1893(1984) Clause F3 (in Appendix F) recommends the following damping coefficient in terms of the critical damping.

(a) Steel and timber structure: 2 to 5 % of critical
(b) Brick and concrete structures: 5 to 10 % of critical
(c) Earthern structures: 10 to 30 % of critical

18.5.5 Natural Frequency of Vibration of a Building

As can be seen in Fig. 18.1, the maximum acceleration coefficient is obtained for a value of the natural frequencies between 0.2 and 0.3 s. It decreases considerably with increase in the period of natural frequency beyond 0.3 s. Thus, the fundamental frequency of the building is a very important component in estimation of seismic forces. The following empirical formulae are used to calculate T in seconds for buildings of simple shapes. (For special buildings with flexible storeys and for irregular arrangements of blocks or for tall towers and water tanks, values of T are calculated by dynamic analysis, as illustrated in the code.)

1. For a reinforced concrete moment-resisting frame building without brick infill panels, the following two formulae [Eqs. (18.4) and (18.5)] are used to give the period in seconds:

$$T = 0.074(h)^{0.75} \qquad (18.4)$$

where h is the height of the building in metres.

$$T = (0.1)n \tag{18.5}$$

as IS 1893 Clause 4.2.1.1, where n is the number of storeys including basement storeys.

2. For all other buildings including moment-resisting frames with brick infill panels and shear walls, we may take

$$T = \frac{0.09h}{\sqrt{\alpha}} \tag{18.5a}$$

or

$$T = 0.05(h)^{0.75} \tag{18.5b}$$

where α is the base dimension in metres along the direction of lateral force and h is the height of the building in metres.

3. Steel building have been observed to have a larger T than concrete building of the same height.

It should be noted that the dynamic analysis of frames ignoring infills, shear walls, etc., can lead to larger T and lower design forces. Usually, the empirical relations give the lower bound safe value and it is mandatory that the values obtained by the theoretical analysis should be always checked with the empirical values.

18.6 DETERMINATION OF DESIGN FORCES

Having determined the horizontal acceleration the next step is to estimate the horizontal forces. Four methods representing different levels of refinement in analysis are available to estimate the design forces produced by earthquakes in multi-storeyed buildings, as base shear V_B. They can be described in the increasing levels of complexity as follows:

18.6.1 Equivalent Lateral Force on Pseudo Static Method (Method 1)

This method is given in IS 1893 (1984) Clause 4.2.1.1. The most commonly used version of the formula for base shear is

$$V_B = KC\alpha_h Mg = KC\alpha_h W \tag{18.6}$$

where

K = Performance factor which depends on the ductility of the buildings as given in Table 18.5, its value decreasing with increasing ductility

C = Flexibility coefficient depending on the natural frequency of the building as given in Fig. 18.5. The shape of this curve is the same as in Fig. 18.3

α_h = Design horizontal seismic coefficient obtained by using the simple seismic coefficient method or the response spectrum method

M = Mass of the building

g = Acceleration due to gravity (Mg will be the weight W in Newtons)

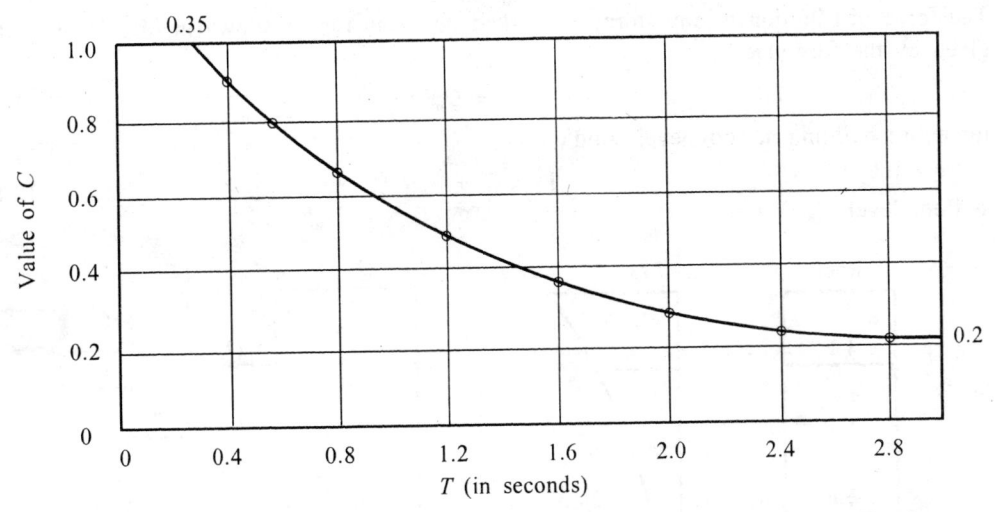

Fig. 18.5 Variation of C with period T stipulated in IS 1893 (1984) Fig. 3.

TABLE 18.5 VALUES OF PERFORMANCE FACTOR K (DUCTILITY FACTOR)
[Table 5 of IS 1893]

Types of structural frame	Value of K
1. Moment resisting frame detailed for ductility	1.0
2. Frames with shear walls detailed for ductility	1.0
3. Frame as in 1 with R.C. infill	1.3
4. Frame as in 1 with masonry infills	1.6
5. R.C. frame not covered in 1 and 2*	1.6

*See also Table 21.1 of Chapter 21.

After finding the base shear V_B for the whole building and taking its distribution forces Q_i along each of the storey, height of the building can be assumed in many ways. One type of distribution used is the inverted triangle. However, IS 1893 assumes the parabolic distribution as indicated by the equation:

$$Q_i = \frac{V_B W_i h_i^2}{\sum_1^n W_i h_i^2} \qquad (18.7)$$

where

Q_i = Lateral force at the required level

W_i = Load (DL + LL) of roof or floor at the required level

h_i = Height measured from base of building (note that for top storeys h_i will be large)

n = Number of storeys

LL is to be 25% of live load if it is < 4 kN/m² and 50% if ≥ 4 kN/m².

The force distribution in any storey is worked out from top as shown in Fig. 18.6(c). It is given by the formulae

$$V_n = Q_n \tag{18.8}$$

for entire building at roof level. And

$$V_i = Q_i + V_{i+1} \tag{18.8a}$$

at floor level.

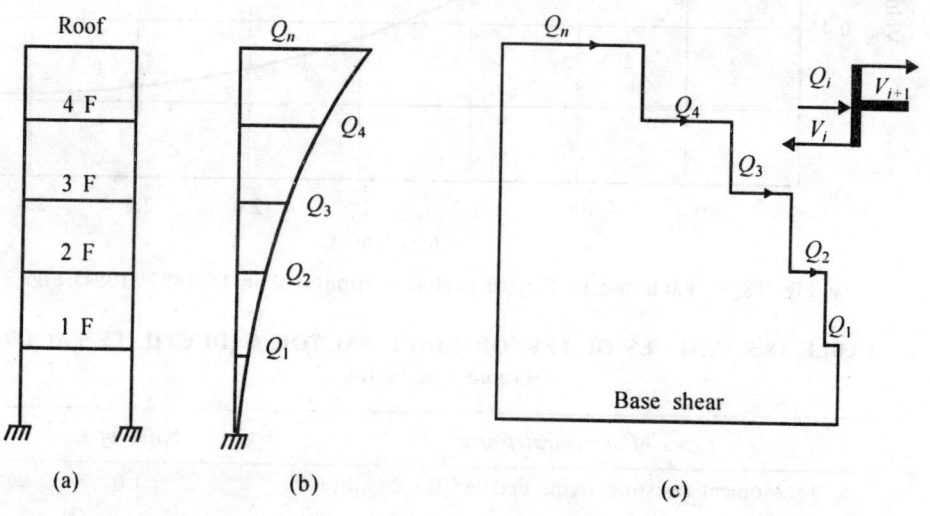

Fig. 18.6 Calculation of base shear: (a) building storeys (b) distribution of forces (c) distribution of shears.

Equations (18.8) and (18.8a) form the basis of the simple equivalent lateral force method which is still the most popularly used method for elastic design.[1] It has been found that the ductility occur only in carefully chosen points. With good detailing of such points, this method has been found to be very satisfactory for design of structures of moderate height. These designs are not very sensitive to earthquake forces. For maintaining the simplicity of the method the natural frequency is determined from empirical formulae.

Steps of the pseudo static method for computation of frequency of an earthquake are as follows:

1. Calculate the dead weight at each floor and roof level, (the floor above the ground is numbered as 1.

2. Calculate the relevant live loads at each floor (except roof level). According to IS 1893 (1984), take 25% of the live load to calculate forces for loading up to 3 kN/m². For the loading more than 3 kN/m² take 50% or the live load to calculate the forces.

3. Calculate the total gravity load W of the building.

4. Estimate the first mode natural period by empirical formula (Section 18.5.54).

5. Choose the appropriate horizontal seismic coefficient from the seismic coefficient method. (The response spectrum method will give the same result for 5% damping.)

---

[1] The steps to be followed for computation of forces are shown in Example 18.1.

EARTHQUAKE FORCES AND STRUCTURAL RESPONSE   323

6. Calculate the seismic design base shear from Eq. 18.6, $V_B = KC\alpha_h W$. (*Note:* In effect, the use of factor C will give the same V_B values as in the response spectrum method with 5% damping and the required natural frequency.

7. Distribute the base shear as component forces acting at different levels of the structure by Eq. 18.6 as shown in Fig. 18.6(c).

8. Analyze the structure under the gravity and lateral loads.

9. Estimate the structural displacement especially the story drifts as indicated in step 10 in section 18.6.2.

This procedure is illustrated by Example 18.1.

18.6.2 Model Superposition Techniques with Response Spectrum (Method 2)

This method is given in IS 1893 (1984) Clause 4.2.2. It is an elastic dynamic approach which assumes that the dynamic response of a structure can be found by considering the response of the building to different modes of vibration independently and then recombining them suitably to study their combined effects. For this method, we use the response spectrum method. We know that a structure like a tall building can vibrate in many modes. The first three modes, of vibration are shown in Fig. 18.7.

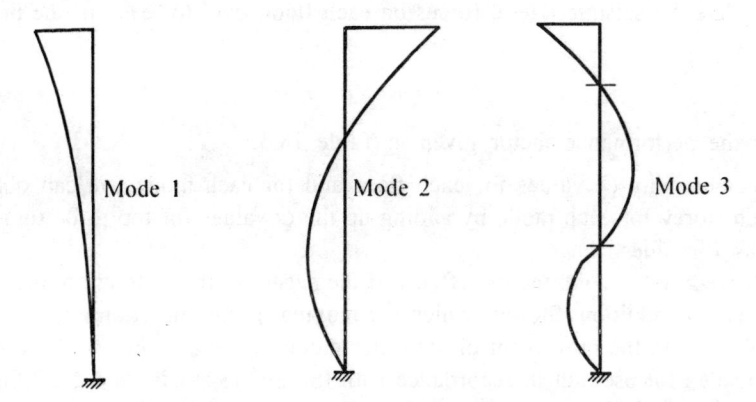

Fig. 18.7   Mode shapes of the first three modes of oscillation of a building.

For dynamic analysis, the mass of a given floor in the building is assumed as lumped at each floor level. For symmetric layout of structures, only one degree of freedom per floor exists. For unsymmetric layouts and eccentric loadings when torsion is also present, three degrees of freedom per floor are possible. Steps of dynamic analysis as recommended in SP 22 are as follows: (For details and worked out example see SP 22. In this example there are a few minor errors which are expected to be corrected in the next revision.)

1. Calculate the dead load and part live load on each floor as in seismic coefficient method as lumped masses.

2. Calculate storey stiffnesses. Each column contributes a stiffness of $K_c = EI/L_c$ and all columns are assumed acting in parallel to each other. The beams are usually flexible.

3. Calculate the natural frequencies T_1, T_2, T_3, etc., of the first three modes (or others if necessary) and its corresponding mode shapes coefficient ϕ' for each storey as well as for

each mode using Stodola Vianello interaction procedure. (Mode shapes give the way the structure deflects during vibration as shown in Fig. 18.6.)

4. Calculate the horizontal seismic coefficient for each mode using appropriate value of T and assuming appropriate damping factor (say 5%) using the formula [modification of Eq. (18.3)]

$$\alpha_h^r = \beta I F_0 \frac{S_a^r}{g} \qquad (18.9)$$

5. Calculate the mode participation factor C_r for each mode as shown in Fig. 18.7. This quantity gives the relative participation of each mode in the vibration system. The value is given by the expression:

$$C_r = \frac{\sum_{i=1}^{n} W_i \phi_i^r}{\sum_{i=1}^{n} W_i (\phi_i^r)^2} \qquad (18.10)$$

where W_i is the corresponding lumped mass and ϕ_i^r is the mode shape coefficient obtained in step 3.

6. Calculate the seismic lateral forces on each floor level for each of the modes by using the formula

$$Q_i^r = K C_y \alpha_h^r \phi_i^r W_i \qquad (18.11)$$

where K is the performance factor given in Table 18.5.

7. After obtaining Q_i values for each floor and for each mode, we can obtain the shear force in each storey for each mode by adding up the Q values for top to bottom floors. These are called as V_i^r values.

8. Now we have to combine the effects of the shears in the different modes which cannot be done by simple addition. Storeys which are moving in one direction and in one mode may be out of phase with the movement of the other modes as seen in Fig. 18.7. Various methods are recommended for use, but in accordance with IS 1893 (84) Clause 4.2.2.2 the shear forces acting in the storey are obtained by superposition for the first three modes as follows:

$$V_i = (1 - \gamma) \sum_{1}^{3} V_i^r + \gamma \sqrt{\sum_{1}^{3} (V_i^r)^2} \qquad (18.12)$$

where γ is to be taken as given below:

Height (in metres)	γ
up to 20	0.40
up to 40	0.60
up to 60	0.80
up to 90	1.00

(*Note:* For intermediate height, the values are interpolated.)

9. Find the distribution of forces from the distribution of shear from Eq. (18.8).

$$Q_i = V_i - V_{i+1}$$

$$Q_n = V_n \text{ (at roof level)}$$

10. Calculate the storey drift which is the relative displacement starting from the base from the relation:

$$\text{Drift} = \frac{V_i}{K_i} \qquad (18.13)$$

where V_i is the shear (in kN) and K_i is the stiffness (in kN/mm) of the frame.

A drift of 1/250 of the interstorey height is allowed in the Indian code for earthquake effects as compared to 1/500 for wind loads. Example 2 given in SP 22 (with correction for flexibility of beams for calculation of T) may be used for details to be followed for the calculation.

It should be remembered that modal analysis without the effect of infill can give a large value of T and unsafe value of seismic forces. Hence the values of T obtained from modal analysis should always be compared with the empirical values and the lesser of the two adopted for design. The dynamic analysis can however give a good idea of the vibration mode of the structure as this is governed by relative stiffness and not by absolute stiffnesses.

18.6.3 Detailed Elastic Dynamic Analysis (Method 3)

This is carried out by more exact dynamic analysis available in books on structural dynamics.

18.6.4 Dynamic Inelastic Time-History Analysis (Method 4)

This is the most sophisticated method for an economical seismic design. However, at present, the method remains only as a research tool in investigating real earthquakes.

18.7 CHOICE OF METHOD FOR MULTI-STOREYED BUILDINGS

IS 1893 (1984) recommends the following analysis for the various types of buildings:

Types of building	Method
1. Taller than 90 m in zones III, IV & V	Detailed dynamic analysis with modal analysis for preliminary design
2. Taller than 90 m in zones I & II	Modal analysis with response spectrum
3. Between 40 & 90 m in all zones	Better to use modal analysis with response spectrum. However, the use of seismic coefficient method is allowed in zones I, II and III.
4. Less than 40 m in all zones	Modal or seismic coefficient method

It should be remembered that complex elastic analysis need not give the right answer for an economic seismic design. It can be obtained only by judicious use of inelastic action using induced ductile behaviour at specific points. Ductile detailing by carefully placing right quantities of extra steel at the right places will give a better structure, than routine design by complex calculations. This will not also considerably increase the cost of the structures.

18.8 DIFFERENCE BETWEEN WIND AND EARTHQUAKE FORCES

Eventhough both wind and earthquake produce lateral forces, basic differences exist in the manner in which they act on the structure. While wind forces are external forces proportional to the exposed area, earthquake forces are inertia forces resulting from distortion of the foundation (along with the surrounding earth) with respect to the superstructure. The inertia force is a function of the mass of the building rather than the area of the exposed surface.

Yet another difference is in the distribution of the forces along the height of the building. Whereas the wind forces are constant for each floor, the earthquake forces vary parabolically (some assume it linearly) with the height as shown in Fig. 18.5. There are considerably larger forces (Q forces) acting on the top storeys rather than on the bottom storeys.

18.9 TORSION IN BUILDINGS

Horizontal twisting takes place in buildings whose centre of mass and centre of rigidity do not coincide, the distance between the two centres being called the eccentricity (e). The IS code recommends a design eccentricity of ($1.5e$) as the computations are only approximate. It also recommends to take into account the positive shears, and the negative shears are to be neglected. The procedure is two-fold as under:

(a) Calculate centre of mass at coordinates x and y
(b) Calculate the centre of rigidities

$$x_r = \frac{\Sigma K_y(x)}{\Sigma K_y} \quad \text{and} \quad y_r = \frac{\Sigma K_x(y)}{\Sigma K_x}$$

where K_x and K_y are the stiffnesses with x and y as the principal coordinates. The procedure that can be used for designs is illustrated is SP 22.

18.10 PARTIAL SAFETY FACTORS FOR DESIGN

The following partial safety factors are used for limit state design of collapse of reinforced concrete structures for earthquake loads (Table 18 of IS 456):

(i) $1.5 (D + L)$ loading 1
(ii) $1.2 (D + L \pm E)$ loading 2
(iii) $1.5 (D \pm E)$ loading 3
(iv) $0.9D \pm 1.5E$ loading 4

where

D = Dead load

L = Live load

E = Earthquake load (which may act from left to right or right to left)

The structure should be analyzed for each of the above loadings and members designed for the worst effects. For this purpose each member has to be analyzed and the beam moments tabulated as in Table 18.6 with earthquake forces in either of the lateral directions.

TABLE 18.6 DESIGN MOMENTS OF BEAMS (Example of presentation of data)

Loading	Beam	Design moments		
		End (A)	Centre (C)	End (B)
Loading 1, no sway	Beam AB	−61	+46	−72
Loading 2, sway to right	Beam AB	+23	+36	−88
Loading 2, sway to left	Beam AB	−55	+35	−12
Loading 3, sway to right	Beam AB	+8	+30	−84
Loading 3, sway to left	Beam AB	−69	+32	−15
Loading 4, sway to right	Beam AB	+8	+18	−65
Loading 4, sway to left	Beam AB	−55	+18	+3
Design moments		−69 +23	+46	−88 +3

where

    Loadings condition 1. $1.5(D + L)$

    Loadings condition 2. $1.2(D + L + E)$ and $1.2(D + L - E)$

    Loadings condition 3. $1.5(D + E)$ and $1.5(D - E)$

    Loadings condition 4. $0.9D + 1.5E$ and $0.9D - 1.5E$

When the lateral load resisting elements are not oriented along the principal directions of the structure, it should be designed for full load in one direction plus 40% of the load in the other direction both acting together and both the directions being considered independently.

18.11 DISTRIBUTION OF SEISMIC FORCES

IS 1893 (1984) Clause 4.2 gives the criteria for distribution of the horizontal forces in multi-storeyed buildings. In case of buildings with floors that can provide rigid horizontal diaphragm action, the building is analyzed as a whole for seismic forces. The total shear at each floor is distributed to various elements of lateral forces resisting system assuming the floors to be infinitely rigid in the horizontal plane. In buildings having frames and shearwalls the frames are designed for at least 25% of the seismic shear (see Table 21.1). Observation of buildings damaged by earthquakes have shown the excellent performance of shear walls. Hence shear walls should always be incorporated in buildings in regions of severe earthquakes.

18.12 ANALYSIS OF STRUCTURES OTHER THAN BUILDINGS

The seismic forces should also be taken into account in structures like bridges, chimneys, retaining walls. They are dealt with in more detail in the latest revised version of IS 1893 as discussed in Section 18.15.

18.13 DUCTILE DETAILING

We have already seen that ductile behaviour is one of the important requirements of seismic resistant design. Indian standard IS 13920 (1993) (code of practice for ductile detailing of reinforced concrete structures subject to seismic forces) specifies that all structures in zones IV and V as well as some specific structures in zone III (structures of importance factor more than unity, industrial structures and buildings above five storeys in height) have to be detailed for ductile failure.[1]

18.14 INCREASED VALUES OF SEISMIC EFFECT FOR VERTICAL AND HORIZONTAL PROJECTIONS

IS 1893 (1984) Clause 4.41 states that towers, tanks, parapets, chimneys and vertical cantilever projections attached to buildings (except ordinary walls) should be designed for five times the normal horizontal seismic coefficient. Similarly, horizontal projections like cornices and balconies should resist vertical forces equal to five times the vertical seismic coefficient (which may be taken as half the horizontal seismic effect) multiplied by the weight of the projection. Hence it is preferable to avoid these projections when planning buildings in seismic areas.

18.15 PROPOSED CHANGES IN IS 1893 (FIFTH REVISION)

In the draft revision of IS 1893, the following changes have been introduced:

1. The code itself is to be divided into seven parts each dealing with special aspects of the problem as follows:

> *Part 1*: General provisions and building
>
> *Part 2*: Liquid retaining structure
>
> *Part 3*: Retaining wall
>
> *Part 4*: Industrial structures
>
> *Part 5*: Stack like structure

---

[1] The principal recommendations for this type of detailing are dealt with in Chapter 21 of this book.

Part 6: Bridges

Part 7: Dams and embankment

2. The approach recommended is more or less the same as that the Uniform Building Code (UBC) of U.S.A. The first step is to calculate the maximum acceleration for elastic response and then use a reduction factor according to the following equation:

$$A = \frac{ZIC}{R} \tag{18.14}$$

Base shear $V_b = AW$

where

A = Design horizontal acceleration spectrum in place of α_h

Z = Seismic zone factor more or less the same as F_0 in the old code

I = Importance factor as in old code

C = Structural flexibility factor depending on period T and expressed as $C = 1.25S/T^{2/3}$. It is not the same as in IS 1893 (84) given in Section 18.6.1

S = Soil profile factor varying from 1.0 to 1.5

R = Response reduction spectrum to design value varying from 4 for a masonry wall to 10 for a moment resisting frame

W = Weight in Newtons

3. Comparing Eq. (18.14) with Eqs. (18.2), (18.3) and (18.6) we find that in Eq. (18.14) the values of seismic coefficient α_h has been omitted and seismic zone factor (Z) has been introduced. The values themselves have been made realistic with zone 1 being omitted. *According to the new code only the response spectrum method is used for design of tall buildings.*

4. Dynamic analysis is compulsory for irregular buildings and some of the procedures in dynamic analysis recommended earlier have also been modified. Thus the purpose of the proposed revision is to bring the IS code along the same lines as Uniform Building Code of U.S.A. which has also been adopted as the standard in many other countries [3].

CONCLUSION

According to the present IS code, the Indian sub-continent has been divided into five seismic zones. Zones I and II are zones of 'low seismic risk'. In these zones, no special design or detailing other than the following general principles of limit state design and detailing given in IS 456 are generally required for adequate safety against these low intensity earthquakes. Zone III are regions of 'moderate seismic risk'. In these zones, only those structures of importance factor greater than 1.0, industrial structures and structures more than five storeys high are required by Clause 1.1 of IS 13920 (1996) to be detailed for ductility. However, all structures located in zones IV and V should be designed for earthquake forces and detailed for ductility [4].

We should bear in mind the fact that the zoning as given in IS 1893 (1984) does not give the frequencies with which earthquake can occur. It gives only the likely magnitude of the earthquake that can occur in a place. Experience has shown that with the present state of our knowledge it is difficult to predict even the approximate magnitude of earthquakes that can occur in the northern regions of India. The Latur earthquake of 1993 of magnitude 6.4 took place in a region which was placed in zone I of IS 1893 (1984). Similarly Jabalpur placed in zone III experienced an earthquake of magnitude 6.0 in 1997. An examination of the seismic map of India shows that at present the majority of areas of India are in zones I to III. Only portions below the Himalayas, North Eastern India, Kutch, Kathiwar and a few other regions come under zones IV and V. These observations indicate that the earthquake map requires revision. Hence, for structures situated in most parts of India ductile detailing should be the most important factor to be complied with. This will ensure that there will be minimal loss of lives even if an earthquake, which unlike cyclones occurs without notice, hits the place.[1]

EXAMPLE 18.1 (Calculation of horizontal seismic coefficient α_h)
A three-storeyed school building is in seismic zone IV. Its foundation is on isolated footings on medium soil of the period of oscillation of the building $T = 0.2$ second. Calculate the horizontal seismic coefficient α_h by

(a) seismic coefficient method (by α_0)
(b) response spectrum method (by F_0 and S_a/g) assuming a damping factor of 2% only (less than 5%)

Reference	Step	Calculations
	1	*Formulae for calculation of α_h*
Eq. (18.2)		$\alpha_h = BI\alpha_0$ (seismic coefficient method)
Eq. (18.3)		$\alpha_h = \beta I F_0 \, S_a/g$ (response spectrum)
		where
		β = Foundation coefficient = 1.2
		(for isolated footings on good soil)
		I = Importance factor of the school building = 1.5
	2	*Calculate α_h by seismic coefficient method*
Table 18.2		For zone IV, $\alpha_0 = 0.05$
		α_h by seismic coefficient method,
		$\alpha_h = 1.2 \times 1.5 \times 0.05 = 0.09$
	3	*Calculate α_h by response spectrum*
Table 18.2		For zone IV $F_0 = 0.25$
		S_a/g depends on the damping and the natural period of oscillation
		Damping = 2% of critical

---
[1]The fundamentals of such ductile detailing are dealt with in Chapter 21.

EARTHQUAKE FORCES AND STRUCTURAL RESPONSE 331

Reference	Step	Calculations
		Period $T = 0.2$ s $$\frac{S_a}{g} = 0.28$$ $$\alpha_h = \frac{\beta I F_0 S_a}{g}$$ $= 1.2 \times 1.5 \times 0.25 \times 0.28 = 0.126$ *Note:* With the response spectrum method, the period T (denotes the flexibility) and the damping factor affects the value of S_a/g. However, for a period of natural oscillation = 0.2s (when S_a/g is maximum) and a damping factor = 5% which is usually assumed if not specified, the value of α_0 and $F_0 S_a/g$ will be the same. In such case, both the methods gives the same value of α_h. For 5% damping $S_a/g = 0.2$ $\alpha_h = 1.2 \times 1.5 \times 0.25 \times 0.2 = 0.09$ (*Note:* For design of real tall buildings, the response spectrum method should always be used with modal analysis. This example is only to show the limitation of the seismic coefficient method compared to the versatility of the response spectrum method.)

EXAMPLE 18.2 (Calculation of earthquake loads by seismic coefficient method)
A four-storey (G + 3) reinforced concrete office building has a ground plan 20 × 15 m and elevation as shown in Fig. 18.7. The imposed load on the roof is 1.5 kN/m² and that on the floors 3 kN/m². Determine the seismic load on the frame by the seismic coefficient method. Assume there is a shear wall in the Y direction. The roof and floor slabs are 150 mm thick. The size of the beam is 250 × 400 mm and columns 400 × 500 mm. The height of floors is 3 m. There is a wall around the building 12 cm thick. Assume location in zone III.

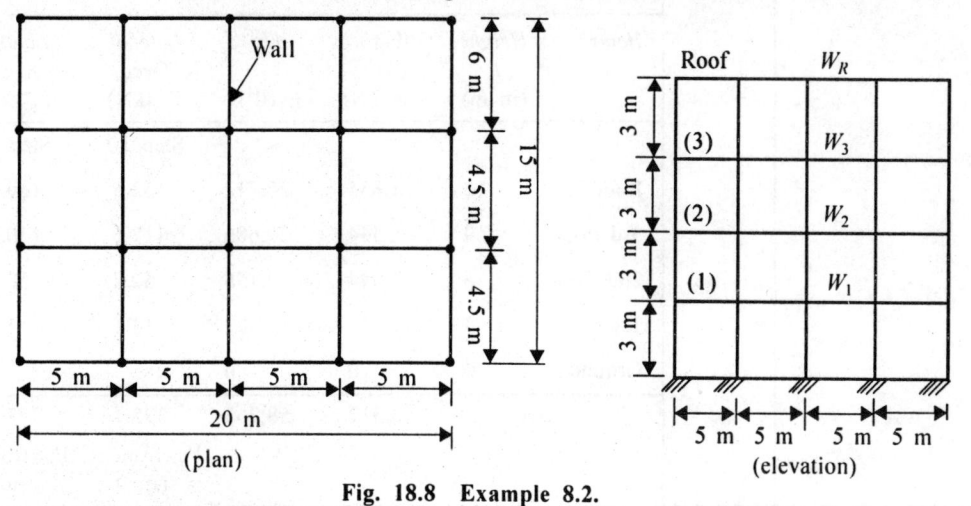

Fig. 18.8 Example 8.2.

Reference	Step	Calculations
From data	1	*Calculate the dead load on roof and floors* 1. Wt. of slab in each floor and roof $(20 \times 15)(0.15)(24) = 1080$ kN 2. Beams in X direction = $4(20)(0.25 \times 0.4)(25)$ (4 beams of 20 m) = 200 kN Beams in Y direction = $5(15)(0.25 \times 0.4)(25)$ (5 beams of 15 m) = 187.5 kN (190 kN) Wt. of total beams = 390 kN 3. Wt of 3 m length = $20(3)(0.4 \times 0.5)(25)$ (20 columns) = 300 kN Taking 1/2 column with each floor = 150 kN 4. Wall at periphery = $70(3 \times 0.12)(20) = 504$ kN 5. Parapet on roof = 170 kN (approx.)
	2	*Determine live load on floors* Up to class 3 kN/m² we take only 25% of the load Live load on floor = $(20 \times 15)(3 \times 0.25) = 225$ kN
	3	*Calculate total gravity loads on roof and floors* Load on roof = $1125 + 390 + 150 + 170 = 1835$ kN Load per floor = $(1125 + 390 + 300 + 504) + 225 = 2544$ kN (with LL)
	4	*Tabulate the results for each floor*

Items	Height h (in m)	Weight w (in kN)	Wh^2 ($\times 10^4$)	Lateral force F_x (kN)	Lateral force F_y (kN)
				Step 10	Step 10
Roof	12	1,835	26,712	152.5	169.5
3rd floor	9	2,544	20,606	117.6	130.8
2nd floor	6	2,544	9,158	52.2	58.2
1st floor	3	2,544	2,289	13.1	14.5
Ground	0	0	0		
Total		9.317	58,765	335.4	373.0
				V_{BX} from step 9	V_{BY} from step 9

Reference	Step	Calculations
Eq. (18.4)	5	*Calculate the base shear* $V_B = KC\alpha_h W$ where $K = 1$ assuming ductility detailing C = Calculated as in step 6 (Fig. 18.5) α_h = Calculated as in step 8 w = As calculated in step 4 in Newtons
	6	*Estimate period for C value* Lateral resistance in X direction by frame only $T = 0.1n$ $T = 0.1 \times 4 = 0.4$ s For lateral resistance in Y direction by the shear walls
Sec. 18.5.5		$T = \dfrac{0.09h}{\sqrt{d}}$ d = Base dimension along force = 15 m h with parapet = 13.8 m $T = \dfrac{0.09 \times 13.8}{\sqrt{15}} = 0.3$ s
IS 1893 Fig. 3	7	*Find value of C from Fig. 18.5* C in X direction (for 0.4 s) = 0.9
(Fig. 18.5)		C in Y direction (for 0.3 s) = 1.0
	8	*Calculate α_h*
Eqs. 18.2 and 18.3		$\alpha_h = \beta I \alpha_0$ or $\dfrac{\beta I F_0 S_a}{g}$ The latter formula is used as shown in Example 18.1. α_0 gives the maximum acceleration for 5% damping. Using the first expression
Sec. 18.5.1		$\alpha_h = \beta I \alpha_0$ where $\beta = 1$ for raft foundation $I = 1$ for office building
Table 18.2		For zone III $\alpha_0 = 0.04$ (maximum value) $\alpha_h = 1 \times 1 \times 0.04 = 0.04$

334 ADVANCED REINFORCED CONCRETE DESIGN

Reference	Step	Calculations
	9	Calculate the base shear $V_B = KC\alpha_h W$ Ductile detailing $K = 1$ In X direction $V_{BX} = 1 \times 0.9 \times 0.04 \times 9317 = 335.4$ kN In Y direction $V_{BY} = 1 \times 1 \times 0.04 \times 9317 = 372.7$ kN (say 373 kN)
	10	Determine distribution V_B in each floor in X directions
Eq. (18.7)		$Q_i = V_B \dfrac{W_i h_i^2}{\sum W_i h_i^2}$ On the first floor $= 335.4 \dfrac{2.29}{58.76} = 13.1$ kN For other floors as in Table step 4.
	11	Distribute V_B in each floor in Y direction (As shown in table)
	12	Draw distribution of shear Draw cumulative distribution as in Fig. 18.6, i.e. IS 1893 (1984) Clause 4.2.1.3. Roof: 152.5 kN in X direction Level 3: 152.5 + 117.6 = 270.1 Level 2: 270.1 + 52.2 = 322.3 Level 1: 322.3 + 13.1 = 335.4 Base shear = 335.4 kN

EXAMPLE 18.3 (Calculation of base shear by the uniform building code)
Determine by UBC, the base shear is the longitudinal direction for a 12-storeyed moment resisting framed building of total weight 13,000 tons for the following data:

Z = Seismic zone factor = 0.4

I = Importance factor = 1.0

S = Soil profile factor = 1.0

R = Structural flexibility factor = 12

Assume $h = 44.4$ metres, where

$$C = \dfrac{1.255}{T^{2/3}}$$

EARTHQUAKE FORCES AND STRUCTURAL RESPONSE 335

Reference	Step	Calculations
Eq. (18.14)	1	Base shear formula in UBC $$V = \frac{ZIC}{R}W$$ where $C = 1.25S/T^{2/3}$ $T = 0.05h^{0.75}$ $S = 1$
	2	Calculate C $T = 0.05(44.4)^{3/4} = 0.86$ s $$C = \frac{1.25S}{T^{2/3}} = 1.40$$
	3	Where $S = 1.0$, calculate base shear $$V = \frac{ZIC}{R}W = \frac{0.4 \times 1 \times 1.40}{12} \, 13{,}000 = 606.7 \text{ kN}$$ [*Note:* This base shear is to be distributed to each storey.]

EXAMPLE 18.4 (Seismic force on cantilevers)

A cantilever portico slab 1 m × 1 m and 100 mm thick projects from a 225 mm thick wall to the outside of a building. The slab is located at 1 m below the roof of the building. Assuming that the weight of the roof on slab is 8 kN per metre length of the wall, examine the stability of the cantilever if it is situated in seismic zone V (*see* Fig. 18.9). (Assume that the wall above the cantilever supports the roof.)

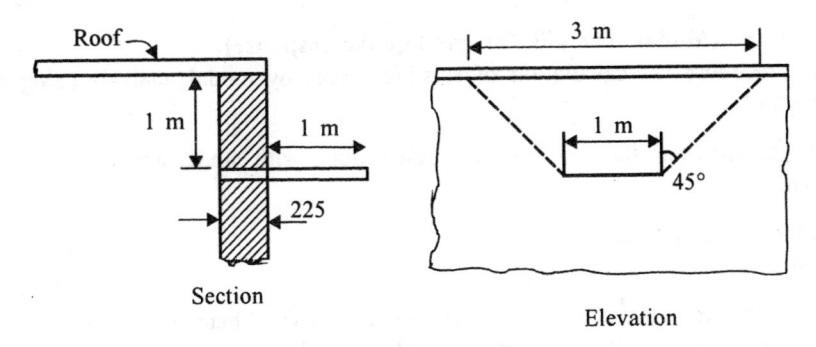

Fig. 18.9 Example 18.4.

Reference	Step	Calculations
	1	*Calculate weight of slab*
		Weight of slab = (1)(1)(0.1)(25) 2.5 kN
	2	*Determine force due to earthquake*
		Acceleration in vertical direction= 1/2 (Horizontal acceleration)
		Factor for cantilever projection = 5.0
		α_h for zone V = 0.08
		Resultant α = (0.08)(0.5)(5) = 0.20
		Force due to earthquake effect = 2.5(0.20) = 0.5 kN
		This acts at the CG of the slab along with weight of the slab.
		Total downward load = 2.5 + 0.5 = 3 kN
	3	*Obtain the overturning moment*
		Distance of forces from wall = 0.5 m
		Overturning moment about face of masonry
		M_0 = 3 × 0.5 = 1.5 kN
	4	*Stabilizing moment*
		M_s = (Load on masonry)(1/2 × Thickness of the wall)
		Assume dispersion of 45 degrees.
		Load from roof = 3 × 8 kN
		Weight of brick-work per metre = 0.225 × 1 × 20 = 4.5 kN
		$M_s = [(3 \times 8) + (2 \times 1 \times 4.5)] \times \dfrac{0.225}{2} = 3.7$ kNm
		As M_s is greater than M_0, it is stable.

EXAMPLE 18.5 (Modal Analysis for earthquake response)

Indicate the procedure for calculation of seismic forces by modal analysis using response spectrum method.

(*Note:* Response spectrum method is usually used along with modal analysis method to get the best results.)

1. Procedure: Refer Section 18.6.2
2. Comments

    (i) Computer analysis will be more convenient than manual analysis
    (ii) Base shear calculated by modal analysis, and response spectrum method can be as much as 10 to 50% less than that calculated by the elementary seismic coefficient method.

REFERENCES

1. Blume, J.A., Newmark, N.M. and Corning, L.M., *Design of Multi-storey Reinforced Concrete Buildings for Earthquake Motion*, Portland Cement Association, Chicago, 1961.

2. Panby, T. and Priestly, M.J.M., *Seismic Design of Reinforced Concrete and Masonry Buildings*, John Wiley and Sons, New York, 1962.

3. Uniform Building Code (UBC) 1994, International Conference of Building Officials, Whittier, California, 1988 (including 1989 supplement).

4. Krishna and Chandrasekaran, *Elements of Earthquake Engineering Design*, Saritha Prakasan, Meerut, India, 1994.

CHAPTER 19

Design of Shear Walls

19.1 INTRODUCTION

Walls can be designed as plain concrete walls when there is only compression with no tension in the section (*see* Chapter 17). Otherwise they should be designed as reinforced concrete walls [1]. *Shear walls* are specially designed structural walls incorporated in buildings to resist lateral forces that are produced in the plane of the wall due to wind, earthquake and other forces. The term 'shear wall' is rather misleading as such walls behave more like flexural members. They are usually provided in tall buildings and have been found of immense use to avoid total collapse of buildings under seismic forces. It is always advisable to incorporate them in buildings built in regions likely to experience earthquake of large intensity or high winds. Shear walls for wind are designed as simple concrete walls. The design of these walls for seismic forces require special considerations as they should be safe under repeated loads. Shear walls for wind or earthquakes are generally made of concrete or masonry. They are usually provided between columns, in stairwells, lift wells, toilets, utility shafts, etc. Tall buildings with flat slabs should invariably have shear walls. Such systems compared to slabs with beams have very little resistance even to moderate lateral loads. Initially shear walls were used in reinforced concrete buildings to resist wind forces. These came into general practice only as late as 1940. With the introduction of shear walls, concrete construction can be used for tall buildings. Earlier tall buildings were made only of steel, as bracings to take lateral loads could be easily provided in steel construction. However, since recent observations have shown consistently the excellent performance of buildings with shear walls even under seismic forces, such walls are now extensively used for all earthquake resistant designs. Surveys of buildings after earthquakes have consistently shown that the loss of life due to complete collapse was minimal in buildings with some sort of shear wall. However the most important property of shear walls for seismic design, as different from design for wind, is that it should have good ductility under reversible and repeated over loads. In planning shear walls, we should try to reduce the bending tensile stresses due to lateral loads as much as possible by loading them with as much gravity forces as it can safely take. They should be also laid symmetrically to avoid torsional stresses.[1] This chapter deals very briefly with the design of reinforced shear walls. Determination of the forces in these walls is not dealt with as it is part of structural analysis.

---

[1]*See* Section 19.3.

19.2 CLASSIFICATION OF SHEAR WALLS

There are many types of shear walls given as under:

1. Simple rectangular types and the flanged walls (called the *bar bell type* walls with boundary elements)
2. Coupled shear walls
3. Rigid frame shear walls
4. Framed walls with infilled frames
5. Column supported shear walls
6. Core type shear walls

A brief description of the behaviour of each type is given below:

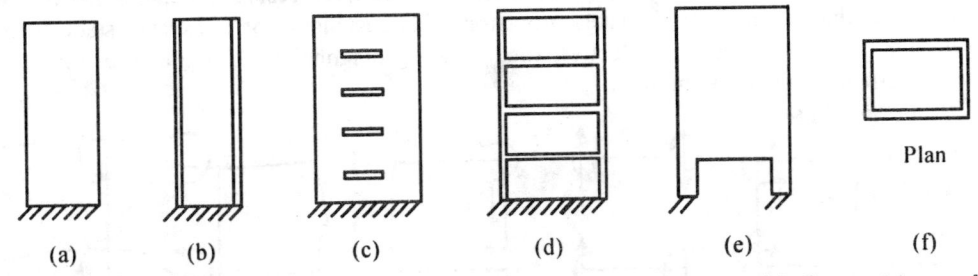

Fig. 19.1 Types of shear walls: (a) plane (simple rectangular) (b) plane with flanges (c) coupled (d) framed with or without infill (e) column supported (f) core type.

19.2.1 Simple Rectangle and Bar Bell Type Free Standing Walls [Figs. 19.1(a) and 19.1(b)]

These simple types were the first to be used in construction. Such shear walls, under the action of inplane vertical loads and horizontal shear along its length, are subjected to bending and shear. Uniform distribution of steel along its length as is used in the simple shear walls is not as efficient as putting the minimum steel over the inner 0.7 to 0.8 length L of the wall and placing the remaining steel at the ends for a length 0.15 to 0.12 L on either side. These latter types are called *bar-bell type* walls which are somewhat stronger and more ductile than the simple rectangular type of uniform section.[1] These walls should be designed in such a way that they never fail in shear but only by yielding of steel in bending. Shear failure is brittle and sudden. One of the disadvantages of this type of shear walls is that as these walls being rigid, during an earthquake it attracts and dissipates a lot of energy by cracking, which is difficult to repair. This defect can be got around by coupled shear walls described below.

19.2.2 Coupled Shear Walls [Fig. 19.1(c)]

If two structural walls are joined together by relatively short spandrel beams, the stiffness of the resultant wall increases and also the structure can dissipate most of the energy by yielding

---

[1] Rules of design of these type of walls are dealt with in Section 19.5.

the coupling beams with no structural damages to the main walls. It is easy to repair these coupling beams than the walls. These walls should satisfy the following two requirements:

(a) The system should develop hinges only in the coupling beam before shear failure.
(b) The coupling beam should be designed to have good energy-dissipation characteristics.

The action of coupling beam is shown in Fig. 19.2A. As the beams are displaced vertically, they tend to bend in double curvature. The consequent shear can reduce the axial force in the up-wind wall by a large amount. Taking M_p as the magnitude of plastic moment in all the beam section the magnitude of N (total reaction) is given by

$$N = \frac{2M_P}{\text{Length of beam}} \times \text{No. of hinges formed} \qquad (19.1)$$

The reduced pressure can lower the shear capacity of the wall. Tests show that under repeated cyclic loads that occur in earthquakes provision of large amount of transverse steel in these beams for ductility is not very effective. Incorporating diagonal steel as shown in Fig. 19.2B is more effective than lateral steel.

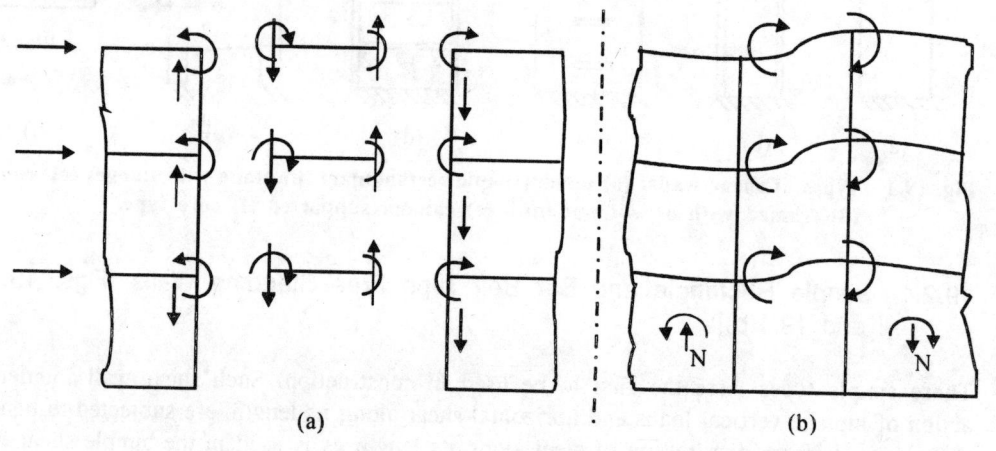

Fig. 19.2A Action of coupled shear walls as energy dissipation devices (coupling beam AB) (a) external forces and reactions of structure (b) action of coupling beams.

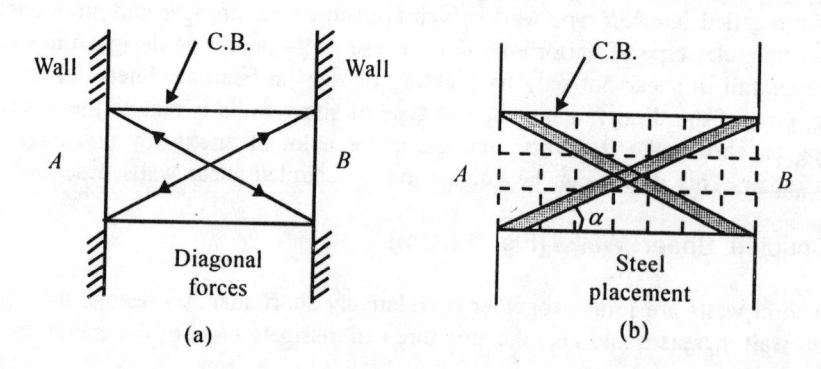

Fig. 19.2B Detailing of coupling beams AB.

19.2.3 Rigid Frames with Shear Walls

The interaction of simple shear walls and rigid frames of a tall building is shown in Fig. 19.3. The deflection of the frame is in the shear mode but that of the wall is in the bending mode. This interaction tends to reduce maximum moments but increases the maximum shears in the shear walls. This increases the tendency of shear failure in the shear walls and this factor should be allowed for in the design.

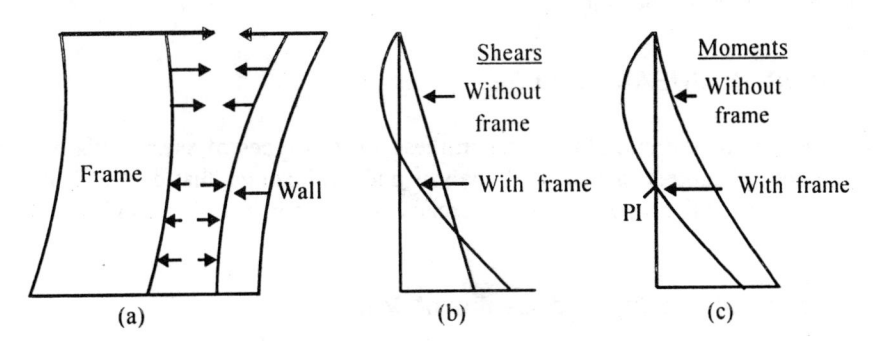

Fig. 19.3 Interaction between structural frame and shear wall: (a) action between frame and wall (b) shears in wall (c) moments in wall.

19.2.4 Framed Wall, Shear Walls and Infilled Shear Walls [Fig. 19.1(d)]

Framed walls are cast monolithically whereas *infilled* frames are constructed by casting frames first and infilling it with masonry or concrete blocks later. A lot of literature is available on the mode of action of these walls.

19.2.5 Column Supported Shear Walls [Fig. 19.1(e)]

When it is necessary for architectural reasons to discontinue shear walls at floor level it becomes necessary to carry the wall to the ground on widely spaced columns as shown in Fig. 19.1(e). In such column supported shear walls, the discontinuity in geometry at the lowest level should be specially taken care of in the design [6].

19.2.6 Core Type Shear Walls [Fig. 19.1(f)]

In some buildings, the elevators and other service areas can be grouped in a vertical core which may serve as devices to withstand lateral loads. Unsymmetry produces twisting and if twisting is not present these walls act as simple shear walls. Cores with designed lintels at regular intervals as in elevator shafts have also good resistance against torsion.

19.3 CLASSIFICATION ACCORDING TO BEHAVIOUR

Shear walls can also be classified according to their behaviour as follows:

(a) *Shear-shear walls* in which deflection and strength are controlled by shear. These are usually low-rise shear walls.

(b) *Ordinary-moment shear walls* in which deflection and strength are controlled by flexure. These are usually high rise shear walls used to resist high winds and cyclones.

(c) *Ductile-moment shear walls* are special walls meant for seismic regions and which have good energy dissipation characteristics under reversed cyclic loads.

Because of the vastness of the subject we confine ourselves to the study of the general aspects of shear walls and details of design of the simple free-standing shear walls (simple rectangle and bar bell type) only.

19.4 LOADS IN SHEAR WALLS

Detailed analysis to determine the forces in these various types of shear walls of a structural system is a special subject in structural analysis and publications that deal with this topic are listed at the end of this chapter [2,3,4,5]. Computer programs for shear-wall analysis are also available.

19.4.1 Centre of Rigidity and Centre of Mass

Lateral stiffness (K) of a shear wall is defined as the force required to be applied at its top in order to displace it by one unit. *Centre of rigidity stiffness* is defined as the point on the horizontal plane through which the lateral load should pass in order that there will be no rigid body rotation. In other words, it is the point through which the resultant of the restoring forces of a system acts. Its location is given by the following expression. Taking K_i as stiffness of individual member and x, y as the distance of the member from a reference point (refer Section 18.9).

$$x_r = \frac{\sum K_i x_i}{\sum K_i} \quad \text{and} \quad y_r = \frac{\sum K_i y_i}{\sum K_i} \tag{19.2}$$

As earthquake-induced lateral forces are proportional to the mass, the resultant force due to earthquake passes through the centre of mass of the floor. This corresponds to the centre of gravity and is given by the equation:

$$x = \frac{\sum m_i x_i}{\sum m_i} \quad \text{and} \quad y = \frac{\sum m_i y_i}{\sum m_i} \tag{19.3}$$

In a building where the centre of stiffness and centre of mass coincide, there is no torsion, and the earthquake loads are taken by the shear walls in proportion to their stiffness. In buildings where they do not coincide, a twisting moment is produced and thus torsion is also analyzed. For wind load, the resultant load acts on the centre of area and it should coincide with the centre of rigidity to avoid torsion.

19.4.2 Principle of Shear Wall Analysis

Analysis of shear walls becomes simple if we assume that all the horizontal loads are taken by the various shear walls without any participation from the frames. In a system where there is no torsion there are a number of shear walls, the load is taken by each wall in proportion to its stiffness as shown below. Representing EL as earthquake load if F_1, F_2, F_3, etc., are the forces on the various shear walls then

$$EL = F_1 + F_2 + F_3,$$

If d is the displacement at the top in the system then

$$F_1 = K_1 \Delta$$

where K_1 is the lateral stiffness. Hence

$$K_1 \Delta + K_2 \Delta + K_3 \Delta + \cdots = EL$$

$$\Delta = \frac{EL}{K_1 + K_2 + K_3 + \cdots}$$

$$F_1 = K_1 \Delta = \frac{K_1}{\Sigma K_1} EL$$
(19.4)

However, if the shear walls are acting along with moment resisting frames, we have to consider the deformations of the frames of the wall both acting together (Fig. 19.3). As already stated, exact solution with computer softwares or approximate solution by use of charts and tables are available for the analysis of various cases of shear walls and they should be used for such analysis.

19.4.3 Stiffnesses of Walls

Let Δ represents deflection of the wall at its top. There are three types of deflection to be considered:

$$\text{Stiffness} = \text{Force required at top for unit deflection} = \frac{W}{\Delta}$$

$$\Delta_1 \text{ bending} = \frac{WH^3}{3EI} \text{ (as a cantilever)}$$

$$\Delta_2 \text{ shear} = \frac{WH}{CAG}$$

where

C = Shape factor (0.8 for rectangle)
W = The load applied

$$G = \frac{E}{2(1+\mu)} \quad \text{(assume } \mu = 0.22\text{)}$$

Let Δ_3 is the part due to foundation rocking and this can be calculated as shown below by assuming the Wrinkler model for foundation (Fig. 19.4). Let modulus of sub-grade reaction be γ (kN/m^3). M due to rotation θ of foundation is given by the following equation:

$$M = \int_{-L_1/2}^{L_1/2} \gamma B x \theta x dx = B\gamma\theta \int_{-L_1/2}^{L_1/2} x^2 dx = \frac{1}{12} BL^3 \gamma \theta$$

Let $R = \gamma BL^3/12$ is the moment to produce unit rotation of foundation. Therefore,

$$\text{Rotation due to moment } WH = \frac{WH}{R}$$

Hence

$$\text{Deflection produced} = (\text{Rotation}) \times (H)$$

$$\Delta_3 \text{ rocking} = \frac{WH^2}{R}$$

Hence

$$\text{Total } \Delta = \Delta_1 + \Delta_2 + \Delta_3$$

Therefore,

$$\text{Lateral stiffness } K = \frac{W}{\Delta} \qquad (19.5)$$

In calculating the stiffness of the wall (i.e. load for unit deflection) all these factors (bending, shear and foundation—rotation) should be taken into account.

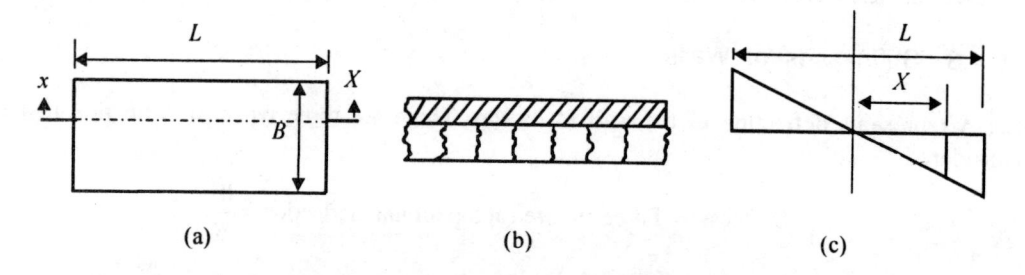

Fig. 19.4 Winkler model for foundation rocking: (a) foundation in plan (b) section along *XX* showing springs (c) rotation due to horizontal forces.

19.5 DESIGN OF RECTANGULAR AND FLANGED SHEAR WALLS

IS 13920 (1993) Clause 9 [7] deals with requirements and design of simple free standings shear walls. Only the main considerations are given below and the code itself should be consulted for fuller details. We should remember the importance of locating shear walls at locations and also directions which are most effective.

19.5.1 General Dimensions

The following factors determine the general dimension [8] of the walls.

1. The thickness of the wall (t) should not be less than 150 mm.
2. If it is a flanged wall, the effective extension of the flange width beyond the face of web to be considered in design, is to be less of the following (Fig. 19.5).

(a) 1/2 distance to an adjacent shear wall web
(b) 1/10th of the total wall height
(c) actual width

3. Where the extreme fibre compressive stresses in the wall due to all loads (the gravity loads and the lateral forces) exceed $0.2f_{ck}$ boundary elements are to be provided along the vertical boundaries of the walls. Boundary elements are portions along the wall edges specially enlarged and strengthened by longitudinal and transverse steel as in columns. These elements can be discontinued when the compressive stresses are less than $0.15f_{ck}$. Boundary elements are also not required if the entire wall section is provided with special confining steel reinforcements.

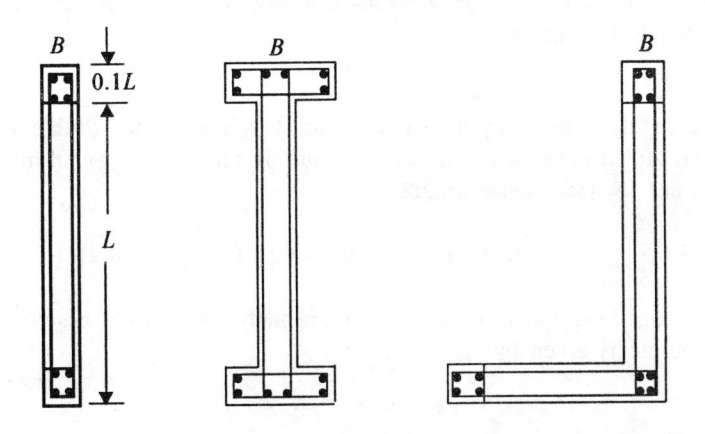

Fig. 19.5 Plain shear walls with boundary elements.

19.5.2 Reinforcements

The following rules are to be observed for detailing of steel:

1. Walls are to be provided with reinforcement in two orthogonal directions in the plane of the wall. The minimum steel ratios for each of the vertical and horizontal directions should be 0.0025.

$$\frac{A_s}{A_c(\text{gross})} = \rho \geq 0.0025 \tag{19.6}$$

This reinforcement is distributed uniformly in the wall.

2. If the factored shear stress (v) exceeds $0.25\sqrt{f_{ck}}$ or if the thickness of wall exceeds 200 mm the bars should be provided as two mats in the plane of the wall one on each face. (This adds to ductility of wall by reducing the fragmentation and premature deterioration on reversal of loading).

3. The diameter of the bars should not exceed 1/10th of the thickness of the part of the wall.

4. The maximum spacing should not exceed, $L/5$, $3t$ or 450 mm, where L is the length of the wall.

19.5.3 Reinforcements for Shear

The nominal shear is calculated by the formula

$$v = \frac{V_u}{td}$$

where

d = Effective width (= 0.8 for rectangular sections)

V_u = The factored shear

The nominal shear should not exceed the maximum allowable beam shear given in IS 456 (2000) Table 20 or the following value.

$$\tau_{c\max} = 0.63\sqrt{f_{ck}} \qquad (19.7)$$

The shear taken by concrete is given by the same value as in beam shear. (Table 9 of IS 456 assuming 0.25% steel) and if necessary increase its value by the following multiplying factor due to axial load as per IS 456 Clause 40.2.2.

$$\delta = 1 + \frac{3P_u}{A_c f_{ck}} \qquad \text{[but not more than 1.5] [see Fig. 26.4]} \qquad (19.8)$$

where P_u is the total axial load and δ is the multiplying factor. Let this stress be τ_c then the shear capacity of concrete is given by

$$V_c = \tau_c t d$$

with

$$V_s = V_u - V_c \qquad (19.9)$$

The steel necessary to resist the shear is determined from the formula

$$V_s = \frac{0.87 f_y A_s d}{S_v}$$

$$\frac{V_s}{d} = \frac{0.87 f_y A_s}{S_v} \qquad (19.10)$$

Table 62 of SP 16 can be used for determining the diameter of shear steel and its spacing.

(5) The vertical steel provided in the wall for shear should not be less than the horizontal steel.

19.5.4 Adequacy of Boundary Elements

The boundary elements should be able to carry all the vertical loads. The boundary elements when provided (as indicated in Section 19.5.1) will have greater thickness than the web. The maximum axial load on the boundary elements due to the effects of vertical load and moments, is

$$P = \text{Sum of factored gravity loads} + \frac{M_u - M_{uv}}{c} \qquad (19.11)$$

where

M_u = Factored moment on the whole wall

M_{uv} = Moment of resistance provided by the rectangular part with the distributed vertical steel across this wall section only, i.e. excluding the boundary elements

c = Centre-to-centre distance between boundary elements

If the gravity loads tend to add to strength of wall the load factor for this load is taken as only 0.8. The boundary elements are designed as a column with vertical steel not less than 0.8% and preferably not greater than 4%. These elements should be provided with special confining steel throughout their height. Detailing of bar bells of these walls is shown in Fig. 19.6.

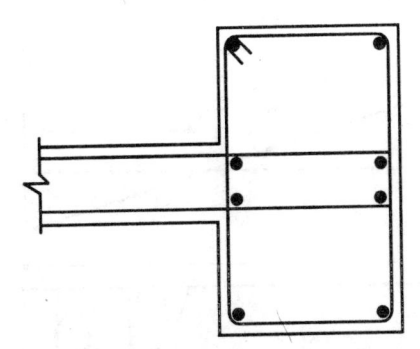

Fig. 19.6 Detailing of barbells in shear walls.

19.5.5 Flexural Strength

The wall section should be safe under the action of combined bending and axial load. This can be determined by interaction diagram. The formula developed in Section 19.6 for rectangular sections may also be used to find the interaction diagram of a wall of uniform section.

The flexural strength assuming a cracked section should be greater than its strength as an un-cracked section to prevent brittle failure. This applies specially to wall sections which, for architectural reasons, are made much larger than that required for strength consideration only.

19.5.6 Required Development Splice and Anchorage

Horizontal steel which acts as web steel for shear should be well anchored near edges of wall or confined to the core of the boundary elements.

Splicing of vertical flexural steel should be avoided as far as possible in regions of flexural yielding which can be taken to extend for a distance of the length of the wall (L) above the base of the wall or one-sixth the height of the wall. If splicing is needed, not more than one-third of the steel should be spliced at such a section and the splicing of adjacent bars should be staggered a minimum of 600 mm.

All continuous steel in shear walls are anchored or spliced as per provision of tension bars. Lateral ties are provided in lapped splices of diameter larger than 16 mm diameter, the diameter of these ties being not less than 1/4 diameter of the bar or 6 mm. Their spacing should not be more than 150 mm.

19.6 DERIVATION OF FORMULA FOR MOMENT OF RESISTANCE OF RECTANGULAR SHEAR WALLS

IS 13920 (1993) Appendix A gives the expressions for moment of resistances of slender rectangular walls with uniform distribution of steel and subjected to axial load and moments. Two formulae depending on whether the section fails in flexural tension or flexural compression are given. These can be derived as follows: Let the strain and stress diagrams are as shown in Fig. 19.7.

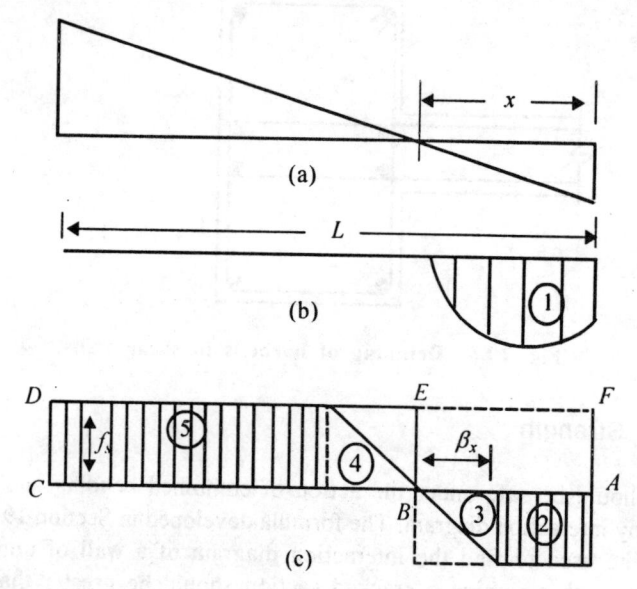

Fig. 19.7 Ultimate flexural strength of rectangular plane shear walls: (a) strain diagram (b) compression in concrete (c) stresses in steel uniformly distributed in the section.

Let

P_w = Axial compression on wall assumed to act at centre of wall.

A_v = Area of vertical steel.

ρ = Vertical steel ratio = $A_v/(tL)$

$\lambda = P_w/f_{ck}tL$

$\phi = 0.87 f_y \rho/f_{ck} = \rho f_s/f_{ck}$

M_u = Moment of resistance of the wall which can be obtained from the strain diagram. For balanced failure we get the following condition:

DESIGN OF SHEAR WALLS 349

$$\frac{\bar{x}}{L} = \frac{0.0035}{0.0035 + 0.87 f_y/E_s} \qquad (19.12)$$

This ratio gives the value of $\bar{x}$ or balanced failure and works out to 0.66 for Fe 415 steel. The forces acting on the section at failure of the section are shown in Fig. 19.7. The magnitude of these forces and their distances from the extreme tension fibre can be written as in Table 19.1:

TABLE 19.1 MAGNITUDE OF THE FORCES ACTING ON THE SECTION AND DISTANCE FROM TENSION FIBRE

	Force	Distance from tension fibre	Part in Fig. 19.7
A.	Compression in concrete $0.36 f_{ck} xt$	$L - 0.416x$	1
B.	Compression in steel (taking steel stress as f_s)		
	(i) $\rho x t (1 - \beta)(t)$	$L - \frac{1}{2}x(1-\beta)$	2
	(ii) $\frac{1}{2} f_s \rho x \beta t$	$L - x + \frac{2}{3}\beta x$	3
C.	Tension in steel		
	(i) $\frac{1}{2} f_s \rho x \beta t$	$L - x - \beta x$	4
	(ii) $f_s \rho (L - x - \beta x) t$	$\frac{1}{2}[L - x - \beta x]$	5
D.	External load on wall P_w	$\frac{1}{2}L$	

The conditions to be satisfied are: $P_w = C - T$ and Moment of forces about mid-point $= M_u$. Taking the first condition, denoting areas by numbers, x/L can be determined as under:

$$P_w = 1 + 2 + 3 - 4 - 5 \text{ (as 3 and 4 are equal)}$$

$$= 1 + (2 - 5)$$

$$= 1 - ACDF + 2ABEF \qquad \text{[from Fig. 19.7]}$$

$$= 0.36 f_{ck} xt - f_s Lt\rho + 2\rho f_s tx$$

$$\frac{P_w}{f_{ck} tL} = 0.36 \frac{x}{L} - \rho \frac{f_s}{f_{ck}} + 2\rho \frac{f_s}{f_{ck}} \frac{x}{L}$$

Taking

$$\frac{P_w}{f_{ck}tL} = \lambda \quad \text{and} \quad \rho\frac{f_s}{f_{ck}} = \phi = \rho\frac{0.87f_y}{f_{ck}}$$

We get

$$\lambda + \phi = \frac{x}{L}(0.36 + 2\phi)$$

$$\frac{x}{L} = \frac{\lambda + \phi}{0.36 + 2\phi} \tag{19.13}$$

Comparing the values obtained for $\bar{x}/L$ from Eq. (19.12) and x/L from Eq. (19.13) we get two cases:

Case 1. If $\dfrac{x}{L} < \dfrac{\bar{x}}{L}$, then tension failure occurs and the value of the moment M_u (the flexural strength) is obtained by taking moment of forces about the mid point. The expression as given in IS code.

$$\frac{M_u}{f_{ck}tL^2} = \phi\left(1+\frac{\lambda}{\phi}\right)\left(\frac{1}{2}-0.42\frac{x}{L}\right) - \left(\frac{x}{L}\right)^2\left(0.168+\frac{\beta^2}{3}\right) \tag{19.14}$$

$$\beta = \frac{0.87f_y}{0.0035E_s} = 0.516 \text{ (for Fe 415 steel)}$$

Taking the particular case of x/L less than 0.5 and neglecting small quantities, we get

$$\frac{M_u}{f_{ck}tL^2} = \phi\left(1+\frac{\lambda}{\phi}\right)\left(\frac{1}{2}-0.42\frac{x}{L}\right) \tag{19.15}$$

Case 2. If $\dfrac{x}{L} > \dfrac{\bar{x}}{L}$, we have compression failure assuming N.A. within section, i.e. $x/L < 1.0$. The value of M_u is given by the following expression:

$$\frac{M_u}{f_{ck}tL} = \alpha_1\frac{x}{L} - \alpha_2\frac{x}{L} - \alpha_3 - \frac{\lambda}{2} \tag{19.16}$$

where

$$\alpha_1 = 0.36 + \phi\left(1 - \frac{\beta}{2} - \frac{1}{2\beta}\right)$$

$$\alpha_2 = 0.15 + \frac{\phi}{2}\left(1 - \beta - \frac{\beta^2}{2} - \frac{1}{3\beta}\right)$$

$$\alpha_3 = \frac{\phi}{6\beta}\left(\frac{L}{x} - 3\right)$$

For a given value of L, t, ϕ, f_{ck}, f_y by varying x/L the values of P_u and M_u can be obtained and can be used for plotting the interaction curve for the wall.

EXAMPLE 19.1 (**Design of simple shear wall with enlarged ends**)
Design a shear wall of length 4.16 m and thickness 250 mm is subject to the following forces (see Fig. 19.8) Assume F_{ck} = 25 and f_y = 415 N/mm² and the wall is a high wall with the following loadings:

Loading	Axial force (kN)	Moment (kNm)	Shear kN
1. DL + LL	1950	600	20
2. Seismic load	250	4800	700

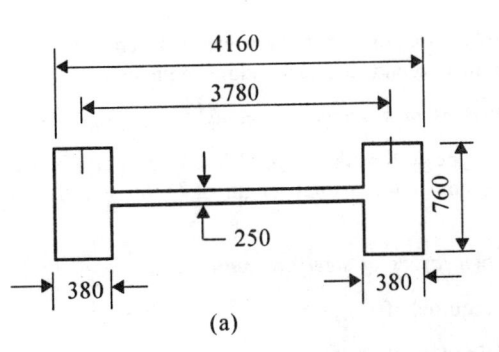

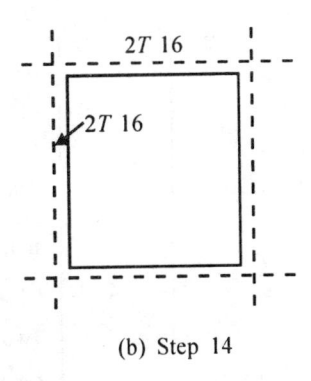

Fig. 19.8 Example 19.1.

Reference	Step	Calculations
	1	*Determine design loads*
IS 456		(A load factor of 0.8, enstead of IS value of 0.9, and 1.2 to be applied to gravity loads depending on whether it assists or opposes stability)
Table 18		
		P_1 (case 1) = (0.8 × 1950) + (1.2 × 250) = 1,860 kN
		P_2 (case 2) = 1.2(1950 + 250) = 2640 kN
		Moment = 1.2(4800 + 600) = 6480 kNm
		Shear = 1.2(700 + 20) = 864 kN
	2	*Check whether boundary elements are required*
		(Extreme stresses are more than 4 N/mm² boundary elements are to be provided)
Sec. 19.5.1		Assuming uniform thickness L = 4160; t = 250

Reference	Step	Calculations
		$$I = \frac{bd^3}{12} = \frac{250 \times 4160^3}{12} = 1.5 \times 10^{12} \text{ mm}^4$$ $$A = bd = 4160 \times 250 = 1.04 \times 10^6 \text{ mm}^2$$ $$f_c = \frac{P}{A} \pm \frac{M}{I} y$$ $$= \frac{(2.64)10^6}{(1.04)10^6} \pm \frac{(6.48)10^9}{(1.5)10^{12}} \frac{4160}{2}$$ $$= 11.52 \text{ and } -6.45 \text{ N/mm}^2$$ $0.2 f_{ck} = 0.2 \times 25 = 5 \text{ N/mm}^2$ As extreme stresses are high, boundary elements are needed. Also there is tension in one end due to bending moment.
Sec. 19.5.1	3	*Adopt the dimensions of boundary elements* Adopt a bar bell type wall with a central 3400 mm portion and two ends 380 × 760 mm giving a total length of (3400 + 380) = 4160 mm.
Sec. 19.5.2	4	*Check whether two layers of steel are required* Two layers are required if (a) Shear stress is more than $0.25 \sqrt{f_{ck}}$ (b) The thickness of section is more than 200 mm (a) Depth of section = Centre to centre boundary elements = 3400 + 380 = 3780 mm $$v = \frac{V}{bd} = \frac{864 \times 10^3}{250 \times 3780} = 0.92 \text{ N/mm}^2$$ $0.25\sqrt{f_{ck}} = 0.25\sqrt{20} = 1.11 \text{ N/mm}^2$ (b) Thickness of 250 is more than 200 mm Use two layers of steel with suitable cover.
	5	*Determine steel required* Let us put minimum required steel and check the safety of wall ($p = 0.0025$). $A_s(\text{min}) = 0.0025 \times 250 \times 1000$ per metre length)
Sec. 19.5.2		$= 625 \text{ mm}^2$ is two-layers

DESIGN OF SHEAR WALLS 353

Reference	Step	Calculations
		Provide T10 @ 250 (314 mm² per metre length)
		Spacing of 250 is less than 450 (max.) allowed.
		Provide the same vertical and horizontal steel.
	6	*Calculate V_s to be taken by steel*
		$v = 0.92$ N/mm²
Step 4		τ_c (for 0.25% steel and $f_{ck} = 20$) = 0.36 N/mm²
		Maximum shear allowed = 3.1 N/mm²
		Designed steel is necessary for V_s
IS 415		$V_s = (0.92 - 0.36)bd$
Table 9		$\quad = 0.56 \times 250 \times 3780 = 529,200$ N
		$\quad = 529.2$ kN
	7	*Calculate steel necessary to take V_s*
		As the wall is high horizontal steel is more effective. Therefore,
		$\dfrac{V_s}{d} = 0.87 f_y \dfrac{A_{sv}}{s_v}$
		$d = 3780$
		Required $\dfrac{A_{sv}}{s_v} = \dfrac{(529.2)(10^3)}{(0.87)(415)(3780)} = 0.388$
		Consider 1 m height = s_v
		Horizontal steel area = 628 mm² = A_{sv}
		Available $\dfrac{A_{sv}}{s_v} = \dfrac{628}{1000} = 0.628$
		The nominal steel provided will satisfy shear requirements.
	7(a)	*Find flexural strength of web part of wall*
		Vertical load on wall (with factor 0.8 on DL)
		$P = 0.8(1950) + 1.2(250) = 1860$ kN
		Assuming it as a UDL over the area the axial load for the central part beams
		beams = P_w
		$P_w = 1860 \dfrac{3400 \times 250}{(3400 \times 250) + 2(380 \times 760)}$
		$\quad = 1860 \times 0.595 = 1107$ kN

354   ADVANCED REINFORCED CONCRETE DESIGN

Reference	Step	Calculations
Sec. 19.6	8	Calculate the parameters λ, ϕ and x/L
		$\lambda = \dfrac{P_w}{f_{ck} tL} = \dfrac{1107 \times 10^3}{20 \times 250 \times 3400} = 0.065$
		$\phi = \dfrac{0.87 f_y}{f_{ck}} = \dfrac{(0.87)(415)(0.0025)}{20} = 0.045$
		$\beta = 0.516$
Eq. (19.3)		$\dfrac{x}{L} = \dfrac{\lambda + \phi}{0.36 + 2\phi}$
		$= \dfrac{0.065 + 0.045}{0.36 + 0.09} = 0.24$
		$\dfrac{x}{L}$ is less than 0.5.
		$\dfrac{M_u}{f_{ck} tL^2} = \phi\left(1 + \dfrac{\lambda}{\phi}\right)\left(0.5 - 0.42\dfrac{x}{L}\right)$
		$= 0.045\left(1 + \dfrac{0.065}{0.045}\right)[0.5 - (0.42)(0.31)]$
		$= 0.041$
Eq. (19.15)		$M_u = (0.041)(20)(250)(3400)^2 = 2370$ kNm
		This is less than 6480 required.
	9	Calculate moment to be carried by the boundary elements
		$M_1 = 6480 - 2370 = 4110$ kNm
	10	Calculate compression and tension in the boundary elements due to M_1
		Distance between boundary elements = 3480 + 380
Eq. 19.11		= 3.86 m (=c)
		Axial load $= \dfrac{M_1}{c} = \dfrac{4110}{3.86} = 1065$ kN
		This load acts as tension at one end and compression at the other end.
	11	Calculate compression due to the axial loads at these ends
Step 7(a)		Fraction of area at each end $= \dfrac{1 - 0.595}{2} = 0.2025$
		Factored compression at compression end. Taking worst case $P_2 = (0.2025)(2640) = 535$ kN

Reference	Step	Calculations
Step 1		Factored compression at tension end (taking P_1)
		= (0.2025) (1860) = 377 kN
		Compression at compression end = 1065 + 535 = 1600 kN
		Tension at the tension end = –1065 + 377 = –688 kN
	12	*Design the boundary elements compression*
		(a) Design one end as column with details as in Fig. 19.10
		(b) Check laterals for confinement
		(c) Check for anchorage and splice length
	13	*Design tension side of shear wall*
		Provide the same steel also on the tension end as in compression end. Check also for the tension (earthquake forces can act in both directions).
	14	*Design the reinforcement around openings, if any, of the wall*
		Openings are provided in the main body of the wall. Assume opening size as 1200 × 1200.
		Area of reinforcement cut off by the opening
		= 1200 (thickness) $\dfrac{\% \text{ steel}}{100}$ = 1200 × 250 × 0.0025 = 750 mm^2
		4 nos. 16 mm bar area = 804 mm^2
Fig. 19.8(b)		Provide 2 nos 16 mm, one on each face of the wall, on all the sides of the hole to compensate for the steel cut off by the hole.

EXAMPLE 19.2 (Lateral stiffness of shear walls)

A bar bell type shear wall with central part 3600 × 150 mm and two 400 × 400 mm strong bands at each ends is supported on a footing 8 m × 4 m which rests on soil whose modules is 30,000 kN/m^3 (Fig. 19.9). Determine the lateral stiffness of the wall. Assume f_{ck} = 20 and height of the wall is 14 m.

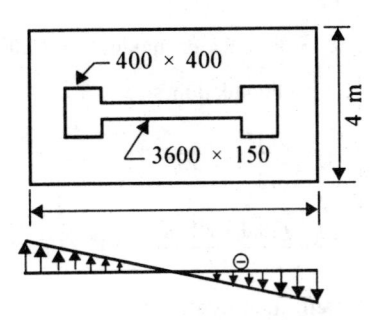

Fig. 19.9 Example 19.2.

Reference	Step	Calculations
Sec. 19.4.3	1	*Calculate deflection of wall due to unit load at top (bending deflection)* $$\Delta_1 = \frac{WH^3}{3EI}$$ $$E_c = 5700\sqrt{f_{ck}} = 25.5 \text{ kN/mm}^2$$ $$I = \frac{150(3600)^3}{12} + 2\left[\frac{400(400)^3}{12} + (400)^2(2000)^2\right]$$ $$= 1.87 \times 10^{12} \text{ mm}^4$$ Deflection due to unit load $W = 1$ kN $$\Delta_{I-1} = \frac{(14000)^3}{(3)(25.5 \times 10^3)(187 \times 10^{12})} = 1.92 \times 10^{-7} \text{ mm/N}$$
	2	*Calculate deflection due to shear* $$\Delta_2 = \frac{WH}{CAG}$$ where C = Constant (=0.8 for rectangle). $$A = (150 \times 3600) + (2 \times 400 \times 400) = 8.6 \times 10^5 \text{ mm}^2$$ $$G = \frac{E}{2(1+0.22)} = \frac{25.5}{2(1+0.22)} = 10.45 \text{ kN/mm}^2$$ Deflection for unit load $\Delta_{I-2} = \dfrac{H}{0.8AG}$ $$= \frac{14000}{(0.8)(8.6 \times 10^5)(10.5 \times 10^3)}$$ $$= 0.194 \times 10^{-5} \text{ mm/n}$$
Sec. 19.4.3	3	*Calculate rocking of foundation due to unit load at top* Let λ = Foundation modules = 30,000 kN/m^3 Foundation size = 8 m × 4 m $$M = \frac{\gamma BL^3}{12}\theta = R\theta$$ $$R = \frac{(30,000)(4)(8)^3}{12} = 512 \times 10^4 \text{ kNm/radian}$$
Sec. 19.4.3		Deflection due to $W = 1$

Reference	Step	Calculations
	4	$\Delta_{I-3} = \dfrac{H^2}{R} = \dfrac{(14)^2}{(512)(10^4)} = 3.83 \times 10^{-5}$ mm/N Calculate lateral stiffness of the wall $\text{Stiffness} = \dfrac{\text{Load}}{\text{Deflection}}$ Total deflection due to unit load $\Delta_I = \Delta_{I-1} + \Delta_{I-2} + \Delta_{I-3}$ $= (1.92 + 0.19 + 3.83) \times 10^{-5} = 5.94 \times 10^{-5}$ mm Stiffness $\dfrac{1}{\Delta} = \dfrac{10^5}{5.95} = 16.8$ kN/mm

REFERENCES

1. Varghese, P.C., *Limit State Design of Reinforced Concrete*, Prentice-Hall of India, New Delhi, 1994.

2. Kazimi, S.M.A. and Chandra, R., *Analysis of Shear Walls Buildings*, Torsteel Research Foundation in India, Calcutta, 1980.

3. Bryan Stafford and Alex Coull, 'Tall building structures', *Analysis and Design*, John Wiley and Sons, New York, 1991.

4. Parme, L. *Design of Combined Frames and Shear Walls*, Portland Cement Association, Illinois, U.S.A., 1967.

5. Iain, A., Macleod, *Shear Wall Frame Interaction—A Design Aid*, Portland Association, Illinois, U.S.A., 1970.

6. Ladislav Cerny and Roberto Lean, 'Column supported shear walls', *Concrete Framed Structures—Stability and Strength*, R. Narayanan (Ed.), Elsevier Applied Science, London, 1986.

7. IS 13920 (1993), Ductile Detailing of Reinforced Concrete Structures Subjected to Seismic Forces, New Delhi.

8. Notes on ACI 318 (1989) with Design Applications, Portland Cement Association, Illinois, 1990.

CHAPTER 20

Design of Cast *In Situ* Beams—Column Joints

20.1 INTRODUCTION

Performance of beam–column joints becomes critical in framed buildings which are subjected to horizontal forces like wind and earthquake. The awareness for proper design of joints started with the publication of the *ACI–ASCE Committee 352 Report* in 1976 titled 'Recommendations for design of beam column joints'. Among the Indian codes of practice, IS code 13920 on ductile detailing of reinforced concrete structures subjected to seismic forces [1] deals briefly with the subject. This chapter describes briefly the fundamental concepts in design of beam–column joints.

20.2 TYPES OF CAST *IN SITU* JOINTS

ACI–ASCE Committee 352 Report divides beam–column joints into two types: Type 1 joints are those subjected to gravity and horizontal loads without any significant inelastic joint deterioration. Whereas type 2 joints are subjected to repeated load reversals (as those due to earthquakes) which cause inelastic deformations. In this chapter, we shall deal with the strength aspects common to both these types of joints but will not deal in detail with type 2 joints subjected to large cyclic loads with inelastic action and dissipation of energy. Information regarding the latter can be found in literature dealing with earthquake engineering.

20.3 JOINTS IN MULTI-STOREYED BUILDINGS

The four different combinations of beam and column joints that can occur in buildings as shown in Fig. 20.1 are the following:

1. Top corner joint where two members meet
2. Exterior joint where three members meet
3. Top interior column joints where three members meet
4. Internal joints where four members meet.

DESIGN OF CAST *IN SITU* BEAMS—COLUMN JOINTS 359

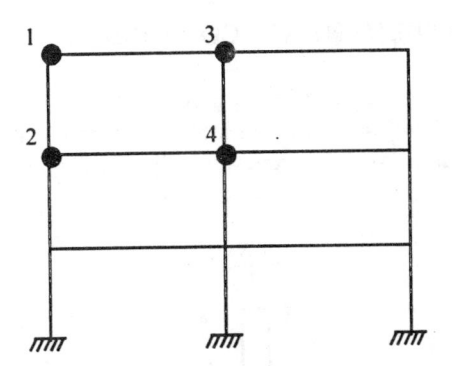

Fig. 20.1 Joints in a building frame: 1. corner joint 2. exterior joint 3. top interior joints 4. interior joints.

20.4 FORCES ACTING ON JOINTS

Moments, shears and direct loads can act on joints discussed in Section 20.3. Reversal of loads can reverse the sign of moments in the beams as shown in Fig. 20.2. Thus in an internal joint, the two beams that frame into the column can have moments acting in the same direction or opposite directions: one due to the gravity loads and the other due to wind or earthquake loads.

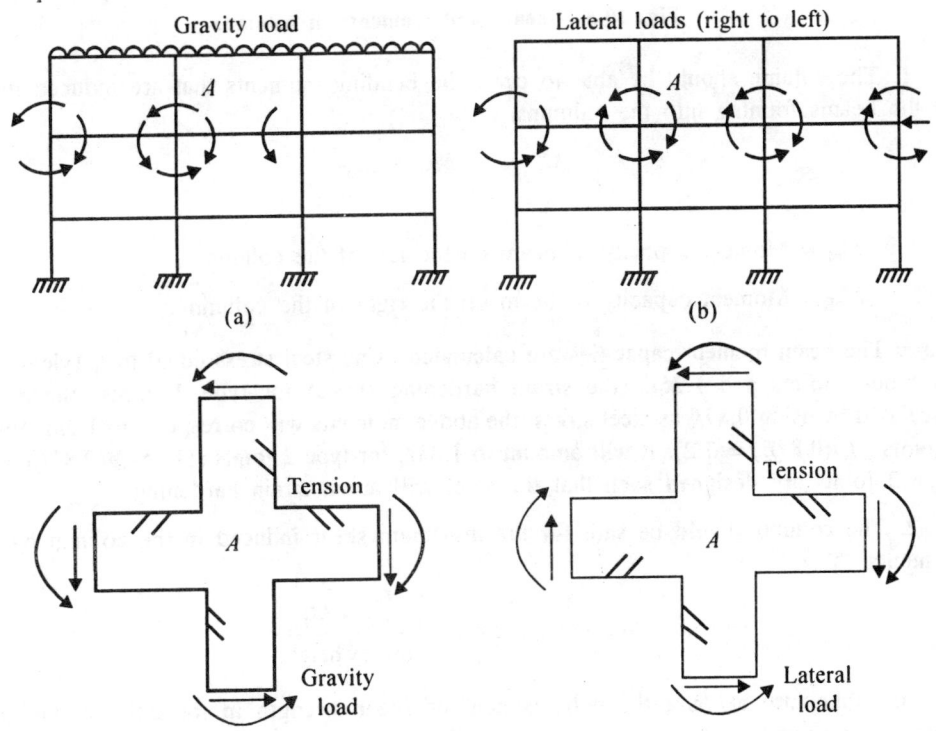

Fig. 20.2 Moments at joints due to gravity loads and lateral loads acting from right to left. (Moments due to lateral loads change in direction when lateral loads are from left to right.)

20.5 STRENGTH REQUIREMENT OF COLUMNS

One of the most important requirements of column design is that under no circumstances failure (either in bending or shear) should take place in columns (Fig. 20.3). This is sometimes referred to as the 'strong column–weak beam design'. The following requirements are necessary for this purpose:

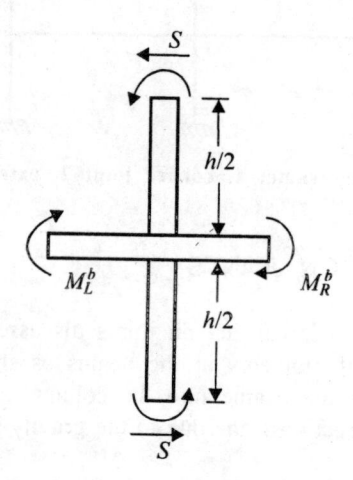

Fig. 20.3 Shears and moments in column.

1. The column should be able to carry the bending moments that are induced in them by the beams framing into the columns.

$$M_{u \text{ (col)}} = M_{bL} + M_{bR} \qquad (20.1)$$

where

M_{bL} = Moment capacity of beam on the left of the column

M_{bR} = Moment capacity of beam on the right of the column

[*Note:* The beam moment capacities are calculated using steel stress equal to f_y (yield stress) for type 1 joints and $1.25f_y$ (the strain hardening stress) for type 2 joints. Since M_u is calculated by using $0.87f_y$ as steel stress, the above moments will correspond to $1.2M_u$ for type 1 joints [$f_y/(0.87f_y) = 1.2$]. It will amount to $1.4M_u$ for type 2 joints [$(1.25f_y)/(0.87f_y) = 1.4$]. Type 2 joints are designed such that the steel will attain strain hardening.

2. The column should be safe for the maximum shear induced in the column by these moments.

$$\text{Max column shear} = \frac{M_{bL} + M_{bR}}{\text{Storey height}} \qquad (20.2)$$

We may take into account the enhancement of shear strength in the column due to the presence of compressive force in the column by using the formula given in IS 456 Clause 40.2.2 [see Fig. 26.4].

$$\tau_c' = \tau_c \delta \quad \text{and} \quad \delta = 1 + \frac{3P_u}{A_g f_{ck}} \not> 1.5$$

where

P_u = Axial compressive force in neutons

A_g = Gross area of concrete section in mm^2

The ACI formula for increased shear due to compression is given in ACI 318 Eq. (11.4) and with a reduction factor of 0.85. It approximates to

$$\tau_c' = 0.13\sqrt{f_{ck}}\left(1 + 0.07\frac{P_u}{A_g}\right) \tag{20.3}$$

(*Note:* These values are much larger than IS 456 values.)

20.6 FORCES DIRECTLY ACTING ON JOINTS

Figure 20.4 gives the forces that can be assumed to act on a joint. There should be sufficient anchorage of steel to meet these forces. The moments from the beams on both sides of the column can be equal or opposite in sign under gravity and wind loads. We have already seen that the moments considered at the failure condition should be due to the yield stress f_y for

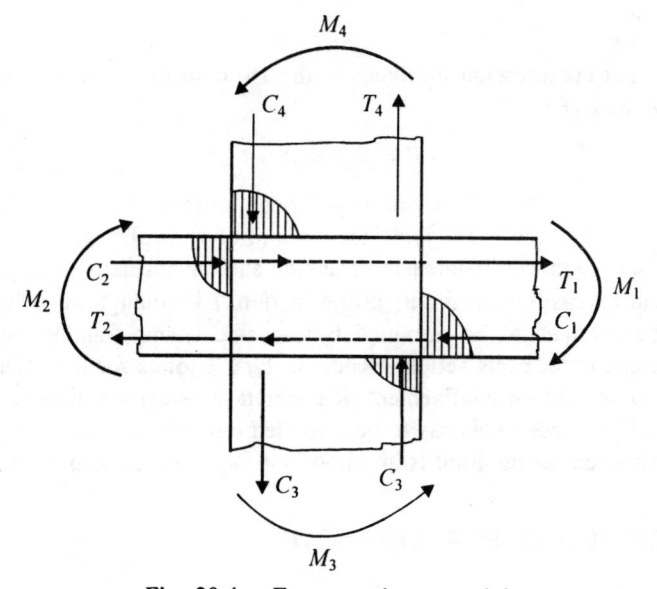

Fig. 20.4 Forces acting on a joint.

class 1 joints and that due to strain hardening stress ($1.25f_y$) for class 2 joints. The resultant shear on the joint as seen in Fig. 20.4 will be the following:

$$V_u = T_1 + C_2 - V_{\text{col}} \tag{20.4}$$

Also

$$T_1 = C_1 \quad \text{and} \quad T_2 = C_2$$

In type 1 joints, there may be no joint degradations but in type 2 joints after a large number of reversals the joint resistance has been found more by the formation of a mechanism as shown in Fig. 20.5, inside the joint. After the elastic action, we can get the diagonal compression mechanism, i.e. the strut mechanism to carry the loads. Further deterioration of the joint can lead to a truss mechanism for equilibrium of forces [2].

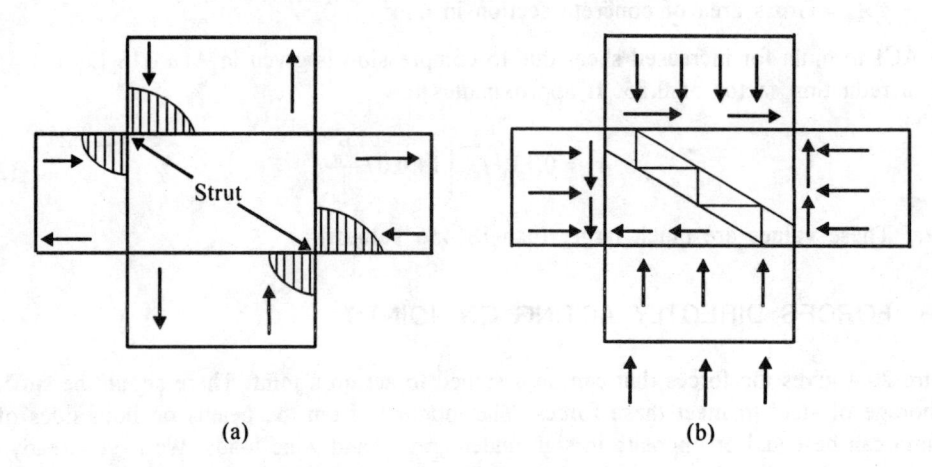

Fig. 20.5 Strut and truss mechanism resisting forces in joints: (a) strut mechanism (b) truss mechanism.

In the strut or diagonal compression mechanism, the shear in the joint will be a component of the strut compression (P).

$$V = P \cos \alpha \tag{20.5}$$

The cross-sectional area of the effective strut will depend on factors such as joint confinement, compression that act on the core of the joint, the amount of vertical steel and the column size. In the truss mechanism, there will be intensive cracking and the absolute necessity of horizontal and vertical steel can be easily seen for its proper action. Eventhough initially the horizontal confinement may be assumed to be provided by the side beams framing into the column, reversal of the load can reduce this action. Hence, in type 2 joints for the safety of the joints it is more sensible to depend on confinement of concrete by steel specially provided for this purpose. In type 1 joints, these steels has to be provided only if there are no beams to confine the joint. Thus confinement of the joint is an important factor for good performance of joints.

20.7 DESIGN OF JOINTS FOR STRENGTH

As seen from the above discussion, three main factors considered in design of joints, in addition to the strength requirement of column already discussed in Section 20.4 are the following:

1. Anchorage of main reinforcement of beams
2. Confinement of the core of the joint
3. Shear strength of the joint

DESIGN OF CAST *IN SITU* BEAMS—COLUMN JOINTS   363

To this we may add ease of construction as another necessary requirement [2]. Let us now consider each of the three requirements in more detail.

20.8 ANCHORAGE

Firstly, we shall consider the anchorage requirements given in ACI and IS codes. These are based on different concepts in the two codes.

20.8.1 Anchorage Requirements in ACI 318

The requirements specified for anchorage in ACI 318 and IS 456 are not the same. The method of providing the development length as shown in Fig. 20.6 in Indian practice and described in IS 13920 Clause 6.25 is not accepted by the ACI code. The ACI requirements for the development length can be summarised as follows:

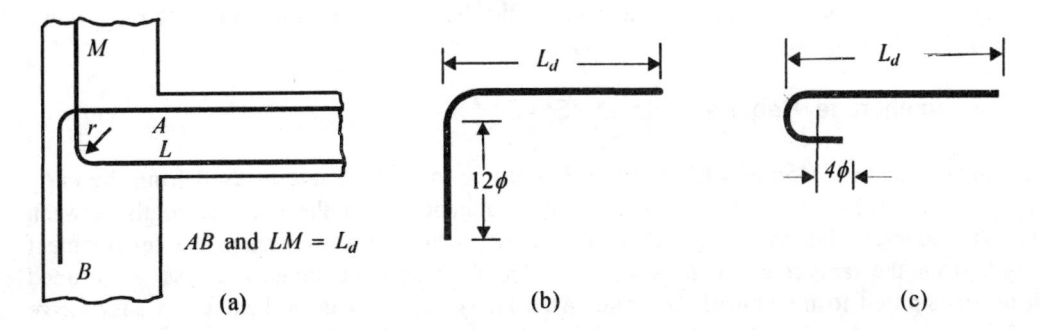

Fig. 20.6 Development lengths: (a) according to IS 456 (b) ACI standard bend (c) ACI standard hook.

1. The basic development length in tension of a straight HYD bar upto 32 mm (#11 bar) given in ACI 318 (83) Clause 12.2 is as follows:

$$L_d = \frac{0.02 A_b f_y}{\sqrt{f_{ck}}} \text{ (in SI units)} \tag{20.6}$$

where A_b is the area of steel.

'Modification factors' of 0.7 for good cover and 0.8 for the presence of confirming stirrups can also be applied to this formula. Larger diameter of bar will have larger bond and the factor corresponding to 0:02 above increases correspondingly.

2. The development length of HYD bars in tension ending in an ACI standard bend or hook is given by ACI 318 Clause 12.5.1. Using the relation $f'_c = 0.8 f_{ck}$, it can be expressed as

$$L_{dh} = \frac{0.27 \phi f_y}{\sqrt{f_{ck}}} \tag{20.7}$$

where ϕ is the diameter of the reinforcement rod.

For joints with load reversals where the bend should end in confined concrete ACI 318 (1983) Clause 21.6.4.1 gives the following formula. The end should have the ACI standard hook and bend

$$L_{dh} = \frac{\phi f_y}{65\sqrt{f'_c}} \text{ in p.s.i.}$$

converting it into SI units. Hence

$$L_{dh} = \frac{0.20\phi f_y}{\sqrt{f_{ck}}} \tag{20.8}$$

where L_d should also not be not less than 8ϕ or 150 mm. Therefore

L_d = Straight length to tangent point + Bend radius 4ϕ + Diameter of bar ϕ

For anchorage to be effective the diameter of the bar chosen should satisfy the L_{dh} value available at the joint.

20.8.2 Anchorage Requirements in IS 456

The anchorage in IS 456 and BS 8110 still continue to be the same as used from the early days of reinforced concrete. The length required depends on the bond strength between steel and concrete. For every 45 degree of the bend or hook, the increase in the development length from the tangest point is assumed as 4ϕ. With such bending no stress is assumed to be transmitted to the end of the bend. Alternately the rod can be bent in an easy curve and the development length provided can be as shown in Fig. 20.6(a). The bearing strength at the bend should also be checked by the following equation derived from IS 456 Clause 26.2.2.5.

$$r \geq 0.456 \frac{f_y}{f_{ck}} \left(1 + 2\frac{\phi}{a}\right) \phi \quad \text{(see page 179 of Reference [3])} \tag{20.9}$$

where

ϕ = Diameter of the rod

r = Radius of the bend for developing $0.87 f_y$ in the bar

a = Centre-to-centre distance between bars or the cover *plus* bar size for a bar adjacent to the face of the member.

With this easy curve we assume that the bar is stressed beyond the bend.

20.8.3 Anchorage Requirements for Joints

The anchorage provided for the tension bars in external joints should satisfy the above code requirements. The anchorage of bars in the internal joint is not a problem if the beam bars are made continuous through the column and adequately spliced outside the column core as usually done. The bond stress developed in the bar need compensate only for the stress in the bar which can be tension on one side and compression on the other thus making the forces cumulative.

The following empirical rule is also usually recommended to satisfy the bond requirements [1] in column bars.

$$\frac{\text{Total depth of column}}{\text{Diameter of beam bar}} \geq 20, \qquad \frac{\text{Total depth of a beam}}{\text{Diameter of column bar}} \geq 20 \qquad (20.10)$$

20.9 CONFINEMENT OF CORE OF JOINT

The second item to be considered is the confinement of joints. Laboratory tests have shown conclusively the necessity of confining the concrete inside a joint subjected to horizontal forces. The confinement can be by 'side beams' or by spirals (laterals) placed inside the joints. In type 1 joints, the presence of beams of dimensions extending to at least 3/4th the size of the column (on both sides of the column) in the direction of the shear forces can be taken as providing the necessary confinement. According to the *ACI–ASCE Joint Committee Report*, type 1 joints which are not confined as above, should be provided with the specified confining steel in the form of hoops or laterals in the direction of shear. Whereas joints which are properly confined by beams need to be provided only with one-half the specified steel. In type 2 joints, confining steel in the concrete is to be provided irrespective of whether side-confinement is provided by beams or not.

The principle of providing spirals or circular hooks for confining joints is based on the assumption that the load taken by spirals should fully compensate the loss of strength due to the spalling of the cover. The per unit volume of the spirals is also assumed to be twice as effective as the equivalent longitudinal steel [2]. Let

A_{sh} = Area of spirals used

S = Pitch of the spirals (not more than 1/4 dimension of column or 100 mm but not less than 75 mm)

A_g = Gross area of column

A_k = Area of the concrete core from outside to outside of the spirals

D = Diameter of the core

Equating forces according to the above assumptions, we get

$$\frac{2A_{sh}\pi D(0.85f_y)}{S} = 0.63 f_{ck}(A_g - A_k)$$

This reduces to the expression given in IS 13920 Clause 7.4.7 for the area of spirals.

$$A_{sh} = 0.09 S D_k \left(\frac{A_g}{A_k} - 1\right) \frac{f_{ck}}{f_y} \qquad (20.11)$$

Since rectangular hoops are considered only half effective, as spirals we require double the steel and the formula for rectangular hoops as given in IS 13920 Clause 7.4.3 is the following:

$$A_{sh} = 0.18 SH \left(\frac{A_g}{A_k} - 1\right) \frac{f_{ck}}{f_y} \qquad (20.12)$$

where

S = Spacing of hoops

H = The larger dimension of the confining hoop measured to its outer faces. In addition, if it exceeds 300 mm it should be sub-divided by additional ties as shown in Fig. 20.7.

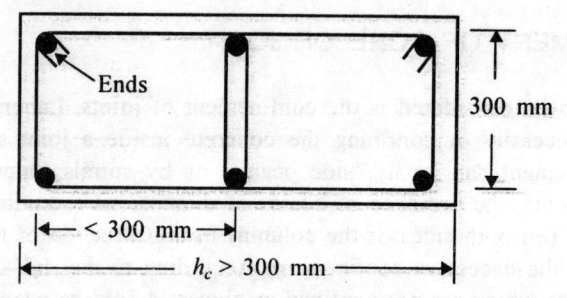

Fig. 20.7 Confining hoop (distances between longitudinal bars should not exceed 300 mm).

These ties should be closed hoops which end in standard 145° hooks and 10ϕ extension (but not less than 75 mm) with ends embedded inside confined concrete. These confining steel in type 1 and in type 2 joints should be continued in the column beyond the joint for a length not less than the following:[1]

1. The largest lateral dimension of the member or
2. 1/6th of the clear span or
3. 450 mm.

20.10 SHEAR STRENGTH OF THE JOINT

Thirdly, we shall examine the requirements for shear strength of the joint. The shear force acting on a joint depends on the probable forces that will be transmitted from the beams to the columns. The shear acting on the joint is given by the following equation with reference to Fig. 20.4. Considering the maximum forces that can be developed in the beams at failure conditions,

$$V = C + T - V_{col} \qquad (20.13)$$

where

$T = A_{s1}$ (steel stress)

$W = A_{s2}$ (steel stress)

A_{s1}, A_{s2} = Areas of tension and compression steels

$$V_{col} = \frac{\text{Sum of moments in the beams}}{\text{Storey height}}$$

The steel stress is f_y for type-1 joints and $1.25 f_y$ for type-2 joints.)

---
[1] See also Sections 21.6.3 and 21.6.4

Having estimated the value of V at the joints, the design can be made by any one of the methods given in Table 20.1.

TABLE 20.1 METHODS OF JOINT DESIGN

No.	Requirement	Park and paulay	Joint committee	ACI 318	IS 3920 (Ductility design)
1.	Anchorage	Diameter of beam bars not to exceed 1/25 column depth in HYD bar	Diameter of beam bars and column bars to be limited $$\frac{\text{Column depth}}{\text{Diameter of beam bar}} \geq 20$$ $$\frac{\text{Beam depth}}{\text{Diameter of column bar}} \geq 20$$	Development length to be satisfied as per ACI code	Development length should satisfy IS 456 practice
2.	Confinement	Transverse steel to be governed by joint shear	Transverse steel controlled by minimum steel volumetric ratio requirement Eqs. (20.11) and (20.12)	Transverse steel governed by column confinement and shear	Transverse steel controlled by minimum volumetric ratio requirements as in the Joint Committee
3.	Strength in shear	Shear in type 2 joints is calculated using a strain hardening factor 1.2 times moment at yield. All shear to be taken by steel ($V_c = 0$). Vertical shear is not considered but column must have at least 8 bars	Shear is calculated by using strain hardening factor for beam moments. Shear to be taken mostly by concrete. Total capacity is limited but not broken down to V_c and V_s. Specified laterals are provided. Vertical shear not specially considered but column should have at least 8 bars	Shear is calculated using moments of beam at yield type 1 joints. Horizontal shear is taken by concrete and steel ($V_c + V_s$) as in beams-shear. Shear strength of concrete depends also on compression in concrete	Shear in column is calculated by using a factor 1.4 times caused by beam to avoid shear failure. Beam moments obtained by using $0.87 f_y$. Minimum transverse steel to be provided as in confinement (see IS 3920)

20.10.1 ACI–ASCE Committee Method

The ACI–ASCE Committee's observations on design of joints for shear need special mention. The Committee 318 made the recommendation based on tests on joints that the shear strength of the joint is to be considered as a function of the cross-sectional area of concrete A_j only, provided the required minimum amount of transverse steel is already provided, Fig. 20.8. The shear strength of the joint is given by the following equation:

$$V_n = \gamma \sqrt{f'_c}\, hb_j \text{ (in p.s.i. units)}$$

where

$\gamma = 20$, for internal joints, 15 for external and 12 for corner joints

h = Depth of column as shown in Fig. 20.8

b_j = Effective joint width = $\frac{1}{2}(b_{beam} + b_{col})$ but $\leq b_{beam} + (1 \text{ or } 2)\left(\frac{1}{2}h_{col}\right)$

$A_j = hb_j$

Converting the above equation into SI units, and with reduction factor 0.85 and $f'_c = 0.8 f_{ck}$, we get

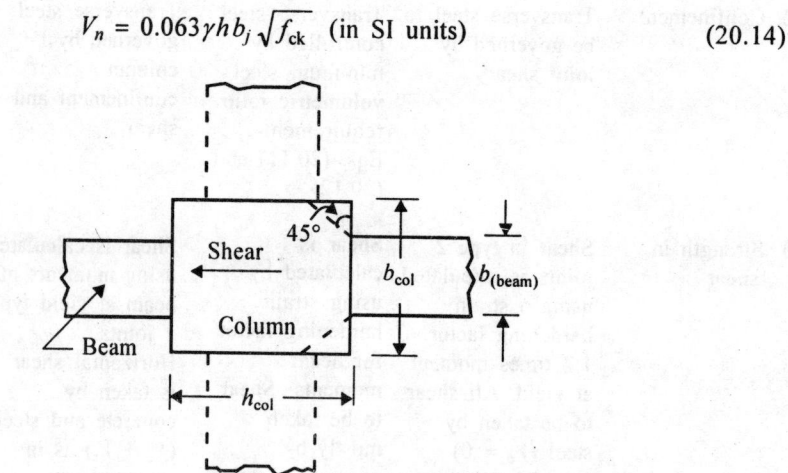

$$V_n = 0.063 \gamma h b_j \sqrt{f_{ck}} \text{ (in SI units)} \qquad (20.14)$$

Fig. 20.8 Plan view of beam–column joint in shear.

When the factored shear in the joint exceeds the shear strength of the joint as given by Eq. (20.14) we can either increase the column size or increase the depth of the beam. The former increases the shear capacity and the latter will reduce the amount of steel and hence the shear to be transmitted to the joint.

20.11 CORNER (KNEE) JOINT

These joints occur commonly in construction and require special consideration. As shown in Fig. 20.9, it can be a closing or an opening corner joint.

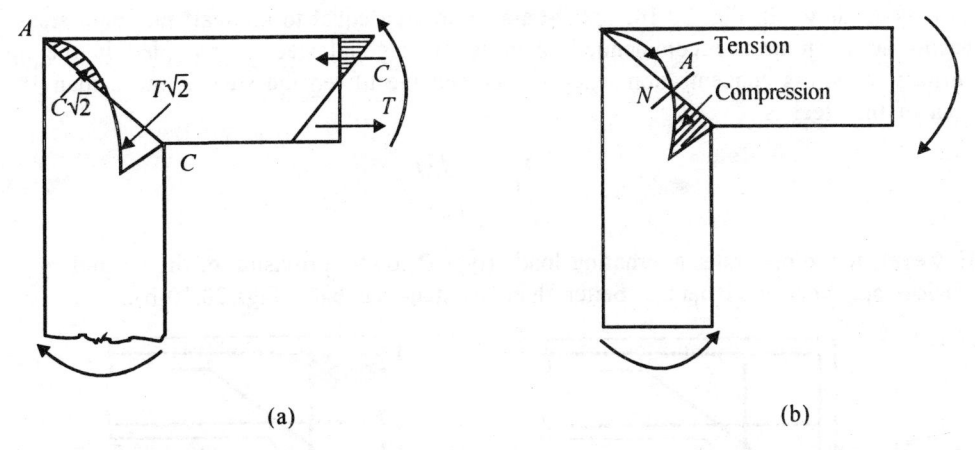

Fig. 20.9 Distribution of stresses in opening and closing joints (a) opening joint (b) closing joint.

The neutral axis and the stress distribution in both the arms at sections away from the junction in these joints are as in the case of ordinary beams. However, at the corner, the stress distribution is different and is, as shown in Fig. 20.9, with zero stress at A and neutral axis at nearly a third point along AC from the inside joint C. There is also a stress concentration at C. In an opening joint at C we have tension stress concentration and since the concrete is weak in tension there will be a tendency of premature cracking unless special steel is provided at this place. If the stress concentration is compressive it does not pose any problem as the concrete is strong in compression. In general, two types of cracking can occur in these joints as discussed below:

1. If the shear in the core is high it can cause cracking due to diagonal tension. If T is the tension in steel transmitted to the joint

$$\text{Diagonal tension} = \text{Shear} = v$$

or

$$v = \frac{T}{bd} = \frac{A_s f_y}{bd}$$

Assuming a maximum shear strength = $0.8\sqrt{f_{ck}}$

$$\frac{A_s}{bd} = \frac{0.8\sqrt{f_{ck}}}{f_y} = \rho \qquad (20.15)$$

with $f_{ck} = 25$ and $f_y = 415$ we get $\rho = 0.0096$.

Thus if the tension steel is more than 1% then the diagonal steel has to be provided to limit the growth of tension cracks due to shear.

2. Cracking can occur due to stress concentration at the joint if it is an opening joint. The total tension in diagonal $AC = \sqrt{2}T$ as shown in Fig. 20.9. If this tension acts on a small area then the cracking will occur.

IS publication SP 34 (1987) [4] p. 110 gives the method of detailing of 90° opening corners as shown in Fig. 20.10. Splay-bars of an area equal to one-half the main steel in the beams at the joint is recommended. Additional diagonal steel is provided in the opening corners for stress concentration when the percentage of tension steel is more than 1%. The area of the steel is

$$A_s = \frac{\sqrt{2}T}{0.87 f_y} \qquad (20.16)$$

However, for joints under alternating loads (type 2 joints) provision of the secondary steel as vertical and horizontal bars is better than the diagonal bars, Fig. 20.10(b).

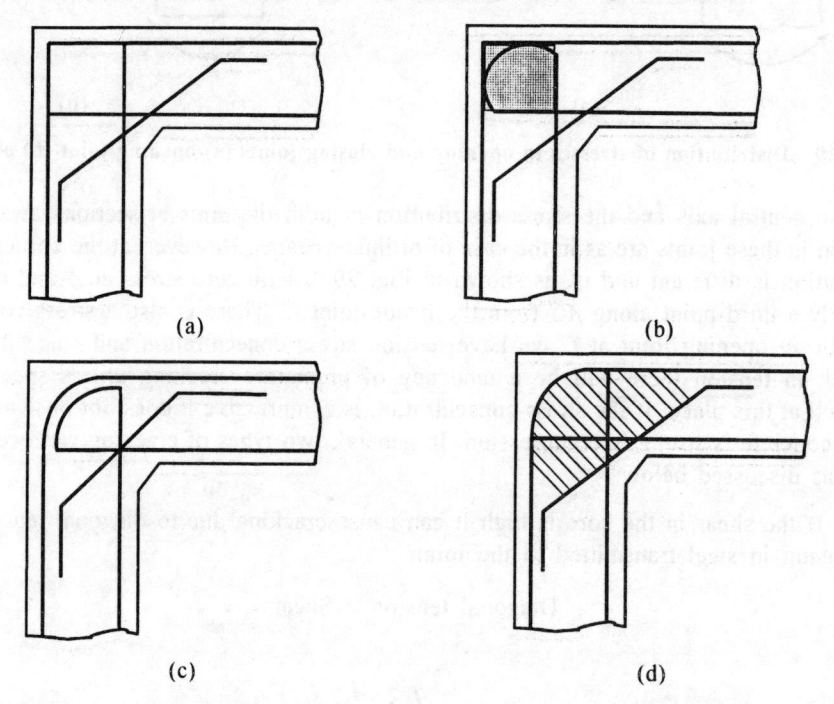

Fig. 20.10 Detailing of knee joints: (a) 90° opening corners of members with percentage of steel ≤ 1. Splay steel to be 50% of main steel (b) joints under alternating loads (c) alternative to (a) (d) members with more than 1% steel (SP 34 page 110).

20.12 DETAILING FOR ANCHORAGE IN EXTERIOR BEAM–COLUMN JOINT

In this three-member joint (item 2 in Fig. 20.1) the tension from a beam is transferred to the joint core by anchorage of bars. Since under cyclic loads in type 2 structures, these anchorages have a tendency to straighten up, their hooked ends should be placed always along the column bars and restrained by stirrups. IS 13920 (93) Clause 6.2.5 requires that in external joints, both the top and bottom bars of the beam shall be provided with anchorage length beyond the face of the column equal to the development length in tension *plus* 10ϕ *minus* the allowance for 90° bend.

20.13 PROCEDURE FOR DESIGN OF JOINTS

The various steps in design of joints can be summarised as follows:

Step 1: Check the capacity of column to take the vertical load as well as the maximum probable moments from the beams joining the column, Eq. (20.1).

Step 2: Check the capacity of column in shear produced by the maximum probable moments from the beams joining the column, Eq. (20.2). (The beam capacities assumed should be $1.2M_u$ for type 1 joints and $1.4M_u$ for type 2 joints.)

Step 3: Check the anchorage requirement of tension bars in the beam inside the core of the columns.

Step 4: Check the confinement of the joint and provide steel for the confinement if found necessary.

Step 5: Check for shear. Calculate the ultimate shear acting on the joint and check the capacity of the joint to withstand the shear.

Step 6: Check the detailing of the steel at the joint. Table 20.1 gives the procedure for checking anchorage, confinement and shear in joints by various methods.

CONCLUSIONS

Recent research has shown that the three important factors that influence strength of joints are anchorage, confinement and shear strength. The topic of joint design specially under conditions of cyclic loadings is still in the development stage and more explicit recommendations for design of joints are expected from research in the near future.

EXAMPLE 20.1 (Design of confining steel at joints)

(a) A circular column is 300 mm in diameter. Find the diameter and spacing of hoops to be used for confinement assuming that the concrete used is M20 and the steel is 415.

(b) What will be the lateral reinforcements if the column is rectangular, 650 × 500 mm in size.

Reference	Step	Calculations
Sec. 20.9	(A) 1	*Circular column with circular hoops* *Choose the spacing of circular hoops* $S > \dfrac{300}{4} = 75$ mm; 100 mm and $\not< 75$ mm Adopt 75 mm.
	2	*Find the dimensions of core* Diameter of the core = 300 − (2 × cover) + (2 × diameter of the hoop) = 300 − 80 + 16 = 236 mm

Reference	Step	Calculations
	3	Calculate the area of hoop required
Eq. 20.11		$A_{sh} = 0.09 SD_k \left(\dfrac{A_g}{A_k} - 1 \right) \dfrac{f_{ck}}{f_y}$
		$= 0.09 \times 75 \times 236 \left[\left(\dfrac{300}{236} \right)^2 - 1 \right] \dfrac{20}{415} = 47.28 \text{ mm}^2$
		Adopt 8 mm (50.27 mm^2) at 75 mm pitch.
(B)		Rectangular column with rectangular ties
	1	Choose the spacing of ties
		$S > \dfrac{500}{4} = 125$ or 100 and $\not< 75$
		Adopt 100 mm.
	2	Find the dimension of core (adopt 10 mm ties)
		$= 650 - 80 + 20 = 590 > 300$ mm
		and $500 - 80 + 20 = 440 > 300$ mm
		Put cross ties to reduce to 590/2 and 440/2 i.e. 295 and 220.
		Larger dimension = 295 mm = H
	3	Determine diameter of ties
Eq. (20.12)		$A_{sh} = 0.18 SH \left(\dfrac{A_g}{A_k} - 1 \right) \dfrac{f_{ck}}{f_y}$
		$= 0.18 \times 100 \times 295 \left(\dfrac{650 \times 500}{590 \times 440} - 1 \right) \dfrac{20}{415} = 64.47 \text{ mm}^2$
		Provide 100 mm (78.54 mm^2) at 100 mm spacing with cross ties in both directions.

EXAMPLE 20.2 (Design of an exterior type 1 joint)
An exterior joint has the following members framing into it

1. Column. 550 × 550 mm with 2% steel with a maximum load on the column 5000 kN, bar diameter 30 mm.
2. Main beam. 500 × 600 ultimate capacity 430 kNm (say) and tension steel 5 nos 25 mm (2454 mm^2).
3. Spandrel beam 450 × 750 mm.

Assume $f_{ck} = 30$ and $f_y = 415$, Storey height = 3 m. If the joint may experience slow reversal of moments due to wind loads, design the joint.

DESIGN OF CAST IN SITU BEAMS—COLUMN JOINTS

Reference	Step	Calculations
	1	*Check column moment capacity from interaction diagram* Column 550 × 550; p = 2%; f_{ck} = 30 $\dfrac{P}{f_{ck}bD} = \dfrac{5000 \times 10^3}{30 \times 550 \times 550} = 0.55$ $\dfrac{p}{f_{ck}} = \dfrac{2}{30} = 0.067;$ $\dfrac{M_u}{f_{ck}bD^2} = 0.067$
SP 16 Chart 31		$M_u = 0.065 \times 30 \times 550(550)^2 = 324 \times 10^6$ Nmm Column, above and below joint have twice this capacity, i.e. 2 × 324 = 648 kNm.
	2	*Check the stability condition of the column with capacity of beam* $\dfrac{\Sigma M_{col}}{\Sigma M_{beam}} = \dfrac{648}{430} = 1.5$ Desirable capacity for class 1 joint $= \dfrac{f_y}{0.87 f_y} = 1.15 \simeq 1.2$ $\left(\text{Class 2 joints should have a ratio } \dfrac{1.25}{0.87} = 1.4\right)$
Text Eq. (20.7)	3 3.1	*Check anchorage of* 25 mm *bars (beam to column)* *ACI requirement with a 90° bend.* $L_{dh} = \dfrac{0.27 f_y}{\sqrt{f_{ck}}}$ (anchored in the core) $= \dfrac{0.27 \times 25 \times 415}{\sqrt{30}} = 511$ mm Length available = 600 − (Cover) − (Diameter of the column bar) = 600 − 40 − 30 = 530 mm Anchorage can be made within the core and will give enough development length.
	3.2	*Find IS requirement for developing* $0.87 f_y$ *in the bar*
SP 16 Table 6.5		When ϕ = 25; f_{ck} = 30 Development length = 940 mm The bar will go straight downward and then bent around a circle and continued down.
Text Eq. (20.9)		Minimum radius of bend $r = 0.456 \dfrac{f_y}{f_{ck}} \left(1 + \dfrac{2\phi}{a}\right) \phi$

374 ADVANCED REINFORCED CONCRETE DESIGN

Reference	Step	Calculations
		Let $a = 100$ mm, $\phi = 25$, then
		$r = \dfrac{0.456 \times 415}{30}\left(1 + \dfrac{50}{100}\right)25 = 236$ mm
	4	*Provide for confinement by minimum transverse steel*
		Spandrel beam is only on one side confinement by transverse steel is needed.
		Procedure as in Example 20.1
IS 13920		$S = 100$, $\phi = 12$ mm (assume), $A_k = 494 \times 494$
Clause 7.4.8		$H = \dfrac{494 + 12}{2} = 253$ mm
Eq. (20.12)		$A_{sh} = 0.18 \times 100 \times 253\left[\left(\dfrac{550}{494}\right)^2 - 1\right]\dfrac{30}{415} = 78.9$ mm^2
		Provide 12 mm (113.1 mm^2) at 100 mm spacing.
	5	*Check for shear in column* (type 1 joint)
Text		Design shear in column $= 1.2\dfrac{M_{beam}}{h}$
Sec. 20.5		$= \dfrac{1.2 \times 430 \times 10^6}{3000} = 172$ kN
		$v = \dfrac{172 \times 10^3}{550 \times 550} = 0.57$ N/mm^2
		(For type 2 joint, we will use a factor 1.4.)
IS 456		Assume half as tension steel (2/2) = 1%
Table 19		Allowed $\tau_c = 0.66$
		Column safe in shear
		(Alternately, we may use the ACI formula. *See* step 5 of Example 20.3.)
	6	*Check for shear capacity of joint as type 1 joint*
		$V_u = T - V_{col}$ (external joint)
		$T = A_s f_y$ (For type 2 joint, we use $1.25 f_y$)
		$T = 2454 \times 415 = 1018.4$ kN
		V_{col} (without factor 1.2) $= \dfrac{430 \times 10^6}{3000} = 143$ kN

DESIGN OF CAST *IN SITU* BEAMS—COLUMN JOINTS

Reference	Step	Calculations
Text Sec. 20.10.4 Eq. (20.14)		$V_u = 1018 - 143 = 857$ kN Calculate the shear capacity of the joint according to the Joint Committee. It is independent of shear steel. $V_n = 0.063 \gamma \sqrt{f_{ck}} \ b_j h_{col}$ $b_j = \dfrac{500 + 550}{2} = 525$ m $h_{col} = 550$ and $\gamma = 15$ $V_n = 0.063 \times 15 \times \sqrt{30} \times 525 \times 550 = 1495.5$ kN This is larger than 857 kN. Therefore, the joint is safe in shear.

EXAMPLE 20.3 (Design of an internal type 1 joint)

The following are the details of an internal beam column of type 1 joint, subjected to reversals which are not due to earthquakes.

1. Column 600 × 600 with 8 nos 25 mm bars. Column factored load is 1400 kN. Storey height is 3 m.
2. Beams on either side are 400 × 500 with 3 bars of 28 mm (1846 mm²) on the top and 3 bars of 25 mm (1473 mm²) at the bottom.

Assuming $f_{ck} = 25$ and $f_y = 415$, design the joint.

Reference	Step	Calculations
	1.	*Check the strength of column* Percentage of steel $= \dfrac{3927}{600 \times 600} \times 100 = 1.09\%$ This is more than 0.8% and less than 6%. Bending capacity (as in Example 20.2) of each column. $M_u = 648$ kNm, $2M_u = 1296$ kNm.
 SP 16, Table 4 SP 16, Table 4	2.	*Check the stability condition of the column with capacity of beams* M_u of beam with 1846 mm² in tension % of steel $= \dfrac{1846}{400 \times 450} \times 100 = 1\%$ $M_{u1} = 3.15 bd^2 = 255$ kNm (approx.) Similarly, M_u with $\dfrac{1473}{400 \times 450} \times 100 = 0.82\%$ steel $M_{u2} = 2.60 bd^2 = 210$ kNm (approx.)

Reference	Step	Calculations
		$\dfrac{\sum M_{col}}{\sum M_{beam}} = \dfrac{2 \times 648}{255 + 210} = 2.79$
		Hence the column failure will not take place.
	3	*Check the anchorage of bars*
		Extend the longitudinal beam bars through the column. The empirical requirement is depth of column/beam bar dianeter ≥ 20.
		$\dfrac{600}{25} = 24 > 20$
	4	*Confinement by transverse steel*
		Assume 10 mm stirrups.
		Dimension of core = 600 − 80 + 20 = 540 > 300
		which introduces cross ties.
		$H = \dfrac{600 - 80}{2} + 15 = 275$ mm
		$S = \dfrac{600}{4}$ or 100 but $\not< 75$
		Adopt $S = 100$.
Eq. (20.12)		$A_{sh} = 0.18 \times 100 \times 275 \left[\left(\dfrac{600}{540}\right)^2 - 1 \right] \dfrac{25}{415} = 70.5$
		Adopt 10 mm (78.5 mm^2) bars with cross ties.
	5	*Check shear in columns* (*type* 1 *joint*)
Eq. 20.2		$V_{col} = \dfrac{1.2\left(M_u^L + M_u^R\right)}{h}$
		$= \dfrac{1.2(255 + 210) \times 10^6}{3000} = 186 \times 10^3$ N
		$v = \dfrac{186 \times 10^3}{600 \times 600} = 0.52$
		Find IS 456 values.
Table 13		Assume $A_{st} = 0.5\%$, $\tau_c = 0.49$.
Text		Factor $= 1 + \dfrac{3 \times 1400 \times 10^3}{600 \times 600 \times 25} = 1.47 < 1.5$
Eq. (20.3)		$\tau'_c = 0.49\,(1.47) = 0.72$ N/mm^2
		Hence it is safe.

Reference	Step	Calculations
Eq. (20.3)		ACI values $$\tau'_c = 0.13\sqrt{f_{ck}}\left(1 + 0.07\frac{P_w}{A_g}\right)$$ $$= 0.13 \times 5\left(1 + \frac{0.07 \times 1400 \times 10^3}{600 \times 600}\right) = 0.83 \text{ N/mm}^2$$
	6	*Find the shear in joint (both beam moments are clockwise)* $V_u = T + C - V_{col}$ $V_{s1} = 1846$ mm^2; $V_{s2} = 1473$ mm^2 $T = 1846 \times 415 = 766 \times 10^3$ N $= 766$ kN $C = 1473 \times 415 = 611 \times 10^3$ N $= 611$ kN *Note*: (For type 2 joints steel the stress will be 1.25f_y.) $$V_{col} = \frac{M_U^L + M_U^R}{h} = \frac{(255 + 210) \times 10^6}{3000} = 155 \times 10^3 \text{ N}$$
Eq. (20.13)		$V_u = 766 + 611 - 155 = 1222$ kN
	7	*Check by Joint Committee recommendation*
Eq. (20.14)		Strength of joint $= 0.63\gamma b_j h_{col}\sqrt{f_{ck}}$ γ for internal joint $= 20$; $o_j = \dfrac{b_b + b_c}{2}$ (assume)
Sec. 20.10		$V_n = 0.063 \times 20 \times 5 \times \dfrac{400 + 600}{2} \times 600$ $= 1890$ kN > 1222 kN (required.) Hence the joint is safe in shear.

REFERENCES

1. IS 13920 (1993), Ductile Detailing of Reinforced Concrete Structures Subjected to Seismic Forces, Bureau of Indian Standards, New Delhi.

2. Ferguson P.M., Breen J.E. and Jirsa J.O., *Reinforced Concrete Fundamentals*, John Wiley & Sons, New York, 1988.

3. Varghese P.C., *Limit State Design of Reinforced Concrete*, Prentice-Hall of India, New Delhi, 1994.

4. SP 34 (1987), *Handbook on Concrete Reinforcement and Detailing*, Bureau of Indian Standards, New Delhi.

CHAPTER 21

Ductile Detailing of Frames for Seismic Forces

21.1 INTRODUCTION

Moment resisting frames with or without shear walls or bracings for resisting earthquake forces can be designed according to the Indian Codes of Practices in one of the following two ways as indicated in Table 21.1:

1. Designed and detailed according to IS 456 as ordinary concrete frames (OCF) as a non-ductile system with value of performance factor $K = 1.6$ (Section 18.6.1).

2. Designed and detailed according to IS 13920 (1993) (*Ductile Detailing of Reinforced Concrete Structures Subjected to Seismic Forces—Code of Practice*) as ductile concrete frames (DCF) [1] with value of $K = 1.0$.

The evaluation of this philosophy in the Indian code took place as follows: In the third revision of the Indian Code Criteria for Earthquake Resistant Design of Structures IS 1893 (1975), the base shear V_b in a building due to earthquake force was calculated by a simple formula and ductile detailing was compulsory for zones IV and V. This code underwent its 4th revision in 1984 which introduced the concept of performances factor (K). According to this concept, the base shear is calculated from the formula $V_B = KC\alpha_1 W$ (*see* also Chapter 18). This factor K, is thus a multiplier of the value of the base shear obtained by earlier codes, bring the result in the line with other international codes. How to apply this factor depends on the structural framing system and the ductility of the construction. According to this principle, designers have the following two options for design of moment-resisting structures. This can be done with or without using shear walls with proper assumption being made in the amount of its participation.

(a) Using 1.6 times the load as obtained by using the old codes (i.e. $K = 1.6$) with the structure detailed as ordinary concrete frame (OCF) according to IS 456.

(b) Using the same load as in the early codes with the value of $K = 1$ with the frame detailed for ductility (DCF) from IS 13920.

The increased load in OCF will lead to an increase in the main steel. On the other hand, the increased load for ductility in DCF will lead to an increase in the transverse steel.

A study of the cost and performance characteristics of typical buildings in various zones in India showed that with the introduction of the performance factor, cost of construction will

TABLE 21.1 LATERAL LOAD RESISTING SYSTEMS

Type	Load resisting system	% Load resisted Vertical	% Load resisted Horizontal	K
Ductile system	Moment resisting frame only	Frame 100%	Frame 100%	1.0
	Vertical load carrying frame with shear wall	Frame and wall 100%	Wall 100%	1.0
	Moment resisting frame and shear wall	Frame 100%	Frame and wall in proportion to stiffness Frame ≥ 25% Wall ≥ 80%	1.0
	Shear wall only	Wall 100%	Wall 100%	1.0
Non-ductile system	Any combination	Combined 100%	Combined 100%	1.6

(*Note*: K values are given in Table 18.5 Chapter 18.)

slightly go up in zones II and III with ductile detailing but in zones III to V it is more economical to design frames as ductile frames than in ordinary frames [2]. This leads to the following recommendations in the Indian code:

(i) In zones IV and V of the seismic map of India, all buildings must be designed as ductile concrete frames with $K = 1$ (i.e. ductile detailing has been made compulsory in zones IV and V using $K = 1$).

(ii) In zone III, the code allows relaxed ductile detailing with $K = 1$ (i.e. ductile detailing is compulsory only in the case of structures of importance factor more than unity and for structure more than five storeys high.

(iii) In zones I and II, design can be made as ordinary concrete frames (OCF) even with $K = 1$ as the seismic risk involved in these regions is very small.

The IS code 13920 (1993) gives the necessary guidance for the ductile detailing. This chapter explains the general principles of this code and reference should always be made to the original code when carrying out the actual design and detailing of a structure.

21.2 GENERAL PRINCIPLES

Earthquake forces sometimes can be much more than the forces against which we design for and reversal of these forces can also occur many times during an earthquake. The members and the connections are designed by ductile detailing to resist the large forces and reversals by inelastic deformation beyond yield without serious reduction in strength or energy dissipation capacity. Details for ductile detailing given in IS 13920 (1993) are very similar to the provisions of ACI 318 (1993) Appendix A. and other international codes of practice. As already pointed out IS 13920 lays down that the special ductile detailing is necessary for the following situations:

1. All structures in zones IV and V
2. The following three types of structures in zone III.

(i) Structures with importance factor greater than 1.0 (i.e. important structures like hospitals, schools water tanks, bridges and as given in Table 4 of IS 1893 (1984)
(ii) Industrial structures
(iii) Structures more than five storeys high. (All multi-storeyed buildings in Cochin, Mumbai, Calcutta and Chennai (by revised code) come in this category. For remaining structures in zone III and all structures in zones I and II we can use value of $K = 1$ and the ordinary concrete frame detailing based on IS:456. The risk involved on such designs is very small.

21.3 FACTORS THAT INCREASE DUCTILITY

The main devices used to improve ductility performance with seismic loading are the following:

1. Using a simple and regular structural configuration of the structure
2. Using more redundancy on the lateral load resisting system
3. Avoiding column failure (or hinge formation in columns) by adopting 'weak beam–strong column' principle in design

4. Avoiding foundation failures

5. Avoiding brittle failures due to shear, bond, anchorage or compression failure in bending

6. Providing special confinement of concrete at critical points by provision of laterals so that the concrete can undergo large compressive strain before failure

7. Using under-reinforced beam sections so that they can undergo large rotation before failure

The method to achieve these objectives as laid down in IS 13920 (1993) are dealt with in the following sections.

21.4 SPECIFICATIONS OF MATERIALS FOR DUCTILITY

In all 'ductile' structures discussed earliars for three storeys in height we should use concrete of minimum grade, M20, and the reinforcements should not be more than the grade Fe 415 so that the steel is ductile. Welded splices should be designed according to Clause 26.2.5.2 of IS 456 (2000). According to IS 13920 Clause 5.3, 500 grade steel is not recommended in earthquake regions.

Splices and mechanical connectors shall not be provided (a) within a joint (b) within $2d$ from the face of a joint and (c) within the mid quarter length of a bending member where yielding may occur.

21.5 DUCTILE DETAILING OF BEAMS—REQUIREMENTS

A member is considered as a beam when it is in bending and the axial force is less than $A_g f_{ck}/10$, where A_g is the gross area of the section.

If the axial force is more, it is considered as a column. The three requirements of dimensions, longitudinal steel and web steel for ductile detailing of beams are separately considered in the following sections:

21.5.1 Dimensions of Beams

The rules for dimensioning the beams are as follows:

1. The width of beams should be greater than 0.3 times its depth and at least 200 mm in size, as most of the test data now available is for such beams only.

2. The total depth D should not be larger than one-fourth the clear span to avoid deep beam action.

21.5.2 Longitudinal Steel in Beams [IS 13920 (1993) Clause 6.2]

We must note that earthquake can produce reversible forces and moments as shown in Fig. 21.1. The following provisions given in IS 13920 (1993) Clause 6.2 are incorporated in ductile detailing:

1. There should be at least two bars throughout the length of the beam: one at the top and another at the bottom to take care of reversal of moments due to earthquakes.

2. The tension steel ratio in beams shall not be less than $r_{min} = 0.24\sqrt{f_{ck}}/f_y$ to avoid a sudden collapse by tension failure (i.e. minimum steel should be about 0.3% for $f_y = 25$ and Fe 415).

3. To avoid congestions, the maximum percentage of steel on any face at any section should not be more than 2.5%.

4. Redistribution of moments due to lateral loads is not allowed. As actual moments of earthquake forces can be more than the estimated values, the steel both at the top and at the bottom face of the member at any section *along its length* shall be at least equal to one-fourth the maximum negative steel provided on the face of either joint (Clause 6.2.4).

5. As the seismic moments are reversible the positive steel *at a joint face* must be at least equal to one-half the negative steel at the face (Clause 6.2.3).

6. In the joints, between beams and columns, detailing should be as follows:

(i) In internal joints, all the bars of the beam at the joints should be continued through the column to the opposite beam. It may also be necessary to check the size of the column for the transfer of forces (see Chapter 18).

(ii) In external joints the full anchorage length should be provided for all the bars in the beams both at the top and the bottom. (The present IS code provision is considered as inadequate by ACI Standards and if necessary it is considered good practice to provide external 'beam' or stub beams for anchoring the bars.) It may also be necessary to adopt a wider column to satisfy side anchorage length requirement of ACI. (Refer also Chapter 18 for details.)

7. As lap splices are not very reliable under cyclic loading, they should be provided with special care as already dealt with in Section 21.4. All lap splices should also have full-development lengths with stirrups for its full length at spacings not more than 150 mm.

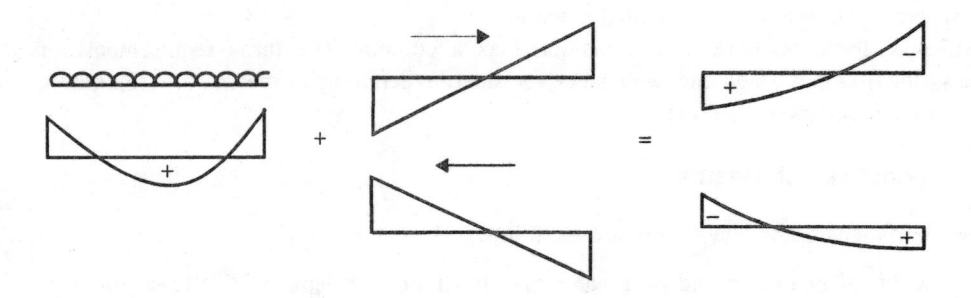

Fig. 21.1 Reversible earthquake effects. Calculation of design moments for beams.

21.5.3 Web Reinforcement in Beams

Clause 6.3 of IS 13920 gives the following rules for detailing web reinforcement:

1. The minimum size of stirrups should be 6 mm for beams of span 5 m but for spans exceeding 5 m it be 8 mm and the 135° bend should extend $10d$ (instead of $8d$ as in ordinary stirrups) but < 75 mm.

2. The magnitude of the shear forces resisted by these reinforcements should be the maximum of the following shear:

(a) The value of shear force calculated as per analysis.

(b) The shear force in beam due to the formation of plastic hinges at both ends of the beam plus the factored gravity load on the span (as shown in Fig. 21.2). This is given by the following formulae:

$$V_u = V_{D+L} \pm 1.4 \frac{M_u^A + M_u^B}{L} \tag{21.1}$$

Fig. 21.2 Calculation of design shear for beams.

It should be noted that for calculating the shear V_{D+L} due to dead and live load, IS 13920 Clause 6.3.2 prescribes a partial safety factor of 1.2 only. Sway to both sides should be considered for design. (*See* Chapter 20, $f_y/(0.87f_y) = 1.2$ approximately and $(1.25f_y)/(0.87f_y) = 1.4$.)

3. The contribution of inclined bars for shear resistance should be neglected as they are ineffective when the direction of shear changes in seismic loading.

4. The spacing of stirrups over a length twice the effective depth at the either end should not exceed one-fourth the effective depth or 8 times the diameter of the smallest longitudinal bar. But this spacing should not be less than 100 mm (Fig. 21.3). The first hoop should start within 50 mm from joint face. The same spacing is to be used for a distance of twice the effective depth where flexural yielding is likely to occur.

5. At other locations, the spacing should not exceed one-half the effective depth.

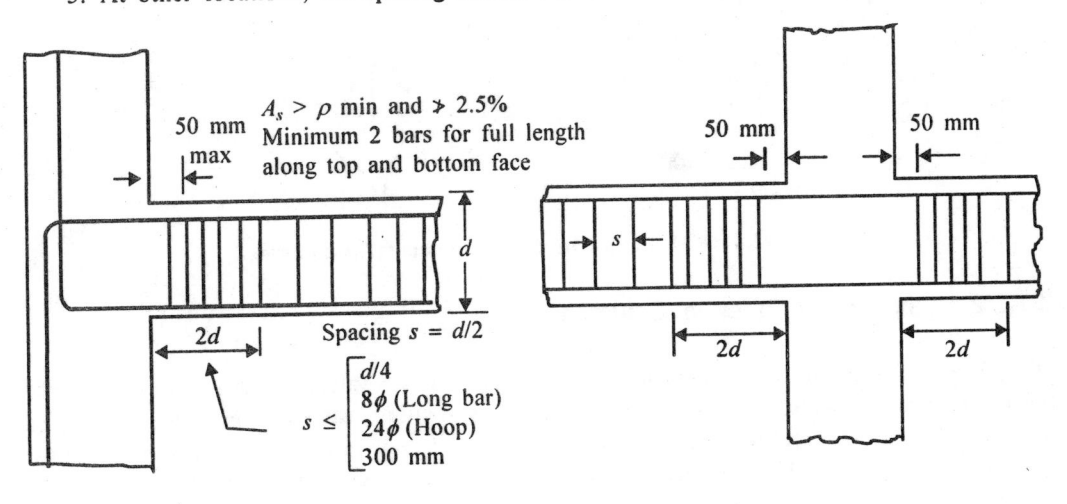

Fig. 21.3 Detailing of reinforcement in beams for ductility.

21.6 DUCTILE DETAILING OF COLUMNS AND FRAME MEMBERS WITH AXIAL—LOAD (P) AND MOMENT (M)—REQUIREMENTS

The four major requirements for columns are given in IS 13920 Clause 7. They are briefly dealt with in the following sections:

21.6.1 General Considerations of Columns

Eventhough it is not specifically stated in IS 13920, the 'strong column-weak beam' concept should be used in a seismic design of columns. As shown in Fig. 21.4

$$\sum M_{col} \leq 1.2 \sum M_{beam}$$

in each principal plane of the joint so that plastic hinges are not formed in columns. The column member should have this moment capacity together with the axial force in the member. Following are some of the other requirements:

1. The minimum dimension of columns should be 200 mm. In beams of frames of span over 5 m (centre-to-centre) and in columns of unsupported height over 4 m, it should be 300 mm. These dimensions ensure confinement of steel.

2. The ratio of the shortest to longest side of the column should not be less than 0.4. (For a 200 mm width the depth cannot be more than 500 mm, i.e. the depth should not be more than 2.5 of its breadth.)

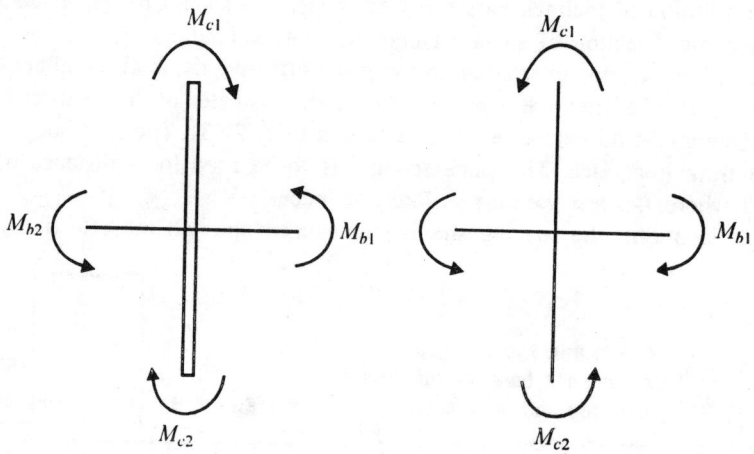

Fig. 21.4 Calculation of design moments in columns.

21.6.2 Longitudinal Steel in Columns

The usual splicing rule of not more than 50% is kept and lap splice-length in columns should be detailed in the same way as tension splicing. As spalling usually takes place at ends of columns during earthquake, splicing should be provided only in the central half of columns. The entire splice length is provided with hoops at spacing not exceeding 150 mm.

Any area of column that extends more than 100 mm beyond the confined core of the column be detailed as shown in Fig. 21.5 to prevent its spalling during earthquakes.

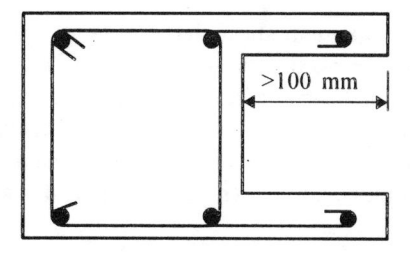

Fig. 21.5 Detailing of parts of columns projecting more than 100 mm.

21.6.3 Transverse Steel in Columns

The third requirement for ductility in columns is the provision of the required amount of transverse steel.

1. Circular columns should be provided with spiral or circular hoops, and rectangular columns with rectangular hoops. The details of hoops should be as per Clause 7.3 of IS 13920. In rectangular hoops, the parallel sides should not be spaced more than 350 mm. If it exceeds 350 mm, a cross should be provided to satisfy this requirement.

2. The spacing of hoops in the middle portion of the column should not exceed one-half the least lateral dimension of the column as shown in Fig. 21.6.

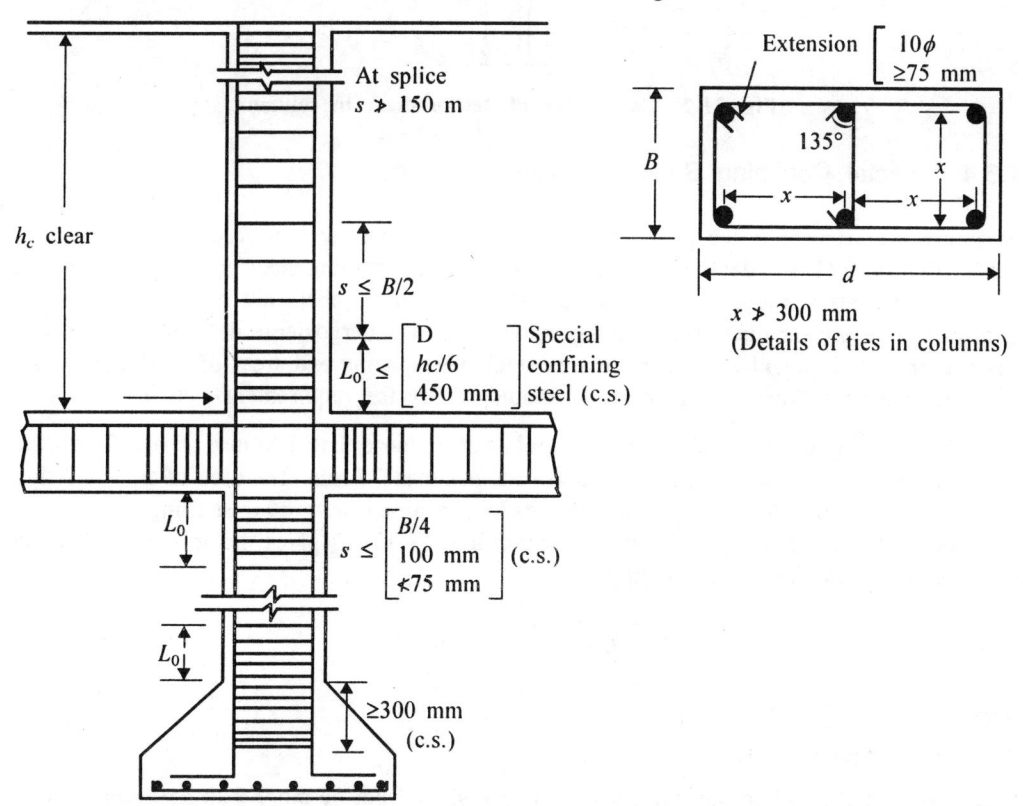

Fig. 21.6 Detailing of columns for ductility (c.s. = confining steel). (Refer also to Fig. 20.7.)

3. The magnitude of the maximum shear that can be developed in the column should be taken as 1.4 times the sum of the maximum moments ($1.4 \sum M_p$) that can be developed by the beams framing into the column from the left and right divided by the storey height or the calculated design shear, whichever is greater. Shear steel should be calculated on this basis and the transverse steel provided should satisfy this requirement (Fig. 21.7) refer Clause 7.3.4 of IS 13920 (1993). Accordingly, the shear V in column design is given by the following equation:

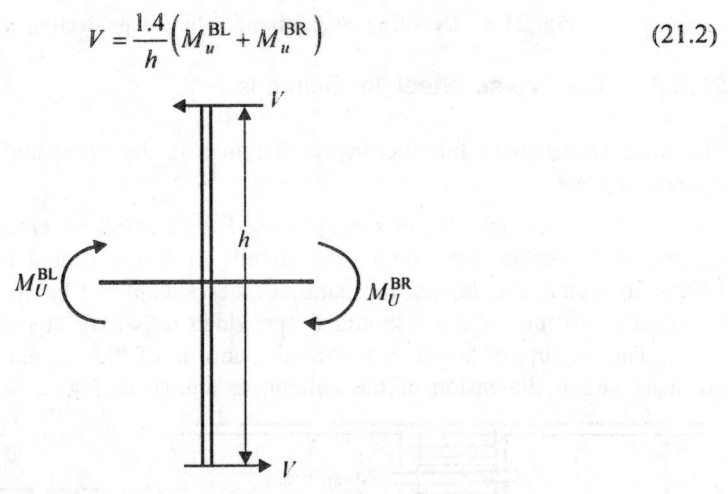

$$V = \frac{1.4}{h}\left(M_u^{BL} + M_u^{BR}\right) \tag{21.2}$$

Fig. 21.7 Calculation of design shear in columns.

21.6.4 Special Confining Steel in Columns

The fourth requirement is the provision of special confining steel in certain places. Clause 7.4.5 of IS 13920 (1993) deals with this subject. Special confining steel in the form of hoops at shorter intervals is required in certain parts of column members not only to resist shear but also to give greater ductility, allowing the section to undergo large deformations. It also assists in preventing buckling of compression steel. The spacing and areas of steel that should be provided is as follows: (In all cases it should satisfy the requirement also for shear).

1. The spacing of the confining steel should not be more than 100 mm or one-fourth the minimum size of the member. It also need not be less than 75 mm. The area of the confining A_{sh} is calculated from the principle that the load carrying capacity after spalling under large strain should be the same as under normal condition. IS 13920 gives the following formula for rectangular ties (*refer* section 20.9):

$$A_{sh} = 0.18 SH \left(\frac{A_g}{A_k} - 1.0\right)\frac{f_{ck}}{f_y} \tag{21.3}$$

whereas

S = Spacing of hoops

H = Larger dimension of the hoop which should not be more than 350 mm (so that if necessary a cross link is provided to satisfy this condition)

A_g = Gross area

A_k = Area of the core measured to outside of the link

A similar formula is also given in the code IS 13920 for spiral hoops. (*See* Chapter 20, Eq. 20.11.)

2. In general, confining should also be provided as in Fig. 21.6 in regions subjected to large deformations at the joints with beams and foundations. They are continued beyond the joint for a distance not less than the largest dimension of the member, 1/6 clear span or 450 mm (*See* Section 20.9).

3. If calculations show that the points of contra-flexure is not in the middle half of the member, the whole column height is provided with confining steel. In such cases, the zone of inelastic deformation may extend beyond the regions given above.

4. Confining steel is also provided in the regions of discontinuity (as illustrated in the code) such as:

    (a) Columns under discontinued walls. These undergo extensive inelastic deformation during earthquake.

    (b) Uncovered portions of columns in basements meant for providing ventilation and columns shorter than others provided for mezzanine floors and lofts. Columns that are short are stiffer than other columns and attract much greater seismic shear force, which may lead to brittle failure.

21.7 SHEAR WALLS

Shear walls should be designed as described in Chapter 19.

21.8 JOINTS IN FRAMES

Column joints can be considered as confined if beams frame into all vertical faces with width at least three-fourth the column face. In such joints, only one-half the confining steel required at the end of the column is provided. Otherwise the joint is also provided with the same amount of steel at the ends of the column. The joint should be designed as in Chapter 20.

CONCLUSIONS

The main requirements for ductile detailing can be summarised as follows:

1. Specification of steel and concrete to be used are indicated
2. In bending members the cross-section sizes, minimum and maximum steels, detailing of longitudinal and transverse steel along beams and joints are specified. The magnitudes of B.M. and S.F. to be used in design are also specified
3. Dimensional constraints on column, magnitudes of moments and shear force to be used for design, detailing of reinforcements in columns, provision of confining steel in places where cyclic inelastic deformation may take place during earthquake, are given
4. Special care should be also taken to detail the joints so that they keep together under earthquake forces.

EXAMPLE 21.1 (Ductile detailing of beams)

Beam AB is to be designed for moments shown in Table 18.6 (Chapter 18).

$$M_A = -69 \text{ kNm and } +23 \text{ kNm}, \qquad M_B = -88 \text{ kNm and } +3 \text{ kNm}$$

The characteristic dead and live loads are 10 and 5 kN/m respectively. The span is 6 m, beams are 300 × 500 mm with 150 mm slab. Assume $f_{ck} = 20$ and $f_y = 415$. The structure is situated in seismic zone IV. Design should be made according to the provisions of IS 13920 (1993).

Reference	Step	Calculations
	1	*Design principle*
		As the structure is situated in zone IV detailed calculations of design are necessary. In Zone III, it may be necessary only to check the detailing of steel as shown in Example 21.2.
	2	*Check for minimum sizes and ratios*
IS 13920		$b = 300$, $D = 500$, $d = 450$ mm
Clause 16.1		$b = 300 > 200$, specified minimum
Text		$\dfrac{b}{D} = \dfrac{300}{500} = 0.6 > 0.3$ (specified)
Sec. 21.5.1		$\dfrac{D}{L_c} = \dfrac{500}{6000} = \dfrac{1}{12} < \dfrac{1}{4}$ specified (beam not too deep)
	3	*Determine negative steel at support B*
IS 13920		Minimum steel ratio $= 0.24 \dfrac{\sqrt{f_{ck}}}{f_y} = 0.24 \dfrac{\sqrt{20}}{415} = 0.0026$ ... (a)
Clause 6.2		Maximum steel ratio $= 0.025$ ... (b)
Sec. 21.5.2		Negative moment at $B = 88$ kNm is
SP 16		$\dfrac{M}{bd^2} = \dfrac{88 \times 10^6}{300 \times 450^2} = 1.45$
Table 2		$p = 0.43\% > 0.3\% < 2.5\%$
		$A_s = \dfrac{0.43 \times 300 \times 450}{100} = 580 \text{ mm}^2$
		Provide 4 nos of 14 mm (i.e. 615 mm²)
	4	*Determine positive steel at support B*
Clause 6.2.3		$M_B = 3$ kNm (positive)
Step 3		Minimum steel ≮ One-half the maximum negative steel
		Provide 2 rods of 14 mm (i.e. 307 mm²)
	5	*Calculate actual moment capacity at M_B*
SP 16		$p_1 = \dfrac{615}{300 \times 450} = 0.00455$; $b = 300$ mm (– ve steel)
Table 2		

DUCTILE DETAILING OF FRAMES FOR SEISMIC FORCES 389

Reference	Step	Calculations
For T beam as calculated		$\rho_2 = \frac{1}{2}\rho_1 = 0.0023; d = 450$ mm (+ ve steel)
		M_B negative = 91 kNm (capacity) (as rectangular beam)
		M_B positive acting as T beam = 41 kNm (capacity)
	6	*Determine negative and positive steel, and capacity of beam at A*
As in steps 3 to 5		Negative steel = 3 of 14 mm (i.e. 461 mm^2)
		Positive steel = 2 of 14 mm (i.e. 307 mm^2)
		M_A negative = 76 kNm (capacity)
Text		M_A positive = 41 kNm (capacity)
Fig. 20.6	7	*Provide positive anchorage (L_d + 10d) in column for steel from beams ends*
	8	*Determine points of cut-off of negative steel*
Clause 6.2.1		(a) Top and bottom steel should be at least two rods throughout the member length. It should be not less than one-fourth the maximum negative steel.
		(b) For reinforcement cut-off, assume the following B.M. specification:
		M_A = 76 kNm, M_B = 91 kNm
γ_f = 0.9		Maximum load on beam = 0.9 (dead load only)
	9	*Specify detailing of splices*
Clause 6.2.6		Splices should not be placed at joints or 2d from the face of the support—splices should have spirals or hoops as shown in Fig. 2 of the code.
	10	*Calculate shear for design of web steel*
Clause 6.3.3		(a) Shear is calculated on the moment capacity of the ends together with a vertical load with a partial safety factor of 1.2 on loads.
Fig. 21.2		(b) ACI code also stipulates that in case of shear due to 'probable flexural strength' at beam-ends is greater than 50% of total design shear the contribution from concrete to shear resistance should be neglected.
		(c) To take care of plastic capacity (strain harding being 1.25f_y) we use a factor $1.25f_y/(0.87f_y) = 1.4$ in capacity calculations. These are taken care as follows:
	10.1	*Shear due to dead and live loads*
Fig. 21.2		$V_A = \frac{1}{2}[1.2(10+5) \times 6] = 54$ kN
	10.2	*For sway to the right*

Reference	Step	Calculations
Fig. 21.2		$V_B = 54 + \dfrac{1.4}{L}[M_A(+ve) + M_B(-ve)]$
		$= 54 + \dfrac{1.4(41+91)}{6} = 54 + 30.8 = 84.8$ kN
		$V_A = 54 - 30.8 = 23.2$ kN
		$V_B = 84.8$ and $V_A = 23.2$ kN
	10.3	*For sway to the left*
		$V_A = 54 + \dfrac{1.4}{L}[M_A(-ve) + M_B(+ve)]$
		$= 54 + 1.4\dfrac{76+41}{6} = 54 + 27.3 = 81.3$ kN
		$V_B = 54 - 27.3 = 26.7$ kN
		$V_A = 81.3$ and $V_B = 26.7$ kN
	9.4	*Final design shear* (kN)
Fig. 21.2		Draw the above two shear diagrams. Design the beam for shear using the envelope of the above shear diagram.

EXAMPLE 21.2 (Ductile detailing of columns)

A block of ten storeyed flats in Chennai has its lower most columns 500 × 700 in size. In order to use the ground floor for car parking the lower columns are made free standing. Comment on the considerations to be given for detailing of these freestanding columns. Assume f_{ck} = 20, f_y = 415 and height of free bay is 4 metres.

Reference	Step	Calculations
	1	*Design principle*
IS 13920 Clause 1.1.1 and Fig. 11		Chennai will be in zone III on the seismic map of India. As the building is more than five storeys high the members should have ductile detailing. Even though detailed calculations for seismic design may not be necessary such detailing is mandatory for this building.
	2	*Check for minimum sizes and ratios*
Clause 7.1.2 Clause 7.1.3		Minimum size > 200 mm is satisfied. Breadth/depth and ratio not to be less than 0.4.
		$\dfrac{b}{D} = \dfrac{500}{750} = 0.67$
	3	*Choose type of stirrups and spacing for confinement (not shear)*
		Assume 10 mm stirrups cover to stirrups = 30 mm
		Breadth of core = 500 − 60 = 440 mm

DUCTILE DETAILING OF FRAMES FOR SEISMIC FORCES

Reference	Step	Calculations
Clause 7.3		Depth of core = 750 − 60 = 690 mm
		Both dimensions exceed 300 mm.
		Provide overlapping loops—otherwise use ties in both directions.
Clause 7.3.3		Spacing of ties at the middle of column should not to exceed 1/2 the least dimension for shear. But for confinement it should not exceed
Clause 7.4.6		1/4 the minimum size of member or 100 mm
		$s = \dfrac{500}{4} = 125$ or 100 mm; $s = 100$ mm
	4	*Find the area of confining steel*
Eq. (21.3)		$A_s = 0.18 SH \left(\dfrac{A_g}{A_k} - 1 \right) \dfrac{f_{ck}}{f_y}$
Clause 7.4.8		$\dfrac{A_g}{A_k} = \dfrac{750 \times 500}{690 \times 440} = 1.235$
		H = Longer dimensons of rectangular hoop = $\dfrac{690}{2} = 345$
		$A_s = 0.18 \times 100 \times 345 \times 0.235 \times \dfrac{20}{415} = 70$ mm^2
		Use 10 mm bar area 78.54 mm^2.
	5	*Specify placing of confining steel.*
		If the column was a normal, column special confining steel was to be placed only at the ends of the column. However as the given columns connect discontinuous stiffness from foundations and the upper stories. We have to attend to the following:
		(i) Provide confining steel for the whole column
Clause 7.4.4		(ii) Continue the special confinement in the column to the next storey for a distance equal to full development of the largest bar in the column.
Fig. 11		
Sec. 21.6.4 (item 4)		(iii) Continue the confinement to the foundation as shown in Fig. 21.4.

REFERENCES

1. IS 13920 (1993), *Ductile Detailing of Reinforced Concrete Structures Subjected to Seismic Forces*, Bureau of Indian Standards, New Delhi.

2. Jain S.K. and Patnaik A.K., *Design and Cost Implications of IS Code Provisions for a Seismic Design of R.C Frame Buildings*. Bulletin Indian Society of Earthquake Technology, p. 307, June 1991.

CHAPTER 22

Inelastic Analysis of Reinforced Concrete Beams and Frames

22.1 INTRODUCTION

The presently available methods for analysis of reinforced concrete structures are the following:

1. Linear elastic analysis
2. Linear elastic analysis with limited allowable redistribution
3. Non-linear analysis with larger redistribution of moments that is generally allowed by the codes. This will require checking of the rotations required for the moment redistribution with the rotation capacity of the sections.
4. Plastic or collapse load analysis of beams and frames as in steel structures.

The first two methods of analysis are allowed in most of the codes such as IS 456, and has been dealt with in Chapter 3. The third method has been permitted in codes like the Euro-code. However, the fourth method of collapse load analysis of reinforced concrete beams and frames has not yet been accepted in the codes of practice. But this method has been used to estimate the capacity of the structures under over loads due to various causes such as earthquakes, bomb blasts, etc.

In this chapter, we will examine in detail the basic principles of the third method and briefly deal with the fourth method. We will limit ourselves with ordinary steadily increasing loadings and will not deal with analysis of structures subjected to large cyclic loads with reversals which involve principles of energy absorption by inelastic action.

22.2 INELASTIC BEHAVIOUR OF REINFORCED CONCRETE

Eventhough concrete by itself is brittle, incorporation of closely spaced spirals or stirrups can be used to give the concrete a lateral confinement. This enables concrete to undergo large strains before failure. Similarly, if we use steel of grade equal to or less than the grade 415 or special steels of higher grade as tensile steel, the section can be made to undergo large strains before failure and thus become more ductile. We will also deal with the material characteristics of concrete and steel to get an idea of the principles of inelastic action of reinforced concrete and then the inelastic analysis of concrete structures as given in the Euro-code.

22.3 STRESS–STRAIN CHARACTERISTICS OF CONCRETE

We use the conservative and safe stress–strain curve as shown in Fig. 22.1 given in IS 456 (as Fig. 21) for the design of the normal concrete section. However, for a study of the non-linear behaviour of a concrete section and its use in structural analysis we have to use the actual stress–strain curve of concrete given in Section 22.3.4, which is to some extent different from that in Fig. 22.1. In the following sections, we will discuss how this curve has been obtained.

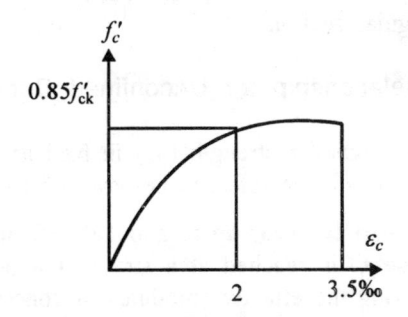

Fig. 22.1 Stress–strain curve of concrete for limit state design.

22.3.1 Historical Review of the Study of Stress–Strain of Concrete

Tests by Richart as early as in 1928 [1] had shown that concrete under combined compressive stresses (under hydraulic lateral compression) gives high strength. The strength was expressed by him as

$$f_c = f_c' + 4.1 F \text{ (lateral)}$$

where f_c and f_c' are the confined and unconfined cylinder strength of concrete and F the confining strength. Careful unconfined compression tests on cylinders at Bureau of Reclamation, U.S.A. around 1940 with the measurement of strains beyond the maximum load showed that the stress–strain curve of concrete is not discontinuous at failure load. The stress–strain curve consists of a rising part and also a drooping part beyond the maximum load point. Around 1950, Hognestad showed by tests on specimens, as shown in Fig. 22.2(a), that in bending compression the stress–strain curve can be represented by Fig. 22.2(b) [2]. Further tests of

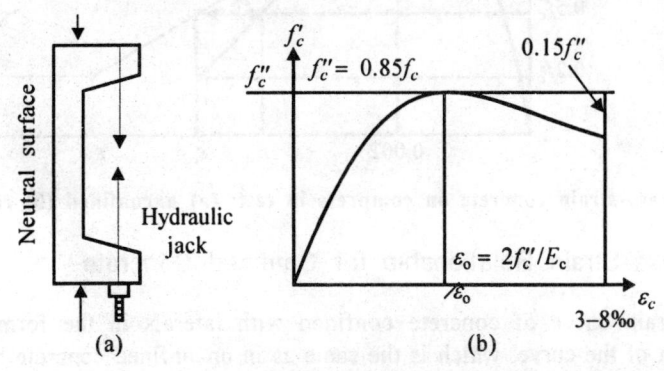

Fig. 22.2 Hognestads stress–strain curve for concrete for limit state design.

concrete in compression with the incorporation of spirals and closely spaced ties or stirrups around 1970 have shown that the failure strains in concrete can be considerably increased by incorporation of lateral ties and thus making the concrete more ductile.

Expressions for the stress–strain curves for such concrete have been suggested by many authors and we will select for study the one similar to that suggested by Kent and Park [3]. Though these investigators based their results only on compression tests of confined concrete in square columns, we assume that the results can be considered as valid also for cases with uniform stress variation. Thus, the results are used in practice for confined concrete in flexure also for square and rectangular sections.

22.3.2 Stress–Strain Relationship for Unconfined Concrete

We will take the unconfined cylinder strength (f'_c) in p.s.i as reference. The nature of the stress–strain curve for unconfined concrete can be assumed as follows (Fig. 22.3):

   (i) The initial upward part is linear up to about $0.45 f'_c$ only.
   (ii) The maximum value f'_c is reached at a strain of about 0.002. This strain is also expressed as follows, by taking the effective modulus of concrete as $E_c/2$:

$$\varepsilon_c = \frac{2 f'_c}{E_c}$$

   (iii) The descending curve is a straight line so that at a strength of $0.5 f'_c$ the strain is given by the following formula (in p.s.i. units):

$$\varepsilon_{c1} = \frac{3 + 0.002 f'_c}{f'_c - 1000} \qquad (22.1)$$

   (iv) The curve droops down to a strength value of $0.2 f'_c$ and concrete crumbles.

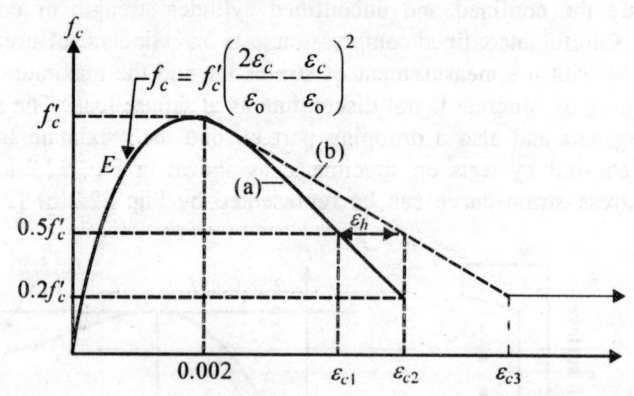

Fig. 22.3 Stress–strain concrete on compress in test: (a) unconfined (b) confined with loops.

22.3.3 Stress–Strain Relationship for Confined Concrete

The stress–strain curve of concrete confined with laterals in the form of hoops has an ascending part of the curve, which is the same as in unconfined concrete but the strain along the descending part at $0.5 f'_c$ is now increased further by a value ε_h.

$$\varepsilon_h = \frac{3}{4}\rho_s\sqrt{\frac{b}{s_h}} \qquad (22.2)$$

so that the total strain on descending curve at $0.5f''_c = \varepsilon_{c1} + \varepsilon_h = \varepsilon_{c2}$, where

ε_{c1} = Corresponding strain at $0.5f''_c$ in unconfined concrete

$\rho_s = \dfrac{\text{Volume of transverse steel in one lateral}}{\text{Volume of confined concrete between laterals}}$

b = Width of confined concrete core between outside to outside of the laterals

ε_h = Spacing-of hoops

The effect of laterals is felt only after the concrete cracks after the peak strength. Also confined concrete maintains a strength of $0.2f''_c$ for very large values of strains as shown in Fig. 22.4.

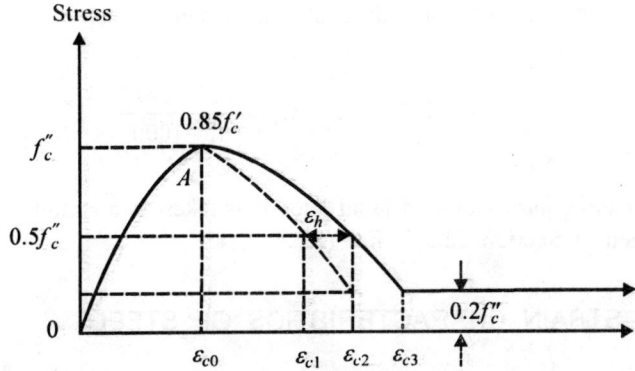

Fig. 22.4 Recommended stress–strain curve of confined concrete for structural calculation.

22.3.4 Stress–Strain Curve for Inelastic Analysis of Structures

Based on the data in Section 22.3.3 the stress–strain curve for bending compression in nonlinear analysis of members can be taken as shown in Fig. 22.4 with the following properties: (This will be obviously different from the stress–strain curve got from the pure compression tests on cylindrical specimen. Many simplified equations for this stress–strain curves have been proposed and the one given below is among them. The aim of the following discussions is only to illustrate the principles involved rather than correctness of the stress–strain formulae. The final equation for this curve has not been accepted by the researchers. Accordingly, any suitable formula for this purpose can be used at the present stage. In our discussions, we assume the following stress–strain curve in bending as shown in Fig. 22.4.)

(i) The maximum safe stress reached in bending in a structure is taken as $f''_c = 0.85f'_c$, where f'_c is the compression strength obtained from cylinder tests.

(ii) The strain at maximum stress is

$$\varepsilon_0 = \frac{2f''_c}{E_c} \qquad (22.3)$$

or can be taken as 0.002 for all practical purposes (E_c is the tangent models).

(iii) The ascending part of the curve is linear up to $0.4f_c''$.
(iv) The ascending part beyond $0.4f_c''$ can be represented by the stress–strain formula. One of the proposed equations by Hognestad is as follows:

$$f_c = f_c'' \left[2\frac{\varepsilon_c}{\varepsilon_0} - \left(\frac{\varepsilon_c}{\varepsilon_0}\right)^2 \right] \qquad (22.4)$$

where

f_c'' = Maximum stress reached = $0.85f_c'$

ε_c = Strain corresponding to stress f_c'

ε_0 = Strain at maximum stress

The full curve OA can also be approximated to a parabola.

(v) The descending part without additional laterals can be taken as a straight line passing through a point.

$$f = 0.5f_c'' \quad \text{and} \quad \varepsilon_{c1} = \frac{3 + 0.002 f_c''}{f_c'' - 1000}$$

in p.s.i. units.

(vi) The descending part with additional laterals is taken as a straight line with additional strains as described in Section 22.3.3, Eq. (22.2).

22.4 STRESS–STRAIN CHARACTERISTICS OF STEELS

Steels are nowadays classified in codes like the Euro-code [4] into two groups depending on their stress–strain curves as shown in Fig. 22.5.

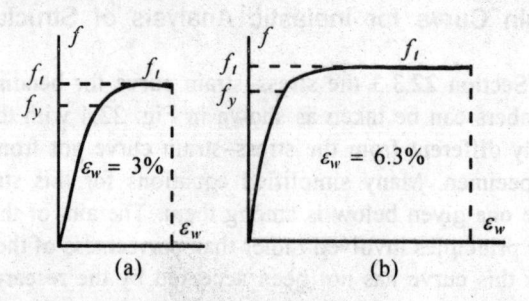

Fig. 22.5 Classes of steel: (a) class N steel (b) class H steel.

1. Steel of normal ductility or Class N steel: These are steels with the ratio of tensile strength to yield strength (0.2% proof stress) between 1.05 to 1.06 tensile strength, i.e. 5 to 6% more than the yield. The elongation at maximum yield load ε_w should be between 2.5 to 3.3%.

2. High ductility steel or class H steel: In this type of steel, the ratio of the strength to yield strength is between 1.08 to 1.1, i.e. tensile strength 8 to 10% more than the yield. The elongation at maximum yield should be between 5 to 6.3%.

INELASTIC ANALYSIS OF REINFORCED CONCRETE BEAMS AND FRAMES 397

In Euro-code, routine design redistribution up to 30% is allowed in reinforced concrete with class H steel and only 15% is allowed in members with class N steel. The comparison among the prescribed mechanical properties of Indian steels are as given in Table 22.1. We can see from the table that only Fe 415 and Fe 250 adequately satisfy the specifications for class H steel. From Chapter 21, we know that only these steels are recommended to be used for ductile detailing.

TABLE 22.1 MECHANICAL PROPERTIES OF STEEL (IS 1786/85)

Property	Grades			
	Fe 415	Fe 500	Fe 550	Fe 250
Minimum yield or proof stress (N/mm²)	415	500	550	250
Minimum elongation % at failure	14.5	12.0*	8.0	30.0
Tensile strength (UTS)	485	545	585	485
(% over yield strength)	(16)	(8)*	(6)	(94)

*Many laboratory tests on Indian steels have shown values less than this prescribed value.

22.5 MOMENT CURVATURE RELATION (M–φ CURVES)

In a bending member we have the following relationship with reference to Fig. 22.6

$$\frac{M}{EI} = \frac{f_c}{E_c x} = \frac{1}{R} = \phi \tag{22.5}$$

Hence

$$\phi = \frac{\varepsilon_c}{x} = \frac{\varepsilon_s}{d-x} = \frac{\varepsilon_c + \varepsilon_s}{d} \tag{22.6}$$

where

ϕ = Curvature (1/m units) (represented also as ψ)
f_c = Maximum stress in compression in concrete
ε_c = Maximum strain in concrete
ε_s = Maximum strain in steel
x = Depth of neutral axis
d = Effective depth of section
E_c = Modules of elasticity of concrete

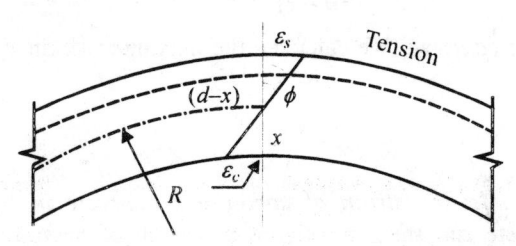

Fig. 22.6 Calculation of curvature in reinforced concrete beams.

The moment curvature diagram of a given section in bending can be theoretically computed from the assumptions of the bending theory by calculating the corresponding values of M and ϕ. The theoretical stress–strain curves for steel and concrete together with the equations of equilibrium and compatibility are used for this purpose. The values of ϕ can be obtained from Eq. (22.5) and (22.6):

The important points along the stress–strain curve at which ϕ values are to be calculated are as follows (*see* also Fig. 22.7):

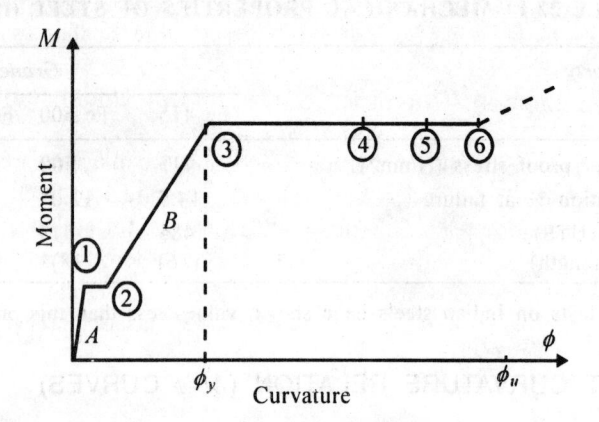

Fig. 22.7 Moment curvature diagram of a reinforced concrete section.

1. *Just before cracking.* In this case, concrete also takes tension and the moment is the cracking moment. Then

$$\phi_1 = \frac{M_{cr}}{EI_{c1}}$$

2. *Just after cracking.* In this case, concrete does not take any tension. We find neutral axis of composite section and I_{c2}. Then

$$\phi_2 = \frac{M_{cr}}{EI_{c2}}$$

3. *At yielding of steel.* In this case, tension steel reaches yield point but concrete does not reach maximum strain under reinforced beams. By trial and error, we can find the maximum strain in concrete from the relation $C = T$.

$$\phi_3 = \frac{\varepsilon_s}{(d-x)} = \phi_y$$

4. *At maximum stress in concrete.* We calculate the maximum strain ε_{c0} from Eq. (22.3) or assume it as 0.002.

$$\phi_4 = \frac{\varepsilon_{c0}}{x}$$

5. *At* $\varepsilon_{cm} = 0.0035$, i.e. *failure strain of unconfined concrete in bending.* Using the stress–strain curve of concrete and steel, determine the depth of neutral axis and ϕ.

$$\phi_5 = \frac{\varepsilon_{cm}}{x} = \phi_u$$

6. At ε_{cm} = Strain hardening of steel

$$\phi_6 = \frac{\varepsilon_{ch}}{x} = \phi_{sh}$$

Such a procedure will give the M–ϕ relationship of the section up to the stage of strain hardening of steel. (We can also note that the value of ϕ in columns subjected to P and M will be small as the depth of the neutral axis will be large so that ε_c/d will be small.)

22.5.1 Simplified Moment–Curvature Curve for an Under-reinforced Beam

For simplicity, we can approximate the moment–curvature diagram of an under-reinforced beam to a bilinear curve. The usual inelastic analysis (method 3 of Section 22.1) uses a linear curve up to ϕ_y, i.e. the linear part of the diagram. However, for collapse load analysis as well as problems dealing with energy absorption beyond elastic deformations we have to make use of the full M–ϕ diagram up to ϕ max which will be bilinear diagram. The capacity for energy dissipation is sometimes expressed as in Fig. 22.8 in terms of curvature ductility as

$$\mu_\phi = \frac{\phi_u}{\phi_y}$$

Use of the properties of unbound concrete for inelastic analysis gives safe values as we know that the value of ϕ can be very much increased by incorporating laterals at the position where rotations should take place for redistribution of moments.

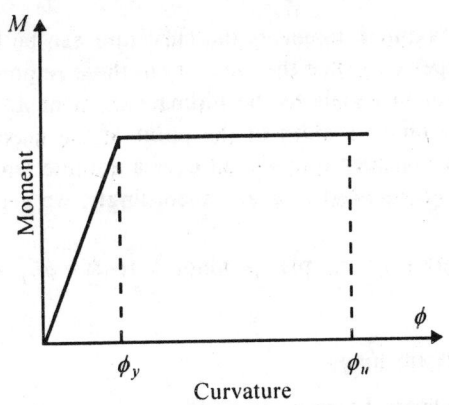

Fig. 22.8 Simplified moment curvature diagram.

22.5.2 Stiffening Effect due to Tension in Concrete

The above discussions on M–ϕ relation is based on the assumption that concrete cannot take any tension. However, on the tension side between cracks the concrete is still active in tension and the effect of this tension is called *tension stiffening effect*. We made use of the action in Chapter 3 dealing with deflection. For accurate evaluation of M–ϕ curve this effect will also have to be considered. Its effect is more in the span length than at the supports and it reduces the conventionally calculated curvature for the same values of the bending moment. For a first approximation, this effect can be neglected. As shown in Fig. 22.9, Eurocode makes correction for this effect for calculation of the M–ϕ relationship [4].

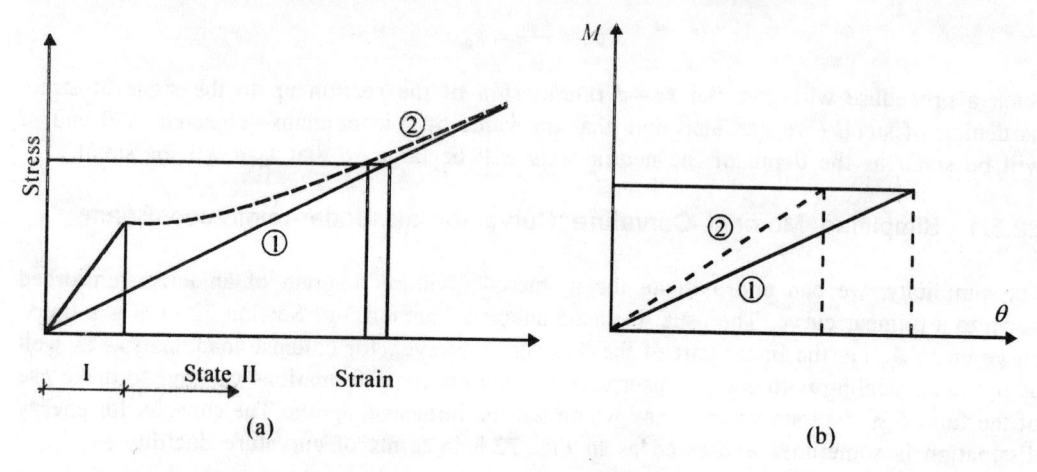

Fig. 22.9 Effect of tension stiffening: (a) concrete stress–strain steel curve (b) moment rotation curve—1. without tension-effect and 2. with tension effect, taken into account.

22.6 CONCEPT OF PLASTIC HINGES (MOMENT ROTATION AT PLASTIC HINGES)

We found that at points of maximum moments the curvature can be high. We can also make the assumption that if we properly confine the concrete in these regions of maximum moments the value of the bending moment equals to the ultimate moment M_u can be reached over a certain length of the member on both sides of the point of the maximum moment. In other words, plastic moment can be considered to spread over a definite length of the member. This length is called as the *length of the plastic hinge*. Accordingly, we get the following relations:

$$\text{The rotation of the plastic hinge} = \int \phi \, ds = \phi L_p = \theta$$

where

L_p = Length of the plastic hinge

θ = Rotation of the plastic hinge in radians

ϕ = Curvature at a point

Many expressions have been proposed for the length of the plastic hinges. It has been found by tests to be approximately equal to one-half the effective depth of the member on either side of the point of maximum elastic moment. Thus the length of a plastic hinge between supports in a beam can be taken as equal to the effective depth of the bending member. (At supports of a fixed beam the length of plastic hinge can be assumed as one-half the depth.)

INELASTIC ANALYSIS OF REINFORCED CONCRETE BEAMS AND FRAMES 401

22.7 EFFECT OF SHEAR ON ROTATION CAPACITY

Inclined shear at supports can, to some extent, affect the plastic rotation capacity. However, it is always better to make the sections safe in shear. Limited tests conducted in 1966 by Beachman and Turliman seem to indicate that shear can be safely transmitted over the plastic zone up to the point of yielding of stirrups.

22.8 INELASTIC OR NON-LINEAR ANALYSES OF REINFORCE CONCRETE BEAMS

The third method listed in Section 22.1 is non-linear analysis. The principles explained above are used for the non-linear analysis of R.C. members. In this analysis of structures, we redistribute more than the nominal amount of bending moment allowed by codes but we also check the following requirement of rotations (denoted by θ).

$$\theta \text{ (required)} \leq \theta \text{ (allowable)}$$

Assuming that the sections are under reinforced, the procedure for this method of analysis as illustrated by Example 22.6 for a two-span continuous beam is as follows:

Step 1: Find the bending moment diagram using factored load and elastic analysis.

Step 2: Reduce the maximum negative moment as required and redraw the distributed moment. Find also the position and magnitude of maximum positive moment. (The redistribution can be more than 30%.)

Step 3: Divide the span into an even (say 4) divisions so that we have an odd member of ordinates (say 1 to 5). Determine the B.M. ordinates from M_1 to M_5. This enables us to use Simpson's rule to find $\Sigma Mm(ds/EI)$.

Step 4: Design the support section and find A_s required. Find x/d and also ϕ_y, ϕ_u values of the section (at yielding of steel and at ultimate failure). Assuming that the diagram is linear from $M = 0$ to $M = M_y$ (Fig. 22.7) draw the M–ϕ diagram.

Step 5: Repeat the above calculation for the span section and determine the M–ϕ diagram for this section.

Step 6: Give unit rotation for the support and draw the (unit moment) diagram. By choosing the apropriate M and ϕ, calculate θ required as

$$\theta \text{ (required)} = \int \frac{M}{EI} m \, ds = \int \frac{1}{R} m \, ds = \int \phi m \, ds$$

This can be found by using the Simpson's rule with odd number of ordinates. Taking s as the intervals the integral is given by the formula:

$$V = \frac{s}{3} [\text{first} + \text{last} + 4(\text{even}) + 3(\text{odd})]$$

product of ordinates.

Step 7: Determine the (θ available) at the centre support which has to rotate by one of the following methods:

Method 1: Using the relation (*see* Sec. 22.6)
$$\theta = \phi L_p = \phi \times \text{(Effective depth)}$$

Method 2: Using Table 22.1 the approximate allowable values of ϕ given in Euro-code for x/d values (for more exact value *refer* Ref. 4) explained in Section 22.8.1.

Step 8: Check the rotation requirement
$$\theta \text{ (available)} \geq \theta \text{ (required)}$$

22.8.1 Allowable Rotation for Inelastic Analysis

The available notation at the plastic hinge depends on many factors. Some of them are the following:

1. Neutral axis depth factor at the section (x/d ratio)
2. Type of steel (yield strength and ultimate tensile strength ratio)
3. Reinforcement index given by $A_s f_y/(bdf_c)$
4. Span depth ratio.

However the major factor is considered in the x/d ratio. The method recommended by CEB-FIP for routine design is to use ϕ values recommended for x/d ratio. Model code CEB-FIP[4] gives a design curve based on results of 350 tests performed in the sixties. These values however neglected the favourable influence of transverse steel on rotation. In addition, the present-day steels are improved versions of steel commonly used in the sixties. Similar experimental values have been published for sections with binders in the members as shown in Fig. 22.10[5]. Euro-code has taken a similar procedure and gives a set of curves (separately for N and H steels). It gives the value of x/d and allowable θ [6]. Approximate values of this curve is given in Table 22.2. This can be used for inelastic design. This method is more popular as they are based on test results.

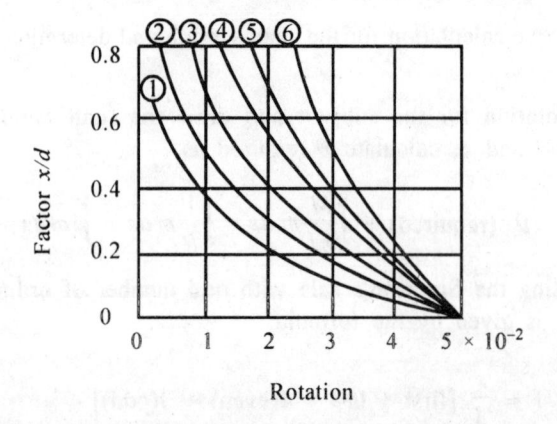

Fig. 22.10 Plastic rotation capacity of reinforced concrete members: 1. CEB-FIP recommendation with no extra binders, 2–6 other results with 0, 0.5, 1.0 1.5 and 2% binders [5].

TABLE 22.2 APPROXIMATE VALUES OF ALLOWABLE PLASTIC ROTATION IN RADIANS [Ref. 4.6]

x/d	Allowable plastic rotation (radians 10^{-3})	
	High ductile steel	Normal steel
0.50	6.0	6.0
0.45	6.6	6.6
0.40	8.0	8.0
0.35	10.0	8.0
0.30	12.0	8.0
0.25	14.0	8.0
0.20	17.0	8.0
0.16	20.0	8.0

Another method suggested to estimate a safe value of allowable rotation is to assume that the curvature beyond yield is spread over the plastic length [7] and the allowable rotation

$$\phi_u \times \text{(Length of plastic hinge)} \qquad \phi_u L_p = \phi_u d$$

This method gives fairly safe values as ϕ_u can be made large by use of extra stirrups. Example 22.6 has been worked out by both of these methods.

22.9 ALLOWABLE ROTATION FOR COLLAPSE LOAD ANALYSIS

Since full plastic analysis does not take serviceability into account but only the strength, extensive cracking and higher rotation of the plastic hinge are allowed in such analysis. Accordingly, as suggested by Mattock and Corley we may assume the hinge rotation can extend up to the ultimate limit so that

$$\theta_p = L_p(\phi_u - \phi_y)$$

22.10 BAKER'S METHOD FOR PLASTIC ANALYSIS OF BEAMS AND FRAMES

The plastic or collapse load of steel structures which are n times statically indeterminate is deemed to be that load that can produce $(n + 1)$ hinges turn into a mechanism. A similar method of analysis is sometimes used also for concrete structures like fixed beams. But, as different from steel structures where the rotation capacity is very large for concrete structures, the rotations necessary for the plastic hinges to form such a mechanism are checked against the rotation capacity of the hinges.

A much more realistic and hence popular method for concrete structures is the method proposed around 1960 by A.L.L. Baker of Imperial College London [8]. The method uses the well-known concepts of elastic analysis which have been modified to suit the plastic analysis of concrete structures.

22.10.1 Assumptions of Baker's Method

The following assumptions are made in Baker's method of analysis:

(i) The collapse load of a reinforced concrete structure is the load that converts a structure which is n times statically indeterminate to a statically determinate structure by forming n hinges. (This is not a mechanism.)

(ii) The load corresponding to the formation of the last n-th plastic hinge is called the *collapse load* of the concrete structure.

(iii) The portions of the structure between the hinges remains elastic.

(iv) Under increasing loads, the relation between rotation and moment of resistance can be taken as bilinear so that when the hinges are formed it rotates under constant moment (Fig. 22.7).

(v) The plastic hinges are concentrated over a definite length at points of maximum moments, called plastic length.

(vi) The safe limiting values of rotation of the plastic hinges should allow for the cracking and the change of EI over the length of the members.

22.10.2 Influence Coefficient Method of Analysis of Structures

The well-known influence coefficient method or matrix method was recommended by Baker for the collapse load analysis of beams and frames. Let $ABCDE$ be a continuous beam which is three-times indeterminate. Let us assume the loading to be a uniformly distributed load and let the indeterminate support moments at B, C and D (which are points of maximum moments) be designed as X_1, X_2, X_3 [Fig. 22.11].

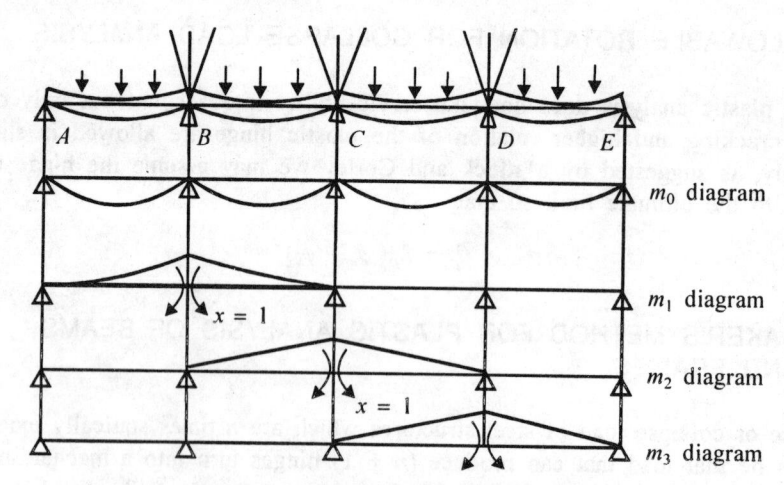

Fig. 22.11 Analysis of continuous beam.

If we have to analyse the structure by elastic analysis using the influence coefficient we proceed as follows:

1. Make the structure determinate by introducing hinges at the points of maximum moments, such as at B, C and D, and assume the moments X_1, X_2, X_3 are acting at these points in the indeterminate structure.

2. Draw (M_0) the bending moment for the determinate system.
3. We introduce unit force at various hinge points and find m_1, m_2 and m_3.
4. Taking point B for an indeterminate structure elastic analysis as there are no discontinuities.

$$\frac{1}{EI}[M_0 m_1 ds + (\int m_1 m_1 ds) X_1 + (\int m_1 m_2 ds) X_2 + (\int m_1 m_3 ds) X_3] = 0$$

Let

$$\int \frac{M_0 m_1}{EI} ds = f_{01}$$

$$\int \frac{m_1 m_2}{EI} ds = f_{11}$$

and so on. Then

$$f_{11} X_1 + f_{12} X_2 + f_{13} X_3 + f_{01} = 0$$

5. For the full elastic solution, we get the following equation:

$$\begin{bmatrix} f_{11} & f_{12} & f_{13} \\ f_{21} & f_{22} & f_{23} \\ f_{31} & f_{32} & f_{33} \end{bmatrix} \begin{bmatrix} X_1 \\ X_2 \\ X_3 \end{bmatrix} + \begin{bmatrix} f_{01} \\ f_{02} \\ f_{03} \end{bmatrix} = 0 \qquad (22.7)$$

6. The zero at the right-hand side in Eq. (22.7) denotes that there will be no discontinuity, which is the case for elastic conditions. However, in a structure which has formed plastic hinges at B, C and D, there will be definite discontinuities due to rotations, θ_1, θ_2 and θ_3. These rotations are opposite in direction to the plastic moments acting at the hinge. Accordingly, for collapse load analysis, Eq. (22.7) reduces to the following equation:

$$\begin{bmatrix} f_{11} & f_{12} & f_{13} \\ f_{21} & f_{22} & f_{23} \\ f_{31} & f_{32} & f_{33} \end{bmatrix} \begin{bmatrix} X_1 \\ X_2 \\ X_3 \end{bmatrix} + \begin{bmatrix} f_{01} \\ f_{02} \\ f_{03} \end{bmatrix} = - \begin{bmatrix} \theta_1 \\ \theta_2 \\ \theta_3 \end{bmatrix} \qquad (22.8)$$

In actual analysis of a frame or a beam, the position of the plastic hinges which occur at points of maximum elastic moments are first assumed on the basis of elastic analysis, Values of X_1, X_2, X_3 are assumed and the values of θ_1, θ_2, θ_3 necessary can be determined. If these are positive (correct sign) and within the permissible values, the values, of X_1, X_2, X_3 assumed are justified. The sections are designed for these values. This procedure is illustrated by Example 22.7.

Example 22.1 (Stress–strain curve for structural analysis)
Determine the salient points on the stress–strain curve of concrete in bending of a unconfined concrete member if cylinder strength of concrete used $f'_c = 25$ N/mm². If such a concrete is confined in a section of breadth 300 mm, total depth 500 mm, and clear cover of 50 mm with 10 mm (78 mm²) stirrups at 100 mm centres, determine the stress–strain curve for inelastic analysis of the structure. Use the relation $f'_c = 0.8 f_{ck}$.

Reference	Step	Calculations
	(A)	*Stress–strain curve of unconfined concrete*
	1	*Estimate stress and strain at the end of linear part*
		Stress = $0.4 f'_c = 0.4 \times 25 = 10$ N/mm^2
		$E_c = 5700\sqrt{f_{ck}} = 5700\sqrt{\dfrac{25}{0.8}} = 3.19 \times 10^4$ N/mm^2
Cl. 6.2.3.1		(According to IS 456 (2000) $E_c = 500\sqrt{f_{ck}}$)
		ε_{ct} = Strain at 10 N/mm^2 = $\dfrac{10.0}{3.19 \times 10^4} = 0.00031$
	2	*Maximum stress attained in a structure (f''_c) and corresponding strain (ε_0)*
Sec. 22.3.4		$f''_c = 0.85 f'_c = 0.85 \times 25 = 21$ N/mm^2
		Theoretically $\varepsilon_0 = \dfrac{f''_c}{0.5 E_c} = 0.0013$
		(Usually a strain of 0.002 can also be assumed for this point.)
	3	*Calculate strain at $0.5 f''_c$ on the descending curve*
		The stress–strain curve on the descending part is assumed as linear.
		$f''_c = 0.5 \times 21 = 10.5$ N/mm^2
		The corresponding strain at $0.5 f''_c$ is given by the formula (in p.s.i. units) as
Eq. 22.1		$\varepsilon_{c1} = \dfrac{3 + 0.002 f'_c}{f'_c - 1000}$
		f'_c (in p.s.i.) = $25 \times 145 = 3625$ p.s.i.
		$\varepsilon_{c1} = \dfrac{3 + 0.002 \times 3625}{3625 - 1000} = 0.0039$
Sec. 22.3.2	4	*Estimate the stress at which the concrete crumbles*
		Stress = $0.2 f''_c = 0.2 \times 21 = 4.2$ N/mm^2
	5	The stress–strain curve is shown as curve *1* in Fig. 22.12.
	(B)	*Stress–strain curve for confined concrete*
	1	The ascending part is the same as in unconfined concrete
	2	*Determine parameters of confinement*
		$b' = 300 - 100 = 200$ mm
		Depth of confinement = $500 - 100 = 400$ mm
		Vol. of transverse steel (with outside dimensions)

Reference	Step	Calculations
		$= 78(2)(200 + 400) = 9.36 \times 10^4$ mm^3 Vol. of concrete within one stirrup spacing $= 200 \times 400 \times 100 = 8 \times 10^6$ mm^3 $\rho_s = \dfrac{9.36 \times 10^4}{8 \times 10^6} = 1.17 \times 10^{-2}$
Eq. (22.2)	3	Calculate enhanced strain over ε_{c1} at $0.5 f_c''$ $\varepsilon_h = 0.75 \rho_s \left(\dfrac{b}{s}\right)^{1/2} = 0.75 \times 1.17 \times 10^{-2} \sqrt{2} = 0.012$ Total strain $\varepsilon_{c1} + \varepsilon_h = 0.0039 + 0.012 + 0.016$ (approx.)
	4	Determine the minimum strength on the ascending curve $0.2 f_c'' = 0.2 \times 21 = 4.2$ N/mm^2 At 4.2 N/mm^2, the curve will be longitudinal.
	5	The stress–strain curves is shown in Fig. 22.12 as curve 2

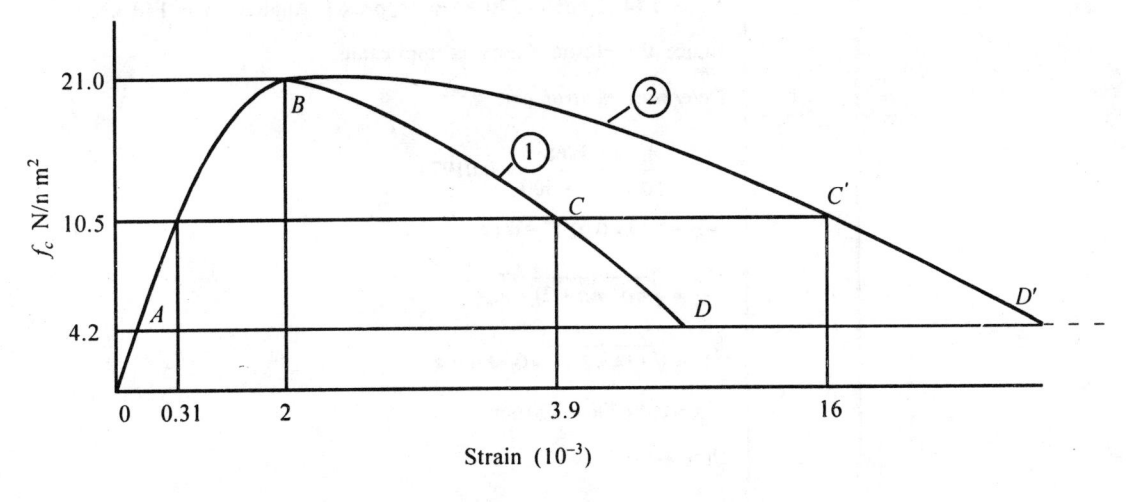

Fig. 22.12 Example 22.1

Example 22.2 (Effect of concrete tension on M–φ curvature)

A reinforced concrete section is 200 mm × 550 mm depth. If the applied moment is 140 kNm determine the instantaneous curvature assuming a tensile stress in concrete at level of steel of 1 N/mm^2 (Fig. 22.13). Assume $E_c = 28{,}000$ N/mm^2 and $E_s = 200{,}000$ N/mm^2 ($m = E_s/E_c = 7.14$); $f_{ck} = 20$.

408 ADVANCED REINFORCED CONCRETE DESIGN

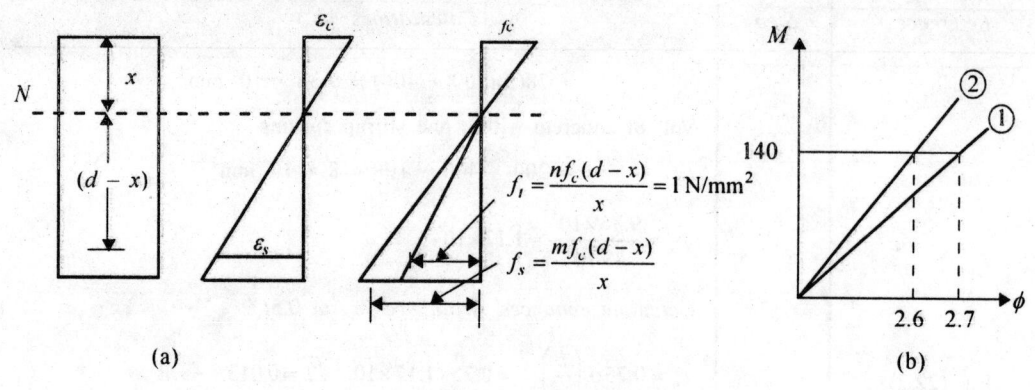

Fig. 22.13 Example 22.2.

Reference	Step	Calculations
		Method 1
		An approximate method is to find the conventional N.A. with no tension in concrete and then reduce the applied moment by ΔM the moment taken by concrete section and recalculate $M/EI = \phi$
		$M_u = 0.14 f_{ck} bd^2 = 270$ kNm (approx.). Applied $M = 140$ kNm.
		Hence the elastic theory is applicable.
	1	*Determine neutral axis x*
		$\rho = \dfrac{A_s}{bd} = \dfrac{1968}{200 \times 500} = 0.0197$
		$m\rho = 7.14 \times 0.0197 = 0.14$
		$\dfrac{x}{d} = \sqrt{m\rho(m\rho + 2)} - m\rho$
		$= \sqrt{0.14 \times 2.14} - 0.14 = 0.4$
		$x = 0.4 \times 500 = 200$ mm
	2	*Determine I*
		$I = \dfrac{bx^3}{3} + mA_s(d-x)^2$
		$= \dfrac{200(200)^3}{3} + 7.14 \times 1968(300)^2$
		$= 5.33 \times 10^8 + 12.6 \times 10^8 = 17.9 \times 10^8$ mm^4
		$= 1.79 \times 10^9$ mm^4

INELASTIC ANALYSIS OF REINFORCED CONCRETE BEAMS AND FRAMES   409

Reference	Step	Calculations
Fig. 22.13	3	*Determine curvature with M = 140 kNm* $$\phi = \frac{M}{E_c I_c} = \frac{140 \times 10^6}{(2.8 \times 10^4)(1.79 \times 10^9)} = 2.79 \times 10^{-6} \text{ mm}^{-1}$$
	4	*Calculate ΔM taken by concrete* With a tension of 1 N/mm² at steel level tension at extreme fibre $$f'_t = \frac{1 \times 350}{300} = 1.17 \text{ N/mm}^2$$ $$\Delta M = \frac{1}{3} \frac{f'_t}{2} b (h - x)^2$$ $$= (0.5 \times 1.17)(200 \times 350)\left(\frac{2}{3}\right)(350)$$ $$= 9.6 \times 10^6 \text{ N·mm}$$ $M - \Delta M = 140 - 9.6 = 130.4$ kNm (net moment)
Fig. 22.13		$$\phi = \frac{130.4 \times 10^6}{(2.8 \times 10^4)(1.79 \times 10^9)} = 2.60 \times 10^{-8} \text{ mm}^{-8}$$ *Note*: This value is less than that calculated without tension effect of concrete. A further correction is needed as, at cracks, the tension is zero. *Method II* (BS 8110 method). The method given in part 2 BS 8110 is to find the exact value of neutral axis by method of successive approximation. The distribution of stresses and strains are as shown in Fig. 22.13(above). $$f_t = \frac{n f_c (d - x)}{x}, \quad n = \frac{f_t(x)}{f_c(d - x)}$$ where $f_t = 1$ N/mm². n is taken as effective tensile modular ratio. Thus the effective area in tension = n (area in tension).
	1	*Find f_c and n* $$f_c = \frac{M}{I} x = \frac{140 \times 10^6}{1.79 \times 10^9} \times 200 = 15.64 \text{ N/mm}^2$$ Taking $f_t = 1$ N/mm² the value of n will be $$n = \frac{(1)(200)}{(15.6)(300)} = 0.042$$

410   ADVANCED REINFORCED CONCRETE DESIGN

Reference	Step	Calculations
	2	Determine the new neutral axis by taking moment of areas $$200\frac{x^2}{2} = 7.14(1968)(500-x) + 0.042\left(\frac{500-x}{2}\right)^2 \times 200$$ $x = 210$ mm
	3	Find I with effective tension concrete $$I = \frac{(200)(210)^3}{3} + (7.14)(1968)(290)^2 + 0.042\frac{(200)(340)^3}{3}$$ $$= 1.91 \times 10^9$$
Fig. 22.13	4	Calculate the curvature $$\phi = \frac{M}{EI} = \frac{140 \times 10^6}{(2.8 \times 10^4)(1.91 \times 10^9)} = 2.6 \times 10^{-6} \text{ mm}^{-1}$$ as in step 4 in Method I. The effect of taking tension in concrete is shown in Fig. 22.2(b). [*Note*: Further correction, for the fact that this tension is available only between cracks, can be made and such a correction is used in Euro-code. This correction tends to decrease the effect of taking full tension into account.]

Example 22.3 (Moment curvature diagram of a concrete slab section)
A reinforced concrete slab is 105 mm thick with 20 mm cover to centre of steel. If the positive steel reinforcement is 424 mm²/m determine the approximate moment curvature diagram. Determine the ductility factor assuming M-25 concrete and Fe 250 steel for reinforcements.

Reference	Step	Calculations
	1	*Data* Effective depth $d = 105 - 20 = 85$ mm Ratio of steel $= \dfrac{424}{1000 \times 85} = 0.005$
	2	Calculate M_r and ϕ before the first crack $$f_{cr} = 0.7\sqrt{f_{ck}} = 0.7 \times 5 = 3.5 \text{ N/mm}^2$$ $$E_s = 5700\sqrt{f_{ck}} = 28.500 \text{ N/mm}^2$$ Strain $= \dfrac{f_{cr}}{E_s} = 1.228 \times 10^{-4}$

INELASTIC ANALYSIS OF REINFORCED CONCRETE BEAMS AND FRAMES 411

Reference	Step	Calculations
		Assume depth of N.A. $x = d/2 = 52.5$ mm
		$\phi = \dfrac{1.228 \times 10^{-4}}{52.5} = 2.3 \times 10^{-6}$ mm^{-1}
		$M_{cr} = \dfrac{3.5 \times 1000 \times (105)^2}{6} = 6.4$ kNm
	3	*Find N.A. depth after cracking*
		$\dfrac{x}{d} = \sqrt{m\rho(m\rho + 2)} - m\rho$
		$m = \dfrac{280}{f_{ck}} = 11.2$ (say)
		$m\rho = 11.2 \times 0.005 = 0.056$
		$\dfrac{x}{d} = \sqrt{0.055(2.055)} - 0.055 = 0.28$
		$x = 0.28 \times 85 = 23.9$ mm
	4	*Take yielding of steel*
	4.1	*Find lever arm and M_y on yielding of steel*
		$d - x = 85 - 23.9 = 61.1$ mm
		Yield strain in steel $= \dfrac{250}{200,000} = 1.25 \times 10^{-3}$ (at $f_y = 250$)
		Strain in concrete $= \dfrac{0.00125 \times 23.9}{61.1} = 4.89 \times 10^{-4}$
		Concrete has not reached maximum strain and may be considered elastic
		$z = d - \dfrac{x}{3} = 85 - \dfrac{23.9}{3} = 77$ mm
		$M_y = A_s f_y z = (424)(250)(77) = 8.16 \times 10^6$ N $= 8.16$ kNm
	4.2	*Determine ϕ on yielding of steel*
		$\phi = \dfrac{\varepsilon_s}{d - x} = \dfrac{0.00125}{61.1} = 20.5 \times 10^{-6}$ mm^{-1}
	5	*Take ultimate failure condition*
	5.1	*Determine x at ultimate condition*
		From IS 456 assume of $\varepsilon_c = 0.0035$

Reference	Step	Calculations
		$x = \dfrac{f_y A_s}{0.36 f_{ck} b} = \dfrac{(250)(424)}{(0.36)(25)(1000)} = 11.78$ mm $\dfrac{x}{d} = \dfrac{11.78}{85} = 0.138$ Lever arm $= d - 0.42x = 85 - 0.42(11.78) = 80.1$ mm $M_u = (424)(250)(80.1) = 8.49 \times 10^6 = 8.49$ kNm
	5.2	Find ϕ at ultimate load $\phi = \dfrac{\text{Concrete strain}}{x} = \dfrac{0.0035}{11.78} = 297 \times 10^{-6}$ mm^{-1}
	6	Write down coordinates of M–ϕ diagram $M = 0 \qquad \phi = 0$ $M_y = 8.16$ kNm $\qquad \phi_y = 20.5 \times 10^{-6}$ $M_u = 8.49$ kNm $\qquad \phi_u = 297.0 \times 10^{-6}$
	7	Determine ductility factor $\text{D.F.} = \dfrac{\phi_u}{\phi_y} = \dfrac{297}{20.5} = 14.5$

Example 22.4 (**Moment curvature curve of a section of an R.C. rectangular beam**) Determine approximate moment curvature values at salient points for a rectangular beam with the following dimensions: $f_{ck} = 37$, $f_c' = 30$ N/mm^2, $f_y = 500$ N/mm^2 (ductile steel) (Fig. 22.14). Assume breadth = 300 mm, total depth = 800 mm, effective depth = 750 mm and $A_s = 1900$ mm^2 (tension steel). [*Note*: This example is given simply to illustrate how a suitable equation to the stress–strain curve of concrete can be used to find M–ϕ diagram.]

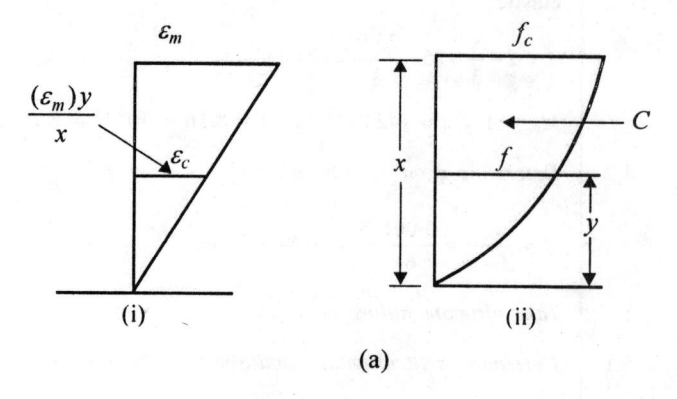

Fig. 22.14(a) Example 22.4.

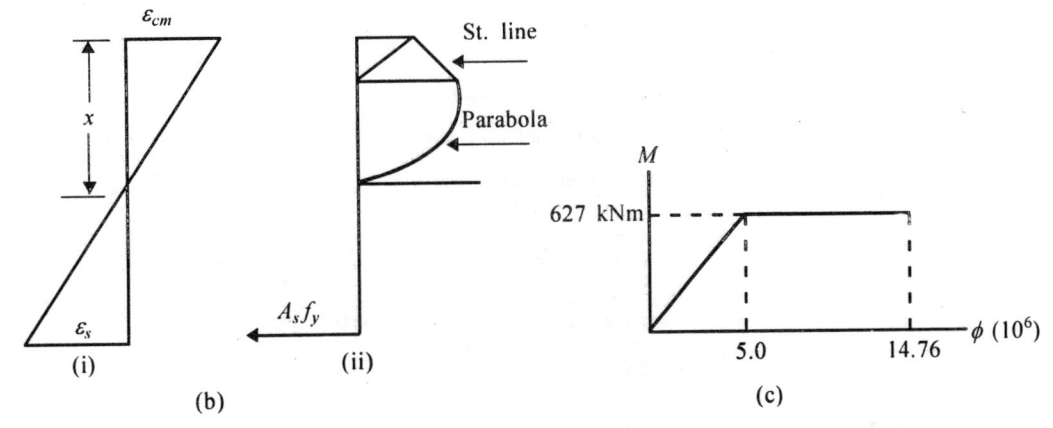

INELASTIC ANALYSIS OF REINFORCED CONCRETE BEAMS AND FRAMES 413

Fig. 22.14(b) and (c) Example 22.4.

Reference	Step	Calculations
	1	Properties of section $$\rho = \frac{1900}{750 \times 300} = 0.0084$$ $E_c = 5700\sqrt{37} = 34{,}671.7 \text{ N/mm}^2$ $E_s = 200{,}000 \text{ N/mm}^2$ m for elastic analysis $= \dfrac{280}{f_{ck}} = 7.57$ $m\rho = 0.0636$ $f_c'' = 0.85 f_c' = 0.85 \times 30 = 25.5 \text{ N/mm}^2$
Chapter 1 Sec. 1.5.1	2	Find ϕ at cracking moment just before cracking $f_{cr} = 0.7\sqrt{37} = 4.26 \text{ N/mm}^2$ Neglecting steel, N.A. at $d/2 = 400$ mm $$I_{gr} = \frac{300(800)^3}{12} = 1.28 \times 10^{10} \text{ mm}^4$$ $$M_{cr} = \frac{f_{cr} I_{gr}}{y} = \frac{(4.26)(1.28)(10^{10})}{400}$$ $= 136 \times 10^6$ Nmm $= 136$ kNm $$\phi = \frac{f_{cr}}{y} = \frac{M}{EI} = \frac{136 \times 10^6}{(34600)(1.28)10^{10}} = 3.08 \times 10^{-7}$$ $= 3.07 \times 10^{-7} \text{ mm}^{-1}$

Reference	Step	Calculations
	3	**Find ϕ just after cracking**
		Concrete does not take tension
		Depth of neutral axis x is
		$$\frac{x}{d} = \sqrt{m\rho(2+m\rho)} - m\rho$$
		$$= \sqrt{0.0636(2+0.0636)} - 0.0636 = 0.3 \text{ (approx)}$$
		$x = 750 \times 0.3 = 225$ mm
		$$I = \frac{300(225)^3}{3} + 1900 \times 7.57 \times (750-225)^2$$
		$= 51.0 \times 10^8$ mm^4
		$$\phi \text{ at } M_{cr} = 136 \times 10^6 = \frac{M}{EI}$$
		$$= \frac{136 \times 10^6}{(34600)(51.0) \times (10)^8} = 7.7 \times 10^{-7} \text{ mm}^{-1}$$
	4	**Find the curvature at yielding of steel** (*using stress–strain curve*)
	4.1	For $\varepsilon_s = \frac{500}{200,000} = 0.0025$ (assuming bilinear curve)
		We use an equation for stress–strain curve to find the value of x at the depth of N.A.
		Let ε_m = the maximum strain is concrete at this stage
Eq. (22.4)		$$f_c = f_c'' \left[\frac{2\varepsilon_c}{\varepsilon_0} - \left(\frac{\varepsilon_c}{\varepsilon_0}\right)^2 \right]$$
		where ε_0 is the strain at the maximum stress at f_c''
		At any point y from N.A. of depth x, we have $\varepsilon_c = \varepsilon_m y/x$
Fig. 22.14(a)		Stress $= f_c'' \left[2 \frac{\varepsilon_m}{\varepsilon_0} \frac{y}{x} - \left(\frac{\varepsilon_m}{\varepsilon_0}\right)^2 \left(\frac{y}{x}\right)^2 \right]$ (i)
		Integrating the stress, we get compression C as
		$C = bf_c'' K$
		K is the integration of the above expression in brackets $y = 0$ to $y = x$. Then

INELASTIC ANALYSIS OF REINFORCED CONCRETE BEAMS AND FRAMES

Reference	Step	Calculations
		$$K = \left[\frac{\varepsilon_m}{\varepsilon_0}\frac{y^2}{x} - \frac{1}{3}\left(\frac{\varepsilon_m}{\varepsilon_u}\right)^2\frac{y^3}{x^2}\right]_{y=0}^{y=x} \quad \text{(ii)}$$ But $\varepsilon_m = \frac{\varepsilon_s}{d-x}x$ Substituting this value in Eq. (ii), we get $$K = \left[\frac{\varepsilon_s}{\varepsilon_0}\frac{y^2}{(d-x)} - \frac{1}{3}\left(\frac{\varepsilon_s}{\varepsilon_0}\right)^2\frac{y^3}{(d-x)^2}\right]_{y=0}^{y=x} \quad \text{(iii)}$$ Now equating forces $C = T$, we get $bf_c''K = T = A_s f_y$ To find the value of Eq. (iii), at the yield of steel, we have $$\frac{\varepsilon_s}{\varepsilon_0} = \frac{0.0025}{0.002} = 1.25$$ $$K = \frac{T}{bf_c''} = \frac{1900 \times 500}{300 \times 25.5} = 124 \text{ (since } T = C\text{)}$$ Accordingly Eq. (iii) becomes with x as depth of N.A. $$1.25\frac{x^2}{d-x} - \frac{1}{3}(1.25)^2\frac{x^3}{(d-x)^2} = 124$$ $$\frac{x^2}{750-x} - 0.417\frac{x^3}{(750-x)^2} = 99.2$$ This gives $x = 250$ mm and $(d-x) = 500$ mm (Theoretically x should be less than the value got in step 3)
	4.2	Find the C.G. of compression by taking moments of forces in Eq. (i). Find corresponding M and ϕ. $$C.G. = \frac{\int f_y y}{\int f}$$ Multiplying Eq. (i) by y and integrating the expression in brackets $$\int\left[\frac{2\varepsilon_m}{\varepsilon_0}\frac{y^2}{x} - \left(\frac{\varepsilon_m}{\varepsilon_0}\right)^2\frac{y^3}{x^2}\right]dy = \left[\frac{2}{3}\left(\frac{\varepsilon_m}{\varepsilon_0}\right)x^2 - \frac{1}{4}\left(\frac{\varepsilon_m}{\varepsilon_0}\right)^2 x^2\right]_{y=0}^{y=x}$$ $$\bar{y} = \frac{bf_c''}{T}\left[\frac{2}{3}\left(\frac{\varepsilon_m}{\varepsilon_0}\right) - \frac{1}{4}\left(\frac{\varepsilon_m}{\varepsilon_0}\right)^2\right]x^2 \quad \text{(iv)}$$

Reference	Step	Calculations
		Now $\varepsilon_m = \dfrac{0.0025 \times 250}{500} = 0.00125$
		$\varepsilon_0 = 0.002; \quad x = 250\text{ mm}$
		$\dfrac{\varepsilon_m}{\varepsilon_0} = 0.625, \quad \dfrac{bf_c''}{T} = \dfrac{1}{124}$
		$\bar{y} = \left[\dfrac{2}{3}(0.625) - \dfrac{1}{4}(0.625)^2\right]\dfrac{250^2}{124} = 160.8 \text{ mm}$
		$M_y = 1900 \times 500 \times (500 + 160.8)\,(A_s \times f_y \times z)$
		i.e. $A_s f_y z = 627.8 \times 10^6$ N $= 627.8$ kNm
		$\phi_y = \dfrac{\varepsilon_s + \varepsilon_c}{d} = \dfrac{0.00125 + 0.0025}{750} = 5.0 \times 10^{-6}(\text{mm}^{-1})$
Fig. 22.14(c)		(The M–ϕ diagram can be assumed as linear form origin to $\phi = 5 \times 10^{-6}$ at $M_y = 627.8$ kNm. This is shown in Fig. 22.16.)
Fig. 22.14(b)	5	*Determine curvature at failure of concrete without special stirrups*
	(a)	We proceed similarly and find ϕ with the help of stress–strain curve.
	(b)	Alternately, we adopt a simpler but approximate method as follows.
	6	*Calculate curvature at the ultimate load according to IS 456*
		$C = 0.36 f_{ck} x b = T = 1900 \times 500 \times 0.8$
		Hence $x = \dfrac{1900 \times 500}{0.36 \times 37 \times 300} = 237.8$ mm
		$\dfrac{x}{d} = \dfrac{237.8}{750} = 0.317$
		Lever arm $= d - 0.42x = 750 - 0.42 \times 273 = 650.5$ mm
		$M_u = (1900 \times 500)(650.5) = 618$ kNm
		(This value is more or less the same as in step 4.)
		$\phi_u = \dfrac{0.0035}{237} = 14.77 \times 10^{-6}$ mm^{-1}
Fig. 22.14(c)	7	*Estimate the ductility factor*
		$D.F. = \dfrac{\phi_u}{\phi_y} = \dfrac{14.77}{5.0} = 2.95$
		This value is much less than that of the slab in Example 22.2, where the steel ratio was 0.005, effective depth only 85 mm and $x/d = 0.138$.

INELASTIC ANALYSIS OF REINFORCED CONCRETE BEAMS AND FRAMES 417

Example 22.5 (**Moment curvature diagram for inelastic design—flanged beam**)
Using the expressions in Example 22.4, determine the moment curvature diagram for inelastic design (assuming it linearly varies from zero to M_{max} at yielding of steel) for the following T section (*refer* Fig. 22.15):

Breadth of flange = 1400 mm, breadth of web = 300 mm, effective depth = 750 mm, depth of slab = 150 mm, area of steel in span = 1700 mm², area of steel over support = 1900 mm² (as in Example 22.3). Assume $f'_c = 30$, $f_{ck} = 37$ and $f_y = 500$ (class *H*), find ϕ and M at yielding of steel.

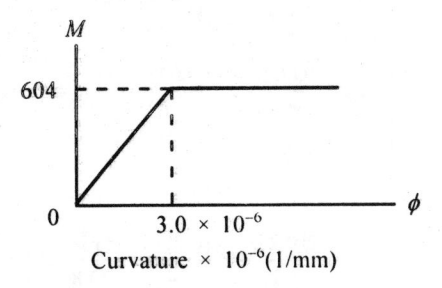

Fig. 22.15 Example 22.5

Reference	Step	Calculations
	1	*Data*
Fig. 22.4		$f''_c = 0.85 f'_c = 0.85 \times 30 = 25.5$
		ε_s at yield of steel $= \dfrac{500}{200,000} = 0.0025$
		Assume $\varepsilon_0 = 0.002$ (strain at maximum stress)
		Equation of equilibrium
Example 22.4 Eq. (iii)		$\dfrac{\varepsilon_s}{\varepsilon_0} \dfrac{x^2}{(d-x)} - \dfrac{1}{3}\left(\dfrac{\varepsilon_s}{\varepsilon_0}\right)^2 \dfrac{x^3}{(d-x)^2} = \dfrac{T}{bf''_c}$
		$\dfrac{\varepsilon_s}{\varepsilon_0} = \dfrac{0.0025}{0.002} = 1.25$
		$\dfrac{T}{bf''_c} = \dfrac{1700 \times 500}{1400 \times 25.5} = 23.8$ (T beam)
		$\dfrac{1.25 x^2}{750-x} - \dfrac{(1.25)^2 x^3}{3(750-x)^2} = 23.8$
		$\dfrac{x^2}{750-x} - \dfrac{0.42 x^3}{(750-x)^2} = 19.04$
		$x = 115$ (N.A. is within the flange.)

Reference	Step	Calculations
Example 22.4 Eq. (iv)		$\phi = \dfrac{\varepsilon_s}{750-115} = \dfrac{0.0025}{635} = 3.9 \times 10^{-6}$ mm^{-1}
	2	*Find C.G. of compression from neutral axis* $\bar{y} = \dfrac{bf_c''}{T}\left[\dfrac{2}{3}\dfrac{\varepsilon_m}{\varepsilon_0} - \dfrac{1}{4}\left(\dfrac{\varepsilon_m}{\varepsilon_0}\right)^2\right]x^2, \quad x = 115$ $\varepsilon_m = \dfrac{0.0025x}{d-x} = \dfrac{0.0025 \times 115}{635} = 4.52 \times 10^{-4}$ $\dfrac{\varepsilon_m}{\varepsilon_0} = \dfrac{0.000452}{0.002} = 0.226$ $\bar{y} = \left[\dfrac{2(0.226)}{3} - \dfrac{(0.226)^2}{4}\right]\dfrac{115^2}{23.8} = 76.6$ mm
	3	*Determine* M_y M_y = (Steel tension) (*L.A.*) $M_y = 1700 \times 500(635 + 76.6) = 604 \times 10^6$ Nmm $\qquad = 604$ kNm (Assume M–ϕ diagram is linear from origin to $\phi = 3.9 \times 10^{-6}$ at M_y = 604 kNm. This is shown in Fig. 22.16.)

Example 22.6 (Non-linear analysis of continuous beam)

A reinforced concrete T beam is continuous over two spans of 8 m and is loaded with a uniformly distributed load of 97 kN/m. If 30% of the central bending moment is to be redistributed to the spans, check the rotation capacity of the beam neglecting the tension stiffening effects. Assume the moment curvature relation are as in Fig. 22.16 taken from Examples 22.4 and 22.5.

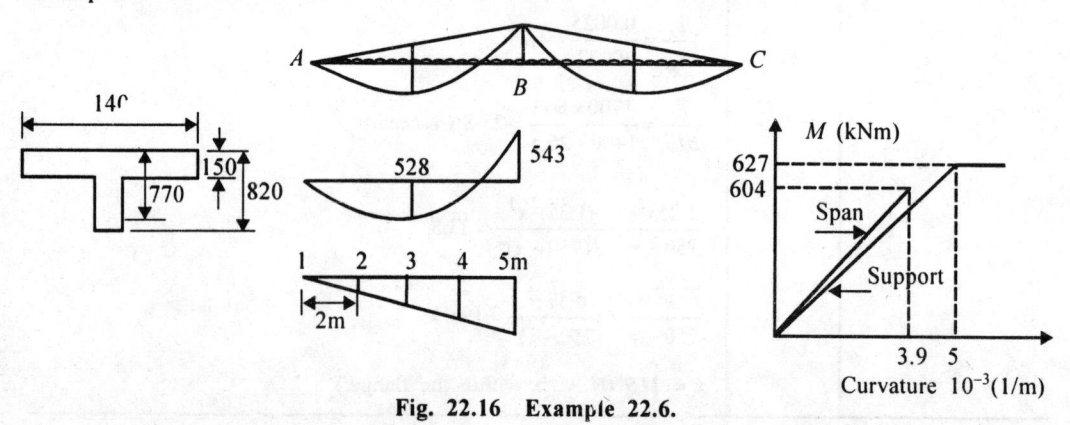

Fig. 22.16 Example 22.6.

INELASTIC ANALYSIS OF REINFORCED CONCRETE BEAMS AND FRAMES 419

Reference	Step	Calculations
	1	*Analyse the beam* Maximum B.M. at support $= \dfrac{97 \times 8 \times 8}{8} = 776$ kNm Reduced B.M. by 30% = 776 × 0.7 = 543 kNm $R_A = \dfrac{97 \times 8}{2} - \dfrac{543}{8} = 320$ kN Maximum positive moment $= \dfrac{R_A^2}{2w} = \dfrac{320^2}{(2)(97)} = 528$ kNm
	2	*Determine M at points 1 to 5 at intervals of s = 2 m from A* M at 1 = 0 (i.e. moment at section 1) $M \text{ at } 2 = (320)(2) - \dfrac{(97)2^2}{2} = 446$ M at 3 = 504 M at 4 = 174 M at 5 = −543 (rectangular section)
Fig. 22.16	3	*Interpolate ϕ for these points from M–ϕ diagrams of span and support (in m^{-1})* Point 1; $M = 0$; $\phi = 0$ Point 2; $M = 446$; $\phi = \dfrac{3.9 \times 446}{604} = 2.87 \times 10^{-3}$ (T beam) Point 3; $M = 504$; $\phi = \dfrac{3.9}{604} \times 504 = 3.25 \times 10^{-3}$ (T beam) Point 4; $M = 174$; $\phi = \dfrac{3.9}{604} \times 174 = 1.12 \times 10^{-3}$ (T beam) Point 5; $M = -543$; $\phi = \dfrac{5.0}{627} \times (-543) = -4.33 \times 10^{-3}$ (Rectangle) [For M–ϕ relationship of a rectangular section, *see* in the end of step 4, Exercise 22.4.]
	4	*Check the rotation required by numerical integration using Simpon's formula.* $\theta \text{ (required)} = \displaystyle\int \dfrac{M}{EI} m. ds$, where m = Unit rotation applied at B $\theta = \displaystyle\int \phi\, m\, ds$ as $\dfrac{M}{EI} = \phi$

420   ADVANCED REINFORCED CONCRETE DESIGN

Reference	Step	Calculations
Table 1, step 4		(*Note*: Strictly, Simpson's rule is applicable only when the same steel is present throughout the section but it is justified in most practical cases.)
		Simpson's formula is as follows (taking product of ordinates)
		$$V = \frac{s}{3}[\text{First} + \text{Last} + 4(\text{Even}) + 2(\text{Odd})]$$
		In this case, $s = 2$m (factors are 1, 1, 4 and 2)
		Table 1 below shows the procedure used.
		Value of θ required by considering both spans
		$\theta = 0.007$ radians
	5	*Calculate θ available at support*
		Method 1: Using Table 22.1 from Euro-code
		$\frac{x}{d}$ at support for $M = 543$ kNm
Exercise 22.4		$\frac{x}{d} = \frac{250}{750} = 0.33 < 0.4$
step 4.1		θ (available) = 0.01 (for ductile steel from Table 22.1)
		Method 2: From the relation $\theta = \phi L_p$. Assume Example 22.4 is valid.
		In step 5, Example 22.4 for the rectangular section, $f_u = 14.76 \times 10^{-6}$ mm^{-1}
Fig. 22.14(c)		Theoretically available rotation $\theta_a = \phi_u \times$ (Depth of the section)
		θ (available) = $14.76 \times 10^{-6} \times 750 \times 0.011$ radians
		(We take L_p as full depth as plange hinge can spread to both sides of the central support.)
		$\theta_y = 0.5 \times 10^{-6}$. From Mattock's equation.
Sec. 22.9		We also note $\theta_p = (\phi_p - \phi_y)d = 0.0089$
	6	*Check compatability; θ (required) < (available)*
		From Table 1
Step 4		0.007 < 0.01 (from step 5)
		The required plastic rotation does not exceed the allowable value. Therefore, the compatability is satisfied.

INELASTIC ANALYSIS OF REINFORCED CONCRETE BEAMS AND FRAMES 421

Reference	Step	Calculations
		TABLE 1

Point	Factor (F)	m	$\phi \times 10^{-3}$	$m\phi F$
1	1	0	0	0
2	4	0.25	2.87	2.87×10^{-3}
3	2	0.50	3.25	3.25×10^{-3}
4	4	0.75	1.12	3.36×10^{-3}
5	1	1.0	−4.33	-4.33×10^{-3}
			Total =	5.35×10^{-3}

Integrating both sides (spans AB and BC)

$$\theta = 2\left(\frac{2}{3}\right)(5.35 \times 10^{-3}) = 0.007 \text{ radians}$$

Example 22.7 (Baker's method of plastic analysis of R.C. structures)
A four span continuous T beam *ABCDE* of 4 m in each span is subjected to a characteristic load of 40 kN/m including its weight. Design the beam so that the beam fails by plastic failure at the supports at an ultimate load with a load factor 1.5.

Reference	Step	Calculations
	1	*Calculate M_0 for each span at ultimate load*
		$w = 1.5 \times 40 = 60$ kNm
		$M_0 = \dfrac{wl^2}{8} = \dfrac{60 \times 4 \times 4}{8} = 120$ kNm (ultimate)
		$= \dfrac{40 \times 4 \times 4}{8} = 80$ kNm (working)
Fig. 22.11	2	*Calculate elastic moments due to working load and ultimate load*
		$M_A = M_E = 0$ (end supports)
From table of		$M_B = M_D = 0.107wl^2$ (penultimate supports)
coefficients		$= 0.107 \times 40 \times 16 = 68.5$ kNm (elastic)
		$= 0.107 \times 60 \times 16 = 102.7$ kNm (ultimate)
		$M_C = 0.071 \times wl^2$ (interior support)
		$= 0.071 \times 40 \times 16 = 45.4$ kNm
		$= 0.071 \times 60 \times 16 = 68.2$ kNm

Reference	Step	Calculations
	3	Choose a plastic moment for design

Assume a plastic moment capacity $M_0/2$. The beam should carry at the collapse load of 60 kN/m and plastic moment of 60 kNm at supports B, C and D: $X_1 = X_2 = X_3 = 60$ kNm. |
| | 4 | Calculate the influence coefficients by using Simpson's formula

Products of ordinates of:

$$\int Mmds = \frac{5}{3}[\text{First} + \text{Last} + 4(\text{Even}) + 2(\text{Odd})]$$

$$f_{01} = \int \frac{M_0 mds}{EI}, \text{ Let } ds = 2 \text{ m from both spans}$$

$$= -\frac{2}{3}[0 + 0 + 4(120)(0.5)]\frac{2}{EI} = -\frac{320}{EI} = f_{02} = f_{03}$$

$$f_{11} = \frac{2}{3}\frac{2}{EI}[0 + (4)(0.5) + 0] = \frac{8}{3EI}$$

$$f_{12} = \frac{2}{3}[0 + 0 + 4 \times 0.5 \times 0.5]\frac{s}{3} = \frac{2}{3EI}$$

$$f_{13} = 0$$

Similarly $f_{21} = f_{12} = \dfrac{2}{3EI}$, $f_{22} = f_{11} = \dfrac{8}{3EI}$

$$f_{23} = \frac{2}{EI}, f_{31} = f_{13} = 0$$

$$f_{32} = f_{23} = \frac{2}{3EI} \text{ and } f_{33} = \frac{8}{3EI}$$ |
| | 5 | Write down the elastic equations in Bakers method omitting EI values

$$\frac{8}{3}X_1 + \frac{2}{3}X_2 + 0 - 320 = -\theta_1 \quad (1)$$

$$\frac{2}{3}X_1 + \frac{8}{3}X_2 + \frac{2}{3EI}X_3 - 320 = -\theta_2 \quad (2)$$

$$0 + \frac{2}{3}X_2 + \frac{8}{3EI}X_3 - 320 = -\theta_3 \quad (3)$$

Calculate EI of cracked section and et

$EI = 1 \times 10^4$ kNm |
| | 6 | Substitute for given X values and solve for θ

As a trial put $X_1 = X_2 = X_3 = 60$ kNm |

INELASTIC ANALYSIS OF REINFORCED CONCRETE BEAMS AND FRAMES 423

Reference	Step	Calculations
		The required rotations θ are calculated $\theta_1 = 120 \times 10^{-4} = 0.012$ $\theta_2 = 80 \times 10^{-4} = 0.008$ $\theta_3 = 120 \times 10^{-4} = 0.012$
	7	*Design concrete section and check whether θ_1 to θ_3 are available for these sections* If so, the values X_1 to X_3 are admissible. Otherwise, change them and recalculate so that admissible values of θ can be obtained.
	8	*Calculate the span moments from the support moments* Span AB; $M_B = 60$ kNm $R_A = \dfrac{60 \times 4}{2} - \dfrac{60}{4} = 105$ kN Maximum span moment $= \dfrac{R_A^2}{2w} = \dfrac{105^2}{2 \times 60} = 91.87$ kNm The span sections being T beams the ultimate moment capacity can be made much more than the required 92 kNm so that the failure does not take place in the span.
	9	*Check the performance of the beam at working load* The bending moments at the span at working load are 68.5 and 45.4, the designed beam can support these loads without much difficulty. The elastic solution for the beam can also be calculated from the following equations. Note: $\int \dfrac{M_0 m \, ds}{EI} = f_{01} = 213$ $f_{11}X_1 + f_{12}X_2 + 0 - 213 = 0$ $f_{21}X_1 + f_{22}X_2 + f_{23}X_3 - 213 = 0$ $f_{31}X_1 + f_{32}X_2 + 0 - 213 = 0$ giving $X_1 = X_3 = 68.5$ kNm, $X_2 = 45.4$ kNm as got from the table of coefficients. Thus the same procedure as used for elastic analysis is used in Baker's method for plastic analysis also.

REFERENCES

1. Richardt, F.E., Brandtzaeg, A. and Brows, R.L. *A Study of the Failure of Concrete under Combined Stresses*. University of Illinois Engineering Experimental Station, Bulletin No. 185, 1928.

2. Hognestad, E., Hanson, N.W. and McHenry, D., *Concrete Stress Distribution in Ultimate Strength Design*, A.C.I. Journal, American Concrete Institute, 1955.

3. Kent, D.C. and Park, R., *Flexural Members with Confined Concrete*, Journal of the Structural Division ASCE, Vol. 97, ST. 7, July 1971.

4. CEB-FIP Model Code for Concrete Structures, Vol. 2 (revised edition), Committee European Beton, London, 1978.

5. Nonlinear Behaviour and Analysis, Ch. CB-6, 'Structural design of tall concrete and masonry buildings', Vol. CB: *Monograph on Planning and Design of Tall Buildings*, American Society of Civil Engineers, New York, 1978.

6. Eibel, J., *Concrete Structures*, Euro design Handbook, Ernst and Sohn, Berlin, 1994.

7. Calvi, G.M., Cantee, E. and Magenes, G., *Evaluation of Rotation Capacity of D Region—Report*, Vol. 62, IABSE Colloquiem, Stuttart, 1991.

8. Baker, A.L.L., *Limit State Design of Reinforced Concrete*, Cement and Concrete Association, London, 1970.

CHAPTER 23

Strip Method of Design of Reinforced Concrete Slabs

23.1 INTRODUCTION

The fundamental concepts of yield-line theory of slabs were originally proposed by a Danish Engineer Ingerslav as early as in 1923 and were developed by Johanssen [1] during World War II. It came to international attention during the IABSE Congress in 1948. It was the only method available for the ultimate load design of slabs till about 1956. It was based on work equation and gave the upper bound values. Structures designed by this theory are safe against total collapse but do not perform well under serviceability conditions.

An alternate method using the equilibrium method representing the safer lower bound solution was developed and presented by Arne Hillerborg of Sweden in 1960 and presented at the IABSE Congress in Stockholm [2]. A simplified version of this method, where the twisting moments in the slab are neglected, came to be known as the *simple strip method* for slabs. Hillerborg also developed an extension of the method called as the *advanced strip method* for slabs with re-entrant corners and slabs supported on columns [2,3].

The 'strip method' has become very popular as a tool for conceptual design of reinforced concrete slabs. This is specially because of the simplicity brought out by the further work done by Wood and Armer of the Building Research Establishment in the U.K. and Kemp [3,4]. They developed the modified strip method described later in this chapter and the concept of strong bands. These strong bands are strips of convenient width and, if necessary, of enlarged depth with concentration of steel in them acting like concealed flexible beams.

The modified strip method along with strong bands give us a very convenient method to design reinforced concrete slabs for which standard elastic solutions are not available. Since IS 456 Clause 23.3.2.3 allows slabs to be designed by any acceptable theory, this method can also be used for routine design.

Johanssen's yield-line theory is described in most of the standard textbooks [5]. This chapter deals only with the basic concepts and introduction to the Hellerborg's simple strip method applicable to simple cases. Deeper understanding of the method and its application to more complex cases can be obtained from a study of the references given at the end of this chapter. The sign convention for indicating the support conditions in this method is shown, by hatching along the fixed edges, dotting at the simply-supported edges and no additional sign at the free ends as indicated in Fig. 23.1.

23.2 THEORY OF STRIP METHOD

The equilibrium equation for a plate can be written as follows:

$$\frac{\partial^2 M_x}{\partial x^2} - \frac{2\partial^2 M_{xy}}{\partial x \partial y} + \frac{\partial^2 M_y}{\partial y^2} = -q$$

This equation can be split into two parts by neglecting the second term, which represents the twisting moment as follows:

$$\frac{\partial^2 M_x}{\partial x^2} = -q_x, \quad \frac{\partial^2 M_y}{\partial y^2} = -q_y$$

and

$$q_x + q_y = q$$

Thus the load is divided into two parts in q_x and q_y in X and Y directions. The original slab is envisaged as a number of independent strips or one-way slabs in X and Y directions carrying the load to the supports. It is assumed that the deflections are negligible so that compatibility conditions need not be considered. Again as torsional moments have been neglected, the simple method should not be applied to slabs without corrections where large torsional moments are present.

23.3 APPLICATION TO A SIMPLY-SUPPORTED SLAB

Let us consider the rectangular slab as shown in Fig. 23.1. The lines *LMN* and *PQR* inside the slab are called the *lines of discontinuity*. They show what load distribution, the designer using conceptual design method, can assume.

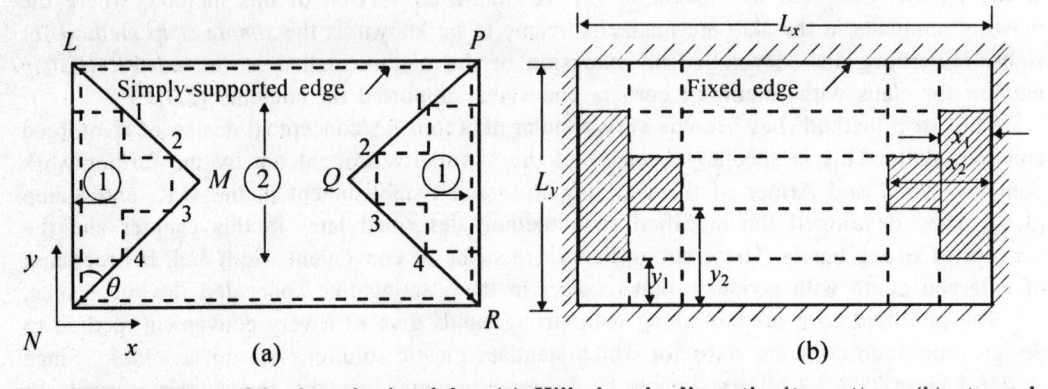

Fig. 23.1 Division of load paths in slabs: (a) Hillerborg's discontinuity pattern (b) stepped discontinuity patterns by Wood and Armer.

All the load in area 1 is carried by strips in the X direction and all the load in area 2 is carried by strips in the Y direction to the nearest supports. The arrows in Fig. 23.1 show the directions of transfer of load and also that of the main steel. The magnitude of the angle θ defining the line of discontinuity between areas 1 and 2 is to be assumed by the designer. Hillerborg suggested that with two adjacent sides both simply-supported or fixed the line of discontinuity will be through the mid angle between adjacent supports. Thus with the two 3,

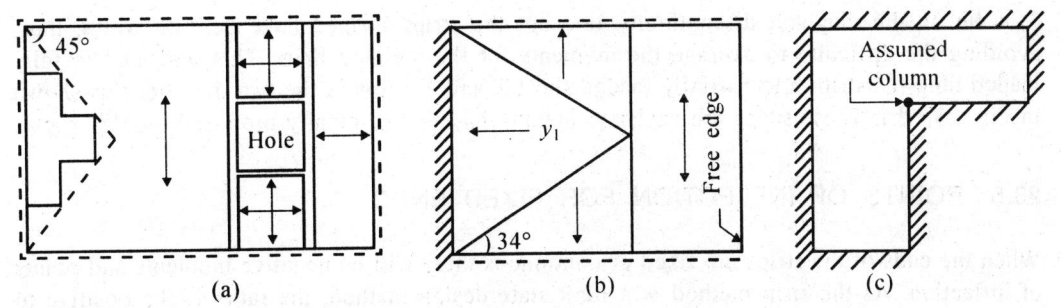

STRIP METHOD OF DESIGN OF REINFORCED CONCRETE SLABS 427

Fig. 23.2 Analysis of slabs by modified Hillerborg's method: (a) concept of strong bands (b) slab with free edge (c) slab with re-entrant corner.

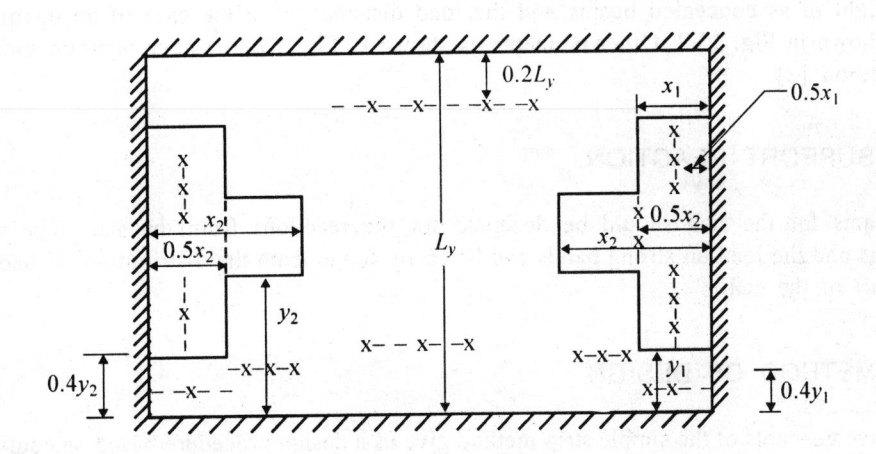

Fig. 23.3 Location of points of inflection (P.I.) in slabs with fixed edges.

3, 4 lie on the diagonal discontinuity line but the strips formed have definite width, thus, avoiding the difficulty to average the moments for the inclined lines. This will lead to fully loaded long (L) strips and partially loaded short X and Y strips in the two directions as shown in Fig. 23.1(b). These strips are easier to handle than that originally proposed by Hillerborg.

23.5 POINTS OF INFLECTION FOR FIXED ENDS

When the ends of the strips are fixed or continuous there will be negative moments and points of inflection. As the strip method is a limit state design method, the ratio of the positive to negative moment can be varied to suit redistribution. However, it has been suggested that these points of inflection (P.I.) may be assumed as 0.2L from the supports for strips loaded the full length and 0.4y_1 for the short (y) strips of loaded length y_1 and 0.5x_1 for the short (x) strips of loaded length x_1 as shown in Fig. 23.3 [6].

23.6 USE OF STRONG BANDS IN SLAB DESIGN

Another concept introduced by Wood, Armer and Kemp is the use of strong bands for design of slabs with openings or other types of discontinuity [6]. These strong bands can be thought of as concealed beams and the load distribution in the case of an opening can be as shown in Fig. 23.2. Other concepts as shown in Fig. 23.2(c) have been developed by this method [2].

23.7 SUPPORT REACTION

The beams for the slab should be designed for the reactions from the slab. The support reactions and the load on strong bands can be easily found from the distribution of loads from each part of the slab.

23.8 METHOD OF DESIGN

The above concepts of the simple strip method give us a design procedure based on equilibrium concept and the designer's intuitive feel of the behaviour of the slabs. Slabs of various shapes and under various types of loadings can be designed by this method. It should however be noted that in the actual performance of the slab under load considerable redistribution of moments may have to take place for the stability of the slab. Hence, adequate rotation capacity and redistribution should be ensured in the design. As slabs have low percentage of reinforcements, this redistribution can easily take place in the structure. The principle of the method is illustrated by Examples 23.1 and 23.2.

23.9 DESIGN OF SKEW SLABS

Skew slabs span over supports which are not at right angles to the centre line of the slab. The skew angle is defined as the deviation of the centre line of the slab from the line prependicular to the supports. Since the present day fast traffic demands straight alignment of roads, culvert—slabs may have to cross waterways at angles other than 90° to their supports. Depending on the skew angle, the following design principle based on theoretical and experimental investigations are used to design such slabs [7].

Case 1 (*Slabs with skew angles less than 15°*). As the torsional moments in these slabs are small such slabs can be treated as normal right angled slabs, the span length being measured between centre to centre of supports, parallel to the centre line of the slab. The main reinforcements are provided in the direction parallel to the centre line of the slab and the nominal distribution steel (amounting to 0.2% of the section of the slab) is provided parallel to the supports.

Case 2 (*Slabs with skew angles more than 15°*). In these cases, there will be considerable torsional moments together with the bending moments and hence a rigorous analysis is required. Ready-to-use influence surfaces for bending and torsional moments for such slabs are available. The Road Wing of the Ministry of Transport, Government of India has also prepared standard designs for such culverts of different spans and different skew angles. The following reinforcements are provided for such slabs [7]:

1. Bottom.
    (a) Main bars placed perpendicular to supports
    (b) Transverse bars placed parallel to supports

2. Top.
    (a) Nominal reinforcement at the top of slab is the direction of the centre line of the slab
    (b) Nominal reinforcement at the top of the slab parallel to the support

3. Corners. Corner reinforcement as for two-way slab action
4. Edges. Top and bottom bars with stirrups to stiffen the edges

23.10 AFFINITY THEOREMS

The theory of slabs usually assumes an isotropic slab with equal reinforcements in perpendicular directions. When considering slabs which are orthotropically reinforced if continuity or fixity exists along one or more edges, or when considering skew slabs, it is possible to transform most of the slabs into their equivalent simple slabs by using the affinity theorems. This topic is dealt with briefly in Reference 8 and in detail in Reference 9.

EXAMPLE 23.1 (**Strip method for slabs fixed along edges**)
A rectangular slab 8 m in the X direction and 5 m in the Y direction has fixed edges and carries a factored load of 20 (kN/m^2)(Fig. 23.4). Determine

(a) the design bending moment diagrams for typical strips for designing the slab by strip method
(b) the design load diagram of the edge beams.

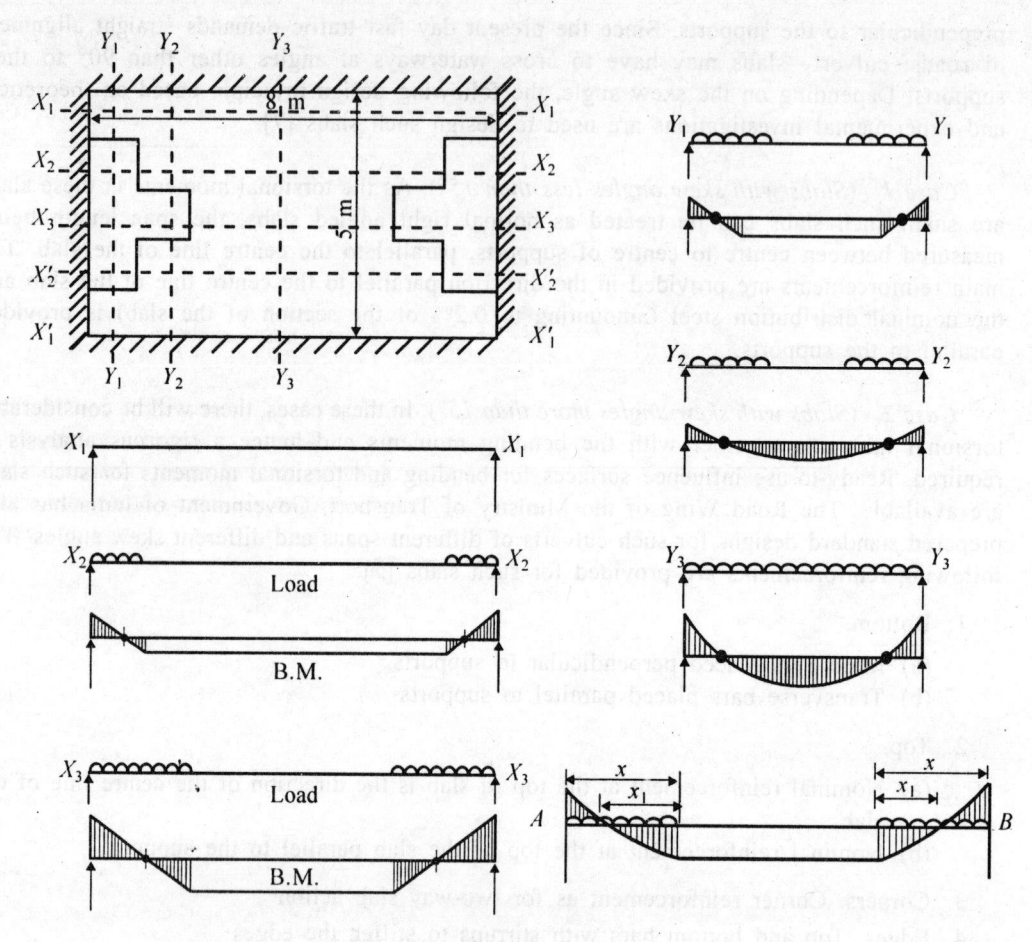

Fig. 23.4 Example 23.1.

Step	Calculations
1	*Draw the discontinuity line*
	Using the step-wise strip, mark the X and Y strips
2	*Determine the loading on the strip*
	Mark the typical XX and YY strips and also the loads on the various strips.
3	*Calculate the bending moment on xx strips*
	Assume ends are fixed. Let the length of loading on the partly loaded strip from fixed end be x. We assume the point of inflection is at half this distance ($x/2 = x_1$) from the fixed end-support. The length from P.I. to the end of the loading point is x_1. Treating the symmetrical beam simply-supported at the P.I. points on each side
	(a) Maximum positive moment $= wx_1 x_1 - wx_1 \dfrac{x_1}{2} = \dfrac{wx_1^2}{2}$

Step	Calculations
	(b) Maximum negative moment $M = \int F dx$
	This is equal to the area of the shear force diagram from the end of the slab to the point of inflection. Using these principles, let us work out the bending moments.
	B.M. on X_1X_1 strips (Fig. 23.4) = No load—no moments
	X_2X_2 strip, $x = 1$ m, $x_1 = 0.5$ m
	Maximum possible B.M. = $\frac{1}{2}wx_1^2 = \frac{1}{2} \times 20 \times (0.5)^2 = 2.5$ kNm
	S.F. at the end = 20; S.F. at P.I. = 10
	Maximum negative B.M. = $\frac{1}{2}(20+10)(0.5) = 7.5$ kNm
4	*Calculate the bending moments on YY strips*
	The points of inflection of YY strips are assumed at 0.4 times the loaded length for partly loaded strips and at 0.2 times the length for fully loaded strips.
	Strip Y_1Y_1 loaded length $y = 1$ m − Distance of fixed-end to P.I. = 0.4 m
	and length from P.I. to the end of loading = 0.6 m
	Positive B.M. = $\frac{1}{2}(20)(0.6)^2 = 3.6$ kNm
	Negative B.M. = $\frac{1}{2}(20 + 12)(0.4) = 6.4$ kNm
	Strip Y_2Y_2 positive B.M = $\frac{1}{2}(20)(1.2)^2 = 14.4$ kNm
	Negative B.M. = $\frac{1}{2}(40 + 24)(0.8) = 25.6$ kNm
	Strip Y_3Y_3 positive B.M. = $\frac{1}{2}(20)(1.5)^2 = 22.5$ kNm (P.I. at 0.2L)
	Negative B.M. = $\frac{1}{2}(50 + 30)(1) = 40$ kNm
5	*Calculate the loads on each short beams (5 m span)*
	Load from X_1X_1 strip = Nil for 1 m length (on each side beam)
	X_2X_2 strip S.F. = 20 kN for 1 m length
	X_3X_3 strip S.F. = 40 kN for 1 m length
	$X_2^1X_2^1$ strip S.F. = 20 kN for 1 m length
	$X_1^1X_1^1$ strip S.F. = Nil for 1 m length
	Hence, the total load on each beam = 80 kN

Step	Calculations
6	Calculate the load on the long beam
	Strip Y_1Y_1 S.F. = 20 kN for 1 m length = 20 kN
	Y_2Y_2 S.F. = 40 kN for 1 m length = 40 kN
	Y_3Y_3 S.F. = 50 kN for 1 m length = 200 kN
	Y_2Y_2 S.F. = 40 kN for 1 m length = 40 kN
	$Y_1^1 Y_1^1$ S.F. = 20 kN for 1 m length = 20 kN
	Total load on each beam = 320 kN (the sum of all the loads)
7	Check for the loads calculated
	Total load on all beams = 2(80 + 320) = 800 kN, which is also equal to the load on the slab = 20 × 5 × 8 = 800 kN.

EXAMPLE 23.2 (Analysis of a propped up slab)

Figure 23.5 shows a slab 4 m in the Y direction and 6 m in the X direction continuous at adjacent sides and unsupported on the other two sides, but propped up at the free corner by a pillar. The slab is to be designed to carry a total factored uniformly distributed load of 10 kN/m². Indicate approximate method to design the slab.

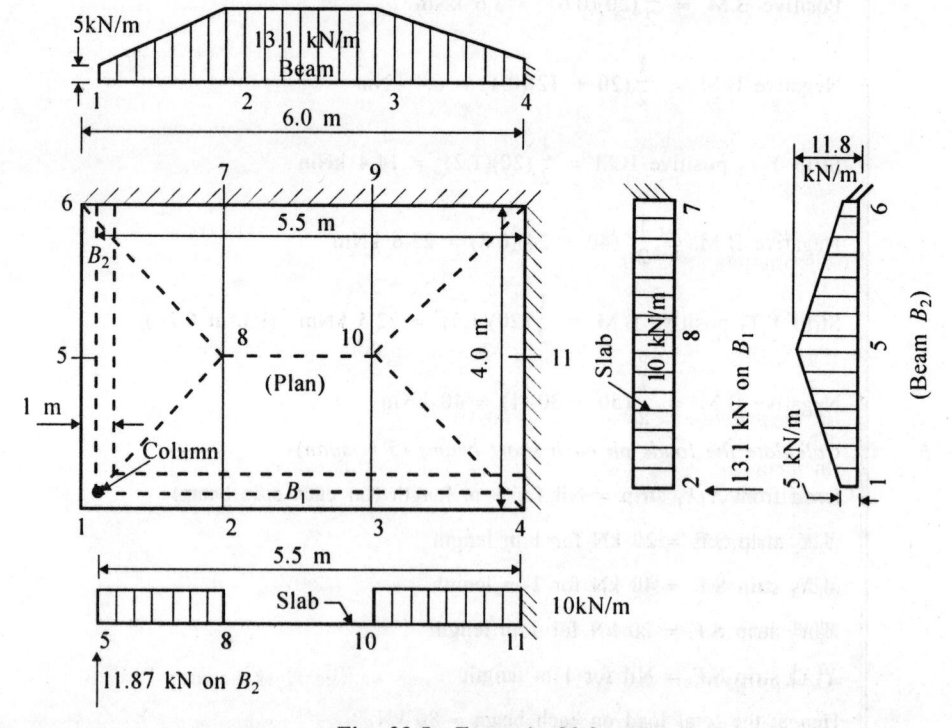

Fig. 23.5 Example 23.2.

STRIP METHOD OF DESIGN OF REINFORCED CONCRETE SLABS 433

Step	Calculations
	A. Approximate method 1. (Strip Method—Rigid Beam)
1	*Assume strong bands along 1,2,3,4 and 1,5,6. Mark discontinuity line along 45°.* It will be more appropriate to use θ with tan θ = 2/3 with more load distributed to the fixed end at points 4 and 6.
2	*Design of beam B_1[1-2-3-4]* (*The strong bands are treated as rigid beams*) Design of beam 1-2-3-4 (beam in X direction). It takes the load from slabs 2-8-7 spanning in Y direction. We may take a width of the slab equal to 1 m adjacent to the propped support and in the X and Y directions as strong bands, which act as beams. Beams act along centre line of these bands. Assuming the slabs in the Y direction as propped cantilevers with uniform load, the maximum reaction on the beam B_1 is given by: $$R_1 = \frac{3}{8}wL_y = \frac{3}{8} \times 10 \times 3.5 = 13.1 \text{ kN}$$ The maximum load intensity is 13.1 kN/m and the loading on beam 1-2-3-4 (6 m span) is as shown in Fig. 23.3.
3	*Design of beam 1-5-6* These are loaded from slabs 5-8-10-11. Again as a propped cantilever, using $\int Mmdx/EI = 0$, the maximum load on this beam is R_2 = 11.87 kN as shown in Fig. 23.5 (*see* the figure at the right hand).
4	*Design of slabs and beams* Design the slabs in X direction and Y direction as in Example 22.1 for the load distribution assumed.
4a	*Design of strong bands* The design loads on the strong bands are as shown in Fig. 23.5
	B. Approximate method 2 (flexible beam) In method I, beam 1-2-3-4 was assumed rigid. However, the support can deflect by a small amount so that the reaction will be (say) 70% of the value with rigid beams. The maximum load = 0.7 × 13.1 = 9 kN This can very much reduce load on the beam 1-2-3-4. The slabs are designed as in method 1.
	C. Approximate method 3 (two-way slab on the rigid beam) The slab can be assumed as two-way slabs with rigid beams on the free ends by using the coefficient for two-way slab on rigid beams. It may be necessary to increase depth

Step	Calculations
	of slab at the free edges to substantiate this assumption (*see* the design of flat slabs for minimum depth of edge beams).
	(*Note*: If we want to use an exact method, we can use tables of moments and shears that are available in publications dealing with various types of slabs under different support conditions.)

REFERENCES

1. Johanssen, K.W., *Yield Line Formulae for Slabs*, (Translation), Cement and Concrete Association, London, 1972.

2. Hillerborg, A., *Strip Method of Design*, Viewpoint Publications, Cement and Concrete Association, London, 1975.

3. Wood, R.H. and Armer, G.S.T., *Theory of Strip Method for Design of Slabs*, Proc. Inst of Civil Engineers, London, Vol. 41, 1968.

4. Armer, G.S.T., The strip method, *A New Approach to Design of Slabs Concrete*, September 1968.

5. Purushothaman, P., *Reinforced Concrete Structural Elements*, Tata-McGraw Hill, New Delhi, 1984.

6. Ferguson, P.M., Breen J.E. and Jirsa, J.O., *Reinforced Concrete Fundamentals*, John Wiley and Sons, New York, 1988.

7. Krishna, Raju N., *Design of Bridges,* Oxford and IBH, New Delhi, 1988.

8. Reynols, C.E. and Steedman, J.C., *Reinforced Concrete Designers Handbook* (*Table 58*), Cement and Concrete Association, Waxham Springs, U.K., 1988.

9. Pannell, F.N., *Nonrectangular Slabs with Orthotropic Reinforcement*, Concrete and Construction Engineering, November 1966.

CHAPTER 24

Durability and Mix Design of Concrete

24.1 INTRODUCTION

Modern portland cement was invented around 1756 and came to be used in Europe to make reinforced concrete in around 1880. In India ordinary portland cement was imported in early days and local manufacture started around 1912. Research in the past two decades has given us a better understanding of cements and today we get a variety of cements in market to choose from.

Reinforced concrete has been in wide use in India for nearly 80 years. In early days, as steel can get rusted under bad environmental conditions, engineers took particular care of durability and in choosing the correct concrete mix with enough cement content and in placing and curing it properly at the site. These structures built with mild steel and carefully placed concrete under proper supervision have withstood all sorts of environments very well in many places in India. On the contrary, a number of important reinforced concrete structures built recently have been found to require urgent repair after only a few years of its completion. This has been due to the fact that soon after World War II (around 1950), emphasis in mix design shifted from durability to high strength. Engineers were interested to get as much strength from as little cement as possible. Consideration for environment was rarely shown. Recent disasters with respect to durability have forced the attention of the engineers to consider environment. Today the durability of concrete and steel, in addition to strength in limit state design of concrete structures, has come back as important subjects to be considered.

This chapter (and Appendix B) examines the factors that affect the durability of concrete and also gives a very brief review of the basic considerations in mix design of ordinary and high strength concrete.

24.2 TYPES OF CEMENTS PRODUCED IN INDIA

Cement is specified by its grade, i.e. the mortar cube strength in N/mm^2 in 28 days. Thus Grade 33 cement (C-33) means cement with standard mortar cube strength of 33 N/mm^2. There are also different types of cements of which ordinary portland cement is the major type used in construction. The principal types of cements available in India are:

1. *Ordinary Portland cement* (*OPC*). About 70% of cement produced in India are of this category and comes in 3 grades, viz., grade 33, 43 and 53 (*see* also Section 24.2.1).

2. *Portland pozzolana cement (PPC) conforming to IS 1489 (1981 Part 1 and 2)*. These are made by blending 10 to 25% reactive pozzolana like fly ash or calcined clay with OPC. These are similar in nature to 33 grade OPC but have slower strength development in the first two weeks. It is very sensitive to curing and require longer curing than OPC.

3. *Sulphate resisting Portland cement (SRPC or SRC) conforming to IS 12330 (1988)*. These are produced in small quantities in India. These are special OPC with less than 5% C_3A and are superior in resistance against sulphates. Cements called Birla Coastal comes in this category. They should not be confused with supersulphated cements (SSC) made from blast furnace slag, calcium sulphate and small quantities of OPC. (SSC is not recommended for use in places with temperatures above 40°C as in India.)

4. *Portland blast furnace slag cement (BFSC or PSC) conforming to IS 455 (1976)*. This constitutes about 10% of cement produced in India. It is made by inter grinding OPC clinker with specially granulated slag from factories. The slag forms 25 to 60% of the cement. Every ton of cast iron produces about 0.3 tons of blast furnace slag which can be usefully used in the cement industry. During its setting the $Ca(OH)_2$ liberated by OPC hydration acts as an activator for the slag. They are also less costly than OPC. Eventhough it is equated with OPC, it behaves more like PPC and has lower heat of hydration and better sulphate resistance. At present, the BFSC cement produced in India is only 33 grade and there are proposals to make 43 grade cements with 45–60% slag content. BFSC with more than 50% slag has good sulphate resistance too.

24.2.1 Grades of Cements Available as OPC

As already stated the bulk of cement used in construction is ordinary portland cement. In the U.S.A. and U.K., OPC is covered by one specification, whereas in Germany OPC is available in 3 grades. The German practice has been accepted also in India and it came about as follows:

Till around 1973, only 33 Grade cement was available in India. However, between 1973–75, the Indian Railways adopted the use of prestressed concrete sleepers in a big way for running the high speed trains. It was soon apparent that the common 33 Grade cement available in the market was inadequate to develop the needed characteristic concrete strength of about 50 N/mm^2 required for the purpose. Hence, the railways developed their own specification for sleeper cements' with a minimum cement strength of 52.5 N/mm^2 in 28 days. Some factories in India came forward to make these cements for the railways, which made them available only to the sleeper manufacturers. Very soon, with the advancement of cement technology, more and more factories found it easy to manufacture higher grade cements with their modernised cement plants. Thus, under OPC grade we have in India:

1. 33 grade as per IS 269 (1989) (modern plants do not produce this grade any more)
2. 43 grade as per IS 8112 (1989)
3. 53 grade as per IS 12269 (1987)
4. Sleeper cements as per IRS-T40-85 (this will be between C 43 and C 53)

The easily available OPC cement today is Grade 43. Grade 33 is available only as PPC. It should be noted that the testing procedures used in India are different from those in U.S.A., where cylinders are used so that the 53 grade cement produced in India would give approximately 25 to 30% less strength as per ASTM standards. The compressive strength developed by the cements with time is shown in Table 24.1.

TABLE 24.1 COMPRESSIVE STRENGTHS OF DIFFERENT GRADES OF CEMENT

Age	Grade			Sleeper cement	
(in days)	Grade 33	Grade 43	Grade 53	Code	Actual
	Recommended values				
3	16	23	27	–	40.3
7	22	33	37	37.5	55.3
28	33	43	53	–	70.3
	Observed laboratory values as % of 28-day strength				
3	30–40	50–60	70–80		
7	50–65	65–80	80–90		
28	100	100	100		
90	100–125	105–115	100–105		
180	115–130	110–120	105–110		

24.2.2 Types of Cements to be Used in Construction

As already stated both strength and durability are of equal importance in civil engineering. In constructions, which may have special problems due to presence of sulphates, chlorides, etc. resistance to these should take the first preference. As regards strength, all OPC can produce strength up to M25 with ease. However, it will be easier (with less quality control) to attain the required strength and to remove shuttering earlier by using a higher grade cement. In all cases the decision to use a particular grade should be made so as to optimise the cost and improve the durability of construction.

[An easy way to check for the adulteration of cement in the field is to burn a sample of cement for 20 minutes on a steel plate by a stove. Adulterated sample (except those containing materials like fly ash) changes colour while unadulterated cement does not do so, as it has already been subjected to high temperature during its manufacture.]

24.3 DURABILITY OF CONCRETE

The various agents or causes of deterioration of concrete have been classified by RILEM in order of their importance as following:

1. Corrosion of steel
2. Effect of sulphate
3. Wetting and drying
4. Freezing and thawing
5. Leeching
6. Acids
7. Internal stresses
8. External stresses
9. Crystallisation of salts
10. Reactivity of aggregate
11. Abrasion

Of these, the most important and common causes of deterioration are corrosion of steel and consequent breaking of concrete by oxidation or chlorides as well as deterioration of concrete due to sulphates. Before we go into the mechanism of corrosion, it is good to have a general idea of the structure of concrete and its influence on durability.

24.3.1 Structure of Concrete, Permeability and Durability

One of the important factors affecting the durability of concrete is the permeability of concrete which is a function of the structure of concrete [1]. Good concrete can be considered as a mixture of coarse and fine aggregates which are completely filled and surrounded by the cement paste. The cement paste itself can be considered as cement particles surrounded by a thin layer of water. The thickness of the water film increases with increase in water cement ratio w. As hydration progresses the products of hydration fills the space occupied by the water. With low water–cement ratio of about 0.3, the hydration products completely fill the water space. The water–cement ratio required for complete chemical reaction is around 0.27. However, in very high strength concrete we many even adopt water–cement ratios around 0.25, leaving significant amount of unhydrated cement which are viewed as very fine aggregates.

24.3.2 Water–Cement Ratio and Permeability

If the water-cement ratio is more than 0.30, some of the water space is not be rilled with products of hydration and the free water space will act as an open capillary when the concrete hardens. These spaces can occur as continuous capillaries allowing easy passage for ions, gas or moisture. The relation of the volume of pores V_p to the volume of concrete V_c has been expressed quantiatively as follows [1]:

When water–cement ratio denoted by $w > 0.40$

$$\frac{V_p}{V_c} = \frac{w - 0.36m}{w + 0.32} \qquad (24.1)$$

where m is the degree of hydration of cement (i.e. volume percentage of cement that has completely hydrated and $m > 0$ but < 1).

If $w < 0.40$,

$$\frac{V_p}{V_c} = \frac{0.14w}{w + 0.32} \qquad (24.2)$$

In Eq. (24.1) with large values of m (i.e. curing being better) and smaller values of water–cement–ratio, the value of V_p decreases considerably. The relation between water–cement ratios and permeability is shown in Fig. 24.1.

24.3.3 Cement Content and Permeability

Cement content (expressed as kg/m^3 of concrete) is also important for impermeability of concrete. A minimum cement content of 300 kg/m^3 (including permitted admixtures) is required for most of the aggregates used in practice for low permeability concrete.

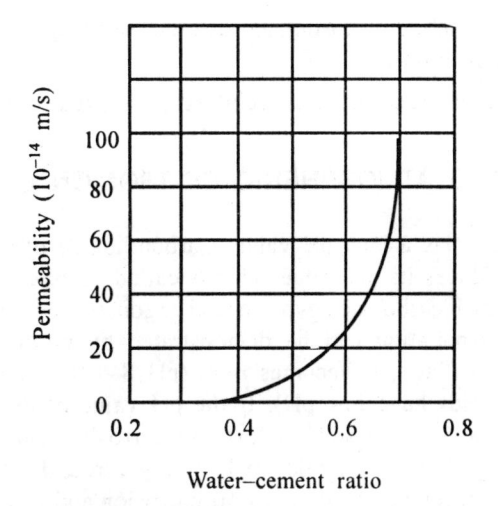

Fig. 24.1 Variation of permeability of concrete with water–cement ratio.

24.3.4 Curing, Nature of Skin of Concrete and Permeability

The nature of the skin formed during the placement and proper curing has a great influence on permeability of concrete. The skin should not crack as happens if alternate wetting and drying is resorted to during curing.

24.3.5 Cracking of Concrete and its Influence on Permeability

Another cause of high permeability in concrete is the occurrence of cracks due to dimensional changes (intrinsic cracks) or due to loads. The intrinsic crack can be formed by the following:

1. Plastic settlement
2. Plastic shrinkage
3. Early thermal changes
4. Drying and shrinkage.

Shrinkage cracks can be reduced by proper curing and by provision of shrinkage steel in the direction of restraint. Alternate wetting and drying of concrete should not be allowed till final curing has been completed. Lack of secondary steel, loading by heavier loads than the designed load during construction, can also lead to cracking.

24.3.6 Summary of Factors Affecting Permeability of Concrete

Summarising the above facts, we find that permeability is the most important factor that affects durability. The factors that affect permeability of concrete, which in turn affects durability, are the following:

1. Water–cement ratio
2. Minimum cement content
3. Maintenance of a proper skin (surface) of concrete by proper curing or other means

4. Reducing cracking of concrete by curing, provision of shrinkage steel and other standard methods prescribed in specifications.

These factors should be attended to in all reinforced concrete constructions.

24.4 CARBONATION AND ATMOSPHERIC CORROSION OF STEEL

Fresh concrete is alkaline and has a high pH value. Carbondioxide of the air reacts with the products of hydration and reduces its pH value. This is called *carbonation* [2]. The products after carbonation are calcium bicarbonate, gypsum and argonate; the first two being soluble in water. The process of carbonation can be demonstrated by putting phenolphthalein on concrete. If it has lost its alkalinity, it becomes pink (pH 8–10) (pH values 12–13 can be considered as alkaline and acids have low pH). If the pH value of the concrete around the steel is only about 10, the passivity of steel is lost and corrosion can take place at a rapid rate in the presence of oxygen and high relative humidity (around 60–75%). At very low humidities, there will not be sufficient moisture to start corrosion and with very high humidities the pores can get blocked thus decreasing the rate of corrosion. Alternate wetting and drying can lead to the acceleration of corrosion of steel.

The notion that ordinary concrete protects steel by preventing ingress of water and oxygen is a mistaken notion. It is now generally accepted that the steel is protected by the passivity induced by the highly alkaline nature, which should be preserved to prevent corrosion. Only special coating can keep off water and air from concrete. In concrete, the carbonation depth (in mm) = $10\beta t$ (approx.), where t is in years. The value of β = 0.6 to 1.0 for low strength concrete; 0.2 to 0.5 for medium strength concrete; and 0.1 to 0.2 for high strength concretes depending on the exposure conditions.

24.5 CHLORIDE PENETRATION AND STEEL CORROSION

Corrosion due to chlorides is entirely different from corrosion due to atmospheric corrosion [3]. Plain dense concrete (without any steel) immersed in water containing chlorides is very durable. In fact, the high concentration of chloride ions present in sea water increases the solubility of some of the expansive components like ettringite that are produced by sulphate attack on concrete and this relieves the severity of sulphate attack in sea water. Concretes above the low tide usually undergo some damage due to wave action and accumulation of salts in the alternate wetting and drying zone.

However, corrosion of steel in reinforced concrete due to chlorides is electrochemical in nature. It takes place irrespective of the pH value and can occur even in uncarbonated concrete. Chloride ions transfer from outside to inside (assuming that there are no chloride ion in the constituents of concrete) by diffusion through the pores and capillary action. To avoid chloride penetration concrete should have the least amount of capillary pores (low permeability) and an impermeable skin. Protection of piles in sea water by coating with bituminous paint is an artificial method of providing an impermeable skin.

It is also important that to start with, the constituents of concrete, viz., sand, coarse aggregate and the mixing water should themselves be free from chloride. The limit of soluble chloride allowed in fine and coarse aggregates are 0.10 and 0.03%, respectively. The total chlorides allowed in concrete including that in cement is 0.025% by weight (IS 456 Table 7).

Estimation of soluble chlorides in sand is made by first washing the sand in distilled water and then estimating the chlorides in water by titration with a standard solution of silver nitrate with a suitable indicator. IS 456 Table 1 gives permissible limits of solids in water to be used for reinforced concrete works. Use of local borewell water (which may be brackish in summer in most of the coastal towns in India) without testing is a dangerous practice that can lead to corrosion of steel in the long run. If during an investigation of chloride corrosion, tests show presence of chlorides at all depths of the member then the chloride would have been in the constituent materials or mixing water and has not got in from outside.

It has been found that blast furnace slag cement (whose setting chemistry is different from OPC) reduces the effective diffusion coefficient of free chloride ions 1/2 to 1/10th of the value obtained with OPC. The fundamental factor for watch reduction in chloride effects is the reduction in capillary pore space. This is specially true in structures in the salt spray zone in coastal areas which are subjected to alternate wetting and drying. Sea water itself has a salinity of over 32 parts per thousand with 9.3 p.p.m. dissolved oxygen and pH around 8.15.

In addition to the individual effect of carbonation and chloride, their combined action should be considered in many situations. The minimum cover specified for R.C. in sea water is 65 mm in tidal zone and 40–50 mm in other areas.

24.6 PRESENCE OF SALTS AND STEEL CORROSION

It is a common observation that steel laid in contact with brickwork (as in reinforced brick work) and G.I. water-supply pipes laid in brick jelly concrete corrode very fast. This is due to the presence of salts in the bricks. Whenever steel is to be placed in such situations it should be separated from bricks with sufficient cover of very dense and impermeable concrete. In the presence of salts in the surrounding materials like concrete, solid or bricks, steel tends to corrode very rapidly. Such details are important in reinforced brickwork.

24.7 SULPHATES AND CONCRETE DISINTEGRATION

Presence of high concentration of *soluble sulphates* can break up even plain concrete. Such sulphates are present in ground water, sewerage, industrial wastes, sea water and in foundation soil. IS 456 (2000) Table 4 gives the precautions to be taken in these cases. The remedy depends on the concentration of soluble sulphate present in the soil or water. The soluble sulphates react with the calcium hydroxide and tri-calcium aluminate (C_3A) present in the cement to produce products of gypsum and calcium sulphate aluminate which occupy more volume than their constituents. This expansion breaks up the concrete exposing the steel in reinforced concrete to corrosion.

The remedy in situations where there are only sulphates and no chlorides is to use the right type of cement. Sulphate resisting Portland cement (SRPC) with C_3A less than 5% is suitable for such situations. However, it should be noted that there are still differences in opinion regarding the necessity of special sulphate resisting cements in reinforced concrete structures in sea water where chlorides and sulphates are present. Any cement with C_3A only 5–8% is desirable to be used in concrete in such cases. But all agree that in marine situations the most important precaution is to use a dense concrete with low permeability of at least grade 30 and sufficient cover for reinforcement. Coatings to reduce permeability also help durability of these structures. IS 456 Clause 8.2.8 deals with concrete in sea water.

24.8 CURING OF CONCRETE (IS 456 Clause 13.5)

Curing of concrete is carried out by water curing or by membrane curing. It is classified as follows:

1. *Good.* When the work is protected from the sun and the relative humidity is kept more than 80% continuously.

2. *Poor.* When the work is protected from the sun and the relative humidity is less than 50%.

3. *Average.* When the curing is between the above.

We should remember that concrete made from OPC is to be cured with good curing for at least 7 days and concrete made from PPC and BFSC should be continuously cured for at least 10 days.

The minimum stripping time of the sides of formwork and the time for removal of props are specified in Clause 11.3 of IS 456 (2000). It is to be clearly noted that curing of concrete in most cases goes on beyond the stripping time.

24.9 SUMMARY OF RECOMMENDATIONS FOR DURABILITY OF CONCRETE

The important steps to be taken to ensure durability of reinforced concrete structures are the following:

1. The cover provided should be according to code of practice (refer Appendix B).
2. The cement content should not be less than 300 kg/m^3.
3. The water–cement ratio should preferably be not greater than 0.4.
4. The aggregates and water should be free from deleterious substances.
5. If the structure is liable to sulphate attack rules regarding sulphate resistance should be followed [Table 4 of IS 456 (2000)].
6. The quality control of concreting (producing, transporting, placing) should be 'good'.
7. Good curing should start immediately after the concrete placed at site hardens and should continue without interruption for the specified minimum period.
8. Stripping of formwork and removal of props should comply with the provisions of the code.

24.10 POLYMER CONCRETE

Polymer concrete are concrete composites in which organic polymers like latex are added. There are three types of polymer concrete that can be produced:

1. Polymer Portland cement concrete (PPCC)
2. Polymer impregnated concrete (PIC)
3. Polymer concrete (PC)

In PPCC, the polymer is added to the mortar and the polymeric network is formed *in situ* during curing. In PIC, the concrete is impregnated with polymer which is polymerized *in situ*. In PC, the polymer is added to the aggregate for better binding properties. All these materials are at present used mainly for repairs, or coatings to reduce the diffusion rate of chlorides and to improve durability. They are widely used in storage of radioactive fuels, desalination plants, ferrocement boats, high strength piling, etc. Its commercial application is rather restricted due to its increased cost factor.

24.11 MIX DESIGN OF CONCRETE

The lowest grade of concrete specified in the revised IS 456 (fourth revision) for general reinforced concrete work is grade 20. For marine work, the grade should Concrete of grades up to 20 are called ordinary grade concrete and those between 25 and 55 as standard concrete. Concrete of grade 60 and above are called *high performance concrete* which are nowadays used for special construction in developed countries. IS 456 (1978) Clause 5.2 recognised an increase in strength of 10% in three months, 15% in six months and 20% in one year and above. The 28-days strength of concrete made from OPC.

When the criteria imposed on a concrete mix is strength, it is called a *designed mix*. When the mix proportions are fixed it is called a *special prescribed mix* and when these proportions comply with any standard it is called a *nominal mix* or *ordinary standard mix*.

24.11.1 Basic Principles of Mix Design

The basic principles of mix design for a designed mix are the following:

1. The major factor that governs the strength of *ordinary grade concrete* is Abram's Law, which states that "the lower the water–cement ratio the higher is the strength of concrete". Thus, concrete, for concrete *below 35 grade concrete* lean or rich mixes with the same water–cement ratio and gives the same strength. The richness of the mix can be neglected in these ordinary grade concretes.

2. Subsequent investigations have shown that there should be a modification for 'the law of strength' (Abram's Law) for concrete, say, above 35 grade. In high strength concrete, the following factors are also important:

   (i) Aggregate cement ratio
   (ii) Maximum size of aggregate
   (iii) Strength of coarse aggregate
   (iv) Workability of concrete
   (v) Fineness of cement or grade of cement.

The effect of these factors are briefly described below:

Aggregate cement ratio

In high strength concrete, the richness of the mix affects the strength. With the same water–cement ratio and equal compaction, the leaner mixes give higher strength than richer mixes.

Maximum size of aggregate

The optimum size of aggregate for a given strength depends on the required strength of the concrete. Grade 20 concrete can be easily made with 75-mm aggregate. In fact such concrete will also require about 35 kg/m^3 less cement than that for 20-mm aggregate.

There seems to be transition near about grade 28 concrete where the maximum size of the aggregate is important. For concrete above 35 grade concrete, the optimum size of the aggregate seems to be around 20 to 25 mm. Thus, "it is just not economically possible to make grade 40 concrete with 40 mm aggregate" [4]. With special concrete like concrete for spinning pipes, poles, etc., the maximum size of aggregate should be further reduced to about 12.5 mm. However, when considering high strength concrete for attaining the same strength and workability, 20-mm aggregates may require less cement than 10-mm aggregate.

Strength of coarse aggregate

With high strength concrete mixes, higher workability obtained by more cement and water reduces its strength. Hence low workability with better placing methods, like heavy vibration with pressure, were used for production of high strength concrete before the invention of plasticizers and superplasticizers. In modern times, better placeability can be easily achieved for low workability concrete by using superplasticizers. Thus in high strength concrete, mix design, we design the mix for low workability and introduce additives to improve workability. Hence the use of additives has become a necessary factor in design of high strength concrete. It is also very much necessary that the additive should match the cement used in the construction.

Grade of cement

The grade of cement has considerable effects on the rate of development of strength and also on the 28-day strength of concrete. The fineness of the cement also affects the workability of the mix; higher fineness requiring larger water–cement ratios for the same strength and workability. Accordingly the grade of cement to be used should match the required strength development of concrete.

24.11.2 Mix Design Procedure

From the above discussion it is clear that the procedures for design of mixes for ordinary grade and high strength concrete have to be different. The two procedures are briefly dealt with in the following sections. The details of the procedures for field application should be obtained from the references given at the end of this chapter.

24.12 DESIGN OF ORDINARY GRADE CONCRETE

There are many methods for design of ordinary grade concrete mixes. The methods used in India are IS method, ACI Method and the D.O.E. (Department of Environment) method developed by Building Research Station, U.K. The booklet *Concrete Mix Design* [4] is a good source of information about these methods.

The most important factor in these methods is the selection of the water–cement ratio applicable to Indian cements. It has been found that the relationship between water–cement ratio and cube strength for ordinary grade concretes using Indian cements can be represented by a non-dimensional curve as shown in Fig. 24.2, which is applicable to all grades of cements [5]. This curve has been obtained from the set of curves for different grades of cements available in SP 23 [6,7] and results of laboratory tests. Figure 24.2 can be conveniently used to select the grade of cement required for a given grade of concrete as well as for finding

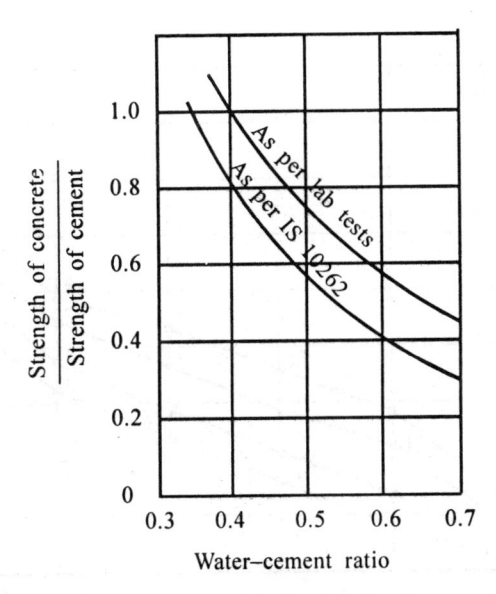

Fig. 24.2 Strength–w/c ratio relation for ordinary Portland cements (curve can be used for selection of w/c ratio for a given strength and given grade of cement).

the required water–cement ratio for a given grade of concrete and cement. Thus, for example referring to Fig. 24.2., we see that with a water–cement ratio of 0.4, the ratio of concrete strength to cement strength is 0.8 based on IS 10262 (and about 0.95 based on actual laboratory tests). With water–cement (w/c) ratio of 0.4 and 33 grade cement, we can get a strength 0.8×33 (say 25 N/mm^2) and with 43 grade cement a strength of 0.8×43 (say 35 N/mm^2). Use of 33 grade cement for a 35 grade concrete will not give the desired results with w/c ratio of 0.4

24.12.1 Other Factors to be Considered in Mix Design of Ordinary Grade Concrete [6,7]

It is convenient to remember the following approximate data about the requirements of ordinary grade concretes:

1. The amount of water that is necessary to maintain a fair amount of workability of ordinary grade cement with 20 mm, maximum size aggregate, is around 160 to 180 kg/m^3 of concrete. Some variation from this amount may be found necessary depending on the maximum size of coarse and fine aggregates. The larger the maximum size the lesser will be the water requirements. The grading of the aggregate also affects the water requirements for a specified workability. The minimum cement content should be specified in the code.

2. Aggregates below 4.75 mm are called *fine aggregates*. The proportion of fine aggregates in the total aggregates depends on the grading curve we choose for the mix. The grading curve usually recommended for 20 mm maximum size aggregate is shown in Fig. 24.3(a). Such curves for different maximum size of aggregates are available in literature and are shown in Figs. 24.3(b) and 24.3(c) [1,8]. Table 24.2 gives the gradings recommended for the 10 and 40 mm aggregates. As can be seen in Fig. 24.3, the percentage of fines to be used

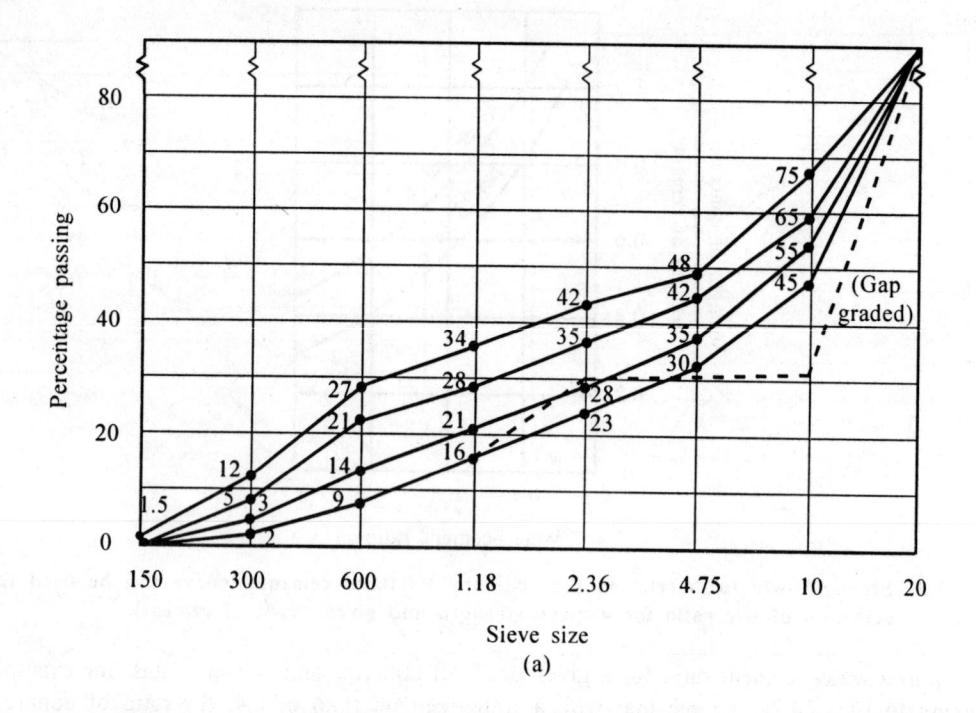

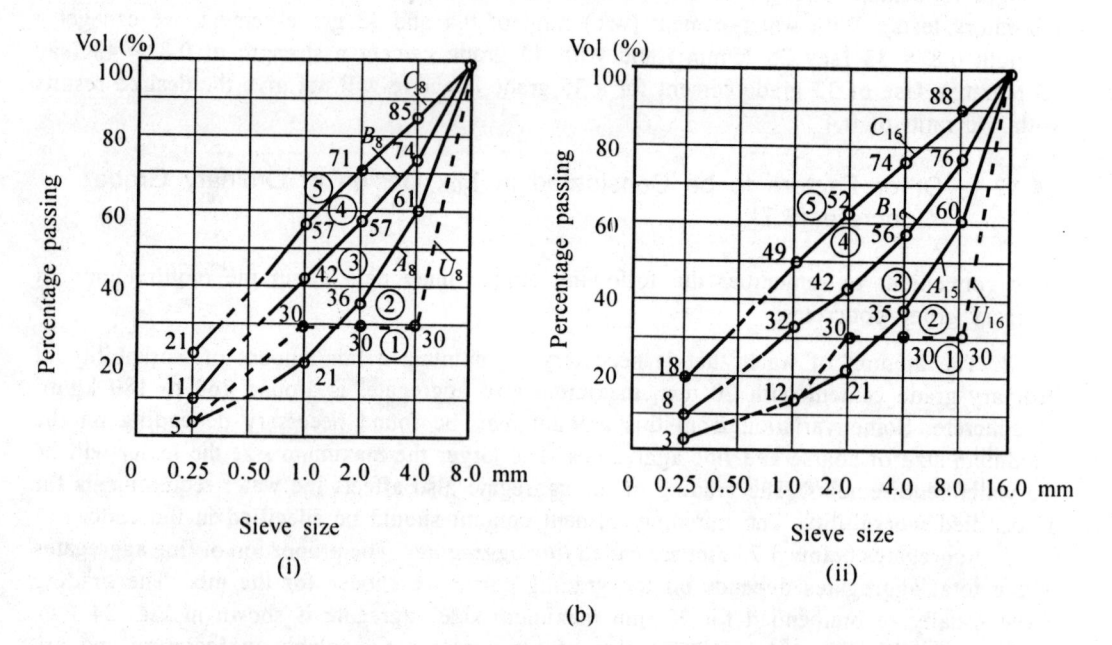

Fig. 24.3 Contd.

DURABILITY AND MIX DESIGN OF CONCRETE 447

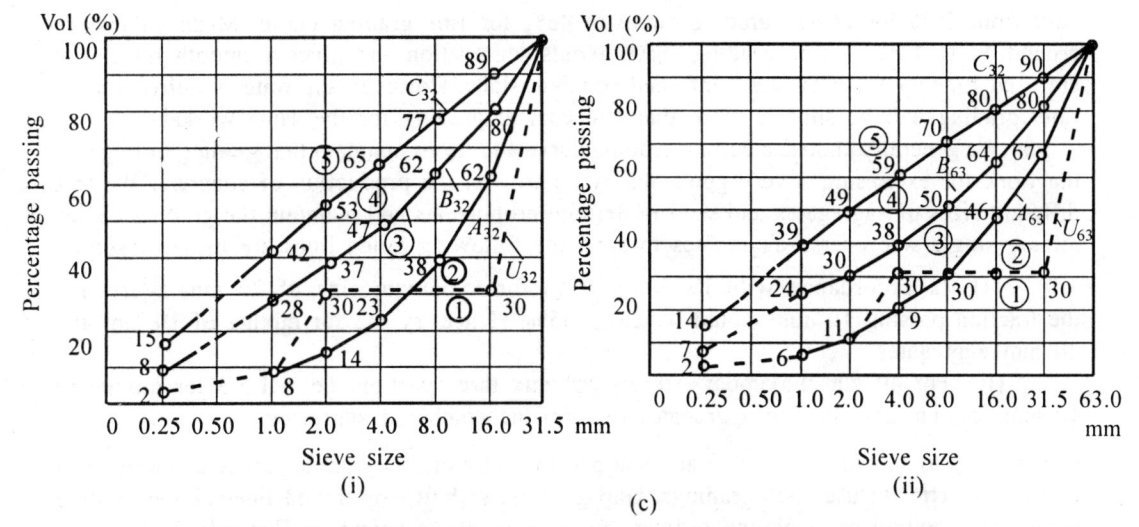

Fig. 24.3 Recommended grading curve for (a) 20 mm [1] (b) 8 mm and 16 mm [8] (c) 31.5 mm and 63.0 mm [8] maximum sizes of coarse aggregates. There are five zones 1–5. Gradings in zone 3 are preferred, although zone 4 is also accepted. *Note:* The essential criterion for composition of aggregate mix is the absolute valume and not the weight of the defferent particles [8].

TABLE 24.2 SELECTED GRADING CHART [COARSE AGGREGATED]

IS sieve	Grading (percentage passing)			
	I	II	III	IV
For 10 mm maximum size aggregate [1]				
10 mm	100	100	100	100
4.75 mm	30	45	60	75
2.36 mm	20	33	46	60
1.18 mm	16	26	37	46
606 microns	12	19	28	34
300 microns	4	8	14	20
150 microns	0	1	3	6
For 40 mm maximum size aggregate [1]				
40 mm	100	100	100	100
20 mm	50	59	67	75
10 mm	36	46	52	60
4.75 mm	24	32	40	47
2.36 mm	18	25	31	38
1.18 mm	12	17	24	30
600 microns	7	12	17	23
300 microns	3	7	11	15
150 microns	0	0	2	5

vary from 30% for coarse grading curve to 45% for fine grading curve. Mean values are around 40 to 44%. A fine grading curve avoids segregation and gives a smooth finish and are used for thin sections like reinforced concrete slabs. However, the water requirement for finer grading will be slightly more than the coarser grading for the same workability.

3. We generally choose a certain grading curve and try to maintain that grading throughout the work for exercising a very good quality control on the production of concrete. We mix different sizes of aggregates and sand in definite proportions and maintain the grading. When mixing single size coarse aggregates to sand the following guide lines are usually used.

(i) In accordance with IS 383 (1970) for maximum size of 20 mm aggregate, the fraction passing 10 mm should be 25 to 55%. Hence try 1:2 for mixing of 10 mm and 20 mm aggregates.

(ii) For 40 mm maximum size we can mix three portion, i.e. 1:1.5:3 for mixing of 10 mm, 20 mm and 40 mm aggregates to form the combined aggregate.

Gap grading as shown in Fig. 24.3 are also possible with ordinary grade concrete and in some cases it is better to use such gradings. Gap grading is shown by dotted lines as the grading curve. The method of designing ordinary grade concrete is shown in Example 24.1.

4. In the field, correction has to be made for the moisture in the aggregate, absorption of water by coarse aggregates and also for the bulking of sand produced by handling of moist sand. Most of the published data in mix design is for saturated-surface dry (SSD) conditions of the aggregate. This is different from the wet or oven dry condition that the aggregates can exist at the site or laboratory.

5. Data necessary for mix design of ordinary grade concrete according to IS 12062 (given in SP 23) is presented in Tables 24.3–24.8 as under.

TABLE 24.3 PROPORTIONS FOR NOMINAL MIX CONCRETE [IS 456 (2000) Table 9]

Grade of concrete	Mass of total aggregate per bag of cement	Proportion of fine to coarse aggregate by mass	Quantity of water per bag of cement (litres)
M10	480	1:2 (generally)	34
M15	330	1:1.5 (upper limit)	32
M20	250	1:2.5 (lower limit)	30

TABLE 24.4 MINIMUM CEMENT CONTENT AND MAXIMUM w/c RATIO FOR DURABILITY OF R.C. WORK [20 mm nominal maximum size aggregate]

Exposure (minimum grade)	Minimum cement content IS 456 1978 Table 19	Minimum cement content IS 456 2000 Table 5	Maximum w/c ratio IS 456 1978 Table 19	Maximum w/c ratio IS 456 2000 Table 5
Mild (M20)	250	300	0.65	0.55
Moderate (M25)	290	300	0.55	0.50
Severe (M30)	360	320	0.45	0.45

Note: For other exposures, see IS 456 (2000) Tables 4 and 5.

TABLE 24.5 APPROXIMATE FREE WATER CONTENT AND SAND CONTENT PER CUBIC METRE OF CONCRETE FOR TRIAL MIXES (Table 42, 43 of SP 23)

Maximum size of agregate	w/c ratio = 0.6		w/c ratio = 0.35 (high strength)	
	Water (kg/m³)	Sand (%)	Water (kg/m³)	Sand (%)
10	208	40	200	28
20	186	35	180	25
30	165	30	–	–

Note: Sand is in percentage of total aggregate in absolute volume. Basic conditions: I. C.F. = 0.8 (slump 10–30 mm) 2. crushed rock and natural sand zone II.

TABLE 24.6 ADJUSTMENT OF WATER CONTENT AND SAND PERCENTAGE
(Table 44 of Sp 23)

No.	Changed condition	Adjustment in	
		Water content	Sand percentage
1	Sand zone I*	Nil	+1.5%
2	Sand zone II	Standard condition	–
3	Sand zone III	Nil	−1.5%
4	Sand zone IV	Nil	−3.0%
5	C.F. (for every change of ±0.1% from 0.8)	±3%	0
6	w/c ration (for every change of ±5% from 0.6)	Slight	±1%
7	Rounded aggregate	−15 kg/m³	−7%

Standard condition: w/c ratio 0.6 and C.F. = 0.8. See Table 24.8 for sand zones.

TABLE 24.7 APPROXIMATE PERCENT AIR ENTRAPPED IN CONCRETE
(Table 41 of SP 23)

Maximum size of aggregate (mm)	Percentage air
10	3.0
12.5	2.5
20	–
25	2.0
40	–
50	1.5
70	1.0
150	0.5

TABLE 24.8 GRADING ZONE OF SANDS [IS 383 (1970)]

IS sieve	Percentage passing			
	Zone I	Zone II	Zone III	Zone IV
10 mm	100	100	100	100
4.75 mm	90–100	90–100	90–100	95–100
2.36 mm	60–95	75–100	85–100	95–100
1.18 mm	30–70	55–90	75–100	90–100
600 μm	15–34	35–34	35–59	80–100
300 μm	5–20	8–30	12–40	15–50
150 μm	0–10	0–10	0–10	0–15

24.12.2 Mix Design as Given in SP 23

The method of mix design recommended by IS and presented in SP-23 is based on calculating absolute volumes by the following two equations.

$$V = \left(W + \frac{C}{S_c} + \frac{1}{\tau}\frac{F_a}{S_f}\right)\frac{1}{1000} \tag{24.3}$$

$$V = \left(W + \frac{C}{S_c} + \frac{1}{1-\tau}\frac{C_a}{S_c}\right)\frac{1}{1000} \tag{24.4}$$

where

V = Absolute volume of 1 m^3 of concrete *minus* the volume of entrapped air in m^3
W = Mass of water in kg/m^3 of concrete
C = Mass of cement in kg/m^3 of concrete
S_c = Specific gravity of cement
F_a, C_a = Masses of fine and coarse aggregates per w/c of concrete
S_f, S_c = Specific gravity of fine and coarse aggregates
τ = Ratio of fine aggregate to coarse aggregate in absolute volume.

24.13 DESIGN OF HIGH STRENGTH CONCRETE

The procedure developed by Erntroy and Shacklock, for concrete of grades M35 and above, is one of the popular procedures used in the UK and India for mix design of high strength concrete [9,10]. Details of the procedures of mix design can be obtained from the reference and only basic assumptions and procedures are given in this section, as follows:

1. As the inherent capacity for maximum strength that can be developed by different grades of cements are different, a suitable cement should be chosen for the production of the required high strength. For example, it is difficult to get M40 concrete with C33 cement. It is preferable to use C53 cement.

2. For high strength concrete, it is preferable to use high strength aggregates like granite aggregate than rounded gravel aggregates.

3. The optimum size of coarse aggregate for high strength concrete is 20 mm.

4. It is very important to note that the aggregate cement ratio (i.e. leanness or richness of the mix) has also a great influence on the strength of the mix. Leaner concrete mixes with the same water–cement ratio give higher strength than richer mixes. (This can be explained by stating that the cement mortar in the mix should be just sufficient to bond all the very strong aggregates together and the thickness of the mortar should also be a minimum.)

5. For a given type of cement, type of aggregate, target strength and workability, there is a fixed water–cement ratio that will give the required strength.

6. Again for a given workability and water–cement ratio there is a fixed aggregate–cement ratio for the type of cement and aggregate used in the mix.

Incorporating all these principles, a set of curves and tables have been derived from tests to determine for a given strength. For a given type of aggregates (granite or gravel) and given type of cement (grade 47 and 53) we can find the following:

  (i) The water–cement ratio and
  (ii) The aggregate–cement ratio

Tests, made by Erntroy and Shacklock for high strength concrete with ordinary Portland cement of UK grade C47 and rapid hardening Portland cement of UK grade C53, have been published. In the absence of other data, even though the grades as specified in UK and India are not the same these may be used for 43 grade and 53 grade cements now available in India for preliminary tests till further data becomes available. The pertinent curves as modified by Lydon [10] for granite aggregates of maximum size 20 mm are given in Figs. 24.4 and 24.5. Similar curves for granite aggregates of 10 mm and gravel aggregates of 20 mm and 10 mm are available in References 9 and 10.

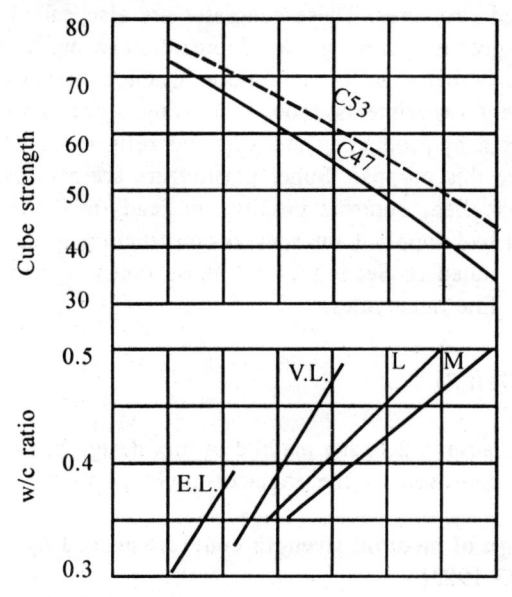

Fig. 24.4 Typical relation between 28 day cube strength curve, water–cement ratio and workability using two types of cements and 20 mm granite coarse aggregate [10].

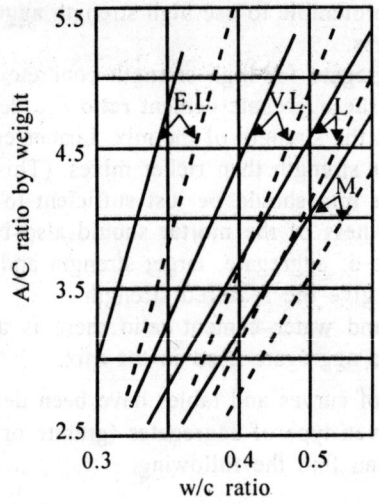

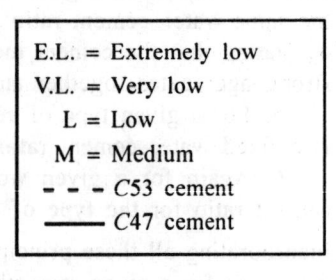

Fig. 24.5 Typical relation between aggregate–cement ratio and workability for two types of cement using 20 mm granite and gravel aggregates [10].

E.L. = Extremely low
V.L. = Very low
L = Low
M = Medium
- - - C53 cement
——— C47 cement

The workability of the resultant mix can be further improved by additives. In fact all modern high strength concretes are made by adding suitable super plasticisers, even though the old practice was to place such concrete by heavy vibration and pressure.

24.14 DESIGN OF VERY HIGH STRENGTH CONCRETE MIXES

Concrete of grade M60 and other higher grades are being used for construction of very tall buildings and other specialised structures. These concretes are also called *high performance concrete*. Eventhough, the subject is still in the development stage, we know that one of the requirements of such concrete is that even the micropores in the concrete should be filled up with reactive materials. Recent experiments show that using silica fumes (very very fine particles of silica obtained as a by-product in industry and refined for concrete-making) as an additive material can serve this purpose. Super plasticizers are also used as an agent to improve workability. With excellent controls possible in ready-mix concrete plants, such concretes are being produced and supplied for special construction in countries like U.S.A. on a large scale. As already stated in Section 24.3.1 these concrete may also have excess unhydrated cement acting as fine aggregates.

24.15 MIX DESIGN METHOD

According to IS 456 (2000) Clause 9.2.1, any method of mix design can be used for arriving at the mix proportions to be approved by the engineer-in-charge of the work.

EXAMPLE 24.1 [Mix design of medium strength concrete according to IS 10262 (1982) and SP 23 1983]

Design a concrete mix according to the procedure given in SP 23 (1983) for the following data and explain how to prepare a mix design report [7].

DURABILITY AND MIX DESIGN OF CONCRETE 453

1. Characteristic strength 20 N/mm² with grade 43 cement
2. Maximum size of aggregate 20 mm (angular)
3. Degree of workability 0.90 compaction factor 50 mm slump
4. Degree of control good
5. Type of exposure mild
6. Specific gravity of coarse and fine aggregate 2.6
7. Sand (by sieve analysis zone III) (Table 24.8)

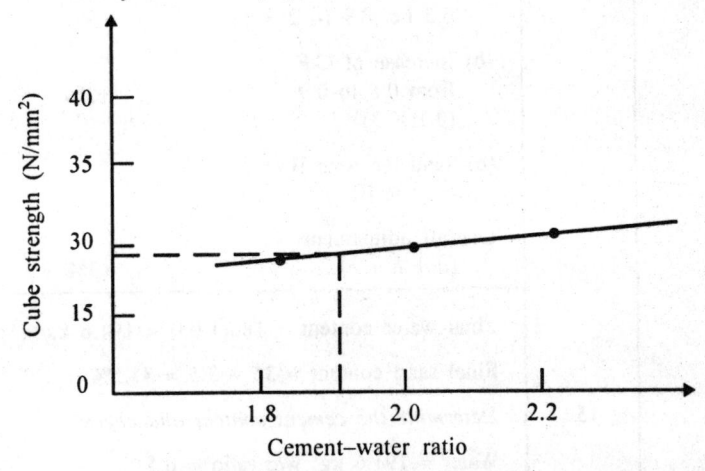

Fig. 24.6 Example 24.1 (step 8).

Reference	Step	Calculations
	1	*Determine target strength*
Chapter 25		Required mean strength = f_{ck} + 1.65s
Table 25.1		Let standard deviation for M20 with good control s = 4.6 N/mm²
		Required strength = 20 + 1.65 × 4.6 = 27.6 N/mm²
	2	*Select water–cement ratio for grade of cement*
		$\dfrac{f_{ck}}{f_c} = \dfrac{27.6}{43} = 0.64$
Fig. 24.2		Required w/c ratio = 0.50 (approx.) [or from IS 10262]
Table 24.4		This is lower than the allowable for mild exposure = 0.65
	3	*Select water and sand cement to be used*
		For 0.6 w/c ratio we have the following data required
		For 20 mm aggregate, C.F. = 0.8 (sand zone II)
Table 24.5		Water content = 186 kg/m³
		Percentage of sand in total aggregate by absolute volume = 35%

Reference	Step	Calculations
	4	*Make corrections required*

Change	± *Water content*	± *Sand*
(a) Decrease of w/c from 0.6 to 0.5 i.e. 0.1 (= 2 × 1)	0	−2.0%
(b) Increase of C.F. from 0.8 to 0.9 (0.1 × 3)	+3%	0
(c) Sand for zone II to zone III	0	−1.5%
Overall adjustment (a + b + c)	+3%	−3.5%

Final water content = 186(1.03) = 191.6 kg/m^3

Final sand content = 35 − 3.5 = 31.5%

	5	*Determine the cement content and check*

Water = 191.6 kg; w/c ratio = 0.5

Requirement of cement = $\dfrac{191.6}{0.5}$ = 383 kg/m^3

Table 24.4   Minimum specified = 300 kg/m^3 only. Sesign O.K.

| | 6 | *Determine the coarse and fine aggregates* |

Assume entrapped air as 2%

Volume of solid concrete = 0.98 m^3 = V

Eq. (24.3)

$$(1000)V = W + \frac{C}{S_c} + \frac{F_a}{\tau S_f}$$

Step 4   where τ = percentage of sand in absolute volume (31.5%).

$$980 = 191.6 + \frac{383}{3.15} + \frac{F_a}{(0.315)(2.6)}$$

F_a = Wt. of fine aggregate = 546 kg/m^3

Eq. (24.4)

$$980 = 191.6 + \frac{383}{3.15} + \frac{C_a}{(0.685)(2.6)}$$

C_a = Wt. of coarse aggregate = 1187 kg/m^3

| | 7 | *Calculate materials per bag for field mix* |

Cement = 50 kg

Water = 50 × 0.5 = 25 kg (litres)

Reference	Step	Calculations
		Sand = $\dfrac{546}{383} \times 50 = 71$ kg
		Crushed stone = $\dfrac{1187}{383} \times 50 = 160$ kg
		Note: (1) The water content should be corrected for absorption of water by coarse aggregates (as mix assumes surface dry aggregates only) and also for free moisture is same).
		(2) If aggregates are to be mixed, follow Section 24.12.1.

Mix Design Reports (*Step 8*)

A mix design report consists of two parts: The first part should give the results of laboratory tests on workability and strength of at least four trial mixes. The second part is meant to check the standard deviation that can be expected in the field. It is based on the analysis of at least 40 trial mixes made under field conditions. The procedure is shown by Fig. 24.7.

Report Part I—Data of Laboratory Tests

Trial mix no. 1. Make a trial mix with calculated proportions including w/c ratio as in Example 24.1 using a small laboratory mixer. Measure its workability and behaviour under a trowel. Take enough quantities for casting six cubes, three for 7-day strength and another three for 28-day strength tests. Examine consistency of mix with a trowel.

Trial mix no. 2. Repeat trial mix no. 1 and adjust its consistency by varying sand and water within 1 to 2% to produce a cohesive mix. For the new mix recalculate the needed water content and cement content for the assumed w/c ratio. Make the second trial mix with the same w/c ratio as in trial mix no. 1 but with the new sand, water and cement content of the adjusted consistency. Cast 6 cubes three for 7-day and three for 28-day strengths.

Trial mix no. 3. Reduce the w/c ratio of trial mix 2 by 10% keeping the water content the same but adjusting cement and sand to keep fineness approximately the same as in trial mix no. 2. Cast 6 cubes three for 7-day and three for 28-day strengths.

Trial mix no. 4. Increase w/c ratio of trial mix 2 by 10% keeping the water content the same but adjusting the cement and sand. Cast 6 cubes as before.

Analysis of the report. Presen data of four tests as in Table 24.10.

The strength results of trial mixes 2.3.4 are then plotted with cement–water ratio as *X*-axis and cube strength as *Y*-axis. This plot should be approximately a straight line as shown in Fig. 24.6. Read off from the graph the required cement–water ratio and derive the required water and cement requirements (see page 453).

Final recommendations will be made per bag of cement (50 kg).

456 ADVANCED REINFORCED CONCRETE DESIGN

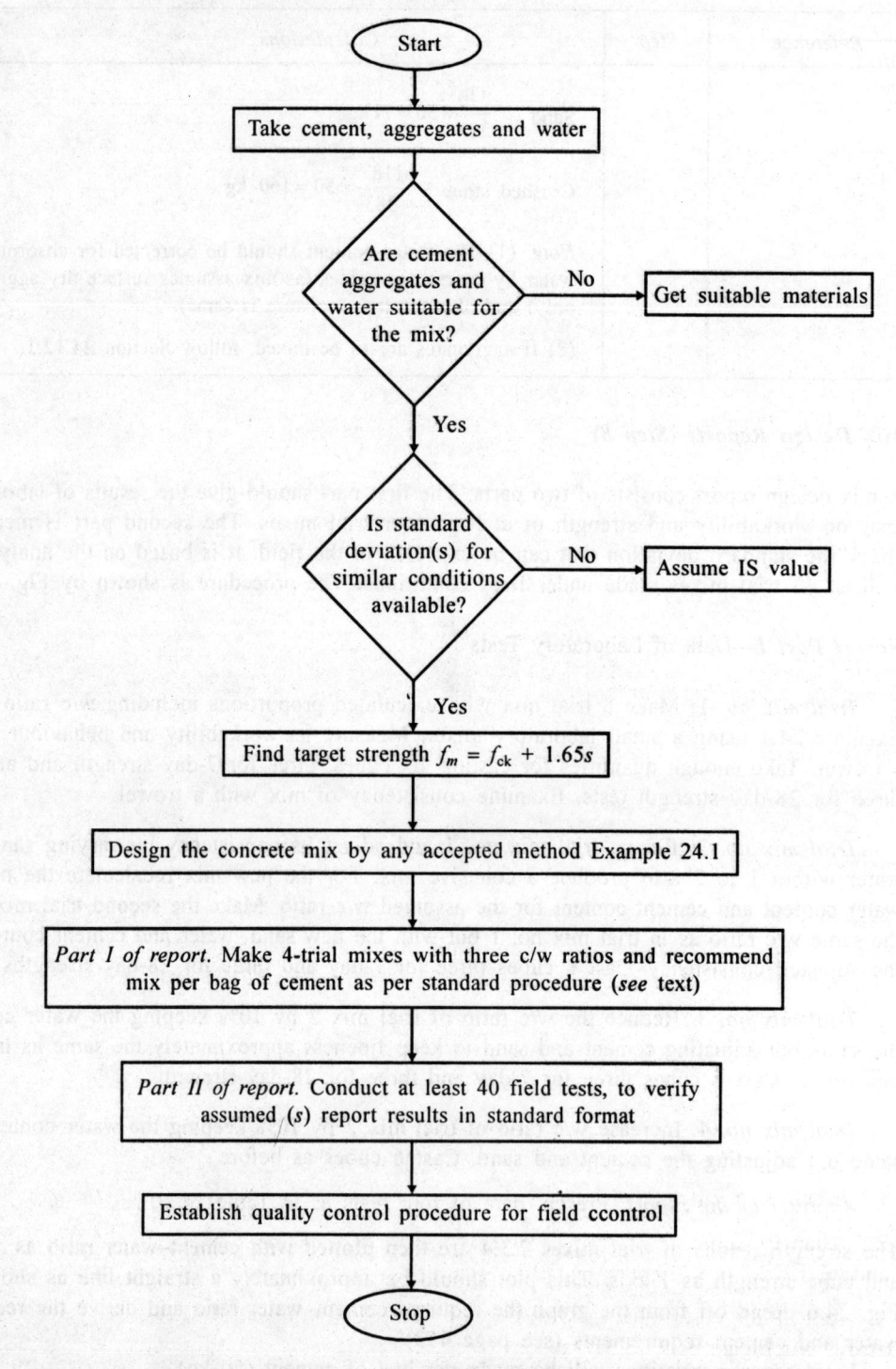

```
                    Start
                      │
                      ▼
        Take cement, aggregates and water
                      │
                      ▼
              ╱────────────╲
             ╱ Are cement   ╲      No
            ╱ aggregates and ╲────────▶  Get suitable materials
            ╲ water suitable ╱
             ╲ for the mix? ╱
              ╲────────────╱
                      │ Yes
                      ▼
              ╱────────────╲
             ╱ Is standard  ╲      No
            ╱ deviation(s) for╲──────▶  Assume IS value
            ╲ similar conditions╱
             ╲  available?   ╱
              ╲────────────╱
                      │ Yes
                      ▼
        Find target strength $f_m = f_{ck} + 1.65s$
                      │
                      ▼
        Design the concrete mix by any accepted method Example 24.1
                      │
                      ▼
   Part I of report. Make 4-trial mixes with three c/w ratios and recommend
         mix per bag of cement as per standard procedure (see text)
                      │
                      ▼
   Part II of report. Conduct at least 40 field tests, to verify
         assumed (s) report results in standard format
                      │
                      ▼
       Establish quality control procedure for field ccontrol
                      │
                      ▼
                    Stop
```

Fig. 24.7 Flow chart for concrete mix design report.

TABLE 24.10 PRESENTATION OF TRIAL MIX DATA

Mix no.	w/c ratio	c/w ratio	Cement (kg)	Water (kg)	Aggregate (kg) Fine	Aggregate (kg) Coarse	Workability Observed	Workability Measured	Strength (in N/mm²)
1	*0.50	2.0							
2	0.50	2.0							
3	0.45	2.2							
4	0.55	1.82							

*This is the theoretical w/c ratio obtained from Fig. 24.2 (0.50 is taken as an example).

1. w/c ratio and litres of water (per bag of cement)
2. Fine aggregate in kg per bag of cement
3. Coarse aggregate in kg per bag of cement
4. Theoretical yield (m^3) per bag of cement
5. Cement content (kg/m^3)

Report Part II—Data of Field tests

The second part of the report should consist of results of actual field tests. Using a field concrete mixer as well as materials and controls, make at least 40 mixes and cast test cubes. Twenty-eight day cube strengths of the 40 results are analysed for verifying the standard deviation, initially assumed in the calculations, for the field conditions. If the assumed standard deviation has not been achieved by the controls at the site, then either tighten the controls or redesign the mix to suit the standard deviation obtained at site.

CONCLUSION

These two reports are to be submitted by the contractor to the engineer in charge before the commencement of any large project to arrive at the final mix proportions. Also continuous field control as explained in Chapter 25 should be maintained throughout the construction period to ensure quality in construction. Mixes so derived are called *designed mixes*.

REFERENCES

1. Eibel, J., *Concrete Structures—Eurodesign Handbook*, Ernst & Sohn, Berlin, 1994.

2. Allan P. Crane, *Corrosion of Reinforcement in Concrete Construction,* Ellis Harwood, London, 1972.

3. ACI SP 65, Performance of Concrete in Marine Environment, American Concrete Institute, Detroit, Michigam.

4. Concrete Mix Design—A.C.C., 3rd ed., Research and Consultancy Directorate, Bombay, 1993.

5. Rao, P.S., 'Influence of Cements on Concrete Mixes', *Proceedings of the Workshop on Quality Assurance for Concrete Mixes for Practising Engineers*, Anna University, Guindy, Madras, 1993.

6. IS 10262 (1982), Concrete Mix Design, Bureau of Indian Standards, New Delhi.

7. SP 23, Handbook on Concrete Mixes Based on Indian Standards, Indian Standards Institution, 1982.

8. CEB-FIP Model Code for Concrete Structures, 3rd ed., Appendix D, Concrete Technology Cement and Concrete Association, London, 1978.

9. Erntroy and Shacklock, Design of High Strength Concrete Mixes—Proceedings of Symposium in Mix Design and Quality Control of Concrete, Cement and Concrete Association, London, 1954.

10. Lydon, F.D., *Concrete Mix Design*, Applied Science Publishers, London, 1973.

CHAPTER 25

Quality Control of Concrete in Construction

25.1 INTRODUCTION

One of the most important requirements in good concrete construction is that the quality of concrete placed in the structure should conform to that specified in the design. The major part of concrete used in developed countries is produced in central ready-mix plants. But in India, most of it is mixed at site and as different from the steel made in the factory, its quality can vary from site to site and at the same site, from day to day. The only method we can use for evaluation of strength of such materials is the statistical method. In this chapter, we shall examine these methods and their uses for design of concrete mixes and also field control of the quality of concrete. The subject will be dealt with under the following heads:

1. Principles of statistics as applied to concrete mix design and concrete quality control
2. Application of statistics in concrete mix design
3. Specification for control of concrete (acceptance criteria)
4. Testing of structures in case of non-conformity to acceptance criteria

25.2 STATISTICAL PRINCIPLES

We shall examine some of the basic principles that are used for the mix design and quality control of concrete.

25.2.1 Measure of Central Tendency

In all statistical observations it is necessary to get an idea of the *central tendency* of the variable being studied. The following are some of the properties used to represent this tendency:

    1. Median. It is the mid value or that value which has as many values greater than itself as there are less than itself.
    2. Mode. It is the most commonly occurring value in the set of values being considered.
    3. Mean value or the arithmetic mean. It is the arithmetic average of all the values being considered.

Of these the 'mean value' is the most important in the study of data occurring with normal distribution. The mean value of a number of data is given by the formula

$$\bar{x} = \frac{\sum x_i}{n} \tag{25.1}$$

25.2.2 Statistical Distribution

If the cube strengths of a large number of samples from a prescribed mix made under the same conditions are taken, it will be found that eventhough a greater number of them will have cube strength near about the average value, a few of them will show strength larger than the average and lower than the average. The curve showing the cube strength and the frequency of occurrence will be as shown in Fig. 25.1. Such curves are called *frequency distribution curves*. In the above case of cube strength, the 'observed distribution curve' is very similar in shape to the theoretical curve called normal distribution curve, which can be expressed by the equation:

$$y = \frac{N}{s\sqrt{2\pi}} \exp\left[-\frac{(x-\bar{x})^2}{2s^2}\right]$$

where

y = Frequency

N = Total number of observation of x

s = Standard deviation[1]

$\bar{x}$ = Mean value

x = Variable

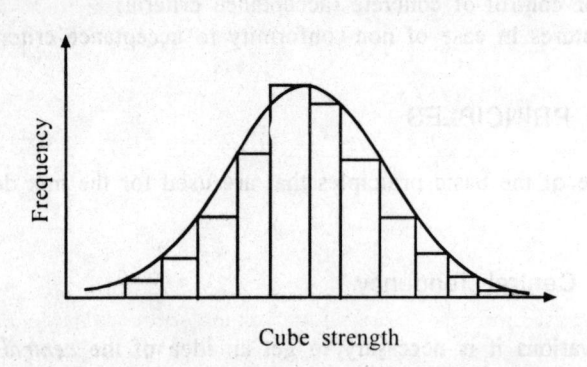

Fig. 25.1 Frequency distribution of cube strength.

By putting the distribution in such a mathematical expression it is possible to study its characteristics more easily by theory. There are many natural occurrences that do not follow normal distribution and they may be then tried to be fitted to other theoretical distribution

---
[1] See Section 25.4.

curves like binormal distribution, Poisson's distribution, etc. However, it has been observed and generally accepted that the strength of concrete, produced in the laboratory or field, follow the normal distribution curve. The difference between various conditions of works being reflected in the spread of the distribution. Under laboratory conditions where the control is good, the spread will be small, and with bad control the spread will be large (Fig. 25.2). The 'normal distribution curve' may itself be skewed but theoretical studies on concrete strength ignore the effects due to skewness. The two properties that define such a symmetrical normal distribution curve are the mean value $\bar{x}$ and the standard deviation s explained below.

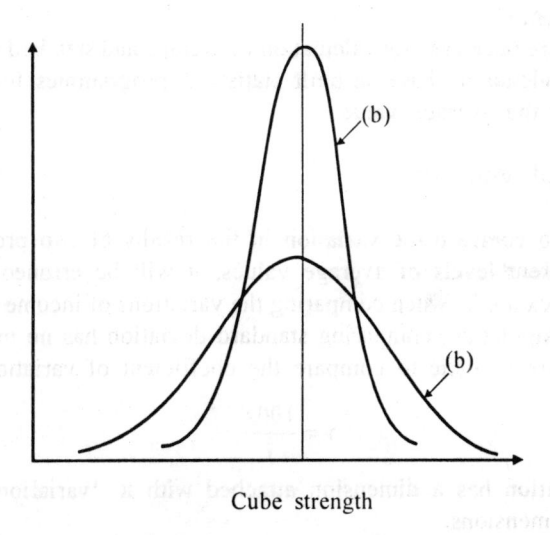

Fig. 25.2 Spread of typical frequency distribution curve: (a) small variation (good control) (b) large variation (bad control).

25.2.3 Standard Deviation

In statistics, the variance s^2 of a distribution is the mean of the sum of the squares of the deviations from the mean value of the samples. It is given by the formula

$$s^2 = \frac{\Sigma(x_i - \bar{x})^2}{n} \qquad (25.2)$$

where $(x_i - \bar{x})$ is the deviation of the individual observation from the true mean value $\bar{x}$. From this value, standard deviation s is derived by taking the square root of the variance.

$$s = \sqrt{\frac{\Sigma(x_i - \bar{x})^2}{n}} \qquad (25.3)$$

where $\bar{x}$ be the 'true mean value' from a large population, which is unknown with limited samples. With small number of samples, the mean value will generally differ from the true mean. Substitution of $\bar{x}$ of a small sample for the true mean $\bar{x}$ will give the minimum value for $(x_i - \bar{x})^2$. Hence, the correction by division by $(n - 1)$ instead of n is made and the standard deviation is expressed for small samples as given in IS 456 (1978) Clause 14.5.2 as

$$s = \sqrt{\frac{\Sigma(x_i - \bar{x})^2}{n-1}} \qquad (25.4)$$

It should be noted that $\bar{x}$ is expressed in the same unit as cube strength (N/mm^2). According to IS and ACI codes, the minimum number of results for calculation of an acceptable standard deviation is 30. BS code specifies that at least 40 samples should be tested to arrive at the probable standard deviation.

Frequency tables were once used for calculation of average and standard deviation. Nowadays, all modern scientific calculators have in-built statistical programmes for easy evaluation of standard deviation and the average value.

25.2.4 Coefficient of Variation

When it is necessary to compare the variation in the results of two products under similar conditions but of different levels of average values, it will be erroneous to compare their standard deviation. For example, when comparing the variations of income of a rural population with that of an urban population, comparing standard deviation has no meaning. Under these circumstances, it is more realistic to compare the coefficient of variation.

$$v = \frac{100s}{\bar{x}} \qquad (25.5)$$

whereas standard deviation has a dimension attached with it, 'variation' is expressed as a percentage, with no dimensions.

25.2.5 Normal Probability Curve

For comparison of different normal distribution curves, it is desirable to reduce all of them to single standard form. The first step to this end is to make all the curve of the same area by using relative frequency as the Y ordinate. Along the X coordinate we plot the quantity $z = (x_i - \bar{x})/s$, i.e. the deviation of x from mean value of $\bar{x}$ in terms of the standard deviations as z. This procedure gives a single standard normal curve with a central value $z = (x_i - \bar{x})/s = 0$ and X-axis in term of the standard deviation as shown in Fig. 25.3. For this standard curve, areas between the ordinates placed at various position of z are as shown in Fig. 25.5. This area is called the *probability integral* and gives the probability that the individual values chosen at random from a normally distributed population will have a smaller value than the mean ordinate. Thus for example, for $(x_i - \bar{x})/s = 1.64$, the probability integral (for one side of the curve only) is 0.05, so that 5% of the cube strength will deviate from the mean value f_m.

In the limit state design, 'characteristic strength of a material' (such as concrete) is defined as the strength of material below which not more than 5% of the results are expected to fall (IS 456 Clause 5.1.1). This simply means that the probability of failure is 5%. To get the normal probability curve from the normal frequency distribution curve, the total number of observations, which is equal to the area under the curve, is made equal to unity as shown in Figs. 25.3 and 25.4. The spread of the curve is practically between $-3s$ to $+3s$, but theoretically it is between $-\infty$ to $+\infty$. From the per cent of areas represented for various value of z given in Figs. 25.3 and 25.4 we can see that the value of z corresponding to 5 per cent failure is given by $z = (f_{ck} - f_{mz})/s = 1.64$ and the failure corresponding to $z = 2.0$ is 2.3%. Similarly, the failure corresponding to $z = 2.33$ is about 1% and that corresponding to $z = 3$ is about 0.1% only.

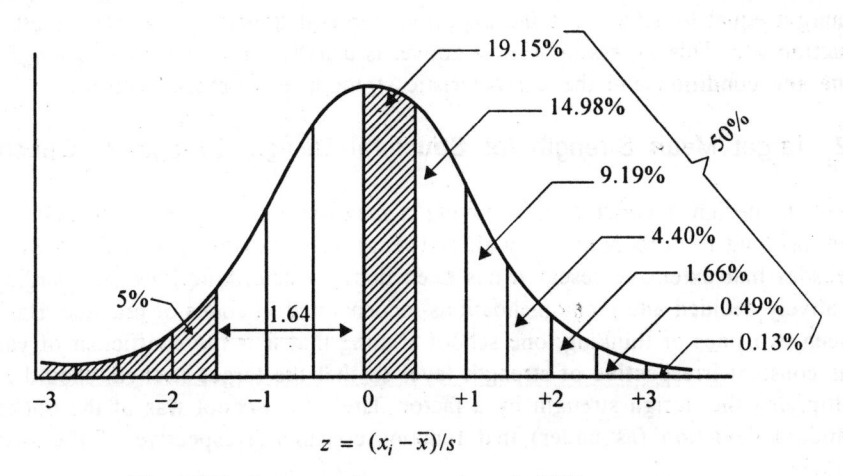

Fig. 25.3 Areas under narmal probability curve.

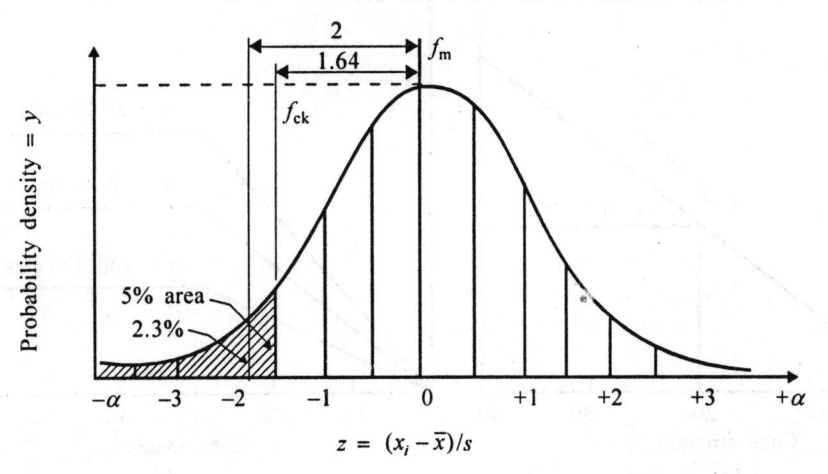

Fig. 25.4 Principle arriving at target strength for a specified characteristic strength.

25.3 APPLICATION OF STATISTICS IN CONCRETE MIX DESIGN

25.3.1 Current Margin

Design of concrete structures are based on the characteristic strength of concrete f_{ck}. However, by definition f_{ck} is the value below which only 5% of the results should fall. From Figs. 25.3

and 25.4, it is evident that the design strength f_{ck} will be less than the mean concrete strength f_m as given by relation $f_{ck} = f_m - 1.65s$ (using 1.65 instead of 1.64).

Hence to get f_{ck} the mean strength should be higher as given by the expression:

$$f_m = f_{ck} + 1.65s = f_{ck} + \text{Current margin}$$

So, assuming the concrete cube strength follows the law of normal distribution, the mean strength for which concrete mix is designed should be in excess of the characteristic strength by a margin equal to 1.65 times the expected standard deviation that can be attained at the construction site. This excess, as shown above, is usually called the *current margin*. It varies with the site condition and the characteristic strength as discussed below.

25.3.2 Target Mean Strength for Structural Design Strength of Concrete

The need to design a concrete mix having a mean strength much in excess of the design strength has long been recognised by all civil engineers. However, it was only during the last two decades that extensive research has been done to determine how the margin should be quantitatively decided and recommendations incorporated in codes of practice. Basically there have been two lines of thinking, one school arguing that it is the coefficient of variation that remains constant irrespective of strength level so that the target strength should be obtained by multiplying the design strength by a factor. The other school was of the option that it is the standard deviation (as under) that remains constant irrespective of the strength level (Fig. 25.5).

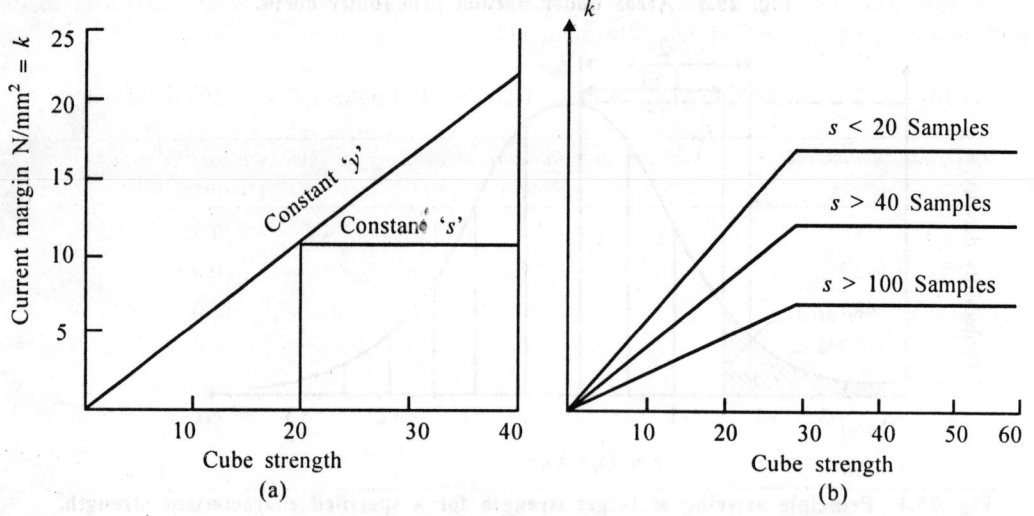

Fig. 25.5 Influence of assumptions on current margin: (a) effect of assuming constant standard s and constant coefficient variation v (b) effect of number of samples on standard deviation.

If one is to take the coefficient of variation as constant the 'current margin' increases as the characteristic strength increases, but if the standard deviation is constant the 'current margin' can remain constant irrespective of specified strength after a certain value as is assumed in BRE publication *Design of Normal Concrete Mixes*. To study the matter in more detail, it is worthwhile to examine the individual factors that make up the total standard deviation [2]. Observations show that they are made up of various components involved in manufacture and testing of concrete. For example it is known that variability of sampling and testing procedure alone can produce a standard deviation of 2.5 N/mm^2. Variation due to cement used in construction can produce a standard deviation of 3.5 N/mm^2 or more. Hence, in British practice it was argued that, under given conditions of work the standard deviation should remain fairly constant especially *above a certain level of characteristic strength*, i.e. for high strength concrete say above grade 25. For lower strength either coefficient of variation or standard deviation may be used for the current margin as shown in Fig. 25.4. It should be noted that this British approach which has also been observed by many ready mix plants in the UK, is not the same as in IS 10262 (1982). In IS 456 (2000), the standard deviation increases with increase in characteristic strength up to grade 25 and then remains constant for the higher grades (Table 25.1). The question always arises what should be the magnitude of the standard deviation be used for mix design. The recommended procedure is to conduct pilot tests under conditions similar to those expected as soon as possible before the work is started. Get 30 to 40 results to derive the standard deviation to be used for mix design. Use can also be made from information obtained from similar previous construction works. In the absence of such data those given in Table 25.1 which is Table 8 of IS 456 (2000) will have to be used to conform to IS specifications. However, it should be noted that these values are high and even though not mentioned in it represents 'fair degree' of control. For good and very good controls they can be much lower. Thus for example, studies show that, for grade 25 concrete with good control the standard deviation can be about 3.6 N/mm^2 and with very good control it can be as low as 2.4 N/mm^2 [1].

TABLE 25.1 RECOMMENDED STANDARD DEVIATIONS [IS 456 (2000) Table 8]

Grade of concrete	Prescribed standard deviations (N/mm^2)	Derived coefficient of variation (%)
M15	3.5	23.3
M20	4.0	20.0
M25	4.0	16.0
M30	5.0	16.6
M35	5.0	14.2
M40	5.0	12.5
M50	5.0	10.0

25.4 EVALUATION OF CONCRETE AT CONSTRUCTION SITES

25.4.1 General Requirements

Specifications required for controlled grade concrete field tests should be made at regular intervals to ensure that the concrete placed in construction conforms to the assumed design strength. This is done by taking enough quantity of sample to cast at least three 15-cm cubes, curing them in the mould under a wet gunny bag for one day after which they are taken out of the mould and 'moist-cured' theoretically at 20°C for 27 days in a field laboratory. The cubes are tested surface dry after a total of 28 days after casting. The mean value of the strength of the three cubes (neglecting those with 15% variation from the mean) is taken as the strength of the sample (IS 456 Clause 15.4). This practise is different from casting only one cube from each sample and determining its strength as applied in some other cases of control. Additional samples may also be taken as required by some organisations for other purposes such as for accelerated curing and testing or checking the field curing adopted in construction. The following information regarding testing of concrete cubes will be found useful in practice:

1. Accelerated curing by heating of cubes in water and testing it as specified in IS 9013 (1978) can be used to estimate the strength of concrete within a few hours after making of concrete.

2. It is usually difficult to test 15-cm cubes of very high strength concrete in the compression testing machine available in the field. In such cases we may use 10-cm cube for testing and the result should be divided by a factor equal to 1.04 to get the strength of 15-cm cube.

3. Concrete specimen which are completely dry before testing can give 10 to 20% more strength than the standard water saturated specimen. Hence cubes should be tested as surface dry only.

4. The rate of loading in standard cube test should be 14 N/mm^2 per minute. Faster loading can produce higher strength and the slower one can produce lower strength.

5. Before testing of cube its exact weight should also be recorded. From it, we can find the density of concrete and also check on the cube strength obtained.

6. As a rule, the minimum dimension of the mould, in which concrete is cast for testing, should not be less than four times the maximum size of the aggregate. Hence, in case we use 15-cm moulds for testing concretes with large size aggregates we cast only the portion sieved through a 20- or 30-mm sieve and apply a correction factor less than unity on the result thus obtained.

Similarly 'production assessment for batching plants' producing controlled concrete can also be laid out when the concrete is produced in central ready mix plants. Concrete should be purchased only from plants which have such control data.

25.4.2 Frequency of Testing

The frequency of testing is given in IS 456 Clause 14.2.2. The number of samples (one sample consists of three cubes) to be taken is shown in Table 25.2.

TABLE 25.2 FREQUENCY OF TESTING OF CONCRETE

Quantity of concrete in the work (m^3)	Number of samples
1–5	1
6–15	2
16–30	3
31–50	4
51 and above	4 plus one additional sample for each additional 50 m^3 or part thereof.

Note: At least one sample should be taken from each shift.

The statistical theory of the method of evaluation of these test results and the different compliance specifications are dealt with in the following sections.

25.4.3 Statistical Theory of Testing Plans

Test plan is a procedure prescribed in specification to provide a continuous assessment of the quality of the product. The testing plan should divide the product into small lots according to the rates of sampling and allow the work to be passed as the work progresses. The risks of using different sampling and testing plans for the producer of concrete and the client using the concrete can be compared by means of the 'Operating Characteristic' (OC curves) described in BS 6000 [2]. According to these methods, cube test results are examined for two conditions:

(i) For the individual values
(ii) For the average value of the cube tests

It should be noted that by definition of characteristic strength, 5% of 'test-results' are allowed to fall below f_{ck}. Hence, such variation should always be allowed in the testing plan as well. This is taken care by the average value of cubes satisfying the prescribed conditions. For practical purposes the sample number for determining the average values cannot be very large, as approval of works has to be made as the work progresses. For this purpose OC curves mentioned above has been made and the British specifications use the 'mean of consecutive four' and the ACI uses 'mean of three consecutive test results' as the bases for approval in their specifications.

25.4.4 Compliance Clauses (Laboratory Cured Specimen)

The compliance specifications in IS, BS and ACI practices are given in the following sections. One test means average of three cubes cast as one sample cured in the laboratory and tested according to specifications.

25.4.5 BS 5328 and IS 456 (Revised) Compliance Clauses

We use the method of moving average values to forecast the trend in the quality of field work. Thus BS 5328 Methods of Specifying Concrete Including Ready Mixed Concrete gives the individual and average strengths to be satisfied for good control of concrete in terms of a given fixed value in N/mm². Both conditions are to be met for compliance [3]. The conditions required by BS and IS 456 (2000) Clause 16.3 are shown in Table 25.3.

TABLE 25.3 BS AND IS COMPLIANCE REQUIREMENT [IS 456 (2000) Table 11]

Specified grade	BS (N/mm²) Individual test results	BS (N/mm²) Mean of four consecutive tests	IS (N/mm²) Individual test results	IS (N/mm²) Mean of a group of four consecutive test results
≤M15	$\geq (f_{ck} - 2)$	$\geq (f_{ck} + 2)$	$\geq (f_{ck} - 3)$	$\geq (f_{ck} + 3)$
≥M20	$\geq (f_{ck} - 3)$	$\geq (f_{ck} + 3)$	$\geq (f_{ck} - 4)$	$\geq (f_{ck} + 4)$ or (for both) $(f_{ck} + 0.825s)$ whichever is greater.

Notes: 1. $(f_{ck} + 0.825s) = f_{ck} + (1/2)\ 1.65s = f_{ck} + (1/2)$ (current margin).

2. If individual test fails only the concerned batch is at risk.

3. In IS, the mean of four tests is *limited to a quantity of 60 m³ of concrete* (see Clause 16.3). It shoud include the first and the last and all the intermediate tests.

4. Each cube strength test is the average of three specimen with individual variation not more than ±15% (Clause 15.4).

5. In continuous production units, frequency of sampling is to be agreed upon mutually by supplier and purchaser (*see* note in IS 456 Clause 15.2.2).

6. For flexural strength compliance, *see* IS 456 (2000) Clause 16.2.

25.4.6 ACI Compliance Clause

ACI compliance Clause is in terms of cylinder strength in p.s.i. Taking f'_c as the specified characteristic cylinder strength in p.s.i., the following conditions should be satisfied:

(i) No individual cylinder strength (average of two cylinders) should fall below the specified cylinder strength f'_c by more than 500 p.s.i. (3.45 N/mm²)

(ii) Average of all sets of three consecutive cylinder strength tests should equal or exceed f'_c.

If the above conditions are not met, the concrete proportions should be adjusted to give a higher mean strength. As can be seen, the second condition about moving averages in ACI is more liberal than the BS condition.

25.4.7 Importance of Compliance Tests

In limit state design, it is very important that the characteristic strength assumed in the design calculations is actually obtained at the construction site. The mean strength of the chosen mix proportions (strength for mix design) should be well above the characteristic strength according to statistical theory explained in this chapter. In addition, while executing the work, samples of concrete should be taken at the construction site according to specifications and tested for 7- and 28-days cube strengths. The cube test values should satisfy the compliance clauses specified in the codes. Unless these conditions are satisfied the strength of the structure will not be the same as that of the designed strength. This is most important for concrete used in columns.

25.4.8 Tests on Field Cured Specimen

According to ACI 318(1989) Clause 5.6.3, the building official may require additional strength tests on specimen that are cured under 'field conditions' to check the adequacy of curing and protection of concrete placed in structure. These are called 'field cured specimen'. Their strength should not be less than 85% of the strength of specimens cured under laboratory conditions. In tropical countries where the curing temperatures and humidities are high, these field cured specimen may give 28-day strengths much higher than the laboratory cured specimen. We should not be misled by these results.

25.5 LOAD TEST OF STRUCTURES

25.5.1 BS and IS Recommendations

As already pointed out in Section 11.2, when cube tests lead to non-compliance it may be necessary to conduct field load tests to prove the worthiness of members like beams, slabs and other members whose deflection can be measured. Loading of a structure to its design ultimate load is neither desirable nor necessary. Clause 9.5.1 of BS 8110 (1985) Part 2 recommends the structure to be loaded to a level appropriate to the serviceability limits and to take measurement of deformations to predict the ultimate strength. The test load usually recommended is the greater of the following characteristic loads.

1. $(DL + 1.25LL)$ [IS]
2. $1.125(DL + LL)$ or $DL + 1.25LL$ [BS]

25.5.2 Load Tests on Structures—IS 456

IS 456 (2000) Clause 17.6 recommends the first of the above loadings. The load should be applied and removed incrementally. It should be applied at least twice with minimum of one hour in between the tests and then applied again for a third time and left for 24 hours.

Deflection measurements should be made 5 minutes after application of each load increment. Signs of cracking should also be noted. A careful study of the deflection readings is made by comparing the measured performance with that expected on the basis of the design calculations. The required criteria for good performance are the following:

1. The initial deflection and cracking if any should be in accordance with design calculations.
2. Where significant deflection has taken place during the loading for 24 hours, the percentage recovery for concrete members should be atleast 75% for reinforced concrete and 85% for prestressed concrete (classes 1 and 2) members. If within 24 hours the reinforced concrete member does not recover atleast 75% of the deflection under superimposed load, the test should be repeated after a lapse of 72 hours. If the recovery is less than 80% in the second loading, the structure is considered unacceptable. The allowable deflections according to IS 456 Clause 17.6.3.1 is $40L^2/D$ under full test load of $(DL + 1.25LL)$, where L is span in metres and D the overall depth of section in mm. In this case of allowable deflection the recovery clause will not apply.
3. The structure should also be examined for unexpected defects, which should be taken into consideration in the evaluation procedure.

25.5.3 Load Tests on Precast Concrete Products

These tests are usually prescribed to control the quality of the precast elements. Usually a rate of sampling is also specified. For tests for assessing 'serviceability and strength' overload tests may not be required. However, it may be necessary to test a small number of specimens also for ultimate strength. It is generally specified that the test results of the ultimate strength of the specimen should exceed the design ultimate load by at least 5% and the deflection of the specimen up to ultimate load should not exceed (span/40) (BS 8110, Part 2 Clause 9.6). Similar tests are specified for concrete sleepers by the Indian Railways.

25.5.4 ACI Recommendations

It should be clearly understood that load tests are not made to settle disputes or litigation over control of construction quality, but they are conducted to test the safety of the structure as a whole and if necessary to permit it to be used for a lower load rating. Usually for factory manufactured components like railway sleepers and pipes a few specimens are also tested to destruction to determine its ultimate strength. In addition, certain performance criteria under non-destructive tests are also made for routine control of quality of manufacture.

It is difficult to test compression members as deflection and strains are difficult to measure and cracking of compression members occurs only just before failure. Accordingly, most load tests in structures are specified for flexural members only.

The following are the conditions usually prescribed for flexure tests. These are give in ACI 318 (1989) Clauses 20.3 and 20.4.

1. Structure should be generally at least 56 days (twice of twenty-eight days) old before testing.
2. Measurements of deflection are taken immediately before and after the loading.
3. The full total test load to be applied on the structure is 0.85 times the 'ultimate design load', including the service load, i.e. $0.85(1.4DL + 1.7LL)$ given in ACI specifications.
4. Full service load should be applied on the structure before 48 hours of load test if it is not already on the structure.
5. The balance of the test load, i.e. load in addition to the service load, that is already on the structure to make up the test load is to be applied in not less than four equal increments.
6. All loading deflection readings are taken at different intervals till it becomes steady. The final load is to be kept on the structure for 24 hours and the final deflection after 24 hours is also to be noted.
7. The test load is removed after 24 hours and the final recovery deflection after 24 hours of removal of load is also to be noted.

The following are the criteria for the acceptance of the structure:

1. There should be no evidence of failure.
2. If maximum deflection of the structure in inches does not exceed $[L^2/(2000h)]$, the requirements for recovery of deflection can be waived. (In the above formula L is the span and h is the depth in inches. This works out to about $50L^2/h$ mm, where L is in metres and h in mm as in IS 456.)
3. When the maximum deflection is greater than the above value, the percentage recovery should not be less than 75% after 24 hours. If found necessary and safe, a lower recovery may be specified for acceptance depending on the importance of the structure.

25.5.5 Core Tests

IS 456 Clause 17.4 also specifies core tests to determine the soundness of the concrete in an already constructed member. The size of the specimen for core tests should not be less than three times the aggregate size in the concrete.

QUALITY CONTROL OF CONCRETE IN CONSTRUCTION 471

EXAMPLE 25.1 (Field control of concrete mixes)
Results of cube tests on a project with *continuous production of concrete* is as follows. Assumed $f_{ck} = 30$ and expected standard deviation = 5 N/mm². Examine compliance criteria according to BS 5328 requirements. Calculate also the mean strength and standard deviation of the results. Indicate requirements for compliance with IS 456 requirements.

S.No.	Test value (N/mm²)	Calculate mean of four	Pass or fail Individual test	Pass or fail Mean of four
1	40.0		✓	
2	41.0		✓	
3	35.0		✓	
4	31.0	36.8	✓	✓
5	28.5	33.9	✓	✓
6	32.0	31.6	✓	F
7	35.0	31.6	✓	F
8	40.1	33.9	✓	✓
9	28.0	33.8	✓	✓
10	35.0	34.5	✓	✓
11	40.0	35.8	✓	✓
12	29.1	33.0	✓	✓

Part 1: Condition to be satisfied for BS compliance.

1. Individual tests $f_{ck} - 3 = 30 - 3 = 27$ N/mm²
2. Mean of consecutive four $f_{ck} + 3 = 30 + 3 = 33$ N/mm²

Pass or fail is shown in the table.

Part 2:
   (a) Calculated value of $f_m = 34.6$ N/mm²
   (b) $s = 4.9$ N/mm² (on 12 samples only). (These are obtained directly by use of a calculator.) At least 30–40 sample results should be used to find an acceptable standard deviation.

Part 3: IS compliance (for 60 m³ concrete)

Condition: (i) Individual, $f_{ck} - 4 = 30 - 4 = 26$ N/mm²
   (ii) Mean of consecutive four, $f_{ck} + 4 = 30 + 4 = 34$ N/mm² or

   (iii) $f_{ck} + \dfrac{1}{2}$ (current margin) $= 30 + \dfrac{1}{2}(1.65 \times 5) = 34$ (Greater of two) Mean of four should be equal or greater than 34 N/mm².

REFERENCES

1. SP 23 (1982), Handbook on Concrete Mixes (based on Indian Standards), Bureau of Indian Standards, New Delhi.

2. Neville, A.M., *Basic Statistical Methods for Engineers and Scientists,* International Textbook Company, London.

3. BS 5328, Methods of Specifying Concrete Including Ready Mixed Concrete, Bureau of Standards, London, 1976.

CHAPTER 26

Design of Structures for Storage of Liquids

26.1 INTRODUCTION

IS 3370 (1967) Code of Practice for Concrete Structures for Storage of Liquids [1] was published before IS 456 (1978) Limit State Design of Reinforced Concrete. Hence, IS 3370 does not include limit state design of liquid retaining structures. The British Code BS 5337 (1976) [2] called the Structural Use of Concrete for Water Retaining Structures allows the use of the limit state method as well as the method recommended in IS 3370 (called as the alternate method in BS) for design of these structures. The design aspects given in BS and IS codes are briefly explained in this chapter.

26.2 DESIGN OF WATER TANKS

In the design of normal structures like buildings, investigations of ultimate strength is enough to ensure their proper performance. However in water tanks it is equally important to limit the crackwidth to specified values which depend on the exposure conditions. In general we can say that the limit state method gives a more economical structure. It gives a thinner section and uses less reinforcements. Structures designed by the alternate method (IS 3370) will be thicker and will have larger ultimate strength and more than the required cracking strength. It will also have a high degree of safety against thermal and shrinkage cracking if these conditions are also specially considered in the design. However structures designed by both methods have been found to give very satisfactory performance in the field. It should be noted that in BS the minimum grade of concrete exposed to water should be M25. The accepted methods of design are indicated in Table 26.1.

26.3 DETAILS OF IS 3370 (1967)

The Indian code IS 3370 is published in four parts:

 Part 1. General requirements
 Part 2. Reinforced concrete liquid storage structures
 Part 3. Prestressed concrete structures
 Part 4. Design tables (the tables reproduced from the publication by P.C.A., USA)

474 ADVANCED REINFORCED CONCRETE DESIGN

TABLE 26.1 METHODS OF DESIGN OF WATER TANKS

Design consideration	Limit state design BS 5337 (1976)	Alternative method in BS 5337 and IS 3370 methods
1. Cracking	One of the following methods: 1. Direct calculation of crack width of an assumed section. 2. Limiting stress in steel and thus indirectly controlling crack-width, by using tables and charts available for this purpose [3]. 3. By 'deemed to satisfy clause' by limiting the tension in steel to specified values. *Note*: Cracking of immature concrete due to constrained shrinkage and change in temperature should also be considered separately.	By limiting the tensile stress in concrete of an uncracked section. (*Note*: According to IS 3370 Clause 5.3 on liquid retaining faces the direct horizontal tension and bending tension in concrete should satisfy the interaction formula $$\frac{t'}{t} + \frac{\sigma'_{ct}}{\sigma_{ct}} \leq 1$$ where t' and σ'_{ct} are the calculated direct and the bending tension, and t and σ_{ct} are allowable direct and bending tension.
2. Strength	One of the following methods 1. Ultimate limit strength with partial safety factors 2. Deemed to satisfy clause by limiting tension in steel by elastic analysis.	Strength of cracked section by limiting the stresses. The stresses in steel and concrete due to the bending moment, shear force in the section due to the loads should not exceed specified values.
3. Deflection	By calculation given in Chapter 1.	Deemed to satisfy by L/d ratios.

Part 4 is very useful for the structural analysis of water tanks and parts 1 and 2 give information about proper design and construction of reinforced concrete water tanks.

26.4 DURABILITY REQUIREMENTS

The exposure classification, allowable crackwidth as well as the minimum grade of concrete, cement content and cover to be adopted for the various exposure conditions of the different members of the tank (like walls, roofs, floor) are given in Table 26.2. The exposure conditions are classified as class A, B and C. Of these, Class C structures can be designed as ordinary concrete structures but exposure conditions A and B require special considerations.

As crackwidth depends to a great extent on the level to which the steel is stressed, we usually limit the allowable stress in steel and there is no advantage in using steel of grades higher than Fe 415 in water tank members subjected to classes A and B exposure conditions.

DESIGN OF STRUCTURES FOR STORAGE OF LIQUIDS 475

TABLE 26.2 EXPOSURE CLASSIFICATION AND THEIR DESIGN REQUIREMENTS
[BS 5337 Clauses 4.9, 4.10 and Table 8]

Exposure		Minimum grade with 20 mm aggregate	Minimum cover (in mm) to all steels	Allowable crackwidth (mm)	Minimum cement content (kg/m³)
Class	Condition				
Class A	Exposed to moist or corrosive atmosphere or subject to alternate wetting and drying	M30	40	0.1	360 for 20 mm aggregate (for 40 mm reduce by 40 kg/m³)
Class B	Almost in continuous contact with liquid	M25	40	0.2	290 for 20 mm aggregate (for 40 mm aggregate reduces by 30 kg/m³)
Class C	Not exposed to liquid or moist or corrosive condition (normal conditions as in ordinary structures	M20	40	0.3 as in IS 456	As in ordinary structures

Notes: 1. When a wall or face is not thicker than 225 mm and one of its face is exposed to class A or class B then both faces are to be classified as exposed to the worst class. For example, roof of tanks thinner than 225 is to be classified as exposed to class A on both faces.
2. Allowable crackwidth of 0.3 mm is based on appearance rather than on durability.

26.5 DETAILS OF DESIGN METHODS

According to BS 5337, water tank can be designed by any of the following methods:

1. Limit state design
2. Alternate method

The second method is recommended in IS 3370. Both of these design procedures in the above two methods are shown in Table 26.1.

26.6 LIMIT STATE DESIGN PROCEDURE

The design should satisfy separately the following requirements under serviceability and ultimate strength conditions.

1. Under serviceability limit state of loading (with load factors given in IS 456 Table 12)

   (a) The width of cracks should be within the limits specified in Table 26.2.
   (b) The deflection calculated by the standard procedure should be limited to 1/250th of the span.

2. Under ultimate limit state conditions (with 1.6 times the service loads according to BS 5337) the bending capacity, shear strength, bond, anchorage, etc., should be safe when calculated by the usual limit state theory of concrete structures.

26.7 CRACKWIDTH ANALYSIS IN LIMIT STATE METHOD

The crackwidths to be considered are those due to flexural tension, direct tension and due to restrained shrinkage and heat of hydration during the initial stages (see Chapter 2). The crackwidth in direct tension may be 'deemed to be satisfactory' if the steel stresses do not exceed those in Table 26.3. The crackwidth in flexural tension can be controlled by adopting one of the two following procedures:

Procedure 1. The above requirements of crackwidth may be 'deemed to be satisfactory' if the steel under service condition does not exceed those in Table 26.3.

Procedure 2. The crackwidth may be assessed by the use of the formula in BS 5337. When the member is under combined flexure and direct stresses, the calculated flexural strain should be modified when calculating the crackwidth. These procedures are explained below.

TABLE 26.3 ALLOWABLE DIRECT OR FLEXURAL TENSION IN STEEL (SERVICEABILITY LIMIT STATE) (BS 5337 Table 1)

Class of exposure	Allowable stresses (in N/mm²)	
	Plain bars	HYD bars
A	85	100
B	115	130

26.7.1 Procedure 1—Limiting Crackwidth by Limiting Steel Stresses

In this method, we analyse a cracked section (in which concrete takes no tension) and ensure that the tension in steel is limited to values given in Table 26.3. The following design formula can be derived for these conditions for a singly reinforced concrete section in bending.

(a) the neutral axis depth (x) is got from the conventional elastic analysis

From modular ratio (elastic) method of design of reinforced concrete, we know that

$$\frac{x}{d} = -m\rho + \sqrt{m\rho(2 + m\rho)} \qquad (26.1)$$

where $\rho = A_s/bd$ (steel ratio). The resisting moment can be expressed as

$$M = A_s f_s \left(d - \frac{x}{3}\right)$$

$$\frac{M}{bd^2} = \frac{A_s}{bd} f_s \left(1 - \frac{x}{3d}\right) \qquad (26.2)$$

$$\frac{100M}{bd^2} = \frac{100A_s}{bd} f_s \left(1 - \frac{x}{3d}\right) = pf_s \left(1 - \frac{x}{3d}\right) \qquad (26.3)$$

where p is the percentage of steel. Equation 26.3 indicates that $M/(bd^2)$ is fairly proportional to p for a given value of f_s. For routine design we can tabulate the following two tables.

(1) Table 26.4. For values of $M/(bd^2)$ the required $100A_s/(bd)$ for values of f_s i.e. the moment resistance factor and the reinforcement percentage for given value of f_s.

(2) Table 26.5. For values of $100A_s/(bd)$ the values of x/d and $100M/(f_s bd^2)$, i.e. for the reinforcement percentage the neutral axis depth and steel stress factors.

TABLE 26.4 DESIGN BY LIMITING f_s VALUES
($f_{ck} = 25$, $m = 15$) (SP 16 Table 70)

$\dfrac{M}{bd^2}$	\multicolumn{3}{c}{$100A_s/(bd)$ for f_s (N/mm^2)}	$\dfrac{M}{bd^2}$	\multicolumn{3}{c}{$100A_s/(bd)$ for f_s (N/mm^2)}
	100	115	130		100	115	130
0.10	0.11	0.09	0.08	0.94	1.10	0.95	0.83
0.14	0.15	0.13	0.12	0.98	1.15	0.99	0.87
0.18	0.19	0.17	0.15	1.02	1.20	1.03	0.91
0.22	0.24	0.21	0.18	1.06	1.25	1.08	0.95
0.26	0.28	0.25	0.22	1.10	1.30	1.12	0.98
0.30	0.33	0.29	0.25	1.14	1.35	1.16	1.02
0.34	0.38	0.33	0.29	1.18	1.40	1.21	1.06
0.38	0.42	0.37	0.32	1.22	1.45	1.25	1.10
0.42	0.47	0.41	0.36	1.26	1.50	1.29	1.14
0.46	0.52	0.45	0.39	1.30		1.34	1.17
0.50	0.56	0.49	0.43	1.34		1.38	1.21
0.54	0.61	0.53	0.46	1.38		1.43	1.25
0.58	0.66	0.57	0.50	1.42		1.47	1.29
0.62	0.71	0.61	0.54	1.46		1.49	1.33
0.66	0.76	0.65	0.57	1.50			1.37
0.70	0.80	0.69	0.61	1.54			1.41
0.74	0.85	0.74	0.65	1.56			1.43
0.78	0.90	0.78	0.68	1.58			1.45
0.82	0.95	0.82	0.72	1.60			1.47
0.86	1.00	0.86	0.76	1.62			1.49
0.90	1.05	0.91	0.79	1.64			1.50

Note: Maximum percentage of steel used = 1.5%

TABLE 26.5 VALUES OF x/d AND $100M/(f_s bd^2)$ FOR STEEL PERCENTAGES
(Refer SP 16 Tables 91–94) (Eqs. 26.1 and 26.3 Cracked Section) ($m = 15$)

$\dfrac{100A_s}{bd}$	$\dfrac{x}{d}$	$\dfrac{100M}{f_s bd^2}$	$\dfrac{100A_s}{bd}$	$\dfrac{x}{d}$	$\dfrac{100M}{f_s bd^2}$	$\dfrac{100A_s}{bd}$	$\dfrac{x}{d}$	$\dfrac{100M}{f_s bd^2}$
0.10	0.159	0.095	0.54	0.330	0.481	1.04	0.424	0.893
0.14	0.185	0.131	0.60	0.344	0.531	1.10	0.433	0.941
0.20	0.217	0.186	0.64	0.353	0.565	1.14	0.438	0.973
0.24	0.235	0.221	0.70	0.365	0.615	1.20	0.446	1.021
0.30	0.258	0.274	0.74	0.373	0.648	1.24	0.449	1.037
0.34	0.272	0.309	0.80	0.384	0.697	1.30	0.459	1.101
0.40	0.292	0.361	0.84	0.392	0.730	1.34	0.464	1.149
0.44	0.303	0.396	0.90	0.402	0.779	1.40	0.471	1.180
0.48	0.314	0.430	0.94	0.408	0.812	1.44	0.476	1.212
0.50	0.319	0.447	1.00	0.418	0.861	1.50	0.483	1.259

A value of $m = 15$ is recommended by BS for all grades of concrete. (The IS 456 value of $m = 280/(3\sigma_c)$ which for grade 20 concrete work out to only 13.3.) The maximum stress in concrete can be obtained as follows:

$$A_s f_s = 0.5 f_c bx$$

$$f_c = 2\rho \frac{d}{x} f_s \tag{26.4}$$

26.7.2 Procedure 2—Method Based on Calculated Crackwidth for Section in Bending

In this procedure we can either check the preliminary design for crackwidth by directly using crackwidth formula or indirectly control the crackwidth by limiting the stress which is derived from the crackwidth formula. Design charts can be made as shown below for the second procedure.

26.7.3 Derivation of Allowable Steel Stresses for Specified Crackwidth

The formula is BS 5337 for crackwidth in water tanks is as follows:

$$w = \frac{4.5 a_{cr} \varepsilon_m}{1 + 2.5 (a_{cr} - c_{min})/(h - x)} \tag{26.5}$$

where

$$\varepsilon_m = \varepsilon_1 - \frac{0.7 b_t h (a' - x)}{A_s f_s (h - x)} 10^{-3}$$

w = Crackwidth at the point considered for crackwidth

a_{cr} = Distance of the nearest bar from the point

c_{min} = Minimum specified cover

ε_m = Average strain with stiffening effect

h = Overall depth of member

x = Depth of neutral axis calculated by assuming $m = E_s/0.5E_c$

ε_1 = Strain at level considered

b_t = Width of section of centroid of tension steel

a' = Distance of compression face from the point considered for crackwidth

A_s = Area of steel

f_s = Service stress of steel

Here the value of E_c is taken as one-half the instantaneous value for calculation of m.

Equation (26.5) is similar to Eq. (2.5) (in Chapter 2 for crackwidth in beams and slabs except that the probability of the calculated crackwidth being exceeded is less than 5% (instead of 20% in beams) and the stiffening effect is related to the service stres f_s.

This expression is true only for singly reinforced section. With compression steel, there will be a reduction in the lever arm and a reduction in tensile stress, leading to a reduction in crackwidth. This change is small and the same formula can also be used for doubly reinforced beams. We have the option either to use this formula directly or find out the value of steel stress f_s to be used so that the width of crack will be below the specified value.

We shall now investigate how the second option of limiting steel stress to indirectly control the crackwidth can be achieved. Let us consider a reinforced concrete member in which

f_s = Stress in steel

ϕ = Diameter of reinforcement bars

s = Spacing of bars

c = Cover to reinforcement

h = Overall depth of member

$$\varepsilon_1 = \text{The strain at the tension face} = \frac{f_s}{E_s}\frac{h-x}{d-x} \qquad (26.5a)$$

If we consider the crackwidth at the surface of the tension zone, then $a' = h$ and $(a' = h) = (h - x)$. With $E_s = 200 \times 10^3$ N/mm², from Eq. 26.5 we get

$$\varepsilon_m = \left(\frac{h-x}{d-x}\frac{f_s}{200} - \frac{0.7\, bd}{f_s\, A_s}\frac{h}{d}\right)10^{-3} \qquad (26.6)$$

Putting $A_s/bd = \rho$ = The steel ratio, we get the crackwidth formula from 26.5 as follows:

$$w = \frac{4.5 a_{cr}}{1 + 2.5(a_{cr} - c)/(h-x)}\left(\frac{h-x}{d-x}\frac{f_s}{200} - \frac{0.7}{f_s \rho}\frac{h}{d}\right)10^{-3} \qquad (26.7)$$

Rearranging, we get

$$\frac{1000w}{4.5a_{cr}}\left(1+2.5\frac{a_{cr}-c}{h-x}\right) = \left(\frac{h-x}{d-x}\frac{f_s}{200} - \frac{0.7}{f_s\rho}\frac{h}{d}\right)$$

Simplifying and reducing the expression to a quadratic in $f_s/100$, we get

$$0.5\frac{h-x}{d-x}\left(\frac{f_s}{100}\right)^2 - \frac{1000w}{4.5a_{cr}}\left(1+2.5\frac{a_{cr}-c}{h-x}\right)\frac{f_s}{100} = \frac{0.007}{\rho}\frac{h}{d} \qquad (26.8)$$

We also know that crackwidth occurs at the tension face between bars, where

$$a_{cr} = \sqrt{\left(\frac{s}{2}\right)^2 + \left(c+\frac{\phi}{2}\right)^2} - \frac{\phi}{2} \qquad (26.9)$$

From Eq. (26.6), we get the maximum value of f_s for a specified crackwidth w, and a_{cr} and x which depends on ρ. However if we adopt particular values of $w, c, s,$ and ϕ, the equation can be used to give a relationship between f_s and d where

$$h = d + c + \frac{\phi}{2}$$

From specified values of s and ϕ we get

$$\rho = \frac{\pi\phi^2}{4sd}$$

with $m = 15$. We can get the neutral axis depth from Eq. (26.1)

$$\frac{x}{d} = -m\rho + \sqrt{m\rho(2+m\rho)}$$

The value of f_s obtained from Eq. (26.8) is the stress in reinforcement in a given section for a specified crackwidth. It can be used to determine the corresponding $M/(bd^2)$ by Eq. (26.3) as

$$\frac{M}{bd^2} = \frac{p}{100}\left(1-\frac{x}{3d}\right)f_s$$

where p is the percentage of steel. We can thus get a set of curves for values of $M/(bd^2)$ and $100A_s/(bd)$ for particular values of w, c, ϕ and s. This curve can be used as a convenient aid for design of the concrete section for specified crackwidth [3]. Such charts for $w = 0.2$ mm, $c = 40$ mm and 60 mm with $\phi = 20$ mm for various steel spacings are shown by Chart 26.1.

[*Note*: The line $f_s = 130$ in Fig. 26.1 corresponds approximately to the relation between the allowable stress 130 N/mm^2 and $M/(bd^2)$ given in Table 26.4. Similar charts for values of $w = 0.1$ and 0.2, $c = 40$, 50 and 60 with $\phi = 12$, 20 and 32 are available in Ref. 3.]

26.8 DEFLECTION ANALYSIS

The empirical method using the modification factors as specified in IS 456 clause 22.2 or theoretical calculation of deflection as indicated in Chapter 1 can be used for this purpose. (The IS and BS methods are very similar for the calculation of deflection.)

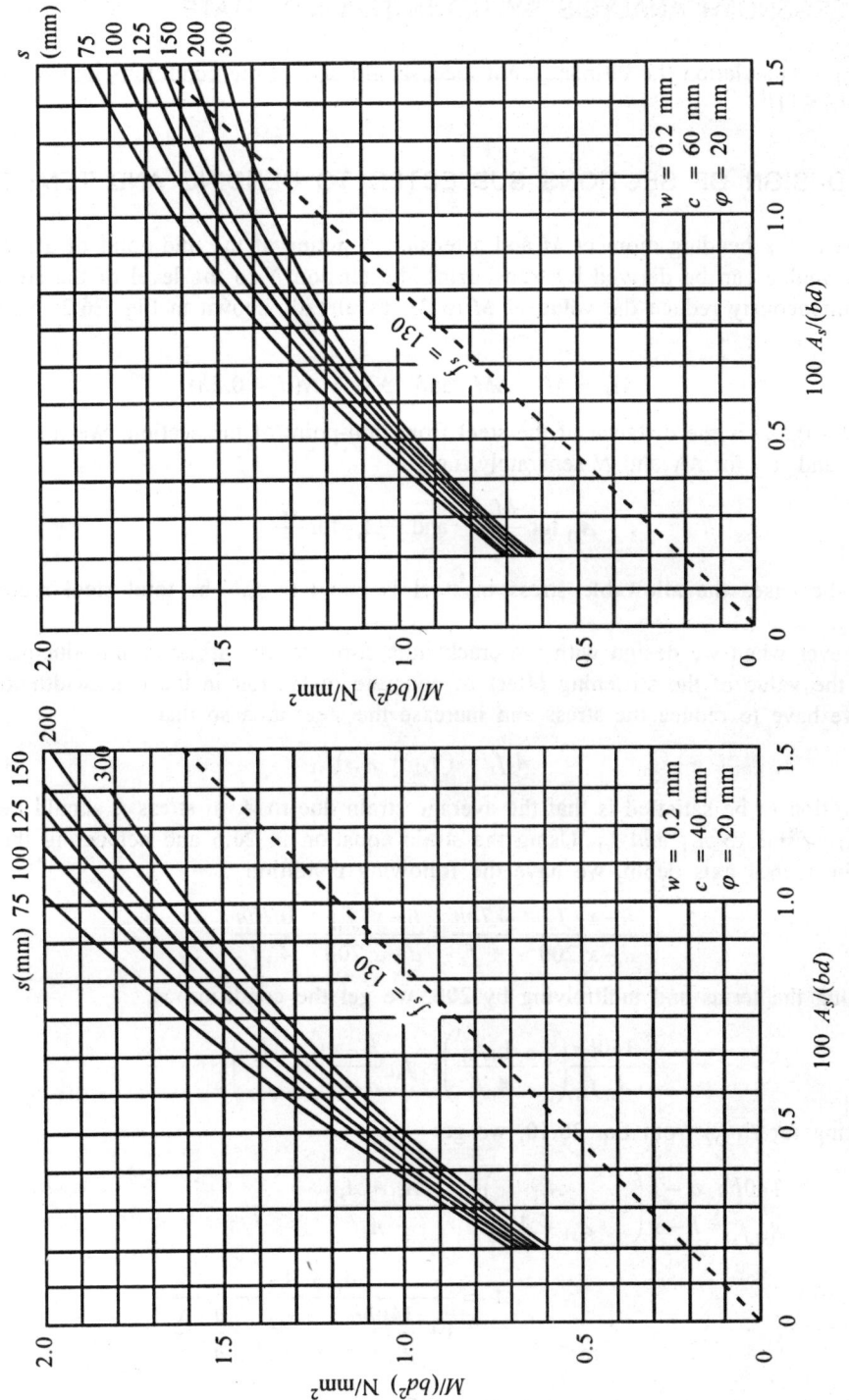

DESIGN OF STRUCTURES FOR STORAGE OF LIQUIDS 481

Fig. 26.1 Percentage of steel required for values of $M/(bd^2)$ with values of crackwidth (w) having cover (c) and diameter of reinforcement (ϕ), spacing of bars (s).

26.9 STRENGTH ANALYSIS BY ULTIMATE LIMIT STATE

For strength calculation the ultimate limit load should satisfy the conditions laid down in IS 456 or BS 8110.

26.10 DESIGN OF SECTIONS SUBJECTED TO BENDING AND TENSION

When there is a bending moment M and a tension N acting at the mid-point of a section, a design formulae can be derived by transferring the tension N to the level of the steel. This will simultaneously reduce the value of M to M_1 as already shown in Fig. 26.2. Now from Eq. 26.8

$$M_1 = M - \Delta M \text{ and } \Delta M = N(d - 0.5h)$$

where $(d - 0.5h)$ is the distance of the steel from mid-point of the section. We then provide steel A_{s1} and A_{s2} for M_1 and N separately, i.e.

$$A_{s1} \text{ for } \frac{M_1}{bd^2} \text{ and } A_{s2} \text{ for } \frac{N}{f_s}$$

In both the cases the allowable stress in steel is equal to f_s. The total steel required is $A_{s1} + A_{s2}$.

However when we design with the crackwidth formula, the arbitrary introducing of A_{s2} violates the value of the stiffening effect of concrete in tension in the crackwidth formula. Hence we have to reduce the stress and increase the steel area so that

$$A_s f_s = (A_{s1} + A_{s2}) f_{s1} \tag{26.10}$$

The condition to be satisfied is that the average strain due to A_s at stress f_s should be equal to the strain due to A_{s1} and f_{s1}. Using the strain equation in 26.6 and neglecting the small change in neutral axis depth, we have the following condition

$$\frac{h-x}{d-x} \frac{f_s}{200} - \frac{0.7bh}{A_s f_s} = \frac{h-x}{d-x} \frac{f_{s1}}{200} - \frac{0.7bh}{A_{s1} f_{s1}}$$

Regrouping the terms and multiplying by 200, we get the condition as

$$\frac{140bh}{A_{s1} f_{s1}} \left(1 - \frac{A_{s1} f_{s1}}{A_s f_s}\right) = f_{s1} \frac{h-x}{d-x} \left(1 - \frac{f_s}{f_{s1}}\right)$$

Substituting for f_{s1}/f_{s2} from Eq. 26.10, we get

$$\frac{140bh}{A_{s1} f_{s1}^2} \frac{d-x}{h-x} \left(1 - \frac{A_{s1}}{A_{s1} + A_{s2}}\right) = 1 - \frac{A_{s1} + A_{s2}}{A_s}$$

$$A_s = \frac{A_{s1} + A_{s2}}{1 + \dfrac{140bh}{A_{s1} f_{s1}^2} \dfrac{A_{s2}}{A_{s1} + A_{s2}} \dfrac{d-x}{h-x}} \tag{26.11}$$

$(A_{s1} + A_{s2})$ is increased to A_s to satisfy the stiffening effect.

26.10.1 Interpretation of the Effect of e = M/N

We have the value of M_1 from Section 26.8 as

$$M_1 = M - \Delta M = M - N(d - 0.5h)$$

Putting $M/N = e$, we get

$$\frac{M_1}{N} = \frac{M}{N} - (d - 0.5h) = e - (d - 0.5h)$$

The following interpretations can be made as already indicated in Section 26.11:

(i) When $e = (d - 0.5h)$, the value of $M_1 = 0$ and only N need to be considered at the level of the steel for design

(ii) When $e > (d - 0.5h)$, the value of M_1 is positive and the section is in bending so that we can apply the BS-crackwidth formula.

(iii) When $e < (d - 0.5h)$ the value of M_1 is negative so that there is tension in the region which was in compression with M only. In this case we should have reinforcements designed on the basis of allowable stresses given in Table 26.3 on both faces. The respective areas will be as shown in Fig. 26.2.

$$A'_{s1} = \frac{0.5N}{f_s}\left(1 + \frac{e}{d - 0.5h}\right) \quad \text{and} \quad A_{s2} = \frac{0.5N}{f_s}\left(1 - \frac{e}{d - 0.5h}\right)$$

Fig. 26.2 Distribution of steel in section subjected to bending and tension.

26.11 MODIFICATION OF CRACKWIDTH FORMULA IN SECTIONS WITH BENDING AND TENSION

The depth of neutral axis x as calculated for bending only will change when the bending is combined with tension. The procedure to estimate the modified value for calculation of crackwidth is as shown in Fig. 26.3.

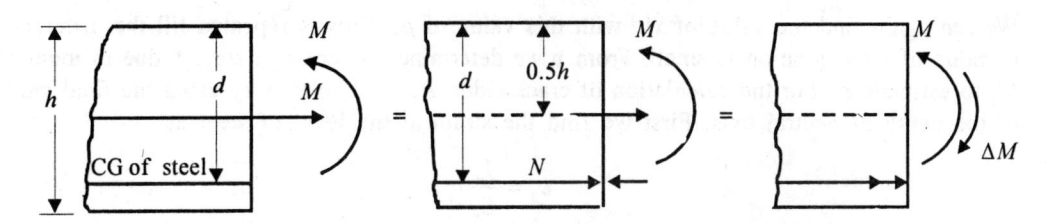

Fig. 26.3 Concrete section subjected to bending and axial tension.

484   ADVANCED REINFORCED CONCRETE DESIGN

Let the bending moment is M and tension at the centre of the section is N. As shown in Fig. 26.2 the combined effect of M and N will be equivalent to another bending moment and a tension N at the level of the steel. From Fig. 26.2, we have

$$M_1 = M - N\left(d - \frac{h}{2}\right) = N\left[\frac{M}{N} - \left(d - \frac{h}{2}\right)\right]$$

If we put $e = M/N$ we will get two cases as already indicated

Case 1. When e is greater than $[d - (h/2)]$ the value of M_1 is positive and the crackwidth calculation by the formula for crackwidth in bending is valid.

Case 2. When e is less than $[d - (h/2)]$, the value of M_1 is negative and the section is under tension. The crackwidth calculations by the formula for cracks in bending is not applicable. The value of ε_m in Eq. (26.5) has to be modified.

The following approximate method can be used to find the depth of the neutral axis and hence ε_m for case 1. The effect of the tension (N) is to reduce the value of the depth of the neutral axis (x). Its effect can be approximated to having lesser steel p_1 instead of p. Then in the expression

$$M_1 = M - N\left(d - \frac{h}{2}\right)$$

Therefore,

$$\frac{M_1}{N} = e - \left(d - \frac{h}{2}\right) \tag{26.12}$$

when $e - (d - h/2) = 0$, $M_1 = 0$, no steel is required. Hence as a first approximation, we assume the reduced value p as p_1, where

$$p_1 = p\frac{(e + h/2) - d}{e + (h/2)} = p\left[1 - \frac{d}{e + (h/2)}\right] \tag{26.13}$$

We now find the value of the depth of neutral axis x with this modified steel area p_1. As moments are calculated about the centre of compression, the exact value of (x/d) can be found by iteration. The second approximation of the area of steel will be p_2 and

$$p_2 = p_1\left(1 - \frac{d - (x/3)}{e + (h/2) - (x/3)}\right) \tag{26.14}$$

We can again find the value of x/d with this value of p_2. This is repeated till the difference in value of x by iteration is small. From x we determine the stress in steel f_s due to moment M_1 to estimate ε_s. For the calculation of crackwidth, ε_m is calculated by using the final value of the depth of neutral axis. First we find the strain at the level of steel as

$$\varepsilon_s = \frac{f_s}{E_s}$$

The maximum strain at the surface of the tension area is ε_1

$$\varepsilon_1 = \frac{f_s}{E_s} \frac{h-x}{d-x}$$

ε_m is determined from Eqs. (26.5) and (26.6) for estimation of crackwidth.

26.12 RECOMMENDED PROCEDURE OF DESIGN

The following recommendations are useful for the design [3]:

1. If the serviceability design is based on allowable steel stresses, we use the alternate method of design which is more or less the same as in IS 3370 (see Section 26.13).

2. If the serviceability design is based on crackwidth, we use the following criteria for preliminary design:

(a) If Fe 250 steel is used, we need design only for ultimate strength unless $M/(bd^2)$ exceeds 1.0 for class A and 1.5 for class B exposures.

(b) If Fe 415 steel is used, for class A exposure, design for ultimate limit state if $M/(bd^2)$, is less than 0.6 and for service load, if $M/(bd^2)$ is greater than 0.8.

(c) If Fe 415 steel is used for class B exposure, design for ultimate strength if $M/(bd^2)$ is less than 0.8 and for service load if $M/(bd^2)$ is greater than 0.8.

26.13 ALTERNATE METHOD OF DESIGN

Designing a Section for limiting crackwidth as well as providing the necessary strength can also be carried by the alternate method given in the following sections.

26.13.1 Designing for Crackwidth by Alternate Method (IS Method)

In this method, we analyse or check the already assumed section as an uncracked section by elastic analysis and restrict the tensile stress in concrete as shown in Table 26.6. We assume that the crackwidth will be limited if the specified tensile stresses are not exceeded. IS 3370 requires the interaction formula to be satisfied as in Table 26.1. The following are the recommended values of the allowable direct and bending tension to be used in the interaction formula.

TABLE 26.6 ALLOWABLE STRESSES IN CONCRETE IN UNCRACKED SECTION
(N/mm²)(IS 3370 Method and Alternate Method in BS 5337)
(IS 3370 Table 1)

Grade of concrete	Direct tension		Bending tension		Shear = $Q/(bjd)$	
	IS	BS	IS	BS	IS	BS
20	1.2	–	1.7	–	1.7	–
25	1.3	1.31	1.8	1.84	1.9	1.94
30	1.5	1.44	2.0	2.02	2.2	2.19

Note: 1. Steel stress = m × Concrete stress
2. m = 15 in BS and $280/f_{ck}$ in IS.

(a) Direct tension t = 1.1 to 1.7 N/mm² for M15 to M40 concrete depending on the concrete strength.

(b) Bending tension σ_{ct} = 1.5 to 2.4 N/mm² for M15 to M40 concrete depending on the concrete strength.

26.13.2 Designing for Strength by Alternate Method of Design

Strength design means safety against overload and other contingencies. For this purpose we use the conventional cracked section analysis and restrict the steel and concrete stresses to specified values given in Tables 26.7 and 26.8. As can be seen from Table 26.3 in BS 5537 the steel stresses are the same as prescribed in the deemed to satisfy method.

TABLE 26.7 ALLOWABLE STRESSES IN CONCRETE IN CRACKED SECTION
(N/mm²)(IS 3370 Method and Alternate Method in BS 5337)
(Ref. Table 15. IS 456)

Grade of concrete	Direct compression IS	Direct compression BS	Bending compression IS	Bending compression BS	Shear IS	Shear BS	Average band (plain bars) IS	Average band (plain bars) BS
20	5.0	–	7.0	–	456	–	0.8	–
25	6.0	6.95	8.5	9.15	Table	0.77	0.9	0.9
30	8.0	8.37	10.0	11.0	(17)	0.87	1.0	1.0

TABLE 26.8 ALLOWABLE STRESSES IN STEEL IN CRACKED SECTION
(N/mm²)[IS 3370 Method and Alternate in BS 5337][Refer IS 3370 Table 2]

Condition	Exposure	Plain bars IS	Plain bars BS	Deformed bars IS	Deformed bars BS
Tension					
Direct	A	115	85	150	100
Bending	B	125	115	190	130
Compression in columns	A and B	125	125	175	140
Shear					
Direct	A	115	85	150	100
Bending	B	125	115	175	130

26.14 DESIGN OF SECTION SUBJECTED TO TENSION ONLY

We have to satisfy the following two equations:

$$A_s f_s = N \qquad (26.15)$$

$$[bh + (m - 1)A_s]f_{ct} = N \qquad (26.16)$$

For balanced design we combine the two equations to get

$$h = \frac{N}{b}\left(\frac{1}{f_{ct}} - \frac{m-1}{f_s}\right) \quad (26.17)$$

A_s is calculated from Eq. (26.16) and we place the steel usually at the midline of the section.

26.15 DESIGN OF SECTION SUBJECTED TO BENDING AND TENSION

We use the same principle as explained is Section 26.10 and finally check the section for stresses.

26.16 DESIGN OF PLAIN CONCRETE TANKS

IS 3370 Part 2 Clause 3.2.3 allows plain concrete members of concrete tanks to be designed by allowing bending tension as specified in IS 456 Clause 5.2.2 with the flexural tensile strength $f_{cr} = 0.7\sqrt{f_{ck}}$ N/mm². By providing a suitable factor of safety (say 1.5) such design can take care of cracking under bending tension. However, the minimum required steel for shrinkage and temperature (see Section 26.18) should also be always provided in plain concrete members.

26.17 STRUCTURAL ANALYSIS OF TANK WALLS

Except for simple cantilever walls and circular rings free at the base, the analysis of tank walls for moments and shear is difficult. The following methods are usually used in India:

1. Reissner method
2. Carpenter's simplifications of Reissner method
3. Elastic analysis as plates with various boundary conditions [4]
4. Approximate methods
5. Tables published by the American Portland Cement Association [5,6] and adopted by IS 3370 as Part 4.

We shall deal only with the last two methods as it is always advisable to use the code for practical designs. We should remember that it is difficult to fix the base of tanks and in design procedures it is advisable to investigate other conditions also and use judgement in determining the required reinforcements.

26.17.1 Approximate Methods of Analysis

For fixed bases of tanks, we assume the cantilever action is limited to the following lengths:

1. For circular tanks let H = height, D = diameter, t = thickness.

Value of $H^2/(Dt)$ of circular tanks	Height of cantilever action (larger of the two)
Less than 6	Full height
6–12	$H/3$ or 1 m
12–30	$H/4$ or 1 m
>30	Full hoop tension

2. For rectangular tanks with length (L) and breadth (B), the following criteria are used for the analysis:

(a) For L/B less than 2, the walls are analyzed as continuous beams.

(b) For L/B more than 2, the long wall is taken to act as a cantilever fixed at the base and the short wall is assumed to bend horizontally with its support on the long wall for a portion above $H/4$ or 1 m from the bottom. Below the above depth the short wall also acts as cantilever.

(c) For tanks like swimming pools, where L and B are both large they should be designed independently as cantilevers.

26.17.2 Analysis by Tables in IS 3370 Part 4

The following tables are available in IS 3370 Part 4 for structural analysis of water tanks. The coordinates are taken as shown in Fig. 26.4 (table numbers refer to those in the code).

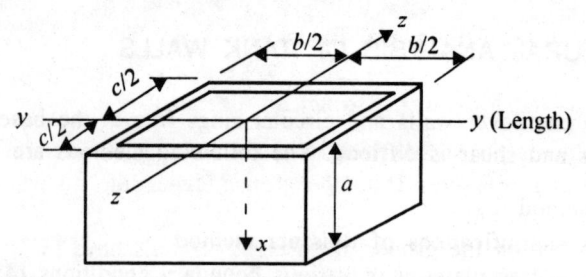

Fig 26.4 Coordinates for tables in IS 3370 part 4.

1. Tables 1 to 4: Moment coefficients for slabs with various edge conditions.
2. Tables 5: Moment coefficient for rectangular water tanks with walls free at top and hinged at bottom
3. Table 6: Moment coefficients for rectangular tanks with walls hinged at top and bottom
4. Table 7: Shear at edges of wall panel, hinged at top and bottom
5. Table 8: Shear at edges of wall panel, free at top and hinged at bottom
6. Tables 9 to 19: Forces in circular tank walls under various conditions
7. Tables 20 and 23: Forces in circular roof slabs.

Triangular loads correspond to liquid pressures, uniform load to gas pressure and trapezoidal load to liquids like petrol which may evaporate and build up a pressure above the surface unless vents are provided for the gas to escape.

26.17.3 Shear Coefficients

In rectangular water tanks the shear force along the edges of the walls is used not only to investigate the shear and bond characteristics but it is also used to determine the moments and tension in the adjacent walls. These shear force values are given in Tables 7 and 8 of IS 3370 Part 4. The shear values in the table are shear per unit length in terms of wa^2. The use of shear coefficient is shown in Example 26.1. The effect of tension in shear is shown in Fig 26.5 (IS 456 Clause 40.2.2 and ACI Eq. 11.9).

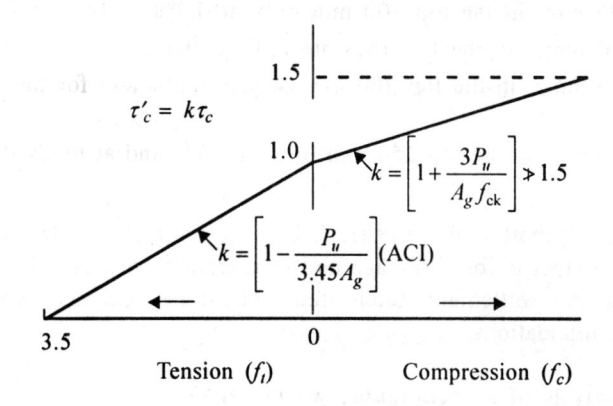

Fig. 26.5 Modification of shear strength of concrete subjected to tension or compression.

With $\tau'_c = k\tau_c$:

$$k = \left[1 + \frac{3P_u}{A_g f_{ck}}\right] \not> 1.5$$

$$k = \left[1 - \frac{P_u}{3.45 A_g}\right] (ACI)$$

26.17.4 Other Forces Acting on Tanks

Many forces such as the following also act on the tank:

1. Vertical moments thrust and weights from the roof of the tank
2. Earth and water pressures from the ground depending on whether tanks are on the ground, or below the ground
3. If the tanks rest on the ground the weight of walls have to act though the edges of the tank which is usually thickened. For moderate size tanks the walls are made continuous with the base but in large tanks like swimming pools the slab and the wall are designed as separate units. The joints are also provided with waterstops to prevent leakage from the tank.

26.18 DESIGN OF GROUND SLABS

Section 16 of Handbook BS 5337 [7] gives details about joint and reinforcement in floor slabs. IS 3370 Part 2 Clause 7 also gives constructional details. Minimum reinforcement in ground slabs is dealt with in Chapter 2.

26.19 MINIMUM STEEL PROVIDED FOR SHRINKAGE AND TEMPERATURE

IS 3370 Part 1 gives the minimum steel to be provided in walls, floors and roof in each direction. The recommendations for Fe 250 steel are as follows:

For section of thickness up to 100 mm, provide 0.3% in both directions

For section 100–450 mm we can reduce percentage linearly from 0.3 for 100 mm to 0.2% for 450 mm.

For section 450 mm and above, a minimum of 0.2% should be provided in both directions. The above steel can be reduced by 20% for Fe 415 steel.

BS Handbook [7] recommends that with Fe 415 steel and grade 30 concrete, the steel provided in ground slabs is as follows:

For slabs up to 200 mm: in the top 100 mm only at 0.3%

For slabs up to 300 mm: in the top 150 mm only at 0.3%

For slabs up to 400 mm: in the top 200 mm only at 0.3% and for the bottom 50 mm at 0.3%

For slabs over 400 mm: in the top 250 mm only at 0.3% and at the bottom 100 mm at 0.3%

In ground slabs the top part is more critical than the bottom part. However we should remember that the above steel is for shrinkage and temperature only, and do not account for the foundation bearing and settlement. Such steel should be designed separately on the principles of design of foundations.

EXAMPLE 26.1 (Analysis of a rectangular water tank)

A rectangular water tank is 10 m × 10 m in plan and in 5 m deep. It rests above the ground with its walls hinged at the base and free at the top. Determine the maximum bending moment and shear forces for design using the tables in Part IV of IS 3370.

Reference	Step	Calculations
	1	Symbols in the table
		a = height, b = length, c = breadth
		Origin is the centre of tank on the top. x-axis downwards, y-axis along the length, z-axis is in the transverse (width) direction (Fig. 26.4)
	–	From the given data find the table to be used
IS 3370		a = 5 m, b = 10 m, c = 10 m, c/a = 2
Part IV		Use tables, $\dfrac{b}{a} = 2$
Table 5		In this table, look for $\dfrac{c}{a} = 2$
		Horizontal moment = $(M_y \text{ coeff})wa^3$
		Vertical moment = $(M_x \text{ coeff})wa^3$
	2	Obtain moment coefficients from Table 5 Page 3 of the code

DESIGN OF STRUCTURES FOR STORAGE OF LIQUIDS

Reference	Step	Calculations

		TABLE 1 COEFFICIENTS FOR MOMENTS

	$\dfrac{c}{a}$	$\dfrac{x}{a}$	Centre of length $y = 0$, $z = c/2$		End of length $y = b/2$, $z = c/2$		Centre of short slab $z = 0$, $y = c/2$	
			M_x	M_y	M_x	M_y	M_x	M_y
IS 3370	2.0	0	0	0.045	0	−0.091		
Table 5		1/4	0.016	0.042	−0.019	−0.094	As in $y = 0$	
		1/2	0.033	0.036	−0.018	−0.089	($b = c$)	
		3/4	0.036	0.024	−0.013	−0.065		
	1	Bottom Hinged						

Note: Negative sign of the moment denotes tension inside the tank and positive sign denotes tension outside the tank.

3 · *Determine the shear coefficients from Table 8*

The horizontal moments are combined with tension produced by shear on the adjacent wall. As there is no thrust from the dome above, there are no forces to be combined with the vertical moment. The weight of cover (if any) acts as vertical compression on the wall. The thickness of the wall will be governed by the horizontal moment combined with shear forces at mid-point of the end of the wall.

IS 3370
Part IV
Table 8
Page 34

We need shear coefficients at $x/2 = 2$.

(1) At mid-point of fixed edge $+0.375wa^2$

The positive sign shows that the shear acts in the direction of the load.

(2) At lower third point of side edge for maximum shear, coefficient $= +0.406wa^2$

4 · *Calculation of moment and shear*

Assume $w = 9.8$ kN/m³

Step 3
Table A

At mid-height of the wall at the end of the long slab (moment is negative)

$M_y = -0.089 \times 9.8 \times 5^3 = -109.0$ kN/m

$V = 0.375 \times 9.8 \times 5^2 = 91.9$ kN/m

Maximum shear at the lower third point

$V_{max} = 0.406 \times 9.8 \times 5^2 = 99.5$ kN/m

The thickness of the wall will be governed by the following:

Reference	Step	Calculations
	5	Horizontal moment M_y = 109 kN/m Shear force N = 92 kN/m *Calculate vertical moment* M_x is maximum at x/a = 3/4 and y = 0 Maximum coefficient = 0.036 M_x = 0.036 × 9.8 × 5^3 = 44.1 kN/m This value is to be combined with the weight of the roof for design of steel in the vertical direction (see Example 26.5 for design).

Example 26.2 (Analysis by approximate method)
A rectangular water tank is 10 × 10 × 5 m deep. Assuming the tank is fixed at the base, determine the design moments by the approximate method.

Reference	Step	Calculations
	1	*Assume cantilever action at base*
Sec. 26.17.1		Height of cantilever $H/4$ or 1 m whichever is larger. $\dfrac{H}{4}$ = 2.5 m which is larger than 1 m. Adopt 2.5 m.
	2	*Determine the cantilever moment (let w = 10 kN/M³)* Average pressure on cantilever $= \dfrac{10(10+7.5)}{2} = 87.5$ kN/m Cantilever moment $= \dfrac{wl^2}{2} = \dfrac{87.5 \times (2.5)^2}{2} = 273.4$ kNm
	3	*Calculate the horizontal bending moments (The frame is a square at 7.5 m depth)* Maximum bending moment at edges $M_y = \dfrac{-wl^2}{12} = -\dfrac{(10 \times 7.5)10^2}{12} = -625$ kN/m Maximum bending at centre $M_y = \dfrac{+wl^2}{16} = \dfrac{(10 \times 7.5)10^2}{16} = 468.8$ kN/m
	4	*Determine the direct tension on the wall due to shear* Assume the wall above the cantilever produces the pressure on the two side wall. Maximum thrust $= \dfrac{10 \times 7.5 \times 10}{2} = 375$ kN/m

DESIGN OF STRUCTURES FOR STORAGE OF LIQUIDS   493

Reference	Step	Calculations
	5	*Maximum shear*
		Maximum shear = Maximum tension = 375 kN/m
		(*Note:* All these values are more than those obtained by analysing the walls as slabs, as can be done by the use of tables.)

Example 26.3 (**Design of tank wall in tension**)
Design a tank wall, which is under a tension of 300 KN (without bending). Assume $f_{ct} = 1.30$ N/mm² (allowable tension in concrete) $f_s = 130$ N/mm² (allowable tension in steel).

Reference	Step	Calculations
	1	*Data*
		$N = 300$ kN, $b = 1000$ mm, $f_{ct} = 1.3$ and $f_s = 130$ N/mm²
		As the whole section is in tension crack calculations are not valid.
		Design the section by alternate method (IS 3370 Part 2).
	2	*Thickness of walls required*
Eq. (26.17)		$h = \dfrac{N}{b}\left(\dfrac{1}{f_{ct}} - \dfrac{m-1}{f_s}\right)$
		Assume $m = 15$. Then
		$h = \dfrac{300 \times 10^3}{1000}\left(\dfrac{1}{1.3} - \dfrac{14}{130}\right) = 198$ mm. Say 200 mm.
	3	*Amount of tension steel required*
		$A_{st} = \dfrac{300 \times 10^3}{130} = 2307$ mm²
		Provide two rows with equal steel in each row of 1150 mm²
		(16 mm at 175 mm gives 1149 mm².)

EXAMPLE 26.4 (**Design of tank wall in bending and tension**)
Design a section for a water tank wall to resist a bending moment of 7.6 kN with an axial tension of 170 kN/m by the conventional method

Reference	Step	Calculations
	1	*Data*
		Under serviceability condition: $M = 7.6$ kN, $N = 170$ kN, cover = 40 mm

Reference	Step	Calculations
		Assume thickness of wall $h = 300$ m
		Diameter of steel reinforcement = 20 mm
		$d = 300 - 10 - 40 = 250$ mm
Sec. 26.11.1	2	Check $e = \dfrac{M}{N}$
		$e = \dfrac{7.6 \times 10^3}{170} = 44.7$ mm
		$d - \dfrac{h}{2} = 250 - 150 = 100$ mm
		As e is less than $(d = h/2)$, section is subjected to tension. Hence the crackwidth formula is not applicable.
	3	*Design of tension steel on both faces*
Sec. 26.10.1		$A_{s1} = \dfrac{1}{2fs}\left(N + \dfrac{M}{d - 0.5h}\right)$
		$A_{s2} = \dfrac{1}{2fs}\left(N - \dfrac{M}{d - 0.5h}\right)$
		$A_{s1} = \dfrac{1}{2 \times 130}\left(170 \times 10^3 + \dfrac{7.6 \times 10^6}{100}\right) = 946$ mm^2
Fig. 26.3		$A_{s2} = \dfrac{1}{2 \times 130}\left(170 \times 10^3 - \dfrac{7.6 \times 10^6}{100}\right) = 361$ mm^2
	4	*Check the maximum tension of uncracked section for safety against cracking*
		Bending tension $= \dfrac{M}{I} y$
		$I = \dfrac{bh^3}{12} + (m-1)A_{s1}\left(d - \dfrac{h}{2}\right)^2 + (m-1)A_{s2}\left(d^1 - \dfrac{h^1}{2}\right)^2$
		Determine the bending and axial tensions and check for the safety by the inter action formula or otherwise as in Example 26.5.

EXAMPLE 26.5 (**Design of tank wall by the limit state and alternte methods**)
Design a tank wall where $M_y = 109$ kN/m (at mid-height $x = a/2$ and $y = b/2$). The shear on the adjacent slab $V = 92$ kN/m [*refer* Example 26.1]. Assume $f_{ck} = 25$, $f_y = 415$ and environment class B. Design by the limit state and the IS 3370 methods.

DESIGN OF STRUCTURES FOR STORAGE OF LIQUIDS 495

Reference	Step	Calculations
		Method 1. Limit state (deemed to satisfy) design BS 5337
Table 26.3	1	*Assume dimensions* Assume h = 400, also assume 20 mm steel is used. Moment produces the tension inside; class B condition Allowable stress = 130 N/mm² d = 400 – 10 – 40 = 350 mm
	2	*Check value of e and examine the section* $d - 0.5h$ = 350 – 200 = 150 mm M = 109; N = 92 (tension) $$e = \frac{M}{N} = \frac{109 \times 10^6}{92 \times 10^3} = 1185 \text{ mm}$$ e is greater than $(d - 0.5h)$. Hence the section is mostly in bending. Both compression and tension occur in the section.
	3	*Determine the value of modified moment and tension at steel level* $M_1 = M - N(d - 0.5h) = 109 - 92(0.15) = 95$ kNm $$\frac{M_1}{bd^2} = \frac{95 \times 10^6}{1000(350)^2} = 0.78$$
Text Table 26.4 SP 16 Table 69	4	*Design the section* Table 26.4 gives the steel percentage = 0.68% for f_s = 130. $$A_{s1} = \frac{0.68 \times 1000 \times 350}{100} = 2380 \text{ mm}^2$$ Additional tension steel $$A_{s2} = \frac{N}{f_s} = \frac{92 \times 10^3}{130} = 707.7 \text{ mm}^2 \text{ (take 710 mm}^2\text{)}$$ Total steel = 2380 + 710 = 3090 mm² Provide 30 mm at 100 mm (3140 mm²) % steel = $\dfrac{3140 \times 100}{1000 \times 350} = 0.9\%$ (Check for the shear using working stress method.)
		Method 2. BS 5337 limit state design based on crackwidth using charts
	1	*Determine the allowable crackwidth* h = 400; Diameter of the steel used = 20 mm

Reference	Step	Calculations
	2	Environment class B crackwidth should not exceed 0.2 mm. From the previous method step 3, $M_1 = 95$ kNm/m; $N = 92$ kN/m *Determine the percentage of steel required* $$\frac{M_1}{bd^2} = \frac{95 \times 10^6}{1000 \times 350^2} = 0.78; \text{ cover} = 40 \text{ mm}$$ $w = 0.2$ mm; $c = 40$ mm and $\phi = 20$ mm
Fig. 26.1		Adopt spacing not to be greater than 300 mm. Percentage of steel = $\dfrac{100 A_s}{bd} = 0.34$ [*Note*: This is much less than the value obtained from method 1.]
	3	*Check the stress in steel* $p = 0.34$, $\rho = 0.0034$ $m\rho = (15)(0.0034) = 0.051$ $$\frac{x}{d} = -m\rho + \sqrt{m\rho(2+m\rho)} = -0.051 + \sqrt{0.051(2.051)} = 0.272$$ This can also be got from SP 16 Table 91. $1 - \dfrac{x}{3d} = 0.91 =$ Lever arm factor $= jd$
Eq. (26.2)		$f_s \dfrac{A_s}{bd} \dfrac{jd}{d} = \dfrac{M}{bd^2}$ i.e. $f_s = \dfrac{M}{bd^2} \dfrac{1}{pj}$ $$f_s = \frac{0.78}{0.0034 \times 0.91} = 252 \text{ N/mm}^2$$ Alternately, using Table 26.5 For $\dfrac{100 A_s}{bd} = 0.34$, we have $\dfrac{100 M}{f_s bd^2} = 0.309$ and $\dfrac{x}{d} = 0.272$
Step 2		$$f_s = \frac{100 M}{bd^2 \times 0.309} = \frac{100 \times 0.78}{0.309} = 252 \text{ N/mm}^2$$ [Hence, this method allows larger stresses in steel than the conventional method.]
Step 4 Method 1	4	*Determine the area of steel required for* M_1 *and* N $A_{s1} = 0.0034 \times 1000 \times 350 = 51190$ mm² for bending $A_{s2} = \dfrac{N}{f_s} = \dfrac{95 \times 10^3}{252} = 377$ mm² for direct tension

DESIGN OF STRUCTURES FOR STORAGE OF LIQUIDS 497

Reference	Step	Calculations
		$A_{s1} + A_{s2} = 1190 + 377 = 1567$ mm²
	5	Make correction for steel area
Eq. (26.11)		$A_s = \dfrac{A_{s1} + A_{s2}}{1 - \dfrac{140bh}{A_{s1}(f_{s1})^2} \dfrac{A_{s2}}{A_{s1} + A_{s2}} \dfrac{d-x}{h-x}}$
		$h = 400$ mm, $d = 350$ mm
		$\dfrac{x}{d} = 0.272$, $x = 0.272 \times 350 = 95$ m
		$d - x = 255$, $h - x = 305$
		$A_s = \dfrac{1555}{1 - \dfrac{140 \times 1000 \times 400}{1190 \times 250^2} \dfrac{365}{1555} \dfrac{255}{305}} = 1820 \text{ mm}^2$
		Adopt 20 mm at 170 (1848 mm²).
		Percentage of steel $= \dfrac{1848 \times 100}{1000 \times 350} = 0.53\%$
		Comments:
		1. In the first method, the steel required was 0.9% and this has been reduced to 0.53%.
		2. As a check we may calculate the crackwidth and confirm that it is limited to 0.2 mm as in Example 26.6.
	6	Check the ultimate limit state
		$M_1 = 95$ kNm/m and $N = 92$ kN/m
BS 5337		Taking partial safety factor $= 1.6$
	6.1	Find steel for ultimate moment and tension
		$M_u = 1.6 \times 95 = 152$ kNm/m
		$N_u = 1.6 \times 92 = 147$ kN/m
		$\dfrac{M_u}{bd^2} = \dfrac{152 \times 10^6}{1000 \times 350^2} = 1.24$ (required $p = 0.369$)
		$A_{s1} = \dfrac{0.369 \times 1000 \times 350}{100} = 1291 \text{ mm}^2$
		Steel for ultimate tension
		$A_{s2} = \dfrac{147 \times 10^3}{0.87 \times 415} = 407 \text{ mm}^2$
		Total $A_{s1} + A_{s2} = 1291 + 407 = 1698$ mm²
		This is less than 1820 mm² (provided). Hence safe.

498   ADVANCED REINFORCED CONCRETE DESIGN

Reference	Step	Calculations
	6.2	*Check for shear*
		$V_{max} = 100$, $V_u = 1.6 \times 100 = 160$ kN/m
		$v = \dfrac{160 \times 10^3}{1000 \times 350} = 0.46$ N/mm^2
IS 456 Table 13		Allowable shear = 0.49. Hence safe.
	6.3	*Check for bond and anchorage*
		Method 3. Design by IS 3370 method
	1	*Assume dimensions*
		Assume a larger $h = 600$ mm. Larger depth assumed satisfies the interaction formula.
		$d = 600 - 10 - 40 = 550$ mm
		$d - \dfrac{h}{2} = 550 - 300 = 250$ mm $= 0.25$ m
	2	*Find eccentricity e*
		$M = 190$ kNm/m, $N = 92$ kN/m
		$e = \dfrac{109 \times 10^6}{92 \times 10^3} = 1184$
		1184 is greater than $d/2 = 250$. Hence the section is in bending.
	3	*Find modified moment and tension steel (strength design)*
Eq. (26.12)		$M_1 = 109 - 92(0.25) = 86$ kNm
Table 26.4		$\dfrac{M_1}{bd^2} = \dfrac{86 \times 10^6}{1000(550)^2} = 0.28$
Table 26.8		Exposure B, $f_s = 130$ N/mm^2
		Required steel percentage = 0.24%
		$A_{s1} = 0.0024 \times 550 \times 1000 = 1320$ mm^2
		Additional steel $A_{s2} = \dfrac{92 \times 10^3}{130} = 707.7$ mm^2
		Total steel = $A_{s1} + A_{s2} = 2027.7$ mm^2
		Percentage of steel $= \dfrac{2027.7 \times 100}{1000 \times 550} = 0.369\%$
		Note: 1. Here with larger thickness of concrete, we get lesser percentage of steel.
		2. Procedures in methods 1 and 2 so far are the same. In method 3, we make the following further checks.

DESIGN OF STRUCTURES FOR STORAGE OF LIQUIDS 499

Reference	Step	Calculations
	4	*Check stress in steel due to M_1 and N*
Table (26.5)		$p = 0.368$ gives values of $\dfrac{x}{d} = 0.285$ and $\dfrac{100M}{f_s bd^2} = 0.343$
		$f_s \text{ bending} = \dfrac{100 M_1}{bd^2} \cdot \dfrac{1}{0.343}$
		$= \dfrac{100 \times 86 \times 10^6}{(1000)(550)^2 (0.343)} = 82.9 \text{ N/mm}^2$
		Direct tension in steel $= \dfrac{92 \times 10^3}{2027.7} = 45.4 \text{ N/mm}^2$
		Total tension $= 82.9 + 45.4 = 128 \text{ N/mm}^2$
		This is less than 130 N/mm² allowed. Hence safe.
	5	*Check for cracking (assuming uncracked section)*
		As the first approximation, moment of inertia of uncracked section
		$I = \dfrac{bh^3}{12} + (m-1) A_s \left(d - \dfrac{h}{2} \right)^2$
		$I = \dfrac{1000(600)^3}{12} + 14 \times 2024 (250)^2 = 19.77 \times 10^9 \text{ m}^4$
		Bending tension $= \dfrac{M}{I} y = \dfrac{109 \times 10^6 \times 300}{19.77 \times 10^9} = 1.65 \text{ N/mm}^2$
		Axial tension $= \dfrac{92 \times 10^3}{(1000 \times 600) + 14(2027.7)} = 0.15 \text{ N/mm}^2$
	6	*Check for interaction near water face*
Table 26.1		Taking axial tension as t and bending tension σ_{ct}.
IS 3770		Interaction formula $\dfrac{t'}{t} + \dfrac{\sigma_{ct'}}{\sigma_{ct}} \leq 1$
Part 2		For liquid retaining faces, i.e. where the bending moments are negative.
Sec. (26.13.1)		$t = 1.3$ and $\sigma_{ct} = 1.9$ for M25 concrete.
Table 26.6		$\dfrac{0.14}{1.3} + \dfrac{1.65}{1.90} = 0.108 + 0.868 = 0.976$
		This condition can be considered as satisfactory.
	7	*Check for shear and bond*
		As in ordinary structures.

500 ADVANCED REINFORCED CONCRETE DESIGN

EXAMPLE 26.6 (Calculation of crackwidth with M and N)

Calculate the crackwidth in Example 26.5 with the following data $h = 400$ mm, Area of steel provided 20 mm rods at 170 mm spacing (1848 mm^2) (step 5 method II), $M = 109$ kNm/m, $N = 92$ kN/m, percentage of steel = 0.53.

Reference	Step	Calculations
	1	*Check whether crackwidth formula is valid*
		$M = 109$ kNm/m, $N = 92$ kN/m
		Modified moment $M_1 = 95$ kNm
Method 1		$e = \dfrac{M}{N} = \dfrac{109 \times 10^6}{92 \times 10^3} = 1185$ mm; $h = 400$; $d = 350$
Step 3		$d - \dfrac{h}{2} = 350 - 200 = 150$ mm
		e is larger than $[d - (h/2)]$. Hence the section is in bending and hence the crackwidth formula in bending applies here.
	2	*Determine the depth of neutral axis with m only*
Table 26.5		$p = 0.51$, Let $m = 15$, $mp = 0.51 \times 15 = 7.65$
SP 16 Table 92		Depth of natural axis with M only can be got from Table 26.4.
		$x/d = 0.32$, i.e. $x = 0.32 \times 350 = 112$ mm
	3	*Find the effect of N combined with M*
		This will cause a shift in the natural axis which can be approximated to a new steel area
Eq. (26.13)		$p_1 = p\left[1 - \dfrac{d}{e + (h/2)}\right]$ (in terms of percentage)
Chapter 1		$= 0.51\left(1 - \dfrac{350}{1185 + 200}\right) = 0.38$; $mp = 5.7$
Table 1.7 SP 16, Table 91		When $p_1 = 0.38$, new value of x/d is 0.286.
		$x = 0.286 \times 350 = 100$ mm
	4	*Carry out further reiteration for finding value of x*
Eq. (26.14)		$p_2 = p_1\left[1 - \dfrac{d - x_1/3}{e + (h/2) - (x_1/3)}\right] = 0.51\left(1 - \dfrac{317}{1352}\right) = 0.39$, with p_2
		$x_2/d = 0.289$. Hence $x = 101$ mm
		As this is close to $x = 100$ we will assume $x/d = 0.289$.
Table 26.5	5	*Calculate tension and strain in steel*
Eq. (26.3)		For $p_2 = 0.39$ we get $\dfrac{100M}{f_s bd^2} = 0.353$

DESIGN OF STRUCTURES FOR STORAGE OF LIQUIDS 501

Reference	Step	Calculations
		Tension in steel due to moment $M_1 = 95$ kNm $$f_s = \frac{100 \times 95 \times 10^6}{1000 \times 350 \times 350 \times 0.353} = 220 \text{ N/mm}^2$$ Strain at level of steel $= \dfrac{f_s}{E_s} = \dfrac{220}{200 \times 10^3} = 0.0011$
	6	*Strain at tension face* Strain at bottom surface of steel $= \dfrac{0.0011(h-x)}{d-x} = \varepsilon_1$
Eq. (26.5a)		$\varepsilon_1 = 0.0011 \times \dfrac{400-101}{350-101} = \dfrac{0.0011 \times 299}{249} = 0.00133$
	7	*Calculate ε_1 and a_{cr}*
Eq. (26.6)		$\varepsilon_m = \varepsilon_1 - \dfrac{0.7bh}{1000 A_s f_s}$ (when $a' = h$) $= 0.00133 - \dfrac{0.7 \times 1000 \times 4000}{1848 \times 221 \times 1000} = 0.00064$
Eq. (26.9)		$a_{cr} = \sqrt{50^2 + \left(\dfrac{170}{2}\right)^2} - 10 = 88.6$ mm
	8	*Calculate crackwidth*
Eq. (26.5)		$w = \dfrac{4.5 a_{cr} \varepsilon_m}{1 + [2.5(a_{cr} - c)/(h-x)]}$ $= \dfrac{4.5 \times 88.6 \times 6.4 \times 10^{-4}}{1 + [2.5(88.6 - 40)/299]} = 0.185$ mm
		This is less than 0.2 mm, hence acceptable.

REFERENCES

1. IS 3370 (1967), *Code of Practice for Concrete Structures for Storage of Liquids*, Parts 1–4, Bureau of Indian Standards, New Delhi.

2. BS 5337 (1976), *Code of Practice for the Structural Use of Concrete for Retaining Aqueous Liquids*, British Standard Institute, London.

3. Threlfall, A.J., *Design Charts for Water Retaining Structures to BS 5337*, A Viewpoint Publication, London, 1981.

4. Keithgreen, J. and Perkins, P.H., *Concrete Liquid Retaining Structures,* Applied Science Publishers, London, 1980.

5. Circular Concrete Tanks without Prestressing (Tables for Design), Portland Cement Association, Chicago, 1972.

6. Rectangular Concrete Tanks (Tables for Design), Portland Cement Association, Chicago, 1969.

7. Anchor, R.D., Hill, A.W. and Hughes, B.P., *Handbook on BS 5337 (1976)*, A Viewpoint Publication, London.

CHAPTER 27

Historical Development of Reinforced Concrete

27.1 CONCRETE IN THE ROMAN PERIOD

The problem of construction of structures that are strong as well as durable under water was faced from early times during the progress of human civilisation. The Romans used brick dust and volcanic ash with lime to produce hydraulic mortar. They also used wooden formwork for casting the structural shapes. In some places, they enclosed hollow earthen pots in this concrete mass to reduce the weight of the structure. These were the fore-runners of modern concrete construction.

27.2 LATER DEVELOPMENT IN CONCRETE

Portland cement as is known today was however first visualised by John Smitten who developed a hydraulic mortar for the Eddystone light house by burning a mixture of lime and clay. This took place in about 1756. From then onwards, rapid developments took place in concrete construction. The following are some of the important dates in the forward march:

1776—James Parker got a patent for producing hydraulic cement by burning modules of clay containing veins of calcareous matter. This was called natural cement. The use of kankar lime obtained by heating of lime modules mixed with clay and use of surkhi in fat lime to produce hydraulic mortar were well known in India also from very early days. However, the basic chemistry of the under-water setting of these hydraulic limes or natural cements were not known those days.

1816—An unreinforced concrete arch bridge was built in France, with concrete in compression.

1824—Joseph Aspden of United Kingdom patented the process to producing cement from lime and clay which when set resembled Portland stone in appearance and this was called Portland cement.

Very soon the practice of strengthening concrete, by embedding steel rails was initiated, eventhough the principles of their action were not known.

1850—J.L. Lambot of France built a concrete rowing boat, strengthened by a mesh of iron rods, for the Paris International Exhibition.

1854—W.B. Wilkinson of England and in 1855 Lambot of France took patents for preparing floors of buildings by burying iron beams or a square mesh of steel rods in concrete, which was cast on falsework.

1861—Francis Coignet published a book describing the many application and uses of reinforced concrete.

1867—Joseph Monier, the owner of a nursery in Paris devised reinforced concrete for making flower pots. He had no knowledge of the behaviour or design of reinforced concrete.

All these developments were made without the knowledge of the principles of action of reinforced concrete. They were more or less only intuitive methods of construction.

27.3 BIRTH OF REINFORCED CONCRETE

In 1872, without theoretical calculations but with a lot of engineering sense, Monier built an R.C. reservoir of 130-cm capacity in reinforced concrete. He can be considered as the world's first builder in reinforced concrete. Reinforced concrete was then known as the 'Monier System'.

Coigner, in 1892, proposed the use of R.C. for the main drainage system in Paris. The larger width of the structure was about 5 m and the walls of the drainage channels were only 80 mm thick. It consisted of reinforced concrete, i.e. concrete reinforced with a steel mesh made out of round bars 8 mm and 16 mm in diameter.

Systematic attempts on theoretical development took place only after the above constructions were successfully made. The German Engineer, Prof. E. Morsh conducted a large number of tests to study the action of reinforced concrete. In 1887, a small book *Das System Monier* (The Monier System) was prepared to back up the Monier's practical method, which caught the imagination of the French, German and Danish engineers.

Prof. Morsch's work can be considered as the starting point of modern theory of reinforced concrete design. The following three fundamental concepts of action of reinforced concrete were clearly initiated in the publication:

1. Concrete is weak in tension so that all the tension can be assumed to be taken by steel.
2. The transfer of stress between concrete and steel takes place through the bond strength developed between steel and concrete on setting of concrete.
3. The volume changes in concrete and in steel due to the atmospheric change of temperature are more or less equal, as the coefficients of expansion of steel and concrete are approximately the same (i.e. $10\text{--}13 \times 10^{-6}$ for concrete and 12×10^{-6} for steel per degree centigrade).

In 1890, Neuman adopted the elastic theory of composite sections to reinforced concrete and pointed out the importance of the relationship between the moduli of elasticity of steel and concrete as well as the effect of the modular ratio on the position of the neutral axis. In 1894, Coignet carried out many experiments and with Tedesco he evolved the 'modular ratio method' to determine the position of neutral axis in reinforced concrete sections. This was the birth of the elastic method also known as modular ratio or working stress method of design of concrete structures which was in use in all national codes till recently.

In 1895, Considere began a series of extensive tests on resistance of R.C. beams and columns and evolved the helically bound R.C. columns, and in 1897, he published a small textbook on the subject of R.C. design. In 1900, Considere also recommended the hooking of the ends of the steel reinforcements with a diameter five times that of the bar. The first teaching course in reinforced concrete design was given in Ecole de Ponts et Chaussess in France in 1897.

Thus, it is important to note that reinforced concrete as is known today is only about hundred years old. From 1900, steady development took place in reinforced concrete construction. In 1903, the first publication of official regulations or codes based on Morsch's recommendations was issued in Prussia. In 1906, the first British Code of Practice for reinforced concrete was published. In U.S.A., considerable work on R.C. was done by Prof. Talbot and his associates at the University of Illinois. The first American Code was drafted by the American Society of Civil Engineers in 1908 and it was published in 1916. Many revisions of these codes have been made by ACI during the last 70 years. From 1920 to 1960, elastic methods were mostly used for design of reinforced concrete structure which have now been replaced by limit state methods. Many developments like use of prestressing, use of ultimate load theory in design, methods for calculations of deflections and crackwidths in concrete structures have been evolved during the last 30 years. As the history of the recent advances in design and analysis of concrete structures are well known only a brief mention of them is made here.

27.4 RECENT DEVELOPMENTS IN REINFORCED CONCRETE DESIGN

Much progress was made in design of reinforced concrete after the Second World War. After 1960, the main progress has been made in evolving refined methods of calculations for design of R.C. structures. Gradually the concepts of Limit States Design have been accepted internationally and the present codes of practice in many countries use these concepts in their design. Concepts on durability under adverse conditions of exposure were also developed in recent times. Similarly improvements in grades of cement and design of concrete mixes were made during this period.

In Great Britain the first code for reinforced concrete was published in 1906. A number of revisions were made on this first code. CP 114 (1948) was revised in 1951, 1957 and 1969 and was in use till 1972. This code mostly used the concepts of elastic design in the earlier years and later 'modified load factor' methods. After the publication of the 'international recommendations for the design and construction of concrete structures', in 1970, by the CEB-FIP, Great Britain was the first country to follow it up with a code based on the principles of limit state design. CP 110 (1972) replaced CP 114 (1969). CP 110 itself was revised in 1985 as BS 8110 (1985). It is accepted as the current code for R.C. design in U.K. It is based on the principles of limit state design. It is also a 'unified code', as the same code of practice covers the practices in reinforced concrete, prestressed concrete and pre-fabricated structures. [It is of interest to note that the first prestressed concrete bridge in India was built over the Palar river in Tamil Nadu in 1954, just four years after the first prestressed bridge was built in U.S.A. in Tennassee in 1950.]

27.5 DEVELOPMENTS IN INDIA

In India before independence, London County Council (L.C.C.) rules and the British codes of practice were extensively used for the design. Many structures have been built in India in reinforced concrete by the British, based on the British codes. The first Indian Standard known as 'Code of Practice for Plain and Reinforced Concrete for General Building Construction' was published in 1953 and it was revised in 1957 as the first revision. This code was further revised in 1964 under the title 'Code of Practice for Plain and Reinforced Concrete'. In 1978, it was published as IS 456 (1978) (third revision). In this third revision the SI units have been introduced for stresses and the concept of limit state which provides a rational approach to design was accepted but working stress method was also allowed as an alternative method. The fourth revision IS 456 (2000) is currently in vogue in India. It has made limit state design mandatory and the status of the working stress method as an alternate method has been discontinued. It gives very detailed recommendations for durability of concrete. Special topics, like concrete walls, also find a place in this revision.

27.6 EVOLUTION OF AN INTERNATIONAL CODE

Limit state design of R.C. members based on IS 456 (2000) is similar to the British Code BS 8110 (1985). Many of the former British colonial countries in Africa and Asia have also accepted BS 8110 as their guides for design till such time as their own national codes are finalised. Codes of other developed countries like U.S.A., and Federal Republic of Germany also have revised their codes of practice to bring it in line with limit state philosophy. The European Common Market countries have evolved a single code called *Eurocode* which will be applicable to all the member countries. Before long a uniform limit states design code is expected to become the international meth od of design of concrete structures. These codes will generally be based on the recommendations of the CEB-FIP organisation, which is the world body promoting knowledge in concrete structures.

APPENDIX A

Calculation of Bending and Torsional Stiffness of Flanged Beams

A1. INTRODUCTION

Two types of stiffness are involved in structural analysis:
 (i) Bending stiffness
 (ii) Torsional stiffness.

Bending stiffness is expressed by the quantity $4EI/L$ and *torsional* stiffness by the quantity $4GC/L$, where

I = Moment of inertia of the section

E = Modules of elasticity (young's modules) of material

C = St. Venants constant (torsional constant) which like the moment of inertia depends on the dimensions of the section

G = Modules of rigidity.

The relationship between E and G is given by the expression

$$G = \frac{E}{2(1+\mu)}$$

If μ is taken as 1/6, $G = 3/7$. If μ is taken as zero, $G = E/2$.

A2. CALCULATION OF MOMENT OF INERTIA OF FLANGED BEAMS

There are many instances where part of the slab is assumed to act with the beam and the moment of inertia of I or L beam has to be calculated.

The following formula for the determination of the moment of inertia of I and L beams is given in *Handbook of Concrete Engineering* by Mark Fintel (Van Nostrand Reinhold Co., New York, p. 18, 1974.

where
$$I = \frac{KbH^3}{12}$$

$$K = 1 + (x-1)y^3 + 3\frac{(1-y)^2 y(x-1)}{1 + y(x-1)}$$

and

    B = Width of flange
    b = Width of web
    H = Total depth
    h = Depth of slab

$$x = \frac{B}{b}$$

$$y = \frac{h}{H}$$

The IS publication SP 16 Chart 88 also gives the necessary curves for quick evaluation of moment of inertia of I section for values of x larger than 5.

For a good estimate of the approximate value of the moment of inertia of T and L beams, the calculations can be simplified if the slab portion of these beams are omitted and the MI of the rectangular part is multiplied by a factor. The values commonly used are the following.

    MI of T beams = 1.8 to 2 (MI of the rectangular part)
    MI of L beams = 1.3 to 1.5 (MI of the rectangular part)

A3. CALCULATION OF TORSIONAL CONSTANT

Value of torsional stiffness is GC/L. Timoshenko has shown that the value of C for a section like T or L composed of rectangles is given by the summation of the value of C of each rectangle to which the section can be divided. These rectangles are so divided that the torsional constant calculated has the largest value. The formula, to be used for each rectangle is:

$$C = \left(1 - 0.63\frac{x}{y}\right)\frac{x^3 y}{3}$$

where x is the shorter dimension of the rectangle and y is the longer dimension of the rectangle.

A4. PART OF SLAB ACTING WITH BEAMS

According to ACI 318(89) Clause 11.6.1.1 the overhanging flange width for the *torsional resistance* shall not exceed three times the flange thickness. Again ACI 318 Clause 13.2.4 in two-way slab design, a beam includes that portion of slab on each side of length not exceeding four times the slab thickness or the projection of the beam above or below the slab whichever is greater.

APPENDIX B

Durability of Structural Concrete

B1. GENERAL REQUIREMENTS

The fourth revision of IS 456 designated as IS 456 (2000) gives special requirements for durability of concrete. These are summarised in this Appendix. The important factors to be considered under durability are described in Clause 8 of IS 456 and are as follows:

1. Planning the shape and size of member to be constructed
2. Choosing an adequate cement content as well as the maximum free water/cement ratio to ensure low permeability of concrete
3. Choosing the right type of cement to suit the environment
4. Ensuring complete compaction of concrete
5. Ensuring sufficient hydration of cement through proper curing
6. Ensuring correct removal of form work and supporting props

Some of these are discussed below:

B2. PLANNING SHAPE OF STRUCTURE

All structures should be so shaped, designed and constructed as to ensure good drainage of water thus avoiding standing pools of water and also run down of water. Extra cover should be provided to steel at corners. In some places, this can be achieved by chamfering. If necessary, surface coatings should also be used to prevent ingress of water or aggressive chemicals and gases. The depth of concrete placed is greater than 600 mm especially with concrete more than 400 kg/m^3 or more, measures should also be taken to reduce the heat of hydration of cement.

B3. CHOOSING DEPTH OF COVER

The old practice of prescribing only the nominal cover (defined as cover to all steels including stirrups) has been replaced as in BS 8110 by linking cover with the properties of the cover concrete such as water/cement ratio, cement content, strength of concrete, type of exposure as shown in Table B1. For this purpose exposure has been classified into five classes [Table 3 of IS 456 (2000)] as follows:

Class 1: Mild—protected concrete surface

Class 2: Moderate—concrete permanently under water or in contact with non-aggressive soils

Class 3: Severe—subjected to alternate wetting and drying: immersed in sea water

Class 4: Very severe—exposure to corrosive fumes, sea-water spray, aggressive soils

Class 5: Extreme—exposure to abrasive action such as sea water carrying solids and flowing water with pH 4.5 or less.

The cover requirements in BS are given in Table B1. We can note that the minimum cement content and the maximum water–cement ratio allowed are also of importance in cover for durability. If the maximum water–cement ratio specified cannot produce the required workability we are allowed to adjust the mix by increasing the cement content (by using blended cement or otherwise) or by using plasticisers or other additives to give the required strength. According to IS 456 (2000) Clause 8.2.4.2, cement contents not including p.f.a (pulverised fuel ash) or g.g.b.f.s. (ground, granulated blast furnace slage) in excess of 450 kg/m^3 should not be used in R.C. members unless special consideration is given to drying shrinkage in thin sections and heat of hydration in thick sections. The minimum cement content specified in the IS 456 revision is 300 kg/m^3. (The cement content in 1:2:4 mix will be roughly 309 kg/m^3.)

TABLE B1 NOMINAL COVER FOR 20 mm MAX. SIZE AGGREGATE
(BS 8110)

\multicolumn{5}{c}{Cover (mm) for exposure conditions}	Lowest grade	Min. cement kg/mm*	Max. w/c ratio
1	2	3	4	5			
20*	20	25	30	50	M50	400	0.45
20*	25	30	40	50	M45	350	0.50
20*	30	40	50	–	M40	325	0.55
20	35	–	–	–	M35	300	0.60
25	–	–	–	–	M30	275	0.65

*Minimum content is independent of grade and includes suitable additions specified for cements.

Note:
(a) 1, mild; 2, moderate; 3, severe; 4, very severe; 5, extreme
(b) *Cover may be reduced to 15 mm using 15 mm maximum size aggregate
(c) Adjustment of minimum cement content for aggregates other than 20 mm is as follows:

For 40 mm aggregate, we can reduce cement content by 30 kg/m^3 but we have to increase the cement content by 20 kg/m^3 for 14 mm and 40 kg/m^3 for 10 mm aggregates [see Table 6 of IS 456 (2000)].

(d) When the w/c ratio can be reduced from the above table the cement content can also be reduced according to the percentage reduction subject to a maximum of 10%

(e) When using cement with additives like p.f.a. on g.g.b.f.s an increase in the total mass of such cements may be needed to obtain the specified grade for the cover being considered. The durability with these cements can be considered as equal to that of OPC provided the same grade as specified is obtained.

B3.1 IS Requirements

1. IS 456 (2000) requirements for nominal cover is similar to but not the same as in Table B1. The IS recommendations are as follows:

(a) The minimum nominal cover for exposure conditions 1 to 5 should be 20, 30, 45, 50 and 75 mm respectively. However for main steel up to 12 mm and in mild conditions of exposure these values can be reduced by 5 mm.

(b) For columns, the cover should not usually be less than 40 mm or main bar diameter. But for columns of minimum dimension of 200 mm or under and where the main steel diameter does not exceed 12 mm the cover can be 25 mm.

2. The minimum cement content and maximum w/c ratio of the concrete should comply with Table B2.

TABLE B2 DURABILITY OF REINFORCED CONCRETE
[IS 456 (2000) Table 5]

Requirement	Exposure condition				
	Mild	Moderate	Severe	Very severe	Extreme
Max.w/c ratio	0.55	0.50	0.45	0.45	0.40
Min:cement (kg/m)	300	300	320	340	360
Lowest grade	M20	M25	M30	M35	M40

Note: Maximum size of aggregate 20 mm, adjustments for other sizes of aggregates can be made as indicated in Table B1 item *c*.

B4. CHOOSING CEMENT FOR STRUCTURES EXPOSED TO SULPHATES

The fourth revision of IS 456 has adopted a similar specification as in BS 8110 for measures to be adopted for sulphate resistance of concrete also. These contain two more classes of exposures than given in IS 456 (1978) Table 20. The requirement are summarized in Table B3. Sulphates are present in most cements and in some aggregates. The total water soluble sulphate content of the concrete mix shall not exceed 4% SO_3 by mass of cement in the mixture.

TABLE B3 CONCRETE EXPOSED TO SOLUBLE SULPHATES
[IS 456 (2000) Table 4]

Class	Sulphate as SO_3		Details of Cement and mix		
	In 2:1 water:soil extract g/L	In ground water g/L	Type of cement	Min: kg/m³	Max: w/c
1	<1.0	<0.3	OPC	280	0.55
2	1–1.9	0.3–1.2	OPC	330	0.50
			SRPC	310	0.50
3	1.9–3.1	1.2–2.5	SRPC	330	0.50
			PPC	350	0.45
4	3.1–5.6	2.5–5.0	SRPC	370	0.45
5	>5.6	>5.0	SRPC with coating	400	0.40

Note: 1. See Ref. 1 Section 3.12 for estimation of soluble sulphates
2. Sea water contains as much as 2.5 g/L sulphates and salinity of over 32 parts per thousand.
3. For structures constantly under sea water which contain both sulphates and chlorides, any cement with C3 A from 5 to 8% can be used in concrete.

B4.1 Allowable Chlorides in Concrete Materials [IS 456 (2000) Table 1]

According to IS 456, the maximum total acid soluble chloride content in concrete expressed in kg/m^3 should not exceed 0.6 in R.C. members [2].

B4.2 Alkali Aggregate Reaction

Aggregates with alkali reactive constituents should not be used in normal practice without taking adequate protective measures.

B5. COMPACTION OF CONCRETE

Special steps should be taken for concreting sloping surfaces. Corners, valleys, junctions should be compacted properly. Porous concrete always leads to leakage and corrosion of steel.

B6. CURING, STRIPPING OF FORMWORK AND REMOVAL OF SUPPORTS

The three important steps of curing, stripping of formwork and removal of props under members should be clearly distinguished. The time schedule as well as the sequences of each of this operations are very important. The order of removal of supports should comply with its structural action. The specifications are given in IS 456 (2000) Clause 11.3.

REFERENCES

1. P.C. Varghese, *Limit State Design of Reinforced Concrete*, Prentice-Hall of India, New Delhi, 1994.

2. Fourth revision of IS 456, BIS, New Delhi, 2000.

3. BS 8110 Part I, British Stands Institution, London, 1985.

Index

A

Acceptance tests, 467
Additional eccentricity, 299
Additives, 452
Affinity theorem, 429
Aggregates, 445
Anchorage of bars, 363

B

Baker's method, 403
Bar spacing
    beams, 44
    flat slabs, 189, 221
    walls, 305
Beams
    column joints, 358
    deep, 50
    deflection, 1
    moment curvature, 44, 397
Bearing strength concrete, 61
Bends in bars, 363
Boundary elements, 345
Buildings natural frequency, 319

C

Cantilever method, 150
Carbonation, 440
Carry-over factors (COF), 246
Cast-*in-situ* joints, 358
Cement
    content, 438
    grade, 436
Coefficient of variation, 462
Collapse load analysis, 392
Columns
    capital, 161
    equivalent, 246
    joint, 358
    shear, 360
    strength, 360
    strip, 171
Compliance test, 467
Concrete
    confinement, 365
    creep, 7
    curing, 442
    fire resistance, 285
    shrinkage, 30
Construction loads, 117
Contraflexure points, 40
Core testing, 470
Corner joints, 368
Corrosion of steel, 77, 437
Cover, 287, 510
Cracking mechanism, 25
Crackwidth, 27
Creep coefficient, 7
Current margin, 463
Curvature, 44, 397

D

Damping, 318
Dead loads, 96
Deep beams
    detailing, 62
    with holes, 58
    lever aim, 52
    shear, 55
    simple type, 51
Deflection
    beams, 1
    cantilever, 14
    imposed loads, 2
    long-term, 7
    requirements, 9
    short-term, 1, 3
    shrinkage, 8
Detailing of steel
    column joint, 370
    corner joint, 368
    deep beam, 50
    fabricated shear head, 225
    flat slabs, 189, 223
    grids, 89
    shear walls, 347
Development length, 363
Drop panel, 161
Ductile detailing, 378
Ductility, 399
Durability of concrete, 437, 474, 509

E

Early thermal cracking, 31, 36
Earthquake
    distribution, 327
    forces, 310, 320
    magnitude, 311
    map, 315

Edge
beam, 263
column, 188
Effective
flange width, 176, 514
length of wall, 297
Equivalent
column stiffness, 246
frame analysis, 241
Exposure classification, 448, 510
External columns, 187

F

Field cured specimen, 468
Fire
classification, 283
effect on concrete, 285
effect on member, 285
effect on steel, 284
spalling, 286
standard, 282
Flat slabs and plates
design of reinforcements, 185
detailing, 189, 190
direct design, 171
equivalent column, 246
maximum steel spacing, 186
minimum thickness, 162
moment in columns, 186
Nicholas method, 174
pattern loading, 180
transverse distribution, 182

G

Gap gradings, 448
Grade
cement, 436
concrete, 444
Grading curves, 446
Grid slabs, 84
Ground slab, 489

H

High-strength concrete, 450
Hill and ridge, 109
Hillerborg's strip method, 425
Historical development, 503

I

Importance factor, 316
Imposed loads, 97
Inelastic
analysis, 392
response spectrum, 318
Integrity bars, 191
Internal ties, 275

J

Johansen, 425
Joints
beam column, 358
in concrete, 118
confinement, 365
corner, 368
detailing, 370
in frames, 359
shear strength, 366

K

Kemp, 425
Knee joints, 368

L

Laboratory-cured specimens, 466
Lateral
load analysis, 147
ties, 275
Liquid storage structures, 473
Load paths, 426
Loads
bottom loading, 50
buildings, 97
construction, 117
dead, 96
deflection curve, 2
earthquake, 313
live, 97
top loading, 50
wind, 101
Load test, 469
Local failure, 52, 61
Longitudinal distribution, 177
Lower bound value, 425

M

Middle strip, 172
Minimum steel
ground slab, 489
water tank, 473, 490
Mix design, 443
Modal analysis, 323
Modified strip method, 427
Moment
curvature, 44, 397
distribution, 39, 41, 253
envelope, 43
of inertia, 507
redistribution, 19, 39, 42
transfer, 214
Monier systems, 504

N

Natural frequency of buildings, 319

INDEX

Neutral axis depth, 13
Nominal mix, 443

O

Openings
   in flat slab, 227
   in strip method, 428
Operating curves (OC), 467
Ordinary grade concrete, 443

P

Parabola, 396
Partial safety factor, 2, 326
Permeability of concrete, 438
Plane concrete walls, 295
Plastic hinges, 394, 404
Plasticisers, 444
Plate theory, 85
Points
   of contraflexure, 40
   of inflexion, 428
Polymer concrete, 442
Portal method, 148
Probability integral, 462
Punching shear, 213

Q

Quality control, 459

R

Raft slab, 241
Redistribution of moment, 39
Reinforced concrete walls, 295
Reinforcements
   spacings (*see* bar spacing)
   types, 396
Ribbed slabs, 73
Richter scale, 311
Rigidity, 183
Rotation, 402

S

Sampling frequency, 466
Shear heads, 225
   with compression, 489
   in flat slabs, 211
   with tension, 489
   in walls, 301
Shear walls, 338
   boundary elements, 346
   design, 338
   detailing, 347
   moment of resistance, 348
Shores, 118
Short-term deflection, 8
Shrinkage steel, 489

Simpson's rule, 401
Skew slabs, 428
Slab beams, 244
Space model, 250
Spalling of concrete, 288
Spandrel beams, 263
Strain hardening, 360
Stress–strain curve
   concrete, 394
   steel, 396
Strip method, 425, 427
Substitute frames, 129
Sulphate disintegration, 441, 511
Superplasticisers, 444

T

Tall buildings, 152
Target strength, 464
T beams, 164, 507
Temperature steel, 31
Tension stiffening, 10, 399
Thermal cracking, 31, 36
Ties in buildings, 275
Top loading, 50
Torsion
   in buildings, 328
   in edge beams, 264
Two-way slabs, 171

U

Ultimate strength of concrete, 393
Unshored construction, 117

V

Variance, 461
Variation coefficient, 462
Vibration, 452
Voided construction, 73

W

Walls
   detailing, 305, 347
   empirical design, 299
   openings, 355
   slenderness, 297
Water–cement ratio, 438
Water tanks, 473
Wind load
   analysis, 112
   cyclones, 115
   gust effects, 102, 114
Wind speeds, 98, 104

Y

Yield-line theory, 425
Yield of steel, 360, 395, 397